Jean / With all my love
Xmas 1997. W.

Flora Britannica

FLORA BRITANNICA

RICHARD MABEY

Supported by Common Ground

With photographs by

Bob Gibbons and Gareth Lovett Jones

Chatto & Windus
LONDON

First published in Great Britain in 1996 by

SINCLAIR-STEVENSON

This edition published by Chatto & Windus
Random House, 20 Vauxhall Bridge Road,
London SW1V 2SA

3 5 7 9 10 8 6 4

A CIP catalogue record for this book
is available at the British Library

ISBN 1 85619 377 2

Design by Ian Muggeridge

Printed and bound in Spain by Cayfosa Industria Grafica

The paper used for this book is an environmentally
responsible product: the mill's liquid effluents present a level
of AOX and COD in accordance with European paper
industry regulations; sulphurous gas emissions into the air are
minimal; the finished paper is chlorine free according to the
rules of the Swedish Environmental Agency; the mill uses
50 per cent of past machine fibre in their paper production.

*Frontispiece: Foxgloves, the 'witches' thimbles' of medieval
herbalists, standing their ground in the landscapes of the
nuclear age (see p. 332).*

Contents

1720. Most-grown flowers were: auricula, carnation, tulip, anemone, pansy, ranunculus.

Introduction

On the 50th anniversary of VE Day, many of the world's leaders gathered in London's Hyde Park for a memorial ceremony and, in the spring sunshine, laid posies of their national flowers around a large globe. It seemed a touching and ingenuous tribute, a rediscovery of one of the oldest spring rituals of reparation. Many of the blooms were traditional: red roses for England, thistles for Scotland and daffodils for Wales; Austria had edelweiss, Canada the sugar maple and South Africa the king protea. There were odd coincidences and improvisations, too, evidence of the momentous changes in the status of many nations over the previous decades. Cornflowers were laid by France, Germany, Estonia, Belarus and the Czech Republic; and formal bouquets of florists' cut-flowers – roses, tulips and carnations – by countries from Albania to Uzbekistan.[1] For many poor and unstable countries, it was clear that the idea of a 'national flower' was unfamiliar, maybe even artificial. But during the ceremony something serendipitous and enchanting happened. In the gusting May breezes, drifts of windblown Japanese cherry blossom – the international flower of peace – blew across Hyde Park like confetti, a reminder that nature makes its own unscripted gestures of renewal.

Plants have had symbolic as well as utilitarian meanings since the beginnings of civilisation. They have been tokens of birth, death, harvest and celebration, and omens of good (and bad) luck. They are powerful emblems of place and identity, too, not just of nations, but of villages, neighbourhoods, even personal retreats. Yet, as at Hyde Park, they are different from any other kind of symbol in having independent existences of their own. Primroses and the last roses of summer still announce the seasons when *they* decide, with no respect for our calendars.

In Britain, wild species have an even more central role in national and local cultures than those from gardens. We pick sprigs of heather for luck, munch blackberries in autumn, remember Wordsworth's lines when the daffodils are in flower, and link hands around threatened trees. Our children still make daisy chains, whack conkers, and stick goosegrass stems on each other's backs. Despite being one of the most industrialised and urbanised countries on earth, we cling to plant rituals and mystical gestures whose roots stretch back into prehistory: holly decoration for the winter solstice, kisses under the mistletoe, the wearing of red poppies to remember the casualties of war. We name our houses, streets and settlements after plants, and use them as the most prolific source of decorative motifs on everything from stained glass to serviettes. From the outside, it must look as if we are botanical aboriginals, still in thrall to the spirits of vegetation.

But is this just the dying stages of an obstinate habit, the outward signs of a longing for the rural life that most of us have lost? Do we really still believe in the bad luck that may-blossom can bring into a house, and in the efficacy of the increasing numbers of herbal nostrums crowding onto chemists' shelves? Or is our seeming respect now a touch tongue-in-cheek? When wild flowers are dragged willy-nilly into shampoo advertisements and state rituals, maybe it is time to ask whether the particular plants themselves have any meaning left for us, or whether they have become purely notional, registers of a fashionably Green 'life-style'.

When work started on *Flora Britannica* in 1992, this was the question which underpinned all others. We were aware of surviving crafts and cottage wisdom, and of the familiarity expressed in our immense legacy of vernacular

The symbolic use of plants is widespread in Europe. This sundial on a farmhouse wall in the Cévennes in France represents a carline thistle – the 'chardon soleil' – whose dried flower-heads are often nailed to doors for good luck.

plant names. But we didn't know whether, as a people (or collection of peoples), we could still be said to have an intimacy with wild plants that was not purely nostalgic and backward-looking. Did people still meet under meeting-place trees? Were children inventing new games for the new, exotic species constantly escaping into the wild, as they did centuries ago for horse-chestnut and sycamore seeds; and was that two-way traffic of wild and cultivated plants over the garden wall still busy? Did plants continue to play any role in our senses of place and season, those fundamental aspects of everyday life that seem everywhere to be under threat from regimentation and the ironing out of local difference? And what names did we use for them now, to ourselves and to each other?

The question as to whether local plant names had survived as real linguistic currency or just as museum curiosities was in one sense the starting point and baseline of *Flora Britannica*. The common names of wild plants are the fullest and most revealing register of the part they have played in our lives. Often they indicate aspects that have touched people's imaginations – a time of flowering perhaps, or a likeness, a use, a scent, an attachment to a particular habitat. So we have Lent lily (wild daffodil) and May-flower (hawthorn); lady's-slipper and foxglove; spindle and self-heal; wood anemones and field poppy. Beyond these is the great lexicon of purely local names. Some species have acquired more than a hundred over the centuries, an extraordinary testament to parish curiosity and inventiveness. Many of these vernacular names record quirks of local geography or custom, or how a particular dialect found its way around a more conventional naming. In south Devon for instance, furze was 'fuzz', wild plums 'bullums', and cleavers 'cliders'. One of the most impressive tallies is for cuckoo-pint or lords-and-ladies (*Arum maculatum*), for which Geoffrey Grigson records some 90 different local tags, all of which say something about the plant's history or associations.[2] Starchwort, for example, recalls the era when the dried and ground-up tubers

Lords-and-ladies or cuckoo-pint has a host of other evocative common names.

were used as a substitute for starch in laundries. The majority are some kind of comment on the appearance of the plant's flowering parts in spring: the long, dull purple or yellow spadix, partially cloaked by a pale green sheath. Many names – even cuckoo-pint itself (pint is short for pintle or penis) – are, perhaps not surprisingly, rather rude: dog's cock (Wiltshire) and priest's pilly (Westmorland), for instance. Others make more genteel reference to the contrasting form and colour of the two parts of the flower shoot, for example Jack in the pulpit (Cornwall) and sucky calves (Somerset). Cuckoo-flower (many locations) probably refers to the time the flower-sheath appears, but may be another euphemistic dubbing and derive from *cucu*, Anglo-Saxon for 'quick' or 'lively'. The commonly used lords-and-ladies is probably a Victorian invention, coined as a polite alternative to this great catalogue of vulgarities.

Most of these names are now confined to books, though, and the worry was that the whole of our plant culture might have become equally moribund. Certainly it has suffered much attrition over the past three centuries. Cromwell's Commonwealth, for instance, suppressed many of the festive parish ceremonies that involved plants (e.g. Beating the Bounds, and May garlands) on the grounds that they were pagan relics.[3] The Victorian era passed on a mixed legacy, too. The practice of using plants as the subject of Christian moral parables added quaint new meanings to a few species but also hastened the end of many of the older and more deep-rooted associations.[4] The so-called 'Language of Flowers' (bay as an emblem for 'glory', acacia for 'platonic love', etc) had no real popular roots at all and was invented from scratch by a group of mid-nineteenth-century French writers.[5]

Yet what we have found in the field research for *Flora Britannica*, and in the multitude of public contributions to it, is that Britain still has a lively popular culture of plants. Although wilder superstitions have faded, and other social groupings – family, friends, schools – are given the loyalty once reserved for the parish, the ancient engagements between plants, people and places continue unabated. What is fascinating is how they are now informed by popular ecology and a sense of social history. The belief, for example, that many of our most interesting plants, from crab-apple to wild garlic, were introduced by the Romans is being replaced by a curiosity about their real origins and uses. And trees with local and historical associations – native lime and black-poplar, yew, wild

Meadow saxifrage and cow parsley in a well-dressing arrangement at Monyash, Derbyshire.

service – are increasingly being planted as landmarks and memorials in preference to exotic species.

The growing affection for trees has doubtless been strengthened by the growing range of threats to them. Around the country they have been symbolic and physical rallying posts for resistance to the 'great car economy'. One in particular, a 250-year-old sweet chestnut on George Green in Wanstead, became the focal point for those fighting the M11 link road through east London. It was occupied by protesters from June 1993 until it was felled amidst scenes of fierce local resistance on 6 December that year. During those months, 400 letters of support were addressed to the tree and its inhabitants.[6] Yet it may be an even sharper sign of how deeply trees have returned to our consciousness that another one could be scapegoated and attacked in an episode that could have come out of the Dark Ages. In Tamworth, Staffordshire, after the funeral of a young man who died when his car crashed into a tree in 1991, ten of the mourners went to the scene of the accident and hacked down the flower-decked tree with axes and a chainsaw.[7] It makes for a rich and sometimes contradictory mixture of science and superstition, communal custom and individual whim; but perhaps this is the shape that folklore is taking on today.

Flora Britannica was launched in the winter of 1991–2, and over the four years that followed it was regularly publicised on television and in the press, as well as through schools, community groups and amenity societies (more than a hundred in total at local and national levels). The many thousands of responses have come in all manner of forms – postcards, tapes of discussions, snapshots and family reminiscences, as well as long and detailed essays on the botanical folklore of individual parishes and individual species. We did not ask for biographical details, so there is no way of statistically breaking down the contributors into young or old, male or female, rural resident or urban newcomer. But on the surface there are no clear biases – except that many contributions came from people who found that talking about their experiences of familiar and commonplace plants enabled them to articulate their feelings about place and nature in general.

Geographically, there are slight biases. Contributions have arrived from all over England, Scotland and Wales, but are densest from areas which have both rich landscapes and a tradition of interest in natural and social history, e.g. Devon, the Welsh Marches, the Sussex Weald and east Suffolk. The big industrial cities are well represented, too, especially Glasgow, Liverpool, Sheffield and Bristol. The only significant gap occurs over central Scotland. There are explanations for this in the history of the region. The infamous Clearances and the long history of sheep and deer ranching impoverished much of the flora, and Presbyterianism frowned on the celebration of what was left. In parts of Scotland, even Christmas trees are still banned from primary schools.[8]

There seems to be little evidence of nationalism in our modern cultural attitudes to plants. What comes across time and time again is the overriding importance contributors attach to their neighbourhood, their local patch. Yet there are feelings shared across Britain which seem to be determined by the kind of place people live in, and sometimes there is more in common between a village in Scotland and a village in Norfolk than there is between two adjacent Welsh settlements. These have been among the most encouraging revelations of the project, echoing as they do the insistence of the 1991 Rio Earth Summit that the future of life on earth depends crucially on local understanding and action.

Many of our contributors would go further than this. For them, an intimate and equal relationship with nature is not so much a path to conservation as its goal. Local plants – which Ronald Blythe once described as 'a form of permanent geography'[9] – are markers not just of their landscapes, but of their autobiographies, as a contributor from Sussex demonstrates:

'Every year on Good Friday we would set off after lunch (boiled cod), each with our basket and a good stock of small balls of wool, for the woods, where we would sink down on the mossy grass and pick bunch after bunch of primroses to decorate the Priory Church on the next morning for Easter Sunday. If Easter was late the woods would be full of the sound of cuckoos and perhaps we might even see a swallow. We might come on a plant of stinking hellebore in the chalky soil but these were rare. White violets had their Special Places. The ones I remember best were at the base of the old flint walls round the churchyard or by the footpath to the Goodwood Dairy which we passed along weekly to fetch our two pounds of butter handed to us by the red-cheeked Mrs Miller, the Scottish dairymaid. Several miles

Left: children explore a fallen plane in Russell Square, London. The damage wrought by the 1987 storm awakened old affections for and new interests in trees. Right: Flora Britannica – high-summer flowers on a Dorset cliff.

of these walls had been built by prisoners from the Napoleonic Wars round parts of the Goodwood Estate. Purple violets were more common but no less cherished, and were followed a few weeks later by masses of pale mauve dog-violets. We did not find wild daffodils in Sussex, though my mother has once bicycled as far as West Dean Woods and seen them there (as I have in the last ten years). Years later, when our children were small and we lived at Bridport in Dorset, we would discover them in the fields near Powerstock and in the Marshwood Vale. The banks of the little Rivers Brit and Asker would be lined with snowdrops here in February, too, and that odd flower butterbur would appear in damp meadows.'[10]

Flora Britannica does not claim to be the last word on the current role of plants in British culture. My hope is that it will trigger as many responses as it records. Yet the popular contributions on which it is based already seem to me to spell out a profoundly important message, an insistence from the grass roots that our vernacular relationships with nature should be taken every bit as seriously as the folklore of less developed areas. The 'post-modern folklore' which they are forming may yet be the best bridge across the gulfs between science and subjective feelings, and between ourselves and other species.

A note on the text

The text includes species of ferns and flowering plants from England, Scotland and Wales. Ireland and the Channel Isles are not included. The species are to some extent self-selected by whether they have figured in local cultures and whether contributors reported this. By British botanists' standards there is an inordinate number of introduced and naturalised species, which are often found more fascinating; but then these *started* with a cultural profile, often by already being in trade or in gardens.

Clive Stace's comprehensive *New Flora of the British Isles* has been invaluable in arranging the text. I have, for the most part, followed his nomenclature in both English and Latin, his ordering of families, and often, where there is doubt or disagreement, his verdicts on the status of species.

The vernacular names (indicated as 'VN') are all ones that were contributed to the project as being in current use, and they are usually printed in the spellings in which they were submitted. I have tried to eliminate obvious copying from previous printed sources, though there are inevitably some borderline cases. Except in special circumstances, I have not indicated particular areas where particular names prevail. The geographical mobility of contributors, who are often writing from one place and remembering another, and the mobility of the names themselves through the mass media would have made this a misleading and potentially inaccurate qualification.

The notes from contributors are printed as they were sent in. Editing has been confined to selection of passages and occasional changes in spelling and punctuation to assist clarity. Editorial additions are indicated by square brackets. As many of the contributions were handwritten, I must apologise if I have made any errors in transcription either in the text itself or in the names of contributors. If notified we will do everything we can to remedy such mistakes in any future editions.

Reference numbers refer to the Source notes section, which is grouped under the same families as the body of the text. First-hand evidence is indicated by name, parish and county of the contributor. Book references appear in the notes under author and year of publication with full references being given in the Select bibliography.

The reference section at the close of the book also contains a supplement on Scottish vernacular plant names and an index in which all places referred to in the text – cities, towns, villages, woods, hills, rivers, churches, etc – are listed, with their counties.

Horsetail family *Equisetaceae*

Field horsetail, *Equisetum arvense* (VN: Mare's-tail, Lego plant). Field horsetail is an abundant plant of waste and disturbed ground, but still a surprising one, seeming like two different species growing in the same spot. The cones are borne on the tops of separate unbranched stems rather like brown sticks of asparagus, which appear in March and April, two months before the branched green stems.

This is the commonest British horsetail, and haunts ground where the soil is compacted or where there is poor drainage. Derelict allotments, canal banks and road-verges are typical sites. In old gravel pits and brickworks, its feathery brushes sometimes grow sharply contrasted against the sculptured flowers of common spotted-orchid.

Children have doubtless always known that horsetail stems can be dismantled in sections – and then put back together again.[1] The famous toy building system may or may not have been inspired by the plant, but it was inevitable that it would provide a nickname sooner or later.

Great horsetail, a plant from an ancient family, whose stems can be dismantled in sections.

'The whole family have been called Lego plants up here [Derbyshire]. I got the name from a botanist nearby, who himself got it from a schoolkid, who for all I know made it up on the spot.'[2]

'I remember it growing by a porch leading to the outside loo, the kind with a high wooden seat. I used to pick one on my journey and sit and pluck it to pieces.'[3]

Horsetails have also been boiled in water to make a fungicide against mildew. 'Several members of my classes have tried this on rose mildew and had better results than with proprietary fungicides.'[4]

Field horsetail, in common with the rest of the family, deposits crystals of silica on its stems and leaves, making them feel like fine sandpaper if you pull them through your hands. The **rough horsetail** or **Dutch rush**, *E. hyemale*, is the most strikingly rough species and, before the days of steel wool, was used for scouring pans – hence the early names of pewterwort and scrubby-grass. It was still being sold for this purpose in Austrian markets in the 1950s.[5] In the seventeenth century John Aubrey recorded the use of 'Horse-taile' by watch-makers and brass-workers for giving an extra finish after filing;[6] it was also used by fletchers for smoothing arrows and as a kind of sandpaper by wood-carvers (see p. 118). Being a scarce species in Britain (and largely confined to the north now), the rush was imported from Holland.

Other conspicuous species include **marsh horsetail**, *E. palustre*, and **water horsetail**, *E. fluviatile*, both quite common tall colonial species of ponds, ditches and marshes. **Great horsetail**, *E. telmateia*, can grow up to six feet tall on damp (though not waterlogged) ground. 'A spectacular plant for indoor decoration – even one stem in a vase – with its lovely ivory stems and grey-green spidery branches.'[7] In London, it is most frequent on railway banks over clay – and in the slightly similar conditions created in cemeteries by gravedigging.[8] **Wood horsetail**, *E. sylvaticum*, mainly an upland species, is the most graceful member of the family, with whorls of delicate feathered branches which droop at the tips.

All the species are widely known as mare's-tails, but the true mare's-tail, *Hippuris vulgaris*, is a flowering waterweed from an entirely unrelated family.

Ferns *Pteropsida*

Ferns are evocative plants, redolent of landscapes of humidity and shade. They conjure up dappled woodland, West Country lanes, old stone walls, even Victorian grottoes. This isn't just romantic fancy. Ferns prosper in moist conditions and reproduce not by seed but by minute spores, which need damp for successful fertilisation.

Their aesthetic and ecological associations mean that ferns have had more than their share of tribulations over the past two centuries. In the Victorian era there was a fashion for collecting them, for the purpose of pressing, producing spore-prints, or especially for 'growing on' in the miniature indoor glasshouses known as 'Wardian cases'. The scale and effects of the 'Victorian Fern Craze' have been graphically documented by David Elliston Allen.[1] Yet, despite the plunder, it is doubtful if any species was made even locally extinct. In fact a number of ferns from rocky places in the north and west have actually expanded their range over the past few centuries, by taking to substitute habitats, particularly walls and the sheltered stonework of old buildings.

But such places are no longer the refuges they once were. The passing of steam has dried out railway cuttings and tunnels. Central heating is doing the same to houses and factories. And everywhere, walls are subject to repointing, weatherproofing and the eradication of hapless plants of any group.

Adder's-tongue, *Ophioglossum vulgatum*, and its relative **moonwort**, *Botrychium lunaria*, are small, scarce ferns of old grassland whose oddity of appearance once made them much in demand by herbalists. They are still special and mysterious plants to discover, barely standing clear of the late spring grass.

Adder's-tongue is the more southerly species. One single short frond grows each year, which divides to form an oval 'leaf' encasing a tongue-like spike (which carries the spore-cases) in something of the manner of lords-and-ladies (see p. 385). In the days of sympathetic magic it was believed to be a cure for snake-bite.

Moonwort is a plant chiefly of dry upland pastures and rock ledges in north and west Britain. Its 'leaf' is fringed with half-moons. It was once believed to be capable of opening locks and unshoeing horses. The seventeenth-century herbalist and astrologer Nicholas Culpeper passed on a Civil War legend about its power:

A depiction of moonwort on a canvas-work panel in Hardwick Hall, Derbyshire.

A screen from Wallington: ferns, including royal fern, by Pauline Trevelyan (and grasses and cornflower by John Ruskin).

'On the White Down in Devonshire, near Tiverton, there was found thirty Horse-shoes pulled off from the Feet of the Earl of Essex his Horses, being there drawn up into a body, many of them newly shod, and no reason known, which caused much admiration ... and the herb usually grows upon Heaths.'[2]

Royal fern, *Osmunda regalis*. A fern which merits its name, growing occasionally up to 10 feet tall, with fronds that are cut into broad and elegant leaflets. It is a species of fens and wet woods, and was one of the most frequently plundered by Victorian collectors. It is now making something of a comeback, in the West Country at least, by escaping from the ornamental lakes and shrubberies where it was introduced a century or so ago, and recolonising woods and river-banks. It is also plentiful in the fens around the Norfolk Broads. An unusual site (1991) was in a crack about seven feet up in a high retaining wall in central Lancaster.[3]

Tunbridge filmy-fern, *Hymenophyllum tunbrigense*, is a delicate, almost translucent species found on damp rock-faces and tree-trunks in shady coombes in western Britain. It was first discovered, outside its main range, near Tunbridge Wells in Kent, in 1696. It is still to be found in the village of Eridge, two miles outside the town (though also over the county border, in Sussex). **Polypody**, *Polypodium vulgare*, is named from the numerous foot-like divisions of its root system. It is typically found growing on the trunks and branches of trees, and is one of those species which helps give an ancient forest feel to the banked and wooded landscapes of the west and north. It is only scattered in middle England. As a native, **maidenhair fern**, *Adiantum capillus-veneris*, is a rare plant of sheltered limestone cliffs near the sea, in west and south-west Britain (excluding Scotland). It has beautiful fan-shaped leaflets on wiry stalks, which rather fancifully suggested female pubic hair to those who named the plant. In the nineteenth century it formed the basis of *capillaire*, a flavouring made by simmering the fronds in water for many hours.

Polypody, often found growing on the trunks of older trees.

Bracken, *Pteridium aquilinum* (VN: Fern). Like all abundant and aggressive plants, bracken has an ambivalent image. Stock farmers resent the way it can take over and sterilise good grazing land. Many naturalists regard it as dull and oppressive, inimical to other species. Yet for those who have lived in brackeny places, its sharp almond scent and the first splashes of yellow on its fronds in autumn can evoke powerful feelings: 'There is a sprig of bracken in the final journal of my late father. To him bracken encapsulated the essence of the countryside. On every walk he went through the same ritual – he would push finger and thumb up the stem of a bracken frond, crush the leaves and inhale the released fragrance ... It evoked for him memories of his first holiday spent by the River Severn. He and his friends slept on the floor of a wooden hut strewn with bracken, listening to the nightingale, wishing he hadn't got to go back to the industrial town of Smethwick.'[4]

Bracken – abundant, durable, versatile and free – has provided many people, from town and country alike, with some of their first *physical* engagements with nature:

'Bracken fronds we plaited, folding mini-leaflets over each other until a strong straight strip design was formed.'[5]

'Another frustrating task [during a childhood in the Lake District] was the making of bracken sandals. Plaits of fern or bracken are, of course, easily made. But joining them together to form soles, using grass or rushes as threads, is more difficult. Sometimes we managed this, but the cool footwear this produced was not very durable.'[6]

'Ideal for thatching a bivouac, if the fronds are laid like tiles on a roof. Has kept generations of tentless scouts warm and dry overnight in bad weather.'[7]

But bracken has also played a more serious role in the rural economy. It has been used for manuring and covering potato beds, for dressing chamois and kid leather, and as fuel and tinder. ('Dead bracken shoved up the chimney and ignited, sets light to the soot and saves bothering with sweeps.')[8] In eighteenth-century Scotland, the naturalist John Lightfoot reported, 'the inhabitants mow it green, and, burning it to ashes, make those ashes up into balls with a little water, which they dry in the sun, and make use of them to wash their linen instead of soap. In many of the western isles the people gain a considerable profit from the sale of the ashes to soap and glass makers.'[9] In the Highlands in the late nineteenth century there were experiments in making bracken silage. Stock ate it greedily, without any apparent ill effects, but the practice never caught on.[10]

Chiefly, though, it was a universal packing and padding stuff. It provided winter bedding for cattle, a cool lining for baskets of fruit and fish, and cushioning for the transport of slate and earthenware.

A glimpse of just how important bracken was in local economies is given by an order made in 1764 for the conservation of fern on Berkhamsted Common in Hertfordshire: 'No person whatsoever shall cut or cause to be cut any fern on the common called Berkhamsted Common from the first day of June until the first day of September yearly under pain of forfeiting and paying for every offence the sum of forty shillings.' (Local legend has it that bracken-cutters used to line up on the common, waiting for midnight to chime from the Parish Church and then staking out their patches, like gold prospectors.) This was not an order imposed by the Lord of the Manor; it was an act of self-regulation drawn up by the commoners themselves, to prevent damage to the recuperative powers of the bracken, and to increase the value of their rights in the future.[11]

Most of these traditional uses became obsolete with the invention of the pneumatic tyre, and later with the development of modern packaging techniques. But farm

Bracken taking on its autumn colours on a Chiltern common.

animals are, here and there, still bedded on bracken, along the Welsh borders for instance. 'I have seen it cut and baled on Titterstone Clee, in Shropshire. I have used it myself instead of straw in my chicken run.'[12] In the Yorkshire Dales the cut fern was 'loaded on a wooden sledge and slid down the fell with the aid of a horse'.[13] In the Forest of Dean one smallholder cuts it for garden mulch: 'I have about one and a half acres of the stuff, some of which I make use of: it is an excellent bedding for the donkey throughout the winter. The soiled bedding, stacked until the spring, is then useful as a mulch around the garden. I also regularly cut fronds in the early autumn. Spread around loosely, about nine inches deep, they make a very effective weed-suppressing mulch.'[14]

But in most areas these practices have died out. And where bracken is no longer needed, it is no longer cut and begins to spread remorselessly. It is reckoned that, by 1990, the extent of bracken cover in Britain was between 1.2 and 2.7 per cent of the total land surface, and maybe up to 15 per cent in areas of rough upland grazing in northern England.[15] This has meant an increase in the incidence of bracken-grazing by animals, in Gwent for instance: 'Bracken is not eaten much by animals, but in recent dry summers when the grass has dried up so much [1993], there has been little else to eat.'[16] The grazing of bracken is not a trivial matter, as it is toxic to all animals, causing serious changes in the composition of the blood. It has also been suspected recently of being carcinogenic in humans if eaten to excess (the young shoots are used as food in the Far East) and even from continued inhalation of the spores.

Despite these problems, bracken is still not regarded as an entirely verminous plant. It has an honourable place on the badges of the Robertson and Chisholm clans, and is figured as one of the defining plants of Ashdown Forest on an embroidered kneeler in Nutley Church, Sussex.[17] Enormous fronds are still measured (13 feet being recorded in Savernake Forest, Wiltshire),[18] and everywhere people still welcome the exquisite young 'fiddleheads' in May and the paper-lacework of fading gold fronds in autumn, picked out by the first air-frosts.

Many of the more familiar wall-ferns belong to the spleenwort family, *Aspleniaceae*. Their native homes are amongst rocks in the high rainfall areas of the west and north. But there have been colonies on old stonework, particularly churches, since botanical records began. And the coming of the railways provided more opportunities for spread. The railway system supplied (and still does to some extent) a range of damp, sheltered habitats for ferns to colonise and a means for ferrying their lightweight spores about. Max Walters, Director of the University Botanic Garden at Cambridge until 1983, has compared how ferns have adapted to church and railway habitats – he calls the two life-styles 'ecclesiastic' and 'ferroviatic' – in the dry, windswept flats of what is probably the least congenial county for ferns in England.

At Old North Road Station, on the disused Cambridge–Bedford line, he discovered (in 1968) a dark fern 'cave' under one of the platforms, where there were plants of **wall-rue**, *Asplenium ruta-muraria*, and 'the largest **Hart's-tongue** (*Phyllitis scolopendrium*) which I have seen in Cambridgeshire'. More remarkable was the appearance of the first recorded colony in the county of **brittle bladder-fern**, *Cystopteris fragilis*, which is entirely dependent on artificial habitats outside its natural range, mainly in the limestone uplands. From about 1920 till 1953 this same platform was also home to one of the very few colonies of maidenhair fern (see above) in inland Britain. 'In Cambridgeshire,' writes Dr Walters, 'the most "continental" part of Britain, a shaded overhang or pit can provide a higher humidity, which may crucially determine the ability of ferns to thrive there.' In more open railside locations, there is **black spleenwort**, *A. adiantum-nigrum*. Of the ten new Cambridgeshire records since 1860 (before which it was found almost exclusively on churches) four were on railway walls, including the one separating Cambridge Station Goods Yard from the Cattle Market.[19]

This is by no means the end of the story of Cambridge's spleenwort familiars. The most celebrated fern patch in the city is an ancient colony of wall-rue on the steps of the Senate House, exactly where it was first recorded by Charles Babington in 1860.[20] (Oxford's more mundane, but far more out-of-place, academic fern is the bracken that grows in small embattled patches around the ancient walls of the Bodleian Library and Sheldonian Theatre.)

Most of these species can be found in similar situations elsewhere in lowland Britain. (And they grow in artificial habitats in the west and north, too. Charles Kingsley's daughter, Charlotte Chanter, reckoned the most luxuriant wall-rue she ever saw was 'growing inside the tower of Morwinstowe Church' in Cornwall.)[21] With them may be

maidenhair spleenwort, *A. trichomanes*, and **rustyback**, *Ceterach officinarum*, so called from the dense, rust-coloured, almost felt-like layer of scales on the underside of the fronds.

Species from the buckler-fern family, *Dryopteridaceae*, are the main contributors to the texture of the woodland vegetation in May and June, after the spring flowers have gone.

Hard and **soft shield-ferns** (*Polystichum aculeatum* and *P. setiferum*) are named from the shape of the spore-cases on the underside of the fronds. The soft shield-fern is one of the most 'sportive' species, and more than sixty sports or varieties were found in the wild and taken into gardens in Victorian times. Many are still in cultivation. In the variety *cristatum*, the tips of the fronds subdivide into tassels or crests; in *decompositum* the leaflets are cut down to the midrib; *abruptum* is remarkable for the way the branches are cut short; *biserratum* has large, broad leaflets; and *proliferum* bears miniature fernlets at the angles formed by the branches with the midrib. 'The handsomest of all,' according to one Victorian guidebook, 'is undoubtedly *plumosum*, in which the fronds will reach nine inches in width and nearly three feet in length. It has a spreading, plume-like habit, but is unfortunately a gem which is "rare" as well as "rich".'[22]

Martin Rickard has traced the origin of this last sport and argues that the collecting fad was not entirely harm-

Above: hard shield-fern, growing in a limestone gryke at Gait Barrows, Lancashire. Left: maidenhair spleenwort is quite common on old walls.

ful. Without it, many remarkable varieties would have gone unnoticed and would quite likely have quickly disappeared, never to be seen again either in gardens or in the wild. 'Plumosum Bevis', for example, 'is sterile in most seasons, and it was not until several years after its discovery in 1876 that any sporangia were noticed. The few spores produced subsequently gave rise to some of the most wonderful hardy British ferns in cultivation today: these are 'Plumosum Drueryi', 'Gracillimum' and the cream of the crop, 'Plumosum Green'. Yet the parent of these marvellous plants was only ever found once, in a lane bank at Hawkchurch on the Devon/Dorset border. It was discovered by a labourer, Jon Bevis, who recognised it as different and pulled it from the hedge and delivered it to a local fern enthusiast, a Dr Wills. Would that hedge-bank and that fern still be there today? Possibly, but I doubt it.'[23]

Broad or **common buckler-fern**, *Dryopteris dilatata*, and **male-fern**, *D. filix-mas*, are the most common and widespread woodland species. The latter's closest relative (with which it often hybridises), **scaly** (or **golden-scaled**) **male-fern**, *D. affinis*, has splendid, shiny ginger scales up its main stems. **Lady-fern**, *Athyrium filix-femina*, is in fact from a different family, but was named in contrast to the male-fern because of its greater elegance and delicacy. It is another 'sportive' species, with some 60 to 70 varieties discovered by the end of the nineteenth century. The first, 'Kalothrix' ('beautiful hair'), was found in the mountains of Mourne at the end of the seventeenth century by Sherard.[24]

'Fiddleheads' of scaly male-fern.

Lady-fern prefers slightly more acid soils than the male-fern, but Walter Scott's rhapsodic description of its haunts could apply to the whole fern tribe:

Where the copse-wood is the greenest,
Where the fountain glistens sheenest,
Where the morning lies the longest,
There the lady-fern grows strongest.[25]

Pine family *Pinaceae*

Norway spruce, *Picea abies*. Across most of northern Europe and America, Norway spruce trees have become known simply as Christmas trees. Ever since Prince Albert and Queen Victoria introduced an old German custom to this country in 1841 and hung lights and decorations on a tree at Windsor Castle, a decked-up spruce has been the centrepiece of northern Christmases. In 1850 Charles Dickens called it 'the new German toy', but there was really nothing new about it. Tree-dressing has been practised around the globe and decorated greenery brought into European dwellings for the winter solstice as far back as the Iron Age. And what is often cited as the prototype English Christmas tree – the branch of evergreen nailed to a board and decorated with gilt oranges and almonds, which a German member of the royal household arranged for a children's party in 1821[1] – is not so different from the ancient kissing bough. But Albert and Victoria's patronage made the notion of decking up conifers at Christmas decisively fashionable; and the Norway spruce (introduced to this country around 1500) quickly proved to be a species which could be grown and marketed on a sufficiently commercial scale to meet the new demand.

Many towns and villages now have communal Christmas trees, erected in market places or on greens. The most famous is the one which is set up in Trafalgar Square, London, close to Nelson's monument. Since 1947, this has

The Trafalgar Square Christmas Tree, first donated by the citizens of Oslo to the citizens of London in 1947.

been donated to the people of London by the people of Oslo, as a gesture of appreciation for the help Britain gave to Norway during the Second World War.

Yet considered purely as a tree, a piece of living greenery, the Norway spruce hasn't much to commend it beyond a symmetrical profile, which makes a kind of conical clothes-horse ideal for hanging decorations on. The twigs are rough, the bark scaly, and the needles hard, spiny and notoriously liable to be shed in centrally-heated rooms. This habit has inspired one nickname for the tree which deserves a wider audience. It appears in some Christmas greetings doggerel from a London industrial cleaning firm, printed on a free Hoover bag and entitled 'On the Trail of the Lonesome Pine Needle':

> *You'll find them in the budgie's cage*
> *And in the baby's cot.*
> *You can afford to leave no stone*
> *Unturned, no tender foot unsocked.*
>
> *It's mid-July, you cry out 'Waiter,*
> *What's this in my soup?'*
> *He replies 'Norwegian Tarragon,*
> *According to the cook'.*[2]

Norway spruce plantations are now a common feature of the British landscape. Some are cultivated solely for the Christmas tree market, but the majority double up as softwood timber plots, with some trees cut for Christmas and the more mature specimens being grown on for pulp or light lumber. It has the advantage over its even more widely planted cousin, the Sitka spruce, of being thoroughly frost-hardy.

The **Sitka spruce**, *P. sitchensis*, is named after the old Russian capital of Alaska on Baranof Island and was introduced to Britain by the arboriculturalist David Douglas in 1831. Since then it has become our most widely planted forest tree – and probably the most loathed, for the intensely regimented and darkly monochrome plantations that were draped across Britain between 1950 and the mid-1980s. In the damp and foggy climate of Alaska, Sitka spruces can grow up to 300 feet tall. In Britain, they rarely grow to half this height – and, in any case, are usually harvested when they are little more than 50 years old.

Sitka spruce self-seeds quite successfully.

European larch, *Larix decidua*. Introduced to Britain from the mountains of central Europe in about 1620, the larch formed the raw material for the first British forestry plantations, on the Duke of Atholl's Perthshire estates in the mid-eighteenth century. (He reputedly planted more

Autumn foliage of hybrid larches.

than 17 million.) The tree has a coarse resinous timber, used these days chiefly for fencing, gates and garden furniture. But larches are peculiar amongst conifers in being deciduous, and the European species has feathery, bright green shoots in spring, turning gold in late autumn. It naturalises freely in the vicinity of planted trees.

A probably unique sport is a prostrate variety: 'Possibly the most remarkable tree in Suffolk is the "Creeping Larch" of Henham Hall, which was planted about 1800, and is the only one known of its vintage. It stands in the pleasure grounds of the Hall (now demolished) and while it is only some 2.7 metres in height with a girth of less than 3 metres, the crown spreads for 26 metres north to south, and for 13 metres east to west. All of this reclines on a raised platform constructed from the railway tracks of the redundant Southwold Line.'[3]

The **Japanese larch**, *L. kaempferi* (introduced 1861), which has vivid blue-green shoots in spring, is also commonly planted. But the most popular variety in forestry at present is a natural cross between this and the European larch, *L.* × *marschlinsii*, which has exceptional vigour and growth rate.

Scots pine, *Pinus sylvestris*. There are Scots pines all over Scotland, but barely a vestige of surviving folklore. They have the status of an endemic subspecies (ssp. *scotica*), a Scots speciality, but do not even have an indigenous name. In England, where the tree has not been native for probably 4,000 years, it nevertheless abounds with associations. It is an ironic twist of history, because it was exploitation by the English that led to the destruction of 'the Old Wood of Caledon' and the virtual elimination of the Scots pine from Scots culture.

In medieval times, the great forest of native pine and birch stretched across most of the Highlands, from Perth to Ullapool. But from the late seventeenth century, it began to be ransacked, first to provide charcoal for the lowland iron foundries, then to support the insatiable timber demands of the Napoleonic Wars. Any chance that the trees might regenerate was dashed by the notorious Highland Clearances in the eighteenth and nineteenth centuries and the blanketing of the denuded hills with sheep and later with deer. By the 1970s it was estimated that little more than 25,000 acres remained, much of it in small scattered clumps.[4]

A massive dead pine, Glen Strathfarrar. One of the essential components of natural pine forest.

But regeneration appears to be winning over the browsers, and the latest Forestry Commission surveys suggest an area of natural pinewood far in excess of this, though the figures are not strictly comparable as the 1994 survey includes woods self-sown from nineteenth-century plantations. In areas fenced off from deer, something approaching the conditions of the boreal pine forests can be glimpsed: rotten trunks still standing, young self-sown saplings, a dense understorey of bilberry and heathers, and the air tangy with the scent of pine resin and juniper.

In England and Wales, the warm period that set in about 5,000 years ago meant that pines were finally driven out by deciduous trees. Since then, all Scots pines have been planted or have self-seeded from planted trees. On many heaths in southern Britain they regenerate almost as vigorously as birch and are regarded as a menace. But many of the older specimens are of great historic interest. The conspicuousness of the tree in the lowlands – the fissured, ruddy-brown bark and rough shelves of evergreen needles amongst the deciduous oak and ash – made it invaluable as a landmark tree. In parts of the country crossed by drove-roads, Scots pines were planted in clumps to mark the way and signal where grazing and hospitality could be had for the night: 'Pines were planted in groups of three or four trees on high ground visible from the previous group ... When were they planted? By whom? Who organised this mass marking? The nearest group to my home [Lower Broadheath, Worcester] is almost directly opposite Sir Edward Elgar's birthplace on the road from Wales via Tenbury Wells, Clifton, Martley, Broadheath to Worcester.'[5]

There are no easy answers to these questions (except that Scots pines live to about 250 years of age), but there is no doubt that the drove-roads across the Welsh border carry one of the great concentrations of waymark pines, from the outstanding group on Bromlow Callow near Minsterley in Shropshire,[6] to the scatter along the roads joining Bewdley, Leominster and Weobley:

'Near Bewdley, many of the farms have two or three Scots pines near the buildings. These we have always believed were markers for drovers bringing cattle and sheep to this area from Wales.'[7]

Native Scots pine on Beinn Eighe, and, in the background, scattered remnants of 'the Old Wood of Caledon' that once covered much of the Highlands.

'The old tale is that the firs were planted to let travelling folk know they would be welcome at the farms [Yatton, Herefordshire].'[8]

There are similar clumps at the entrances to the ridings in Salcey Forest, Northamptonshire, and around farms on the old Brownlow estates at Ashridge (now National Trust land).[9] In Yorkshire, pine-marked fields where cattle could be rested were known as 'Halfpenny Fields'.[10] In Oxfordshire: 'Many farms have plantations of Scots pines around them, or as an avenue leading to the house. There is a theory that these pines had been planted by Jacobite supporters who had moved south when times were hard for Scottish farmers, as a sign of a safe house for Jacobite farmers.'[11]

But the most extensive 'pine-ways' are to be found on the southern chalk downs – rather surprisingly, given that this is not the tree's favourite soil. Kenneth Watts has made a detailed study of the droveways in Wiltshire, and some of the best examples of pine waymarking he has found include: Stock Lane from Aldbourne to Marlborough; Swayne's Firs on the Wiltshire–Dorset border, north-east of Martin Drove End; at Four Barrows on Sugar Hill, on the drove down into Aldbourne. There are also large enclosures surrounded by Scots pines ('drove closes') on, for example, Horse Down west of Tilshead, presumably associated with the Yarnbury Fair, which was held five miles to the south. A single sentinel pine marks Trowle Common Junction near Trowbridge. And there are pine-fringed drove ponds on Golden Ball Hill on the north scarp of Pewsey Vale, associated with the droveway from Tan Hill to Marlborough, and on Pertwood Down on the drove between Monkton Deverill and Tytherington.

'Perhaps the best example is Limmer Pond beside the Chute Causeway in the extreme east of the county. This pond – which is surrounded by massive pines – stands near the line of a drove running south from near Scots Poor to Weyhill Fair site.'[12]

During the enclosure of the East Anglian Breckland in the nineteenth century, large numbers of pines were planted as hedges and shelter-belts. (They were known as 'Deal Rows' locally.) On the light and often windblown soils of this region, pines were felt to stand a better chance of surviving than the more usual hawthorns or hazels. But, remarkably, they were often managed as deciduous hedges – lopped off at the top and then trimmed to encourage denser, wind- and sand-proof foliage. When cutting and management ceased around the beginning of this century, the region's strong winds took over the trimming process. (The geographical location of fields in Breckland was jokingly regarded as dependent on the direction of the wind: 'Sometimes that's in Suffolk, sometimes that's in Norfolk.') As a result Breckland is characterised by miles of dwarfed, contorted Scots pine belts. One of the most extensive runs for several miles alongside the A11 near Elveden, Suffolk. On Barnham Cross Common, nearby, one pine has become so contorted that part of the trunk has looped round itself, forming a hole. It is called the 'Trysting Tree' locally, and lovers link hands through the hole.

In the Lake District, there are pockets of naturally twisted Scots pine, presumably naturalised but with a wild cast about them:

'Between Penrith and Appleby there is a low hill of infertile sandstone called Whinfell, planted mainly with Scots pine. Here and there, I think always self-sown, there used to be odd trees which were also Scots pine but clearly of a different race. The boles were short and stubby, coarsely and very heavily branched and the timber full of big knots. They were the sort of tree no forester would want to encourage and I called them "Whinfell Rogues". Though we are often told that Scots pine became extinct in England and Wales, I have often wondered whether pockets survived, of which the self-set Whinfell Rogues are one.'[13]

There are plenty of small-scale domestic uses for pines. The cones (sometimes known as 'dead apples') are used for weather-forecasting and as kindling. The resin (which seeps through and hardens on the outside of the trunk) is the source of an antiseptic oil and has occasionally been used as a rough-and-ready medicinal chewing-gum for throat infections. Steeped in white wine, it also makes a passable imitation of Greek retsina. Pine shoots can add a resinous flavour to cooking oil and vinegars, and may have been the coniferous bittering agent used in the ancient Scots recipes for 'spruce beer' (cf. heather ale, p. 161) – though true spruce could have been imported from Scandinavia.[14]

Corsican pine, *P. nigra* ssp. *laricio*, comes from south-

A lopped Scots pine windbreak, Breckland.

Representations of oak and pine from a Tudor pattern book, early sixteenth century.

ern Europe and is commonly planted in shelter-belts and forestry plantations on sandy soils. **Lodgepole pine**, *P. contorta*, from western North America, is one of the commonest forestry species. **Monterey pine**, *P. radiata*, from California, is a distinctively domed tree which has been much planted in south-western Britain, where it occasionally self-seeds: 'We have a row of huge Monterey pines on the north side of the Rectory garden. They and the church are visible from high points all over the county [Cornwall]. But the most interesting thing is that the trees are visible from the sea, and sailors and fishermen still use them as a landmark (literally) when they are out at sea.'[15]

'Many of the coastal resorts also use this species for shelter where it gives a very distinctive character to the area.' (Cornwall County Council, Landscape Strategy, 1993)

Maritime pine, *P. pinaster*, from the Mediterranean, is regarded in much the same way along a strip of coast straddling the Dorset–Hampshire border. 'The current landscape started in the 1800s when pines were planted on the heathland. The Scots Pine was originally planted, but was later replaced by the Maritime Pine, which seemed more resistant to drought. Such large numbers of Maritime Pine were planted that this tree became characteristic and received the alternative popular name of the Bournemouth Pine.' (Branksome Park Conservation Policy, Poole, 1989)

Weymouth pine, *P. strobus* (from central and eastern North America), however, is named not after the town but after an early Lord Weymouth, who planted many at Longleat in the eighteenth century.

Juniper family *Cupressaceae*

Juniper, *Juniperus communis*, provides the essential flavouring for one of the country's favourite spirits, gin. But native juniper has probably not been used by British distillers since the last century, and the berries are now mostly imported from eastern Europe. Home-grown fruits, however, are increasingly used as a flavouring, especially with game: 'The local limestone hill, Arnside Knott [Lancashire], provides us with juniper berries to cook with venison. Used whole they give a bitter, crunchy bite to savouries.'[1]

Limestone hills are just one of the favoured places of this evergreen shrub, which tends to grow in colonies and

The aromatic berries of juniper are used for flavouring gin.

have a striking impact on local landscapes. It is widespread over much of Britain, but eccentrically local, occurring only on comparatively well-lit and well-drained rocks and soils. And its two classic habitats could hardly be more different. In the north it prefers cold, rainy sites on acid soils, growing with heather and bilberry on moorland and as an understorey in the Highland birch and pine woods. In the south, it is a species of hot, dry, calcium-rich soils and haunts the parched downlands of chalk country.

The form of individual bushes is also very varied. They can be low and prostrate at one extreme, and conical or cylindrical at the other, sometimes up to 17 feet in height. Bushes can also be bent and trimmed by wind and browsing, and change shape spontaneously with age, the older ones having a tendency to die out from the centre and collapse. (This is evident in one of the most long-standing populations in Britain, on the Taynish peninsula in Argyll, where there are almost no young bushes.) From a distance a large colony of juniper can look like a fantastic piece of topiary, a landscape of tapers and sprays, dark shelves and swells.

In some regions, juniper seems to thrive best in places where there have been cycles of change in land-use, where, for example, a period of grazing, which produces well-lit bare ground and short turf ideal for the germination of juniper seeds, is followed by a period of abandonment, which allows the seedlings to grow on ungrazed. In the Lake District and Northumberland, for instance, it is possible that juniper's survival is related historically to the alternations between pasturing and small-scale mining that characterised the rural economy here.[2] In Teesdale, Durham, it often grows around farms and on the boundaries between the enclosed land and the common grazing, where the effect of such changes would be most strongly felt.

Until the end of the last century, the juniper on the fells of Upper Teesdale was put to a variety of uses. Boughs were cut for firewood (which burns with a cedar-like fragrance) and for making the bases of haystacks. Even its insidious prickles were turned to advantage: 'Juniper was often used as a substitute for barbed wire by

Juniper wood, Little Langdale, Lake District, showing the variety of forms in which the shrub occurs.

being placed on the tops of stone walls ... 120 years ago, the berries of juniper, or "junifer" as it was called, were collected by families who travelled from Weardale, who would grind the berries down to flavour bread and cakes.'[3] In England, juniper is distinctive enough to have places named after it. On the Surrey downs, for example, there are a Juniper Hill, Hall and Bottom.[4] Sometimes the name outlives the plant that inspired it. In north Oxfordshire there is a single large bush outside the Fox pub at Juniper Hill, the village immortalised in Flora Thompson's *Lark Rise to Candleford*.[5] It is a relic of what was presumably a much larger colony on Cottisford Heath. This is where, before the late eighteenth-century enclosures, squatters built the settlement that was to become 'Candleford' hamlet.

In southern England, juniper occurs patchily in many chalk areas, especially in Wiltshire, where small bushes can sometimes be seen on road embankments. The largest population in England is inside the Ministry of Defence's Chemical and Biological Defence Establishment at Porton Down, Wiltshire. Here, in buffer-land barricaded by high-security fencing, there are more than 14,000 juniper bushes, growing amongst heather in a community that is known from nowhere else on the southern chalk. As on the northern fells, though, the youngest bushes seem to date from a release from grazing pressure, in this case the rapid decline in rabbit numbers caused by the myxomatosis epidemic.[6]

The oil extracted from juniper has an ancient reputation as an abortifacient (which may have echoes in the Victorian belief in the effectiveness of gin for the same purpose). In Lothian, in the medieval period, giving birth 'under the savin tree' was a euphemism for a miscarriage or juniper-provoked abortion.[7] Until at least the mid-1980s juniper pills (still on the market in 1993) were being advertised as 'The Lady's Friend' in the small ads in ladies' journals.[8]

The name 'savin' does not seem to have survived in Britain, but in Herefordshire, where the species does not grow in the wild, some kind of juniper was known as 'the savage tree' and was used as a horse medicine (Grigson, interestingly, quotes the name 'horse saving' from Cum-

An overgrown hedge of Leyland cypress.

berland)[9]: 'It was crushed up and put in very small quantities in horse feed, and was said to "ginger them up".'[10]

It is possible that this was a cultivated variety of a European juniper, *J. sabina*, which is known as 'savin', and which yields oil of savarin, more potent – and toxic – than that from common juniper.

Leyland cypress, × *Cupressocyparis leylandii*. Exceptionally rapid growth – up to six feet a year – and dense evergreen foliage have conspired to make this the most popular of all garden hedging shrubs. It first appeared at Leighton Hall, Powys, in 1888, as a cross between Monterey and Nootka cypresses, and by 1995 was reckoned to have close on 60 million descendants in Britain. They are even beginning to appear in farm hedges in open countryside, where, being more public, they will be more contentious than they already are in gardens. Their furious expansion and funereal foliage have made *leylandii* hedges notorious for inflaming disputes between neighbours, though occasionally they are used to pre-empt them. A Milton Keynes contributor heard a friend say, 'I wasn't getting on too well with my neighbour, so I planted a hedge of Leyland cypress.'[11]

Monkey-puzzle family *Araucariaceae*

Monkey-puzzle, *Araucaria araucana*. This striking tree was introduced to Britain by the botanist Archibald Menzies after an expedition to Chile in 1791. The story is that, while he was dining with the Viceroy of Chile, Menzies pocketed some of the kernels of the nuts they were offered for dessert, and succeeded in germinating them in a frame on his ship. The seedlings were subsequently donated to Kew, where they proved to be the previously unrecorded 'Chile Pine' or monkey-puzzle.[1] Much greater quantities of seed were brought back in the nineteenth century by the horticulturalist William Lobb, and the tree became a popular and conspicuous feature of Victorian gardens. But more specimens than was previously realised were planted out in the interiors of nineteenth-century plantations and are now beginning to show themselves. Monkey-puzzle self-seeds very rarely in this country, but the instances are increasing, which may be some small compensation for the ravages the tree has been suffering in its native cloud-forest habitat.

Most ancient British yews are associated with churchyards.

Yew family *Taxaceae*

Yew, *Taxus baccata* (VN: Hampshire weed; Snotty-gogs (for the berries)). A mature yew is a compelling tree whatever its situation. It has the densest, darkest foliage of any evergreen and a buttressed trunk that comes close to the colour of mahogany. Its wood reputedly outlives iron. A 250,000-year-old yew-spear found at Clacton in Essex is the world's oldest known wooden artefact.[1] Yet what sets yews most decisively apart from other trees in Britain is the remarkable and probably unique association they have with ancient churches. At least 500 churchyards in England and Wales alone contain yew trees which are certainly as old as the church itself, and quite likely a good deal older. Yews of great ages are rare outside churchyards, and no other type of ancient tree occurs so frequently inside the church grounds. I do not know of any similarly exclusive relationship between places of worship and a single tree species existing anywhere else in the Western world.

It is obviously a meaningful association, however cryptic, and when you contemplate yews of extreme age it is hard not to believe that the meaning is a profound one. In the village of Fortingall, Perthshire, at the geographical heart of Scotland, there are living fragments of the shell of a stupendous yew. It stands in the corner of a churchyard where there has been a building for worship since at least pre-Reformation times. Nearby there are groups of ancient, possibly Druidical stones. In 1769 Daines Barrington measured its girth at 52 feet, but it was already a hollow ring of wooden pillars, like a wood-henge, and funeral processions reputedly passed *through* the trunk. There is a legend that Pontius Pilate (whose father was supposedly a legionary stationed in Scotland) played under its branches.[2] Guesses about its age range from 2,000 to 9,000 years.

Even modest, middle-aged yews can have a powerful presence. The tree which stands on a mound at the southern corner of St Peter's church in Berkhamsted is probably no more than 350 years old. It was a local tradition for townspeople to gather under it on New Year's Eve, which I can remember persisting until the early 1960s. On windy nights in the heart of winter, the twigs still stream above the High Street like ceremonial bunting.

It is no wonder that ancient yews have been the subject of all manner of theories and myths about their origins, age and meanings. At school we were taught that the mound on which the Berkhamsted tree stands contained our town's plague victims; and yews were certainly once planted over graves to protect and purify the dead. This was a business which could create perils of its own: 'if the Yew be set in a place subject to poysonous vapours, the very branches will draw and imbibe them, hence it is conceived that the judicious in former times planted it in churchyards on the west side, because those places, being fuller of putrefaction and gross oleaginous vapours exhaled out of the graves by the setting sun, and sometimes drawn by those meteors called *ignes fatui*, divers have been frightened, supposing some dead bodies to walk, etc.'[3]

There have been more mundane explanations for yews' presence in churchyards. They were planted in these protected plots to provide wood for long-bows and to keep their poisonous foliage out of the reach of browsing cattle; to provide decoration for the church, or as a *memento mori*.

The distribution of old yews in old churchyards (they are concentrated in south-east and central England, Wales and the Lake District) is reflected in the tree's distribution in the wild. Yew is principally a species of well-drained chalk and limestone soils. In ancient woods it grows in the company of beech, maple and ash, and on sheer slopes such as Stoner Hill in Hampshire it can look dramatic in winter, silhouetted against the white plumes of old-man's-beard. But yew's sticky red berries are popular with birds, and bird-sown seedlings will colonise open chalk downland as well, forming dark thickets under which nothing else can grow. The trees live so long that woods formed in this way can become at least a temporary 'climax' vegetation and persist for centuries. 'Nunton Ewetrees' in Wiltshire, first described by John Aubrey in 1685,[4] still survives near Downton in Wiltshire (though none of the individual trees look very old or large, and some appear to have been coppiced). The famous horseshoe of chalk at Kingley Vale in Sussex has been invaded by yew largely over the past hundred years. But in the heart of the thickets of younger trees is a group of more venerable yews, probably 500 years old. They are spectacularly split and sinuously interwoven, and in places they have welded together to form multiple trunks. Many of the larger boughs have sagged down to ground level, and the experience of scrambling through this twilit wooden labyrinth, over damp and musty leaf litter, is not unlike being in a series of subterranean caverns.

An early theory about the link between yews and churchyards occurs in Sir Thomas Browne's *Hydriotaphia* (1658). He wrote, 'Whether the planting of yewe in

The extraordinary texture of the wood inside a hollow yew.

Yew, *Taxaceae*

Churchyards, hold not its originall from the ancient Funerall rites, or as an Embleme of Resurrection from its perpetual verdure, may almost admit conjecture.'[5] A century later, when Gilbert White wrote a detailed description of what was to become England's most famous (and most measured) yew, he took a slightly different view of its symbolism:

'In the church-yard of this village [Selborne, Hampshire] is a yew-tree, whose aspect bespeaks it to be of a great age: it seems to have seen several centuries, and is probably coeval with the church, and therefore may be deemed an antiquity: the body is squat, short, and thick, and measures twenty-three feet in girth, supporting an head of suitable extent to it's bulk. This is a male tree, which in the spring sheds clouds of dust, and fills the atmosphere around with it's farina.

As far as we have been able to observe, the males of this species become much larger than the females; and it has so fallen out that most of the yew-trees in the church-yards of this neighbourhood are males: but this must have been matter of mere accident, since men, when they first planted yews, little dreamed that there were sexes in trees ...

Antiquaries seem much at a loss to determine at what period this tree first obtained a place in church-yards. A statute passed AD 1307 and 35 Edward I. the title of which is "Ne rector arbores in cemeterio prosternat." ['To prevent the rector from felling trees in the graveyard.'] Now if it is recollected that we seldom see any other very large or ancient tree in a church-yard but yews, this statute must have principally related to this species of tree; and consequently their being planted in church-yards is of much more ancient date than the year 1307.

As to the use of these trees, possibly the more respectable parishioners were buried under their shade before the improper custom was introduced of burying within the body of the church, where the living are to assemble ...

One of the giant yews – probably five centuries old – in Kingley Vale, Sussex. This remarkable valley on the southern chalk has yews of all kinds: a central core of ancient trees and spreading new woods of bird-sown specimens up to a hundred years old.

The farther use of yew-trees might be as a screen to churches, by their thick foliage, from the violence of winds; perhaps also for the purpose of archery, the best long bows being made of that material: and we do not hear that they are planted in the church-yards of other parts of Europe, where long bows were not so much in use. They might also be placed as a shelter to the congregation assembling before the church-doors were opened, and as an emblem of mortality by their funereal appearance. In the south of England every church-yard almost has it's tree, and some two; but in the north, we understand, few are to be found.

The idea of R.C. that the yew-tree afforded it's branches instead of palms for the processions on Palm-Sunday, is a good one, and deserves attention. See Gent. Mag. Vol. L. p. 128.'[6]

Some of White's hypotheses do not really hold water. Individual yew trees would not provide much protection from the wind; and, in any case, this function would hardly have been compatible with their harvesting for long-bows, which were cut from the trunks, not the branches. In fact English yew was regarded as being too brittle for use in long-bows, and the wood was usually imported from Spain and Italy.[7] (This also meant that there was little point in conserving yews – and thus separating them from stock – within the confines of the churchyard walls, especially as animals were frequently allowed to graze in churchyards.)

But the use of yew branches as 'palm' was certainly widely practised, especially when Easter occurred too late for gathering sprays of pussy-willow, the usual English substitute for real palm and olive branches. In the churchwarden's accounts of some parishes, presumably those without yew trees of their own, there are entries for payments made for the purchase of yew 'Palme'.[8] At Kington, in Herefordshire, yew was also used in Whitsuntide celebrations and brought into churches to decorate the tops of the pews.[9]

Since White's time it has generally been presumed that yews were planted in churchyards not as emblems of mortality, but, because of their evergreen foliage, of *im*mortality and resurrection. Yet there have been difficulties in relating this theory to a specifically Christian tradition. As more and more ancient yews have been examined by naturalists and antiquarians, the more it has seemed that many are not just 'coeval' with the church, but vastly older. Circumstantial evidence in the form of earthworks, local legends and the sheer physical bulk of many of the trees has suggested ages of up to at least 2,000 years.

In the 1940s Vaughan Cornish surveyed many of the yews in British dioceses and parishes, and concluded (though without a great deal of solid evidence) that the oldest were not Christian plantings at all. They were the sacred trees of ancient religions, some Druidic, some Celtic, and a few, maybe, relics of pre-Celtic Iberian settlers (hence their frequency in Wales). And, like many pagan icons and practices, they were retained and pragmatically sanctified by the Christian church. Moreover, from the medieval period it became the custom to plant two yews in churchyards, one close to the pathway which leads between the principal entrance of the church and the funeral gateway, and the other beside a path which leads to a second and lesser doorway. Coffins would probably pass both trees during funerals.[10]

The problem with the historical basis of Cornish's theory is getting any accurate confirmation of the ages of the trees. After 400 or 500 years almost all yews begin to lose their heartwood and become hollow, making dating by ring-counts impossible. They also enter long periods of suspended growth, when they put on virtually no extra girth at all. The great yew at Crowhurst in Surrey apparently grew only nine inches in girth in two and a half centuries, from 30 feet in 1630, to 30 feet 9 inches, recorded in both 1850 and 1874.[11] The Selborne yew seemingly *shrank* between 1950, when Sidney Scott reckoned it 'nearly 28 feet',[12] and 1981, when it was more precisely measured at 25 feet 10 inches.

The last record was made by Allen Meredith, who has made a long study of the sizes, positions and archaeological associations of ancient British yews. From this, and the limited documentary evidence that is available about the dates when some church yews were planted, he has drawn up a tentative table linking girth and age. Young trees, whose ages are verifiable by ring-counts, raise no problems, and a yew 12 feet in girth, for instance, is given the uncontentious age of 300 years. The laboriously slow growth of extreme old age, though also partially quantifiable by ring-counts in the trunk's shell, is altogether more speculative. Meredith puts trees with a girth of 30 feet at 2,400 years, and of 33 feet at 3,000 years. This would put three trees in three churchyards in Powys (Defynnog, Discoed and Llanfaredd), all of which exceed 35 feet, at more than 4,500 years of age. (A younger tree – a mere 22 feet in girth, at Strata Florida, over the Powys border – is the burial site of the medieval poet Dafydd ap Gwilym.)[13]

The extrapolation of the growth-curve, back not just into the pre-Christian period but into prehistory, is done with not much more solid evidence than Vaughan Cornish had access to. There is no living heartwood from British yews older than 400 years, no datable ancient timbers (as there are for most other species). No one is sure, yet, at what age they begin to go hollow, or when or why they go into slow-growth mode.

And there is one more complication. The planting of wild tree species was very rare before the Middle Ages – which suggests that the original sacred sites were situated close to existing yew trees, rather than vice versa. Yet can the architects of at least 500 Christian churches really have wedged their buildings into already established ground-plans?

Hollow yew in Much Marcle churchyard, Herefordshire. There is a cluster of ancient churchyard yews in the area and this one is believed to be over 2,000 years old.

But there is no doubt that old yews have an irresistible aura of extreme antiquity, and it is hard not to believe that they antedate their attendant churches. At Hambledon, Surrey, there is a tree with a girth of 35 feet and an enormous spread, so that the lower branches are propped up on tombstones. The hollow trunk is covered with burrs, which have the look of green pin-cushions from their bristling epicormic shoots. At Stedham, West Sussex (12 miles south of Hambledon in this concentrated zone of churchyard yews), the yew is slightly thinner (30 feet in girth), shorter and squatter, and the trunk is held together by wire hawsers. Yet its interior has a quality missing in the Hambledon tree, and found in other hollow yews with dryish interiors. The shelved surfaces of the dead wood have a lustrous, satiny finish, close to the texture of wasp-nests. In places they are bleached like driftwood; but here and there patches of colour break through – the orange of living yew-wood, a violet sheen of the kind sometimes seen in mother-of-pearl, small invasions of green algae.

These brooding, gothic trees frequently became landmarks when they grew in the wider countryside. A venerable yew stands over the site of the ancient holy well of Glangwenlais, in the Carmel Woods near Ammanford, Dyfed. Wordsworth's yew in Lorton Vale, Cumbria ('Of vast circumference and gloom profound/ This solitary Tree! a living thing/ Produced too slowly ever to decay;/ Of form and aspect too magnificent/ To be destroyed …'),[14] still stands by White Beck, High Lorton.[15] The Celtic 'iw' is one of the oldest tree names and, transmuted into 'ew' or 'ewe', is an occasional component of place names, as in Ewhurst (Hampshire, Sussex and Surrey), meaning 'yew-tree wooded hill', and Ewshott (Hampshire). In Cheshire Jill Burton surveyed yew place names, and found 74, mostly attached to isolated farms or cottages: 'The majority of Yew places are in a central band up the Weaver/Dane valley, from the southern border with Staffordshire. There is only one in the Wirral, one in South West Cheshire and one on the extreme East border with Derbyshire. 67 per cent are actually on or very near roads, with 22 per cent being near "old roads". Some 52 per cent are within 500 metres of a boundary. Surprisingly, only 6 per cent are near a church … 28 per cent are at junctions of roads and footpaths, with the inference that the roads and paths led to them, or even that they could have been used as markers for crossings. Less than 12 per cent are in hamlets *and* near roads.'[16]

There are many 'Yew Tree' inns, too (though the trees are invariably much younger than those in churchyards), as, for instance, in Lower Wield, Hampshire, Odstock, Wiltshire, and near Newent, Gloucestershire. A gateway yew at the North Star pub at Steventon in Oxfordshire has been divided into two for access and has a lamp attached.[17]

'On the old drove road across Ashdown Forest two yews were a sign for travellers of the availability of overnight accommodation. Three yew trees indicated additional provision for animals. Three such yews can be found at Duddleswell Crossroad on Ashdown Forest. One of these old yews was uprooted during the 1987 gale; but before the Forest Rangers had an opportunity to right it, most of the wood was cut up and it disappeared.'[18]

'When the very ancient yew in The Lee Graveyard blew down in the 1990 storm, the bellringer told me that it was very unfortunate for the vergers, etc, as it was always used as a WC. Its huge, hanging branches concealed all.'[19]

'A famous Derbyshire yew is in Shining Cliff Wood near Ambergate – a yew supposedly the inspiration for the nursery rhyme "Rock-a-bye-baby". It is known as the Betty Kenny Tree ... Apparently one of the boughs was hewn out to create a cradle. A family used to live in it, hence Rock-a-bye-baby ... Also in Derbyshire, one at Churchtown, near Matlock, is probably nearer 2,000 years old. The font is made from its wood.'[20]

'A wood carving of a dove in flight, as the handle for the font cover for Hastingleigh Church [Kent], was made from one of the yews blown down in the storms.'[21]

'On their smallholding in Torver near Coniston, my uncles used to make walking sticks and shepherding sticks from the branches of yew trees. They would choose a small branch which had grown from the underside of the yew, and which therefore had to turn up towards the light, forming a ready-made handle for the stick, which was then smoothed and finished.'[22]

The story of the yew in St Mary's churchyard, Selborne, encompasses all these aspects, sacred, secular and commonplace. Its abrupt collapse during the great gale of 25 January 1990 has been graphically described by the vicar, James Anderson: 'The massive trunk lay shattered across the church path and a disc of soil and roots stood

Clare Roberts's sketch of the living lich-gate of yew at St Margaret's, Warnham, West Sussex. It is well over a hundred years old and was previously much larger, but had to be clipped back to avoid obstruction to the pavement.

vertically above a wide crater. The bench around the trunk was still in place, looking like a forgotten ornament on a Christmas tree. A stormy sea of twisted boughs and dark foliage covering the churchyard was pierced here and there by a white tombstone like a sinking ship.'[23]

There were also white bones showing through, brought up to the surface from ancient burials or lying tangled in the root-ball. They were quickly taken into custody in the nearby Field Studies Centre, to protect them from dogs and other predators, and permission for an emergency archaeological dig was obtained from the Diocese, on the understanding that all human remains would eventually be reburied in the churchyard. In the course of the next week, two archaeologists from the Hampshire Museums Service uncovered the remains of about 30 individuals, several of which were complete burials in their original positions. They were all apparently Christian burials, in shallow graves *beneath* the root-mass. The earliest, and the deepest, was dated by pottery from the grave-fill to about AD 1200. A number of nails suggested that the man had been buried in a coffin. He had been placed right against the south side of the yew when it was probably about 10 feet in girth. The original site of the young yew was estimated by the archaeologists to be a patch of undisturbed soil with no remains of any kind, just north-west of the centre of the old tree.[24]

Meanwhile, the fallen yew had become the scene of extraordinary activity. People from all over Britain who had once lived in Selborne, or just visited it, came to pay their respects and to buy or beg a piece of the wood. One man remembered having his daily lunch-break in its shade. Another came to retrieve a fragment for his parents, who had become engaged under it. The source of the wood (some of which was sold for the church funds) was the vigorous lopping the tree received before an attempt to winch it back into the vertical and, in effect, replant it. By mid-February almost the entire crown had been removed, and on the 13th the tree was ready to be raised. A time capsule (containing, amongst many other things, a paperback edition of Gilbert White's book, with its early history of the tree) and a probably superfluous load of tree-planting compost were inserted in the root-hole amongst the resettled medieval skeletons, and a three-ton crane began laboriously to winch the tree upright. By dusk it was done. Later, the vicar and children from the local school linked hands round the tree and said prayers for its survival. And providentially (or so it seemed at the time) a water-main promptly burst close by and bathed the yew's roots in water for the next 36 hours.

The prognosis looked good to start with. That summer, the tree put out a bristle of new shoots on the west side, both on its pollarded branches and from the base of the trunk. For once it looked remarkably like the illustration of the tree that Hieronymus Grimm had made for the first edition of White's book, suggesting that the tree may well have been pollarded in the past.[25]

But in the hot summer of 1991 the shoots withered, and by 1992 it was clear that the tree was dead – the result, some pundits believed, of a surfeit of water. Later that year, in a touching ceremony on 28 November, a cutting taken whilst the tree was still alive was planted in the churchyard by the youngest and oldest citizens of the parish together.

Yet the old hulk lives on in its own way. Its hollow shell (which still has patches of that lustrous, layered, satiny deadwood) has been colonised by young hazel and foxgloves, and a honeysuckle of more exotic origins is beginning to cloak the fluted exterior. Some of the wood taken from the larger branches has been made into artefacts for the church, and a yew font-cover and altar screen now join the rough yew cross which has long hung over the nave. A lute has also been constructed from the wood (a very traditional use), with the curved back of the instrument using alternating strips of the dark heartwood and paler sapwood.

Most yew plantings these days are not of the wild variety, but of the tidier but blander fastigiate variety, or Irish yew, whose branches all sweep evenly upwards, as if they had been bound into a bundle. They are mostly descendants of two trees found on a limestone crag in Fermanagh in the 1760s, and presumably ousted 'normal' wild trees because of their resemblance to Mediterranean funereal cypresses, and for the ease with which they could be clipped – and even 'topiarised' – into order. (The clippings have recently become commercially valuable. An alkaloid named taxol, which seems effective against ovarian cancer, has been discovered in yews, and research laboratories and drug companies are offering to buy the foliage in bulk.)

Churches and churchyards

In England (though much less so in areas of Scotland and Wales) wild plants figure conspicuously in both churches and churchyards. There are good historical reasons for this. The parish church was not only the focal point of the community but one of the main custodians of its continuity. Christian churches often developed on the sites of pre-Christian and Celtic holy places, inheriting some of their nature-worship icons – old yew trees, for example – and Christianising practices such as the hanging-up of winter greenery. They were also centres of culture and craftsmanship, where wood-carvers could work, where medicinal herbs were cultivated, and where ceremonies and rituals involving plants – weddings, funerals, harvest festivals, beatings of the bounds – were centred.

Over a period of more than a thousand years, all this activity has left a rich legacy of plants, both real and representational, inside the territory of the church. From the late thirteenth century, carvings of recognisable flowers began to appear on the capitals of pillars, especially in the larger churches and cathedrals, where skilled masons (sometimes itinerant workers from the continent) could be employed.[1] Misericords, bench-ends and pulpits were favourite sites for local carvers to add distinguishable wayside flowers amongst the more formal rose and vine motifs. Since then wild plants have increasingly figured in stained-glass windows, pew-ends, altar-cloths and kneelers. In tiny churches in the Norfolk Broads, wild plants – sawn-off tussock sedges – *were* the kneelers (see p. 391).

The churchyard has also become a sanctuary for plants. At a time when unimproved grassland has all but disappeared across much of agricultural Britain, these small patches of turf – 'God's Acres' – are in many parishes the last refuges for species such as meadow saxifrage, green-winged orchid and hoary plantain.[2]

There are also churches in woods, hedged churches (for instance All Saints, South Elmham, in Suffolk, which has an abundance of pyramidal orchids and sulphur clover), and ditched churches (including a tenth-century example at Hanbury, Warwickshire, with a churchyard surrounded by an eighteenth-century ha-ha).[3] At Tilbury-juxta-Clare in Essex, the circular churchyard of St Margaret's rises above the vast arable fields like an oasis and is surrounded by a hedged medieval bank-and-ditch. In it can be found scarce woodland plants such as stinking iris and stinking hellebore. There is a remote possibility that they are native relics on this ancient site; but more likely they are relics of herbal cultivation or of winter grave decorations.

Many other species have become naturalised from memorial posies and wreaths: lily-of-the-valley, snowdrop (also planted for Candlemas, see p. 421), primroses (on young children's graves especially, see p. 168), including the pink 'churchyard primrose', garden forget-me-nots, and even rosemary 'for remembrance'. Occasionally ornamental plants such as teasel (and even woad), used for flower arrangements inside the church, seed themselves near the porch.

Wild flowers ornament churches both inside and out. Above: a children's Easter flower arrangement in Okeford Fitzpaine, Dorset. Right: alexanders, naturalised in St George's churchyard, Isle of Portland.

Birthwort family *Aristolochiaceae*

Birthwort, *Aristolochia clematitis*. An intriguing perennial from southern Europe, once in wide use as a medicinal herb because of a fancied resemblance between its funnel-shaped yellow flowers and a uterus. Birthwort was given to speed up labour, and it would have been a standard herb in the gardens of abbeys where the nuns had midwifery duties, though it has never been accepted by orthodox medicine. But it also has more potent pharmacological properties. The distinguished Oxford botanist Professor E. F. Warburg was fond of scandalising audiences by describing birthwort as 'a good abortifacient, only found in England in nunneries, where it is an introduced plant'. Whether it was ever used to cover up rare indiscretions will probably remain a monastic secret. But Warburg's widow, Primrose, writes that a botanical friend heard from a Mother Superior that 'she'd got it round her nunnery and she'd needed it the previous year. History does not, unfortunately, say whether she *used* it.'[1]

Birthwort, a relic of old monastic herb gardens, on the site of Godstow Abbey, Oxfordshire. It was used to speed up childbirth and occasionally as an abortifacient.

In Northumberland it was also used by dairy farmers for expelling the afterbirth after a calf had been born.[2]

Most of the naturalised colonies of birthwort are – or were – on the sites of old abbeys or ecclesiastical establishments. Those at Bury St Edmunds Abbey in Suffolk, the twelfth-century almshouse of St Cross in Winchester and the Benedictine Carrow Abbey in Norwich have now gone. But the plant clings on amongst the nettles at the ruins of Godstow Nunnery outside Oxford, and in a few places in and near Cambridge, where it may once have been commercially cultivated.

Asarabacca, *Asarum europaeum*, is a very scarce and declining species, brought here from mainland Europe as a medicinal herb (though some believe there is a native population as well) and naturalised in a few shady places. The old church at Swyncombe in Oxfordshire and Limebrook Priory in Worcestershire are typical sites,[3] as is the much larger colony in Wiltshire, which has been known since 1820. 'One large patch of it was found by Mr Popham ... away from any house, in the left-hand hedge of the lane going from Standlynch Down to the large chalk-pit at Redlynch, near Salisbury.'[4] The slight air of mystery about asarabacca's origins, its secretive habits and darkly glamorous appearance – dark brown bell-flowers hidden amongst creeping, cyclamen-like leaves – have made it a favourite quarry amongst botanists.

Water-lily family *Nymphaeaceae*

White water-lily, *Nymphaea alba*; **Yellow water-lily**, *Nuphar lutea* (VN: Brandy balls). Both species are relatively widespread in lakes, ponds, dykes, canals and slow-moving rivers. The yellow water-lily has smaller flowers held above the water on a stalk, and smells slightly of wine dregs – hence 'brandy balls' and the older name of 'brandy bottle'. There are carvings of it in Bristol Cathedral, Westminster Abbey and the Angel Choir at Lincoln.[1]

But both species are also ornamental enough to have suffered from collection for garden ponds. William Cowper was an eighteenth-century lily-scrumper and wrote

White water-lilies at Little Langdale in the Lake District.

one of his lighter verses about how his dog, Beau, made up for the poet's inept attempts to pick a water-lily with his walking stick:

> *... Beau trotting far before*
> *The floating wreath again discerned,*
> *And plunging, left the shore.*
>
> *I saw him with the lily cropped,*
> *Impatient swim to meet*
> *My quick approach, and soon he dropped*
> *The treasure at my feet.*[2]

Buttercup family
Ranunculaceae

Marsh-marigold, *Caltha palustris* (VN: Kingcup, Mayflower, May-blobs, Mollyblobs, Pollyblobs, Horse-blob, Water-blobs, Water-bubbles, Gollins, the Publican). This is one of the most ancient native plants, probably surviving the glaciations and flourishing after the last retreat of the ice, in a landscape inundated by glacial melt-waters. Until two centuries ago, before the extensive draining of the landscape, kingcups must have been the most conspicuous plant of early spring, blooming at the edges of cattle wallows, in water-meadows and damp flashes on village greens, and growing straight from the dark mud amongst willow and alder roots in wet woods. In the Isle of Man, where plant rituals survived until very recently, it was held in high regard as a spring omen, and flowers were strewn on doorsteps on old May Eve. Now the custom of bringing 'mayflower' – as it is called in English on the island – into the house is enjoying something of a revival, and improvised vases of blooms have been seen on counters and in shop windows.[1]

Marsh-marigolds are in decline as agricultural land continues to be drained, but they are still the most three-dimensional of plants, their fleshy leaves and shiny petals impervious to wind and snow, and standing in sharp relief against the tousled brown of frostbitten grasses. Most of the plant's surviving local names – water-blobs, molly-blobs, water-bubbles – reflect this solidity, especially the splendid, rotund 'the publican' from Lancashire.[2]

Marsh-marigolds provide the first show of spring colour in meadowland.

Globeflower, here flowering by the River Tees in Durham, is chiefly a plant of northern limestones.

The white forms found in gardens are a foreign variety. But Francis Simpson distinguishes a native form with lemon-yellow flowers and taller habit.[3]

Globeflower, *Trollius europeaeus* (VN: Locker gowan, London bobs). A much scarcer but equally striking flower of northern and Welsh woods, pastures and stream-banks, chiefly on limestone in the English part of its range. Its large, butter-yellow petals are tightly bunched, like old-fashioned roses – hence the 'locker' component of some local names. In England it comes no farther south than the Derbyshire Dales and the uplands near Oswestry in Shropshire.

Stinking hellebore, *Helleborus foetidus*, is a perennial of woods and scrub on calcium-rich soils, with lime-green flowers edged with claret and held in sprays above stiff, fingered, evergreen leaves from late January till May. The southern chalk-hills are the place to see it, where the flowers glow amongst tangled hazel shoots and the dark foliage of yew and holly. This is where the eighteenth-century naturalist Gilbert White knew it, in the hanging woods above Selborne in Hampshire (where it still flourishes). He thought it 'very ornamental in shady walks and shrubberies' and transplanted it to his garden. He also noted that the local women 'give the leaves powdered to children troubled with worms', but that it was a 'violent remedy'.[4] There is some clue to this in the smell, not stinking exactly but reminiscent of the unpleasant mousy tang of hemlock. Nonetheless 'setterwort' has remained a garden favourite (though most plants come from continental stock these days) and has become naturalised some way beyond its natural range. A yellow-foliaged form that has occurred in the wild in three different sites recently looks a likely garden plant of the future.[5]

Green hellebore, *H. viridis*, has a similar range and taste in habitats. It is a less spectacular plant, with dull green flowers and sprawling, jagged leaves which die back in winter. But it can often form quite large colonies. The

Stinking hellebore stays evergreen even in the hardest winters.

species most often planted out in gardens, *H. orientalis*, the **Lenten-rose**, is naturalised here and there in southern England.

Winter aconite, *Eranthis hyemalis*, from southern Europe, is often the very first flower to bloom in gardens in mid-January, and is widely naturalised in plantations, roadsides and churchyards. The yellow flowers have been called 'choirboys' in Suffolk, from the ruffs that surround them.[6] **Love-in-a-mist**, *Nigella damascena*, escapes occasionally to waste ground and rubbish-tips.

Monk's-hood, *Aconitum napellus*, is also widely naturalised, but may be native in a few shady places by streams in the south-west. It is probably the most virulently poisonous of all British plants, yet its hooded, bonnet-like blue flowers have made it a favourite border plant. In the Middle Ages it was taken into cultivation, possibly as a hunter's poison (across Europe *A. napellus* and its relatives are also known as 'wolf's-banes') and certainly as a potent pain-killer and a liniment for rheumatism. A famous site is at Roche Abbey, near Rotherham in Yorkshire, where it was probably introduced by the Cistercian monks in the twelfth century.[7] But it killed as often as it cured. Anne Pratt's book on native poisonous plants (1857) is full of warning tales about the dire casualties resulting from nibbling garden refuse or misidentified parsnips (including the tale of the philosopher Van Helmont, who, 'in the confused state of mind consequent on a dose of Aconite, thought that he saw his soul in his stomach').[8] Even skin contact can be dangerous. In 1993, there was an epidemic of poisoning at a florist's in Wiltshire: 'A flower seller was treated for heart palpitations in intensive care after handling bunches of a poisonous flower ... staff at a flower shop in Salisbury suffered shooting pains after poison from a monkshood entered their bloodstreams. The shop's owner bought 150 bunches from a wholesaler, who has now withdrawn them. "I wondered what was wrong – all of a sudden everyone was lethargic and getting pains." '[9]

Baneberry, *Actaea spicata*, is another poisonous member of this toxic family. It is a local plant of limestone regions in the north of England, most often found sprouting from cracks in limestone pavements. Its sprays of

Winter aconite, known as 'choirboys' in Suffolk.

white flowers give it the look of a ground-hugging *Spiraea*. Later in the year it bears shiny green berries which change to black. A cryptic alternative name is Herb Christopher, for which Geoffrey Grigson has one possible, if somewhat fanciful, explanation: 'Perhaps it suggested St Christopher the ferryman by carrying its flowers, so to say, in a shoulder raceme, as the saint carried the infant Christ over the river.'[10]

Wood anemone, *Anemone nemorosa* (VN: Windflower, Grandmother's nightcap, Moggie nightgown). Wood anemone is one of the earliest spring flowers, and one of the most faithful indicators of ancient woodland. Its seed in Britain is rarely fertile and, even when it is, does not stay viable for long. Instead, the plant spreads at a snail's pace – no more than six feet each hundred years – through the growth of its root structure.[11] Wood anemone is consequently a very confined plant, rarely extending its territory beyond its anciently traditional sites. These are usually in long-established woodland, though in the West Country it is also abundant in hedge-banks. Elsewhere it is frequent in ancient meadowland; and in the Yorkshire Dales, for instance, in limestone pavements. In many of these places the colonies may be relics of previous woodland cover, but the windflower's liking for light suggests that it may not be a plant of purely woodland origins. It will not grow in deep shade and opens its blooms fully only in sunshine.

On warm days in early April, a large colony of anemones can fill the air with a sharp, musky smell, which is hinted at in some of the old local names such as 'smell foxes'. Most of these names are now obsolete, but there are at least two comparatively new ones – 'moggie nightgown' in parts of Derbyshire ('In Stanley Common a 'moggie' is a mouse, not a cat'[12]) and the delightful, if not especially appropriate, children's mis-hearing, 'wooden enemies'.

Colonies of wood anemones with purple or purple-streaked petals are quite frequent, e.g. in Wayland Wood, Norfolk, the site of the Babes in the Wood legend. But the sky-blue form, var. *caerulea*, is much rarer, and may have been lost. It was a great favourite of the nineteenth-century pioneer of 'wild gardening', William Robinson,

Wood anemones, flowering in old woodland in early spring, before the leaf canopy closes.

who was careful to distinguish it from the occasionally naturalised European **blue anemone**, *A. apennina*:

'The most beautiful form of our wood Anemone ... is the large sky-blue form. I first saw it as a small tuft in Oxford, and grew it in London where it was often seen with me in bloom by Mr Boswell Syme, author of the Third Edition of Sowerby,[13] who had a great love for plants in a living state as well as in their merely "botanical" aspects, and we were often struck with its singular charm about noon on bright days. There is reason to believe that there is both in England and Ireland a large and handsome form of the wood Anemone – distinct from the common white of our woods and shaws in spring, and that my blue Anemone is a variety of this. It is not the same as the blue form wild in parts of North Wales and elsewhere in Britain, this being more fragile looking and not so light a blue.'[14]

Yellow anemone, *A. rapunculoides*, is a southern European species naturalised here and there as a garden throw-out.

The pasqueflower – the anemone of Passiontide – is now reduced to a scatter of sites on old grassland in chalk and limestone country.

Pasqueflower, *Pulsatilla vulgaris*. One of the most beautiful of our native flowers. Its purple petals, held in the shape of a bell, surround a tuft of yellow stamens, and are cushioned on greyish, feathery leafage. It blooms around Eastertime – hence the name 'Pasque', meaning, like 'Paschal', of Easter. But it has never been common enough to have much of a place in local culture. There was a legend in a few areas that it sprang from the blood of Danes or Romans, because it seemed to haunt old earthworks such as barrows and boundary banks. But this association was more likely due to the pasqueflower's need for the kind of undisturbed chalk grassland often found at antiquarian sites. Such places have always been comparatively safe from ploughing, and the pasqueflower can have a long occupation on them. Until the 1970s it grew on the Fleam Dyke in Cambridgeshire, where the East Anglian poet Edward Fitzgerald knew it in the mid-nineteenth century and commented on it in an annotation to his *Omar Khayyam*. Not many miles away, another nineteenth-century poet, John Clare, found it around the village of Helpston in Northamptonshire:

'You have often wished for a blue Anemonie the Anemonie pulsitilis of botanists & I can now send you some for I have found some in flower to day which is very early but it is a very early spring the heathen mythology is fond of indulging in the metramorphing [*sic*] of the memory of lovers & heroes into the births of flowers & I coud almost fancy that this blue anenonie [*sic*] sprang from the blood or dust of the romans for it haunts the roman bank in this neighbourhood & is found no were else it grows on the roman bank agen swordy well & did grow in great plenty but the plough that destroyer of wild flowers has rooted it out of its long inherited dwelling it grows also on the roman bank agen Burghley Park in Barnack Lordship.'

Letter to his publisher, 25 March 1825[15]

It continues to grow in some quantity at Barnack, in the old stone quarry known as the Hills and Holes, which is just a couple of miles down the road from Burghley Park.

Pasqueflower is now a nationally scarce plant. Its largest colony is on the steep banks of Barnsley Warren in the Cotswolds; the smallest probably the few flowers that cling on in short turf on the Magnesian limestone in West Yorkshire – its most northerly station in Britain.[16]

Traveller's-joy, *Clematis vitalba* (VN: Old-man's-beard, Father Christmas, Baccy plant, Smokewood, Woodbine). In the months of fog and grey skies that precede Christmas, old-man's-beard drapes hedges and banks in chalk country with billows of feathery seed-heads. No wonder the sixteenth-century writer John Gerard called it traveller's-joy. He wrote that it 'is called commonly *Viorna quasi vias ornans*, of decking and adorning waies and hedges, where people trauell, and thereupon I haue named it the Traueilers Ioie ... These plants haue no vse in Phisicke as yet found out, but are esteemed onely for pleasure, by reason of the goodly shadowe which they make with their thicke bushing and clyming, as also for the beautie of the flowers, and the pleasant sent or sauour of the same.'[17]

Nearly two hundred years later, Gilbert White glimpsed another layer of meaning in Gerard's coining. In his Journal for 23 November 1788, he writes: 'The downy seeds of traveller's joy fill the air, & driving before a gale appear like insects on the wing.'[18]

Feelings about the plant – and its uses – find their way into many of its popular names. 'Father Christmas' is comparatively new, and, like 'old-man's-beard', refers to the fluffy seed-heads. 'Woodbine' is a general name applied to climbers (e.g. honeysuckle and bindweed). The dry winter stems have also been cut for smoking – hence old names such as 'boy's bacca' and 'shepherd's delight' (and can this custom have been the origin of Woodbine cigarettes' brand name?). In parts of Sussex, where people can remember it being smoked up until the 1950s, it is called 'smokewood'.[19]

Virgin's-bower, *C. flammula*, is the common fragrant Mediterranean species, which is naturalised on sea-cliffs and sand-dunes in some places in the south and west.

Meadow buttercup, *Ranunculus acris*; **Bulbous buttercup**, *R. bulbosus*; **Creeping buttercup**, *R. acris*. There is little discrimination made culturally between these three common species. They are still universally used in the children's game of holding a flower under the chin: if there is a yellow glow reflected from the skin, the subject likes butter. They are frequently featured in medieval church carvings, perhaps because of their fingered leaves.

Old-man's-beard by a downland track in Oxfordshire. It was called traveller's-joy by John Gerard in 1597, from 'decking and adorning waies and hedges, where people trauell'.

Buttercups in an old orchard. They are amongst the most resilient species and thrive in all kinds of grassland.

Buttercups of some sort can be seen, for example, in Bristol Cathedral (where they are painted with ivy leaves and berries), and carved on the capitals of Southwell Minster Chapter House in Nottinghamshire.[20]

The name 'buttercup' seems obvious enough, linking the yellow of butter with the colour of the flowers amongst which the dairy cows grazed. But it did not come into common use until the eighteenth century. Before that there were a host of equally expressive local names – including goldweed, soldier buttons and kingcup. Most of these are now obsolete, but 'crowpeckle' survives in Northamptonshire in local vernacular and in field names such as Crowpightle.[21]

All three species are common in grassland in most parts of Britain. Meadow buttercup prefers slightly damper, calcareous sites, and can be so dense in meadows in the Derbyshire and Yorkshire Dales that the light from the flowers is dazzling under May sunshine. (In Derbyshire, buttercup petals are a frequent component of well-dressings.) Bulbous buttercup prefers drier soils. Creeping buttercup is a common weed of wet woods, gardens and waste places. A double-flowered form of this species was found in an arable field in Suffolk in 1993.[22]

Goldilocks flowers often lack a few petals.

Corn buttercup, *R. arvensis*, was once a widespread but unloved cornfield weed. The seed-heads are viciously spiny and earned the plant local names such as devil's claws and hellweed.[23] It is now very scarce. **Goldilocks**, *R. auricomus*, is quite frequent in ancient woods and grassland. It has the curious habit of not always forming a full quota of petals, sometimes having none at all. **Celery-leaved buttercup**, *R. sceleratus*, is a tenacious plant of marshes and pond-edges. Sometimes it will cling on at the site of old ponds long after they have dried out.

Adder's-tongue spearwort, *R. ophioglossifolius*. This small-flowered, mud-loving annual has the honour of having Britain's smallest nature reserve – the 300 square yards of Badgeworth Pool in Gloucestershire – dedicated to its conservation. Adder's-tongue spearwort was first noticed in marshy ground close to the pool in 1890. It then vanished for 20 years, but reappeared after the hot summer of 1911, and in 1912 the whole marsh was covered with a froth of yellow flowers in June and July.

It has had fluctuating fortunes ever since, depending on climatic conditions. There were no plants at all, for instance, in 1942, 1959 and 1969. They were drowned out in 1959, and frozen to death during the terrible winter of 1962/63. But the following spring, the buried seeds sprouted abundantly, and there were over a thousand plants, a figure reached again in 1965, 1968 and 1973. It is, as you might expect from a species on the very northern edge of its range in Britain, exceedingly finicky in its requirements. It needs bare soil and moist ground from August to October so that the seeds can germinate. (It is one of a disparate group of species, including starfruit, pennyroyal and small fleabane, that actually prosper in the churned-up mud at the edge of cattle ponds.) It needs a mild, frost-free autumn and sufficient rain to keep the ground moist enough for the seedlings to develop sturdy basal rosettes. (If not, they can be killed by trampling stock or uprooted by birds.) It needs enough rain in early winter to fill the pool with water and submerge the plants. This is essential to protect the seedlings and allow the proper development of the young plants. At this stage the leaf-stems grow rapidly and allow the pale, yellowish-green leaves to reach the top of the pond where they spread out and float on the surface of the water like miniature water-lily leaves. Finally they need freedom from the kind of hard frosts that can freeze the pool solid and kill the growing tips of the plants.

The Badgeworth buttercup has also had an eventful

social history. In 1932 the site was almost destroyed when a new owner began to fill in part of the marsh with rubbish, in the hope of building on it. Fortunately what was left was purchased by a public-spirited local botanist, G. W. Hedley, who presented it to the Society for the Promotion of Nature Reserves. John Moore used this parish drama as the basis for his fictional *Midsummer Meadow* (1936). It is almost a morality play – the green fields of St George's England against the Dragons of development and greed – a conflict which has been repeated across the length and breadth of Britain ever since. It was acted out a second time in Badgeworth in 1973, when a plant-hire firm that had bought a patch of land adjacent to the marsh applied for planning permission to construct a wash-down facility for their vehicles, which would almost certainly have led to oil and chemical pollution of the marsh. A stalwart campaign mounted by local people made the defence of Badgeworth a national issue, and eventually the County Planning Authority refused the planning application and the contested land passed into the benign ownership of a beekeeper.

For a while the village and its tiny buttercup reserve had their moment of international fame. The 'Badgeworth Battle' was reported in places as far afield as Ozark, Missouri. TV teams came to make films, and a coach firm specialising in old people's outings made the reserve its first stop by popular request. In Badgeworth itself, the Women's Institute produced a tie with the buttercup as its motif, and the crowning of the 'Buttercup Queen' became the main event at the village church fête.[24]

Things are quieter now. There is no longer a Buttercup Queen. Sonia Holland, the plant's greatest local champion, died in 1993. But the buttercup itself survives, in its traditional, nerve-rackingly erratic way, and has now been joined by a second Gloucestershire colony.[25]

Lesser celandine, *R. ficaria* (illustrated on p. 381). This is one of the first woodland flowers of the year. More than two hundred years ago Gilbert White noted that the average first flowering around his Hampshire village of Selborne was 21 February.[26] A century later, the Hertfordshire botanist John Hopkinson gave precisely the same date for the years between 1876 and 1886. Another hundred years on, and this is still the time celandines begin to bloom across much of southern England in a typical year – justifying one of celandine's defunct local names, 'spring messenger'.

Another obsolete name is pilewort – the herb given for haemorrhoids. This ancient prescription is based on the Doctrine of Signatures (see p. 380), and on a fancied resemblance between the knobbly tubers and piles. These tubers – which readily break free from the fibrous roots – are one of the means by which celandines are able to spread rapidly in the open and disturbed soil of damp woodland tracks and to colonise stream-banks, ditches and shady gardens. A subspecies, ssp. *bulbilifer*, is an even more aggressive spreader. It has tiny bulblets – 'bulbils' – in the junctions between the leaf-stalks and the main stem, and these are readily strewn about by birds, walkers, car tyres and flowing water.

The most curious name is celandine itself, which derives from the Greek *chelidon*, a swallow. The sixteenth-century herbalist Henry Lyte suggested that this was because it 'beginneth to springe and to flowre at the comming of the swallows'.[27] But most celandines are in

Water-crowfoot, an integral part of the river landscape. The River Wye, at Ross-on-Wye.

flower long before the swallows arrive, and it looks as if the lesser celandine may have been confused with the greater celandine, another yellow-flowered but quite unrelated species, whose connection with swallows I discuss on p. 55.

But perhaps the flower's name indicates something less literal than a coincidence between blossoming and a bird's arrival. Perhaps lesser celandine was seen as a kind of vegetable swallow, the flower that, like the bird, signalled the arrival of spring. Wordsworth certainly thought so. It was his favourite flower and he wrote three poems about it. Before the first, 'To the Small Celandine', he adds a little field-note: 'It is remarkable that this flower, coming out so early in the Spring as it does, and so bright and beautiful, and in such profusion, should not have been noticed earlier in English verse. What adds much to the interest that attaches to it is its habit of shutting itself up and opening out according to the degree of light and temperature of the air.'

The second poem ('To the Same Flower') includes a stanza about children's enjoyment of the celandine's 'glittering countenance':

> *Soon as gentle breezes bring*
> *News of winter's vanishing,*
> *And the children build their bowers,*
> *Sticking 'kerchief-plots of mould*
> *All about with full-blown flowers,*
> *Thick as sheep in shepherd's fold!*
> *With the proudest Thou art there,*
> *Mantling in the tiny square.*[28]

When Wordsworth died in 1850, it was proposed that a celandine would be the most fitting decoration for his tomb in the Lakes. But, ironically, the plant carved on the monument at Grasmere seems not to be 'the Small Celandine' at all, but, because of the old confusion or inept carving, the greater, *Chelidonium majus.*

Perhaps, after celandine's long history of ambivalent medicinal and classical names, a modern, serendipitous tag, accidentally coined by a child, deserves the last word: 'When my daughter was a toddler she mispronounced celandine as "lemon-eye", and since that is what the flower looks like that is what we call it.'[29]

Common water-crowfoot, *R. aquatilis*, the commonest of a large group of water-plants, which can cover ponds, ditches and streams with white flowers in spring and summer. Two other species of rather quicker-flowing water, *R. fluitans* and *R. penicillatus* ssp. *pseudofluitans*, can often be indicators of the sites of old fords or collapsed bridges.[30]

Pheasant's-eye, a cornfield weed once common enough to be sold as a cut flower.

Pheasant's-eye, *Adonis annua*, is one of the many colourful annuals to have been driven out of arable fields by herbicides and modern cultivation methods. It has brilliant scarlet flowers, a little like small anemones in shape, held over feathery foliage. In the eighteenth century it was common enough in the chalky cornfields of southern England to be gathered for sale in Covent Garden Market as 'Red Morocco'.[31] Today it is frequent only in Wiltshire, especially around Salisbury – though it rarely persists in any one place. Plants appeared during the construction of the M4 through Wiltshire in 1971, in a newly constructed car park at Odstock Hospital in 1987, and on several set-aside fields during the early 1990s.[32]

Columbine, *Aquilegia vulgaris* (VN: Granny's bonnet, Granny's nightcap). This cottage-garden favourite, a true British native, is named after two rather incompatible birds. The Latin name, *Aquilegia*, is derived from *aquila*, an eagle, because of a fancied similarity between the petals and eagles' wings (perhaps because of the upturned tips?). Columbine is from *columba*, a dove, and from the more striking resemblance of the bases of the petals to five pigeons perched in a ring. Columbines have been popular subjects for church carvings. They are, for instance, figured on misericord supports in Manchester Cathedral (*c.* 1506) and Ripon Cathedral (*c.* 1490); on a bench-end at St Petroc's in Lydford, Devon; and in a window border with fritillaries at St Bartholomew's, Yarnton, Oxfordshire.[33]

Columbines are scarce but widespread flowers of woods, fens and damp grassland on calcareous soils. Truly

Wild columbines are usually blue or purple, though occasionally white.

wild specimens are usually blue or purple (though I have found colonies with pure white flowers in the Derbyshire Dales). Other colours, and double- and large-flowered forms, increasingly common on road-verges and the edges of woods, are usually garden escapes.

Common meadow-rue, *Thalictrum flavum*, is a local species of fens and wet meadows chiefly in eastern England. The shape of the leaves is similar to the Mediterranean herb, rue.

Barberry family *Berberidaceae*

Barberry, *Berberis vulgaris.* A native shrub or one anciently introduced from continental Europe, which was at one time used for hedging because of its densely packed spiny branches. In the nineteenth century it was found to be an alternate host for the black rust fungus that attacks

Barberry, occasionally found as a hedgerow shrub.

wheat, and many hedges were grubbed out. But it still survives in scattered localities across Britain, especially in pasture country and along old trackways, e.g. at Little Heath, Hertfordshire, and by Chalk Lane, Brandon, and Shaker's Lane, Bury St Edmunds. In Wiltshire there is a long-established barberry hedge at Kington St Michael, on the site of a medieval Benedictine Nunnery.[1]

The oblong berries hang in brilliant red clusters and have been used for making tart pickles and jelly. 'This was the scarlet jelly [once] poured on top of little mutton pies.'[2] A related species, **great barberry**, *B. glaucocarpa*, is widely planted for ornament and hedging. It is 'a peculiarity of the farmland around Minehead, especially in the Porlock Vale'.[3]

Oregon-grape, *Mahonia aquifolium*, is a sprawling, evergreen shrub from western North America, widely naturalised from gardens, and often planted out in woods to provide cover and food for pheasants. In the United States the dark, white-bloomed berries have been used as human food, too. One Wiltshire child used to squeeze the berries to 'produce a very realistic blood to trick the unwary'.[4]

Poppy family *Papaveraceae*

Common poppy, *Papaver rhoeas* (VN: Corn-poppy, Field poppy). Feelings about the meaning of plants can run deep, in many ways. Once, writing about the persistence of plant symbolism, I mentioned that the red poppy, which we wear on Remembrance Day, had been 'an emblem of blood and new life since the Egyptians'.[1] A few days later I received a letter from a London man who had lost two cousins in the First World War and who thought it strange that I 'should apparently be unaware of the English habit of wearing poppies on 11th November each year by way of remembering the Englishmen who died'.[2]

I do understand the English origins of this tradition and my correspondent's sense of grievance that a powerful symbol of personal loss had been seemingly diluted or appropriated. Yet the scarlet poppy's association with death and new life, with corn and harvest, is as old as agriculture, and maybe as civilisation itself. It has been one of the world's most successful 'weeds', and has followed and exploited the spread of farming across the globe so comprehensively that no one is sure of its native home. It is a plant which belongs not so much to a particular home as to a way of life – to the tilling and disturbance of the soil, and to the building (and razing) of communities. The archaeologist Flinders Petrie found poppy seeds mixed up with grains of barley in relics from the Twelfth Dynasty at Kahun in Egypt, which prospered before 2500 BC.[3] There, and maybe across much of the Middle East and the Mediterranean, poppies must have gone through the same evocative cycle: growing up unbidden in the fields, their blood-red petals cut down with the corn, only to spring up again, in numberless quantities, the following summer. No wonder they became such complex symbols of growth, blood and new life. The Assyrians called them 'the daughters of the field'. For the Romans they were the sacred plant of their crop goddess Ceres. Garlands for her statues were made from poppies interwoven with barley or bearded wheat, and poppy seeds were offered up in rituals to ensure the fertility of the crops.[4]

Corn-poppies probably reached Britain mixed up with the seed-corn of the first Neolithic settlers, and even here were soon regarded as ambivalent signs of fertility and death. In the late medieval period they were called 'corn-roses', but often confused with the opium poppy (see below) and believed to induce sleep or headaches (though they have no such properties). Many of the early vernacular names – 'thundercup', 'thunderflower', 'lightnings' – reflect the ancient belief that poppies must not be picked, for fear of provoking storms; and conversely, perhaps, that whilst they were unpicked the crops were safe from summer downpours. And two centuries before the slaughter on Flanders Field, the scarlet troops massed

Field poppies, east Suffolk, now largely confined to the edges of fields.

amongst the wheatfields of southern England were being nicknamed 'soldiers' and 'redcaps'.[5]

Some of these old beliefs found a Victorian echo in the brief but heady fashionability of the cliff-top landscapes of Cromer and Overstrand in Norfolk, which were immortalised as 'Poppy-land' by the *Daily Telegraph*'s drama critic, Clement Scott. Scott had taken to visiting these new seaside resorts in the 1880s and had fallen in love not just with the local miller's daughter, Louie Jermy, but with the sight of waves of scarlet blossoms in fields and lonely churchyards, sweeping down to the very edge of the cliffs, and set against the sparkle of the North Sea in high summer. He began to write ecstatic columns about Poppy-land in August 1883, and started a fad that brought thousands of visitors to the little villages on what the Great Eastern Railway rapidly renamed 'The Poppy Line'. Scott also wrote a popular but painfully sentimental poem about his East Anglian Arcadia, entitled 'The Garden of Sleep', which recalled the flower's soporific reputation: ' 'Neath the blue of the sky, in the green of the corn/ It is there that the regal red poppies are born!/ Brief days of desire, and long dreams of delight,/ They are mine when my Poppy-land cometh in sight.'[6]

The obscene sea of mud and broken bodies that stretched around Ypres and the Somme thirty years later was a very different kind of 'Garden of Sleep'. Millions of soldiers and animals were simply churned into the earth, like so much compost. The men who had been persuaded to fight 'in order to preserve and somehow possess the beauties of the English countryside' were busily engaged, as the poet Ivor Gurney saw, in turning France, a 'darling land ... blessed with a merciful spirit founded on centuries of beautiful living', into a wasteland 'of mud and swamp and brimming shell-holes'.[7] Edmund Blunden saw the terrible irony clearly, and in his bitter poem 'Rural economy' he becomes a farmer, planting seeds of iron, which, manured with 'bone-fed loam/ Shot up a roaring harvest home'.[8]

Yet the real harvest had an ironic and paradoxical healing power. Not everything could be killed. The war artist William Orpen visited the battlefield in the summer of 1917, six months after the carnage of the Somme, and was mesmerised by it: 'No words could express the beauty of it. The dreary dismal mud was baked white and pure – dazzling white. White daisies, red poppies and a blue flower [probably cornflowers], great masses of them, stretched for miles and miles. The sky a pure, dark blue, and the whole air, up to a height of about forty feet, thick with white butterflies: your clothes were covered with butterflies. It was like an enchanted land, but in the place of fairies there were thousands of little white crosses, marked "Unknown British Soldier" for the most part.'[9]

The explosion of the poppies, seemingly coloured by blood but also healing the land, had struck writers at the front since the first summer after the war's outbreak. In the early winter of 1915, Colonel John McRae, a Canadian academic and volunteer medical officer, was treating the appalling casualties after the second battle of Ypres. He found time to write a poem about the poppies' imagery which he sent anonymously to *Punch*. The magazine printed it on 15 December:

In Flanders fields the poppies grow
Between the crosses, row on row,
That mark our place: and in the sky
The larks, still bravely singing, fly
Scarce heard amid the guns below.

We are the Dead. Short days ago
We lived, felt dawn, saw sunset glow,
Loved and were loved, and now we lie
In Flanders fields.

The poem, 'In Flanders Fields', was reprinted around the world. In Georgia in the United States, Moina Michael, a worker with the YMCA, was so moved she vowed to wear a poppy evermore. In 1921 the British Legion was formed and was approached by a YMCA friend of Moina Michael's with samples of artificial poppies she had made. The idea of a 'Poppy Day', marking the anniversary of the Armistice, took off immediately. That year, on 11 November, the British Legion raised more than £100,000 from the sale of poppies.

The Second World War, ironically, almost spelt the end of the real poppy as a field flower. The intensive agriculture necessary during the war years, followed by the development of powerful agricultural herbicides, virtually eliminated them from cornfields. I last saw them on my home patch in 1959. I had worked on the harvest that year as a holiday job, stacking behind a reaper and binder, and poppies had brightened every sheaf, caught up amongst the wheat stalks like shreds of scarlet bunting.

For the next two decades they were more or less confined to hedge-banks and untended gardens, but every so often, when a farmer went broke or a new road was driven through old arable land, there was a glimpse again of that red blaze. Some poppy patches became briefly famous, like those that lined new bypasses at Bury St Edmunds in Suffolk and Steeple Langford in Wiltshire. These poppies sprang from seeds which had quite likely lain dormant in the soil since the advent of chemical herbicides. Poppy plants produce on average 17,000 seeds per plant, hundreds of millions for every field – and about a sixth of these are capable of lying dormant, but viable, for at least 40 years.[10] 'The smithy just 10 yards down the road was demolished two years ago. I had a barrowful of soil from the site, and I was delighted to have poppies and corncockles on the piece of garden the soil had been put on. The cockles need the support of wheat or corn as they grow rather spindly. I am not sure when the smithy was built but I guess these seeds had lain dormant for over 100 years.'[11]

The poppy 'is the most transparent and delicate of all the blossoms of the field … [it] is painted glass; it never glows so brightly as when the sun shines through it' (John Ruskin).

Poppies were sorely missed during the two decades when they were sprayed almost into oblivion, and during the 1980s and 1990s they were put on the market in wildflower seed mixtures and grown in 'wild gardens'. Perhaps we began to *look* at them for the first time as flowers, not simply blotches of red amongst the corn, as Ruskin had a century before:

'We usually think of the poppy as a coarse flower; but it is the most transparent and delicate of all the blossoms of the field. The rest – nearly all of them – depend on the *texture* of their surfaces for colour. But the poppy is painted *glass*; it never glows so brightly as when the sun shines through it. Wherever it is seen – against the light or with the light – always, it is a flame, and warms the wind like a blown ruby … Gather a green poppy bud, just when it shows the scarlet line at its side; break it open and unpack the poppy. The whole flower is there complete in size and colour, – the stamens full-grown, but all packed so closely that the fine silk of the petals is crushed into millions of shapeless wrinkles. When the flower opens, it seems a deliverance from torture.'[12]

It is this wrapped-up, silk-like quality that made poppies such popular flowers for making dolls from – another custom that was revived in the 1980s: 'The flower petals were folded down to reveal the black hairy "head". The "skirt" was kept in place by tying round it a piece of fine grass to form a belt. Mature poppies gave red dollies, but for pink ones and for white ones, the immature buds were opened.'[13]

The policy of set-aside (taking surplus arable land temporarily out of cultivation) that began in the late 1980s gave corn-poppies a chance to return to the landscape on a scale that had not been seen for forty years. On the dry chalk of the north Chilterns they covered not just whole fields but whole hills. They filled valleys in the Scottish lowlands and speckled the route of the M4. They returned not just to Norfolk's 'Poppy-land', but to parts of the great arable prairies in the centre of the county, nicknamed 'England's granary'. And returning in such numbers they carried oddities and sports with them, pink, blue-black, mauve, white: 'In June 1992, I was surprised to come across a corn poppy with petals which were pure white, except on the basal areas which are normally black – and they were a deep pink. All neighbouring plants in the arable field in which it occurred had flowers of a normal poppy colour.'[14]

This aberration is not so different from what has come to be known as the Shirley poppy – perhaps the favourite of all cultivated poppies. Its ancestor was discovered by the Revd William Wilks, of Shirley in Surrey, 'in a wilderness corner of my garden' (once part of Shirley Common) in the summer of 1879 or 1880.[15] It was a single flower in a population of otherwise normal corn-poppies, which had a narrow edge of white on the petals. Wilks saved the seeds, and, out of the 200 plants he raised from them over the next year, four or five had white-edged petals. He continued selectively breeding them for many years, producing paler flowers and eliminating the black blotch at the base of the petals. The final strain contained a range of flowers in all shades from pure scarlet to pure white, with all manner of streaked, patched and edged blooms in between, but all bearing the characteristic white or yellow centres.

A correspondent who lives in the road where Wilks bred his poppies tells us that the occasional 'Shirley' could be found in surrounding fields until the late 1980s, despite the village now being part of greater Croydon, and the fields turned into golf-courses.[16] But there is a permanent bed of Shirley poppies in the large garden of what was once Wilks's vicarage (now an old people's home), and a pub named 'The Poppy' (the Shirley was dropped from the name quite recently) not far down the road.

Shirley poppies bring the poppy's story, always full of ironies, full circle. White Shirley poppies, or white artificials, are worn by peace movement supporters on Armistice Day, to honour the dead without condoning war.

Oriental poppy, *Papaver orientalis*, a robust, tall poppy with very large red flowers, which is quite widely naturalised from gardens. It originated in south-west Asia. **Opium poppy**, *P. somniferum*, is another Asian species, with glaucous, grey-green foliage and white to purple petals, sometimes variegated. It is very widely naturalised

Elizabeth Blackadder's study of Shirley poppies with their typically pale centres.

Opium poppies, widely naturalised, but not producing opium in our climate.

Yellow horned-poppies on the shingle beach near Dungeness Power Station, Kent.

in waste places, and occasionally in arable fields. This is the species whose latex is used for the production of opium, but it produces very little of this in the British climate. The seeds are the poppy seeds of cooking and are free of narcotic chemicals. **Welsh poppy**, *Meconopsis cambrica*, is a yellow-flowered member of the same genus as some of the exotic Himalayan poppies. It is not exclusively Welsh, and occurs in woods and shady places among rocks in south-west England, and, as an escape, much further north.

Yellow horned-poppy, *Glaucium flavum*, is a showy plant of seaside shingle banks. The leaves are silvery-grey, fleshy and covered with fine hairs. The golden yellow flowers are followed by the 'horns', curling seed-pods which can be up to a foot long. The whole plant exudes a yellow latex when broken and is poisonous. Very occasionally, specimens are found inland. In 1991, 25 plants appeared at Greenhithe in Kent, on the sides of a new drainage ditch. Although the location is half a mile from the current coastline, it is approximately where the estuary beach would have been before the marshes were drained, and the poppies almost certainly sprang from long-buried seed.[17]

Greater celandine, *Chelidonium majus*. The flowers of greater celandine would not immediately make you place it in the poppy family. They are custard-yellow and about the size of buttercups. But cut the stalk or leaves, and the latex characteristic of the family (orange in this species) oozes out. It is this that accounts for the 'chelidon' ('swallow' – see lesser celandine, p. 48) in its scientific name, and the plant is still called 'swallow-wort' in parts of North America, as it was by Lyte and Gerard in the sixteenth century.[18] There is a complicated myth, dating from classical writings and compounded in the medieval period, that swallows used the herb as a restorer of eyesight: 'It is called Celandine, not bicause it then first

Greater celandine, from the collection of flower paintings prepared between 1828 and 1851 by women of the Clifford family in Frampton on Severn.

springeth at the comming in of the Swallowes, or dieth when they go away: for ... it may be founde all the yeere, but bicause some holde opinion, that with this herbe the dams [female swallows] restore sight to their yoong ones when their eies be out, the which things are vaine and false.'[19] But sceptical though he might have been, Gerard was still slightly in thrall to sympathetic magic, and was not against recommending the highly corrosive latex for eye disorders in humans, 'for it clenseth and consumeth awaie slimie things that cleaue about the ball of the eie, and hinder the sight'. What it certainly did do was cause severe conjunctivitis in any unfortunate patient treated with it!

The latex also has a somewhat safer – and highly successful – role in herbal medicine as a wart-remover.[20] (Only comfrey and feverfew have more cures attested by our contributors.) This may have always been its role in folk medicine; and the reason it is so often found in rough ground close to buildings (it is especially fond of the foot of stone walls) may be because it was once a common plant in cottage physick gardens. It may even be a native, brought into gardens from open, disturbed habitats.

Oxford is greater celandine's heartland. It grows everywhere in this anciently stony city, at the edge of car parks, on old walls, in cloisters and at the foot of exclusive Fellows' staircases. Its lobed leaves are also unmistakably carved on the shrine to St Frideswide, which dates from 1289 and now sits in Christ Church Lady Chapel. Its presence is probably no coincidence, for Frideswide, as well as being the patron saint of the University, was also a benefactress of the blind. She was the daughter of a twelfth-century Mercian princess, and went into hiding for three years to avoid an arranged marriage. Her luckless suitor subsequently went blind, and, in an act of contrition, Frideswide became a nun. Not long afterwards, she summoned up a holy well in the village of Binsey, just upriver from Oxford. Its water was reputed to have miraculous powers, especially for eye and stomach problems, and this seems to have been the reason for her sanctification. So, when her shrine was carved, the prime eye-herb was added to the magnificent and precisely carved frieze of oaks (both species), hawthorns, ivy, hop, maple and sycamore – the only site in Britain where its image is recognisably carved in a sacred building.

Yellow corydalis on a ruined church tower, Woodrising, Norfolk.

Fumitory family *Fumariaceae*

Common fumitory, *Fumaria officinalis* (VN: Earth smoke, Red-tipped-web), is a common weed of gardens, arable fields and waste places. Both scientific and English names stem from the Latin *fumus terrae* – 'smoke of the earth'. The delicate, grey-green leaves do have a slightly smoky appearance, enough to persuade one seventeenth-century herbalist that 'it appeareth to those that behold it at a distance, as if the ground were all of a smoak'.[1] **Climbing corydalis**, *Ceratocapnos claviculata*, is a cream-flowered scrambler more or less confined to heaths and ancient woods on acid soils. **Yellow corydalis**, *Pseudofumaria lutea*, from the southern Alps, is increasingly naturalised in cracks in walls and pavements throughout the British Isles. A survey of south Essex walls found it on over 18 per cent of the 650 studied.[2]

Plane family *Platanaceae*

London plane, *Platanus × hispanica*. The London plane is unquestionably the city's most dominant tree, but there is no evidence that it is a native Londoner, bred by metropolitan horticulturalists within the sound of Bow Bells. It has traditionally been regarded as a hybrid between the oriental plane, *P. orientalis*, a native of the eastern Mediterranean, and the American plane, *P. occidentalis*, which arose in Spain or southern France in the early seventeenth century, and was first planted in England in about 1680 (though, in 1919, a Dr Augustine Henry claimed that it had appeared at Oxford in 1670).[1] There is little hard evidence for theories about the tree's origins, however, and arguments still simmer about its pedigree. Even its status as a hybrid has been challenged (it produces fertile seeds even in Britain). But it has never been found growing truly wild, so it is unlikely to be a separate species. And though there is a possibility that it is simply a vigorous sport of one of its suggested parents, it is now more usually consigned to that convenient category 'of unknown origin'.

But there is no doubt about its qualifications as a city tree. It is tough and resilient, rarely sheds branches, is indifferent to the most ruthless pruning, and will flourish even in compacted or paved-over soil. Its leathery leaves are easily washed clean of urban grime by rain (though its habit of regularly sloughing off large flakes of old bark, to leave a mottle of pale green, yellow and fawn underbark, is no longer reckoned to have a similar cleansing effect). As a result it has been planted along streets and in squares and parks throughout urban England, and, following the great success of John Nash's plantings in the early nineteenth century, across Europe and North America too.

In London, planes account for more than half the trees along streets and in public spaces, and shape the character of many districts – especially the squares in the West End, where some trees date from the early eighteenth century. But it must be said that, *en masse*, they have their drawbacks. They look marvellous in the sunshine – especially amongst the white buildings and fierce light of southern Europe. But in London on a grey day, against sombre municipal concrete, their heavy foliage can be oppressive. And planted in such numbers, often in close-packed rows, they have proved vulnerable to fungal disease in recent years.

Many of the London park and square planes were blown down in the great storm of October 1987, and this led to the rediscovery of the tree's exquisite pinkish and fine-textured timber, which was once valued as 'lacewood' veneer. Some more robust carpentry was also done with the fallen trees. In St James's Park, one of the rangers chain-sawed a whole tree where it lay into a kind of rustic

The dappled shade of London plane trees in a London square. Berkeley Square, early spring.

Plane leaves in autumn.

children's climbing frame. But many large trees remain, not just in London (e.g. Berkeley Square), but in Oxford, where a specimen at Magdalen College, planted in 1801 (and described as a scion of one raised in the Botanic Garden in 1666), is 23 feet in girth, and at Woolverstone Park, Suffolk, where one has the same girth.[2]

The London plane has a very long flowering and fruiting season. The globular flower-heads – 'bobbles' – dangle on long stalks in May and June (male and female being separate but on the same tree), and develop into a spiky brown fruit-cluster resembling a small mace-head. These hang on the leafless trees throughout the winter, to break up early the next spring, releasing the individual seeds. Both flower-pollen and seed debris seem to be allergenic to many people – which may have encouraged children from a school in Beckenham, south London, to use the hairy seeds as an itching powder, an urban substitute for rose-hip seeds.[3]

The seeds are fertile, and in hot summers will germinate in great numbers even in smidgens of dirt in inner-city gutters. These rarely survive long before they are squashed, swept up or weeded out. But in the outer suburbs, especially on walls by the Rivers Thames and Lea, the plane is now becoming at least a naturalised Londoner, with some of the self-sown saplings growing into tall trees.

Elm family *Ulmaceae*

Elms, *Ulmus* species. In many parts of Britain, elm trees have already become a memory, and sometimes not even that. Only 25 years after the virulent new strain of Dutch elm disease struck, it is proving increasingly hard to remember where big elms grew, what shape they were exactly, and what the English farming landscape looked like when it was still full of those heavy, fulsome, towering trees.

Elms have always been tinged with nostalgia and melancholy. In literature and popular mythology they are trees of drowsiness and brooding, dark reflections, it sometimes seems, of the piled cumulus clouds of late summer. At their most sombre, they have been constant reminders of mortality, dropping vast branches without warning and providing the wood in which human remains are finally laid to rest. 'Elm hateth Man and waiteth' is an old saying.

It is John Betjeman's poignant poems that best catch the tree's image (Tennyson's memorable 'The moan of doves in immemorial elms' excepted). Betjeman's elms are emblematic but always exactly placed. In Highgate he sees 'rich elms careering down the hill/ Full billows rolling into Holloway ...' In Hertfordshire, 'shadowy cliffs' with 'pale corn waves rippling to a shore'. In 'Dear Old Village', 'The elm leaves patter like a summer shower/ As *lin-lan-lone* pours through them from the tower.' His elm epithets themselves sound the complex depths of the tree's presence. They are 'guardian', 'waiting', 'shadowy' and 'whelming' – that extraordinary, perfect adjective from the funeral poem 'In Memory of Basil'. One of his oddest poetic images, at the start of one of his strangest poems ('The Heart of Thomas Hardy'), is also enacted before a gallery of elms:

The heart of Thomas Hardy flew out of Stinsford churchyard
A little thumping fig, it rocketed over the elm trees.[1]

All of these poems were written before the wave of Dutch elm disease that began in the late 1960s, whose impact is hauntingly documented in Gerald Wilkinson's book, *Epitaph for the Elm* (1978). The book's opening echoes the elegiac way elms were talked about in their prime: 'The elms are dying. Gaps appear in familiar lines of trees that we never bothered to think of as elms. Half the tall trees, and many smaller ones, in the roadside hedges seem to have been elms, we notice, now they are so unhappily conspicuous ... Rich corners of rustic England are one year a little yellowed, the next as bare as battle-fields. Those dusty summer lanes in the heart of England that were half black shadow, half tattered sleepy sunlight among cobwebs, nettles and leafy elm shoots will soon be exposed to an unnatural glare.'[2]

Since then elm disease has fulminated, waned and returned again – the pattern you would expect from a virulent fungal disorder. The spores of Dutch elm disease are carried from tree to tree by bark-beetles, but the lethal agency itself is a fungus, *Ceratocystis ulmi*, which damages trees chiefly by blocking their water-conducting channels. Its effects may be mitigated in several ways. As more elms die, the distance to uninfected trees may be greater than the beetles can fly. And with time, even aggressive new strains of the fungus catch infections of their own, which make them less virulent, or able to grow only in sizeable elm branches. This was almost certainly one of the processes that was happening during the 1980s, when, throughout the country, shoots and suckers were reported growing from apparently dead elm stumps and often getting some way beyond the bush stage.[3] Many reached as high as 30 feet and began to flower again. Then, in the early 1990s, the disease struck again – a consequence of these now sizeable young trees being more attractive to both beetles and fungus, and the fungus having had time to evolve strains less vulnerable to viral attack. This see-saw is likely to continue, with sometimes the fungus in the ascendency and sometimes the virus (and young trees), until a *modus vivendi* is reached.

But the single most important reason why the elms have not been driven into extinction by the current wave of disease (as many prophesied they would be) lies in the

Dead elms at Dallinghoo, Suffolk, their almost identical profiles showing their common ancestry as suckers or cuttings from a single tree.

nature of the elm family itself. Although four to six species are conventionally recognised, in the real world elms refuse to be be confined to them. Over the past millennia they have formed a bewildering range of races and hybrids, so that there are now almost as many kinds of elms as places in which they grow. The elms' saving grace has been to reproduce both by seed and by sucker. Seedling elms, with all the variation that cross-breeding brings, are common on the continent and probably were so in southern Britain when the climate was warmer, six or seven thousand years ago (cf. small-leaved lime, p. 116). With the exception of the **wych elm**, *Ulmus glabra*, elms do not often produce seedlings in Britain now. But the majority of kinds produce suckers from their roots, forming the genetically identical clusters of trees known as clones. These differ in height and bark texture, in their degree of uprightness and symmetry, in their vulnerability to disease, and in the size and shape of their leaves. In Buff Wood, at East Hatley in Cambridgeshire, Oliver Rackham has identified at least 29 distinct elm clones in just 40 acres.[4]

A similar diversity exists outside woods. Elms have always been closely associated with human settlements,

'Elms in Old Hall Park', East Bergholt, Suffolk, a pencil and wash drawing by John Constable.

and 'elm' is one of the commonest components in Anglo-Saxon place names. The foliage was cut for cattle and maybe even for human food in prehistoric times, and later the trees were widely planted, or encouraged to spread sideways, as hedges and boundary markers. The suckers were an abundant and convenient source of young trees, and sometimes whole parishes would be hedged from a single clone, the character of the particular tree perpetuating itself through each new generation of suckers. (And even after they were dead: the profiles of the stark skeletons in many East Anglian hedges during the late 1970s matched up like receding images in a mirror.)

Between 1955 and 1967 R. H. Richens conducted an extensive survey of the village elms of East Anglia.[5] He visited and took representative leaf samples from trees growing in the ancient boundary hedges of more than 500 parishes. He found such a variation in leaf-shape (echoed by differences in bark texture and the shape of the whole tree) that it was possible to name types of elm distinctive to small groups of settlements and sometimes even to individual villages. (In Essex alone he identified 27 village types, and the litany of their names, from Frinton, Springfield and Layer de la Haye, to West Hanningfield and North Weald Bassett, is like something from a Betjeman poem itself.)

Richens's explanation for this intensely local distinctiveness was that prehistoric settlers in different river valleys brought suckers of their favourite or local elm type over from the continent. There is no evidence that this happened, and it does seem an unlikely scenario, given the highly forested nature of the countryside (which included plenty of wych elm then) when these early farmers were creating their settlements.

Rackham's alternative explanation is that these varieties of elm were naturally present in Britain, having migrated back with our other forest trees in the post-glacial era.[6] It has now been established by pollen analysis that the species which shows the most variation – the **small-leaved** or **East Anglian elm**, *U. minor* – was present in England before Neolithic settlers arrived. So was the wych elm, which is really a species apart, being both non-suckering and rarely growing naturally outside woods. The **English elm**, *U. procera*, the tall, billowing, traditional tree of the Midlands and southern England, and the most severely hit by Dutch elm disease, is possibly a human introduction. It may be a variety of *U. minor*, as is the narrow, erect-branched **Cornish elm** (*U. minor* ssp.

angustifolia or *U. stricta*). The **Wheatley elm**, fashionable in the 1920s, is a variety of the Cornish, and the **Huntingdon elm** a variety of the hybrid between wych and smooth-leaved elm known (confusingly, in elm disease days) as **Dutch elm**, *U.* × *hollandica*. Mapping the elms' family tree is a Byzantine business ...

But its complexity is the group's best hope of survival. Almost all the suckering types can survive the disease, albeit as bushes; and two – **Boxworth elm** (Cambridgeshire) and **Dengie elm** (Essex) – seem promisingly resistant even as adult trees. There are also many surviving trees in Brighton. This has long been a city with a huge variety of elms in streets and parks, and the local authority made great efforts to protect what was in effect an elm gene bank by burning dead trees promptly, before beetles had emerged, and inoculating living trees: 'Elms dominate so many streets in Brighton, including the London Road from Preston Circus, and indeed the Level, which were saved from the ravages of Dutch elm disease only to be decimated by the Great Storm [of 1987]. Plenty are left and the Council replanted the Level.'[7]

Most of the famous landmark elms – the Nine Elms in south London, the Tenor, Bass and Alto Elms in Sigglesthorpe, Humberside, the Palmers Elm at Hewish in Somerset, the Watch Elm in Avon, for example – have gone.[8] So have the Dancing or Cross Elms of Devon round which May Day dances were performed. Sometimes these were planted close to churches – though one of the hazards associated with elms growing in churchyards was that they were liable to send suckers under the church, as at Ross on Wye in the late nineteenth century, where some sizeable trees grew *inside* the church. (Many of these historic elms are catalogued, along with elm paintings and literature, in R. H. Richens's *Elm*.)[9] But a few elm customs survive:

The Wicken Love Feast in Northamptonshire commemorates the union in 1587 of '"Wikehamon and Wikedyve into one church and again called Wicken" ... Ever since, on Holy Thursday (Ascension Day), after morning service in the church, the 100th Psalm is sung under an elm tree near the Parsonage, where the rector has given cake and ale to all in the parish that assembled or came to it. The elm was blown down some years ago, and a young one is growing in its place. On Ascension Day we still process to the spot after a service in the church and sing the Old Hundredth and pray for the village. Then we go to the Rectory for Holy Thursday cake and Ruddles Ale.'[10]

East Anglian village elms (near Little Wigborough, Essex) regenerating from suckers. They may reach 20 feet in height before succumbing to Dutch elm disease again.

'In Lichfield, Staffordshire, there is an old custom of carrying twigs in procession round the Cathedral Close on Ascension Day. It happens during Rogationtide and the Beating of the Bounds, and, even the church acknowledges, has fertility overtones. (When there was a theological college in the close, it was "believed" that a student's wife observing the procession would conceive!) We end by throwing all the twigs in the font.'[11]

'This tradition has been taken over during the past few years by the younger choristers of the [Lichfield] Cathedral School, who place elm twigs and branches on the doorways round the Cathedral Close early on the morning of Ascension Day. The elm was cut from the nearest available tree, in the Palace gardens ... The twigs carried in the procession later in the day by clergy and congregation are now usually lime – elm having vanished from the bulk of the close.'[12]

'Another nasty habit that we had was rubbing elm leaves between the hands and then rubbing the hands on the face of an unsuspecting victim. It was almost as if one had been stung by nettles – except that, fortunately, the effects did not last more than a few seconds. Hardly surprising since elm and nettle are closely related.' [13]

When elm timber was plentiful it was widely used where durability in wet conditions was needed – for instance in wooden pipes, floorboarding and coffins. The intricate and irregular grain pattern found in some trees (and especially in elm burrs) also made it popular for furniture. These days, the dwindling remains of dead elm trees, which bleach to the colour and texture of bone when dry, are chiefly used by sculptors in wood.

But there is every chance that mature elms may one day return to both commerce and the landscape. Although the current wave of Dutch elm disease may be the worst in historical times, the disease itself is nothing new, and the tree has always crept back. There was an epidemic in southern England between 1819 and 1864, another in Oxford in the 1780s. The dead elms common in Italian paintings between 1450 and 1530 are almost certainly stricken with it. And it may even be implicated in the phenomenon known as the Elm Decline (*c.* 3000 BC), when the pollen deposits from a tree which covered something like one-eighth of the British Isles fell dramatically to half their previous level. Commenting on these events (and echoing elm's morbid mythology) Oliver Rackham concludes, 'Why was it possible in the eighteenth century to insure one's elms against death? Elm was evidently well known to be the tree that specially shared man's fragile tenure of life, and it is difficult to suggest any other explanation than Elm Disease.' [14]

Hop family *Cannabaceae*

Hemp, *Cannabis sativa* (VN: Marijuana, Grass, Pot, Ganja, Hash). Hemp is the source of the listed drugs marijuana (the dried leaves) and cannabis (the dried resin which exudes from the leaves and stem), and these days can be legally grown in this country only under Home Office licence. Previously it was cultivated both as a medicinal herb and as a source of fibre. Cannabis seeds have been found in the remains of the medieval Augustine Monastery at Soutra, East Lothian, and the plant was probably introduced there from the Middle East as a sedative.[1] It continued to be officially prescribed in this country as a tranquilliser (usually as 'Tinctura Cannabis Indicae') up until the 1930s, though this was prepared from imported plants. Naturalised plants usually sprang from spilt hemp seed, which was used as a bird food and in anglers' ground-bait mixtures (which explains the plant's occcasional appearance along canal towpaths).

Now it is being cultivated again, for its fibrous qualities. By 1993 two farmers, one just outside Oxford and the other in Essex, had been granted licences and were together growing more than 1,500 acres of cannabis. The crop is used chiefly in the making of high-grade paper products, such as Bibles and, ironically, cigarette papers, and 'archive quality' paper, which doesn't degrade like that based on wood pulp.[2]

But any smoker wondering if potent material could be scrumped from these farms is likely to be disappointed, as cannabis strains grown for fibre develop little of the active narcotic ingredient (tetrahydrocannabinol) in our climate. Nonetheless, hopeful users have sometimes resorted to

Marijuana sometimes sprouts from bird-seed in waste places.

Hemp and hops from a Tudor pattern book.

clandestine outdoor cultivation, as in a plot discovered in Ashdown Forest and perhaps the four plants found in a woodland clearing near Shrewsbury in 1980;[3] or they have picked the plants which still spring occasionally from birdseed. Hemp seed is, strictly speaking, banned from such products. But now and then the odd unscreened seed slips embarrassingly through, as in a Hexham garden: 'An elderly friend feeds the wild birds regularly on her terrace outside a picture window. When we were visiting she pointed out a magnificent plant growing in the paving and said how excited she was that this lovely plant must have arrived in the birdseed. We told her what it was. Sadly for such a decorative plant she had to pull it up because she had the vicar and several ladies of the church coming to tea that afternoon.'[4]

Unfamiliarity with the fingered leaves of real cannabis plants no doubt lies behind some farcical cases of mistaken identity, in which, for instance, young conker trees have been impounded:

'During the summer of 1977, the Wiltshire Police carried out a drugs raid at my secondary school, looking for cannabis plants. They confiscated and left with a horse chestnut seedling in a pot, obviously not believing its owner that they would find a conker attached to the end of it.'[5]

'A Drug Squad descended on the Sussex Trust for Nature Conservation headquarters at Woods Mill without warning and without a search warrant. The squad had been tipped off that the Trust were growing cannabis. Many plants on the Trust's nature trail are labelled and no doubt someone with insufficient understanding had reported the inoffensive Hemp Agrimony, *Eupatorium cannabinum*, quite unrelated to cannabis.'[6]

Hop, *Humulus lupulus*. Hop's thick vines and bunches of cone-like fruiting heads are most familiar scrambling about hedgerows or – a favourite site – up telegraph poles and their stays. Often they are relics of old hop-gardens. But hop is almost certainly a native and haunts wilder habitats such as fens and river-banks too. In undisturbed and well-lit places the woody stems can stretch up to 20 feet in length. One plant in Kings Bromley, Staffordshire,

Wild hops scrambling over a hedge in Hampshire.

Hop foliage and 'cones', carved in Woolland church, Dorset.

was heavy enough to pull down a holly tree.[7]

Hop has the kind of deeply-lobed leaves beloved of stone- and wood-carvers, and it figures alongside white bryony, grape-vine, maple and buttercup on the medieval capitals of the Chapter House of Southwell Minster in Nottinghamshire.[8] The female flowers are carved there too, tucked under the leaves.

Hops are dioecious, bearing male and female flowers on different plants. The male flowers are small and green, but the female – likened by John Gerard to 'scaled Pine apples'[9] – are large and unmistakable, and the source of the hop's essential oils. Each lobe of the cone-like structure is studded near its base with yellow glands, which exude a mixture of aromatic oils and resins known as lupulin. Their scent is quite distinctive, with hints of garlic and ripening apples and yeast, and it is no surprise that hops were used in herbal medicine long before they were employed in making beer. They were recommended as an appetite stimulant, a mild pain-killer and a sedative. It was a pillow stuffed with hops that finally cured George III's insomnia and popularised this sleep-aid: 'It was remarkable that the first favourable change was due to Mr Addington, not indeed in his political capacity, but rather in his filial capacity. He remembered to have heard from his father, an eminent physician, that a pillow filled with hops would sometimes induce sleep when all other remedies had failed; the experiment being tried on the King was attended with complete success.'[10] More recently, infusions of hops, used as mild tranquillisers, were included in British and American Pharmacopoeias well into the 1900s, and hop pillows are still widely made and sold.

Hops were not introduced as a bittering agent in brewing until the end of the Middle Ages. Even then other wild plants were still often used to preserve and flavour ale – notably bog-myrtle, wormwood, costmary or 'alecost', and ground-ivy or 'ale-hoof' (see p. 317). But hops enabled the brew to be kept longer, and 'beer', flavoured with hops, began to be distinguished from 'ale', made without them. By the end of the sixteenth century beer was the more popular English drink, and hop-growing spread over many areas of southern England and the Midlands. It may have been this expansion that gave rise to another use for the plant – eating the young shoots. But it was in the late nineteenth century, when large numbers of families from London's East End used to migrate down to the Kent hop-fields to help with the pruning and picking, that hop shoots earned a wider popularity. Bundles of the young pruned shoots were boiled like thin asparagus.

There is a custom in hop-growing areas of hanging a bough of hops on a wall or over the fireplace.[11] In Hampshire I have seen boughs hung over the bar in pubs.

Hop boughs are still hung on walls, especially in pubs in hop-growing areas. It is a purely decorative custom now, but echoes the old practice of hanging corn dollies in churches and farmhouses after harvest.

Mulberry family *Moraceae*

Fig, *Ficus carica.* The fig, though it came to Europe from south-west Asia and, like the olive, is irrevocably associated with the languid warmth of the Mediterranean, will fruit perfectly well in windswept Britain. The sweetest and most succulent fig I have ever eaten was grown outdoors on the very top of Harrow on the Hill in Middlesex.[1]

But setting seeds in the wild that will grow successfully into mature trees is another matter; and most wilding figs are isolated specimens which have had some fairly obvious head start, from warmth or inadvertent composting: one close to a lime-kiln near Abergavenny; others on railway embankments, resulting perhaps from discarded food – on top of a tunnel in Oakleigh Park, Hertfordshire, for instance, and 'the substantial population on the embankment between Castleford and Leeds'.[2] The tree which sprang from a grave in Watford in 1913 probably also originated in a snack taken by the unfortunate occupant; but there is a local legend (still well known) that it sprang from the tomb of an atheist, who asked for a fig to be placed in his or her hand in the coffin and who said that if there was life beyond the grave the fig would sprout.

Figs in the fencing, Petworth House, West Sussex.

There have also been car park figs, additionally nurtured by the warmth of car exhausts. A famous tree in the NCP park on Ludgate Hill in London was, at 35 feet, one of the tallest in Britain before the site was developed. There is also 'a large fig tree growing on waste land (at present a car park) in the centre of the one way system in Colliers Wood, south-west London'.[3]

But the most interesting and most plentiful colonies of fig trees are on the banks of urban rivers. Perhaps the specimen spotted by a sharp-eyed fan of the TV series *Last of the Summer Wine* growing on the banks of the Holme 'practically opposite Nora Batty's cottage' is not quite in this category.[4] But most of the individuals or small groups of fig trees found on the banks of the Avon in Bristol, where one is rooted in vertical stonework near Temple Meads Station, the Thames in London, and a number of northern waterways (including a canal in Liverpool close to Tate and Lyle's sugar refinery) do fit into some kind of pattern, with nearby industry and sewage being the usual common factors.

'Figs occur on the banks of the Huddersfield Narrow Canal. The species is thought to have come from the unloading of canal barges carrying groceries. One specimen found by the side of a tiny stream is said to have come from a batch of rotten figs thrown out by my communicator's aunt! Two trees on the banks of the River Colne possibly came from sewage works formerly situated higher up the valley, and a third fig on the banks of the Calder came from a biscuit factory which used to be there.'[5]

'There is a large specimen growing out of the canal side at Brookfoot Mills, Brighouse, Yorkshire. The fig is growing opposite an outfall from the dyehouse which discharged hot water into the canal. A name painted on the wall of the building indicates that a sweet factory was there at one time. I have a fig tree in my front garden grown from a cutting from the Brookfoot plant, now about 10 feet tall.'[6]

Glasgow has a celebrated riverside specimen known as the Dalmarnock Fig. It has many thin trunks about 10 feet tall and sprouts from the vertical stonework of the south-facing bank of the Clyde just downstream from Dalmarnock Bridge and close to the local sewage works.[7]

Fig trees by the River Don, in the old steelworks quarter of Sheffield: 'as much a part of its industrial heritage as Bessemer converters'.

But the most substantial population so far discovered in Britain grows along the banks of the River Don in Sheffield. Isolated fig trees have been known there for some time, but it was not until there was a thorough survey of the river-bank vegetation, organised by Dr Oliver Gilbert of Sheffield University in the early 1990s, that it became clear that there was a healthy colony beside the Don. Some 35 specimens were found during the survey itself; more have been uncovered since, and in at least one place they form a small riverine wood.

Most of the figs grow in the base of retaining walls in the derelict industrial quarter, amongst a dense scrub of Indian balsam and Japanese knotweed. (See p. 108 for an account of the remarkable ecology of this and the other herbaceous plants of this urban river.) The trees are all well-grown specimens, up to 25 feet high and with multiple stems. Like all figs they fruit twice a year, once in May – pear-shaped fruits which swell and soften until mid-July – and again in September, harder fruits which fall off unripe in October. The wild fruits are rarely fit to eat.

There is a curiosity about these Sheffield fig-trees. Why do they all seem to be of much the same age, roughly 70 years? And why are there so many more of them than along other urban rivers? There is no reason to think that Sheffield was a city of obsessive fig-eaters, flushing exceptional quantities of well-manured seeds into the river (though it had good numbers of pickle and cake factories). Dr Gilbert believes that the answer lies with the steel industry. The trees are concentrated in the steelworks area, and the period during which the trees seemingly became established, in the early 1920s, was when the Sheffield steel industry was at its height. The river water was used as a coolant in the factories, and as a result the Don ran at a fairly constant temperature of 20°C – warm enough for fig seeds washed into the river from sewage outfalls to germinate and thrive. When the steel industry began to decline in the city, river temperatures returned to normal, and though established trees were able to survive no new seedlings were able to sprout. As Dr Gilbert says, this 'means that the Wild Figs of Sheffield are as much a part of its industrial heritage as Bessemer converters, steam hammer and crucible steel'.[8]

This isn't just a botanist's view. There is a real interest in the trees developing in Sheffield, and Tree Preservation Orders have been placed on some of them. When I visited the Don in the autumn of 1992, I called into a riverside pub amongst the decaying warehouses and empty foundries and found it had a splendid, spreading fig-tree on its own stretch of bank. The landlord knew the figs' story (as did most of the regulars present) and told me how they had dug in their heels to save their tree when the brewery had threatened to grub it out in a modernisation programme. And if any proof of the 'rightness' of these trees were needed, whilst we were on the bank admiring the pub's specimen a kingfisher came and perched on one of the lower branches.

Black mulberry, *Morus nigra*, is occasionally naturalised from bird-sown fruit, for example on a wall by the Thames at Chiswick in west London.

Nettle family *Urticaceae*

Common nettle or **Stinging nettle**, *Urtica dioica* (VN: Devil's plaything, Hokey-pokey, Jinny nettle). The nettle has given its name to nettle rash (urticaria), to a state of general pique or irritation, and – out of exasperation or respect at its profusion – to at least half a dozen villages in England, including Nettlebed (Oxfordshire), Nettlecombe (Dorset), Nettleham and Nettleton (Lincolnshire), and the Nettlesteads (in Kent, Surrey and Suffolk).[1] A plant which grows in such close, copious and aggressive proximity to human settlements was never in danger of being ignored.

Stinging nettle in flower.

Stinging nettles' natural habitat is fertile, muddy, slightly disturbed ground, especially amongst the lush herbage of silt-rich river valleys and woodland glades manured by feeding animals. Above all, it needs soils rich in phosphates, which is why it has flourished in the wake of human and agricultural colonisation. Human settlements provide phosphates in abundance, in cattle-pens, middens, bonfire sites, refuse dumps and churchyards. Soil-borne phosphates can endure for exceptionally long periods, and ancient nettle clumps (often with other phosphate-lovers such as elder and cleavers) mark the sites of many deserted villages, Little Gidding in Huntingdonshire for instance, and isolated Scottish crofts: 'In the Highlands, ruined crofts occur in very remote places, miles from anywhere. One thing they always have in common – a patch of nettles, even though there are no other nettles for 20 miles in any direction.'[2]

The wooded sites of Romano-British villages on the Grovely Ridge near Salisbury are still dense with nettles subsisting on the remains of an occupation that ended 1,600 years ago.[3] And modern Wiltshire is scarcely any different. Fertiliser run-off from its vast arable prairies has made the county a paradise for nettles, especially along the upper reaches of the River Kennet. In summer a thirteen-and-a-half-mile stretch is an almost continuous double ribbon of nettles (or a broad single swath where the channel has dried out), up to eight feet high. They spread by seed and by means of their tough yellow roots, which as well as extending up to 20 inches a year, are broken off and churned into the mud by cattle, and sluiced further downstream by the river water.[4]

Such an abundance of tender greenery, stings notwithstanding, has been a tempting source of food since prehistoric times. The Romans certainly ate nettles as well as manuring them. On the Celtic fringe they have traditionally been eaten in a kind of soup with oatmeal;[5] and a recipe for St Columba's broth (the sixth-century Irish monk and poet) survives to this day and tastes remarkably like soup made from young peas: 'Pick young stinging nettles before the end of June, when they are 4 or 5 inches high – one handful for each person. Boil, drain, chop and return to pan with water and milk. Reheat, sprinkle in fine oatmeal or oats, stirring until thick. For present-day tastes eat with toast and grated cheese, or peeled soft-boiled egg.'[6]

Nettles have frequently been returned to as a subsistence food, for example during the Irish potato famine of the 1840s and during the Second World War. Ambrose Heath, in his *Kitchen Front Recipes and Hints* (1941), had extravagant praise for both nettle and dandelion leaves: 'A poached egg on a bed of dandelion or nettle purée covered with cheese sauce is an almost perfect meal, containing every one of the foods which we are being told to eat, body-building, protective and energising.' Modern cooks have deep-fried nettle leaves to the texture of green crisps, and used them raw in purées. The distinguished chef Anton Mosimann blends his with fromage blanc, new potatoes and nutmeg into a 'nettle nouvelle'. A woodman from Bedfordshire has a more earthy technique: 'I sometimes make a sort of green vegetable pâté by picking tender nettle tops and smashing them up in a pestle and mortar (which in my case means half a brick and an oak log). I add a mixture of herbs including crushed garlic ... wrap it in a lettuce leaf and eat it as a snack.'[7]

A Devon boy ate untreated nettle leaves at school, a trick that echoes 'grasping the nettle' (in which the stem or leaves are gripped quickly and tightly, so that the toxin-laden hairs are crushed before they can pierce the skin): 'The trick is to roll them up in a special way with the tongue, making sure that there is plenty of saliva to coat them.'[8]

'Many years ago a gypsy arrived at my farm asking for scrap ... After he had loaded his van his hands were filthy, covered with grease, dust and more than a little cow-dung. Without a word he went to the nettle patch and grabbed a plant and stripped off the leaves – just in one movement. I was fascinated, because of course nettles sting. "How can you do that?" I asked. "Finest thing in the world for cleaning the hands," he said. "And what's more you'll never see a gypsy with arthritis." '[9]

The idea of using nettle stings as a counter-irritant, to 'warm away' inflammations, has some roots in sympathetic magic, but it also has a degree of practicality. According to the Elizabethan antiquary William Camden, the Romans, well aware that Britain would be cold, brought their native nettle with them to rub on their skins. ('Roman nettle', *U. pilulifera*, occurred as a casual in some parts of Britain until the 1950s, but if the Romans did resort to therapeutic self-flagellation they would have found English nettles just as efficient and rather more available.) In the Isle of Man the same custom persisted until quite recently: 'It is widely known as "Jinny Nettle" on the island. The Manx name *Undaagah* comes from an old Gaelic word meaning "flaying", because of the plant's blistering effect on the skin. It was used on the island to restore circulation by beating the skin with it.'[10]

'Both my mother and I use this plant on any joint which gives painful symptoms. The treatment is simple – sting the joint liberally with the plant, and if possible move the joint well immediately after application, e.g. go for a walk. Some relief is felt within 30 minutes, but the gently tingling warmth is felt for many hours. The following day, should any pain in the joint persist, sting again, and eventually full and pain-free movement returns.'[11] The use against arthritis has been to some extent vindicated by the modern medical practice of using bee-venom for inflamed joints.

Using nettle switches rather more recklessly was part of a children's ritual on Oak Apple Day, 29 May (perhaps echoing pro-Royalist sanctions – see p. 76), which seemed to be confined to villages close to the Derbyshire–Staffordshire border between Mayfield and Tansley: '[When I was a girl] in Derbyshire in the 1930s, Oak Apple Day was marked by a special ritual. On the way to school in the morning, the boys armed themselves with sprays of nettles, and custom licensed them to use the nettles to sting the bare legs of girls unless they carried a safety talisman of a sprig of oak leaves. Such a sprig guarded the girl completely.'[12] In Mayfield itself, boys were just as liable to get 'nettled': 'Stinging nettles, usually about 30 to 40 cm long, were plucked, either using your handkerchief to wrap round the stalk to protect your hands if you were "soft", or using your bare hands if you were "tough", and then used to lash other children across their bare legs (all primary school boys wore short trousers in those days). There was no particular Royalist sentiment in this at that time. It was just a rather sadistic game, although we were aware of the origins.'[13] (Also in Derbyshire, I have heard of a similar spartan application by a farmer's wife who used a nettle switch to get a brood of children out of bed.)

Nettles have also been used to make green manure, by steeping them in rainwater (cf. comfrey, p. 309),[14] and have a long history of use as a fibre in string- and cloth-making. In Britain a flint arrowhead has been found attached by nettle fibres to fragments of its shaft; and remains of nettle fabric were discovered in a Danish grave from the late Bronze Age, wrapped around cremated bones.[15] Nettle-cloth (the fibrous stems treated in much the same way as flax) was certainly manufactured in Scotland and Scandinavia into the early nineteenth century. When Germany ran short of cotton during the First World War, it resorted to nettles to make military clothing. Something like two and a half million kilograms were gathered from the wild, though it took 40 kg to make a single shirt. During the Second World War, some work was done in Britain on the possibility of using nettles in the same way, but in the end the plants were chiefly used for extracting chlorophyll, and as a dye for camouflage nets.[16]

Pellitory-of-the-wall, common on old stonework, and once prescribed (by sympathetic magic) for kidney stones.

The **Stingless** or **fen nettle**, *U. galeopsifolia*, considered by some to be a distinct species, has been found at Wicken Fen, Cambridgeshire, by the river at Wylye in Wiltshire, at many stations in Norfolk (especially in the Broads), and by the River Kennet near Woolhampton in Berkshire.[17] It is probably much commoner than realised. **Small nettle**, *U. urens*, is a small but stinging annual of cultivated and waste ground, chiefly on light soils.

Pellitory-of-the-wall, *Parietaria judaica*, is a frequent plant of steep hedge-banks, old pavements and walls – especially damp church walls – mainly in the south. It used to be an important plant in herbal medicine, given (because it grew on stone) for kidney and bladder stones. One Cornish farmer has found his horses dosing themselves with it when they are ill: 'There are several old buildings on this farm, which was known to be established in 1393. Growing on them is a profuse amount of pellitory-of-the-wall. When leading my Connemara stallion past one wall he would always stop to eat this plant, and obviously liked it. When another pony was ill, it was the only food she would eat and I am sure that it helped restore her health.'[18]

Mind-your-own-business or **Helxine**, *Soleirolia soleirolii*, is an occasional naturalised escape, mainly in southern England. It is most frequently seen in or near greenhouses, covering damp floors, walls and flowerpots with multitudes of small rounded leaves. It is native in the western Mediterranean.

Walnut family *Juglandaceae*

Walnut, *Juglans regia*. A southern European tree which has been widely planted in Britain. Self-sown trees occasionally spring up in warmer places, in hedgerows and chalk-pits for instance. There are seemingly self-sustaining colonies in a deep Chiltern coombe near Chequers in Buckinghamshire and on Ministry of Defence land at Purfleet, Middlesex. I have seen numbers growing on a railway embankment at Hemel Hempstead, Hertfordshire. Seedlings appearing some distance from parent trees usually spring from nuts carried off and buried by rooks.

Bog-myrtle family
Myricaceae

Bog-myrtle, *Myrica gale* (VN: Sweet gale, Gold withy). Bog-myrtle is a shrub of wet, acid heathland and moors, chiefly in Scotland, North Wales and north-west England, but with surviving populations in, for instance, the Devon and Surrey commons and the Norfolk Broads. It is rather drab in appearance, enlivened by the stubby orange-brown catkins in spring and early summer. But the whole plant emits a resinous, balsamic fragrance, especially when in flower. And in sites where there are large colonies, for example in damp valleys and by streamsides in the New Forest, the scent is astonishingly pervasive. It can carry hundreds of yards, and the aromatic substances responsible are occasionally blamed for tainting milk. In fact sweet gale is rarely browsed and flourishes best at the junction between grazed areas – such as the New Forest 'lawns' – and the heath proper, where competition from grasses is less.

'When my family first came to the New Forest in 1923, we found that the local name for bog-myrtle was "Gold Withy" (literally golden willow). The plant, common in Forest bogs, was blamed for giving a peculiar flavour to the milk of cows which ate it.'[1]

More often bog-myrtle has been both admired and used for its scent. 'Gale' is commemorated in several place names, for instance Galsworthy in Devon, 'the slope of the bog-myrtle'.[2] The aromatic resins have been used for scenting candles, and the whole plant is still used in brewing and cooking and as an insect repellant.

'Gale beer is a drink traditionally made from bog-myrtle. But the tradition of using the fragrant wax in candles is lost with the art of making tallow candles.'[3]

'Grows in wet places on the edge of the North York Moors. The very pleasant aromatic odour of this plant is added to home-made beer by mixing the leafy branches with the hot liquid in the early stages of the beer-making process.'[4] (The Gale Inn at Littleborough, over the border in Lancashire, is apparently named from the plant's use in beer, though the plant itself has vanished from the region.)[5]

Bog-myrtle in damp heathland on Hartland Moor, Dorset.

'In Islay and Jura it was used as a garnish for food and stored with linen as a means of driving away moths and finely scenting the cloth.'[6]

'In Sutherland we are proud to still have small areas of bog-myrtle. When put in sachets with clothes, laundry, or blankets, its scent is as delightful as that of lavender. Insects do not feel attracted to it, especially moths. Some people, especially anglers, even wear a sprig in a buttonhole against midges.'[7]

In 1995 a commercial midge-repellant based on bog-myrtle appeared on the market under the name 'Myrica'. It was produced from wild myrtle gathered by crofters on the Isle of Skye, which was then steam-distilled to produce a volatile essential oil. The initial trials vindicated the plant's folk reputation. Eight volunteers each had one arm covered in a gel made from the essential oil and the other left untreated. Over 10 minutes, the untreated arms recorded 155 bites while the treated arms received just 13.

Beech family *Fagaceae*

A note on trees

With the exception of the much-depleted elms (see p. 58), the beech family is the first of the big woodland tree families of Britain (not all closely related), which include the oaks, birches, alders, limes, poplars, willows, maples and ashes. Trees have size, longevity, economic usefulness and a profound impact on the landscape – which means that they have entered our culture more thoroughly than most smaller flowering plants. Some – oak, elm and willow especially – could have whole anthologies of poetry and paintings collected about them.

Yet to an extent their individual cultural histories have been overshadowed by the roles they have played in the general business of growing and harvesting wood. Although each has its own unique set of meanings and uses, they are treated more as trees than as individual species. They have all been cropped for fuel-wood and used as rough timber. Their names are found as frequent components in place names, and most, in their maturity, have contributed to our considerable legacy of landmark trees. These subjects have been elegantly and exhaustively covered in Oliver Rackham's books (see Bibliography), which have transformed our understanding of the way that trees and woods 'work', socially and ecologically. *Flora Britannica* concentrates more on modern perceptions and meanings.

But there are common threads running through the social and cultural use of trees which are worth summarising at the outset. Traditionally, woodland trees were harvested by *coppicing*, which involves cutting them back to ground level once every 10 to 20 years. All our native deciduous trees (given enough light) will send up sheaves of straight new poles indefinitely when treated in this way. The cut poles were chiefly used as fuel-wood, but also as small-scale domestic or farm timber, often regardless of species.

Outside woods, in parks, pastureland and hedgerows, trees were usually managed by *pollarding*. This involves cropping the branches six to 15 feet above ground level, again on a regular rotation. Pollarding produces less reliably straight poles than coppicing, but means that the trees can coexist with cattle or deer, as the new shoots appear above the browse level. Pollarding also prolongs the life of trees by reducing their top-heaviness and making them more wind-resistant, and over the centuries can produce exceptionally gnarled, characterful individuals. Uncut trees – grown for higher-quality timber – are often referred to as *maidens* or *standards*.

Pollards and, to a lesser extent, uncut trees can attain great ages and can become landmark trees, symbols of continuity in the landscape that can outlive whole dynasties of humans. There are still large numbers of landmark trees in parishes throughout Britain. Some are nationally famous, such as the Tortworth Chestnut, the Major Oak in Sherwood Forest, and the Tolpuddle Martyrs' Sycamore. There are lost trees, too, such as Gerard Manley Hopkins's Binsey Poplars, the Selborne Yew, and, most recently, the Aston on Clun Flag Tree. Many are more local trees which mark ancient boundaries and meeting places or where, according to legend, Queen Elizabeth picnicked or men were hanged. There are also increasing numbers of deliberately planted memorial trees, commemorating coronations, jubilees, centenaries and battles (some even planted out in the *formation* of battles). And there is an unrecorded host of privately loved neighbourhood trees – hidden in by children, watched as weather-vanes from kitchen windows, chatted under or just nodded at on the journey to work.

'There is a large oak in Quinta Drive, Barnet, and although not unusual, it is surely a landmark in this area. It is the place where if we have a coach to meet, it's always at the oak tree, [it] is used for directions in finding your way round our estate, it's the turn round for our Hopper bus.'[1]

Pedunculate oak or **English oak**, *Quercus robur* (VN: Sussex weed). The English have a loving but proprietorial relationship with the oak. Since at least 1662, when John Evelyn lamented to the Royal Society the 'notorious decay' of Britain's '*Wooden-walls*', as 'nothing which seems more fatally to threaten a Weakning, if not a Dissolution of the strength of this famous and flourishing *Nation*',[2] the qualities of the oak and the character of the nation have been linked. Edmund Burke described the aristocracy as 'the great oaks which shade a country' (and two centuries later, in 1995, the felling in Windsor Great Park of a number of 150-year-old oaks was compared by one pundit to 'chopping down the Queen Mother'). In the mid-eighteenth century David Garrick wrote a shanty, two of whose lines were to enter the national folk memory: 'Heart of oak are our ships/ Heart of oak are our men'. They in turn were plagiarised by Samuel James Arnold in his poem on the death of Nelson. The prolific botanical writer and arbiter of Victorian taste John Loudon pronounced the oak 'the emblem of grandeur,

Heart of oak: an ancient pollard, encrusted with lichens, moss and ferns, at Croft Castle, Herefordshire.

strength and duration; of force that resists, as the lion is of force that acts ...'

This is a fair précis of the oak's qualities, though we might be more diffident about claiming them for ourselves these days. But we can have legitimate pride in our legacy of ancient oaks, which vastly exceeds that of any other western European country. There are not many lowland parishes that don't have at least one oak over 250 years old – marking a cross-roads, commemorating an accession, sheltering a village seat, or just marooned out in the fields because no one has the heart to cut it down. Old oaks can live well over 500 years, especially when they are pollarded, and become recognisable as characters long before that. They age with more craggy and eccentric individuality than beeches, with dead branches drying to driftwood grey, the trunks developing burrs and dark rain-trickles and tufts of epiphytic fern, and often collapsing in places to a powdery red mass of rotting heartwood. The survival of trees which are so alien to orthodox forestry standards is a marvel, given the way we have treated so many of our other ancient natural monuments.

The following are accounts of some surviving landmark oaks, or their surviving fragments. Their image tends to be more homely, vernacular and even irreverent than that of some of the regal trees of the past. The archway oak in Sladden Wood, Kent, is typical, commemorating as it does the illegal flattening of this wood in one morning by Hughie Batchelor, a notorious 'agricultural improver' of the 1970s and 80s.

'Of the few mature oaks that were not felled before 1977, only one still survives. It is dubbed the Hughie Batchelor memorial tree. The felling of this oak had been so inexpert that it snapped at about four feet above the cut, and now forms an archway across a path. It still lives, though last year [1994] beefsteak fungus was found growing at its base, which may shorten its life.'[3]

'An ancient oak tree in the centre of Caton village [Lancashire] was believed to be becoming dangerous because of an overhanging branch. Villagers have over the years adopted the tree as a symbol of Caton. It is used as an emblem for the uniform of the local primary school, and of the Caton Sports Association. Tree surgeons, as usual, suggested felling it. But villagers thought otherwise and rallied round to protect the landmark. One parish councillor, Mick Jackson, and another villager volunteered to do the work, and made a metal post and prop to support the falling branch. Mr Jackson said, "The tree is the most important landmark in Caton and we should try and preserve it as long as there is any life in it at all." '[4]

'The vast oak at Eardisley near Hereford, described by Loudon in 1867 (34 feet in girth at three feet high), is still standing. It is said to be well over one thousand years of age and is very much alive. This tree is hollow and children used to use it as a play den. Some years ago the tree caught fire but it was saved, thank goodness. I have been inside this massive oak and looked right up into the sky with ease.'[5]

'I remember weekend jaunts to Epping Forest when living in Walthamstow as a child. Some time leading up to 1956 the Old Oak Tree in Epping was burnt down by vandals. This was a famous old tree and it served as monument and symbol to my family. My dad made a journey to Epping that weekend to say farewell to the tree. We brought home a small piece of charred wood and carved a

A modern (1990) carving of an oak tree in oak wood, in the tower doorway of St Mary's, Ludgershall, Buckinghamshire.

paper-knife from it, which he still has to this day.'[6]

'The Wishing Tree of Isle Maree, Wester Ross, a long-dead oak, is perhaps the most celebrated in all Scotland. The tree stands on an island in Loch Maree ... The original significance of the site lay in a wishing well, long since dried out, where rag (cloutie) offerings were hung from nearby trees. At some stage the clouties were replaced by coins [cf. hawthorn, p. 214], hammered edgewise into the tree. After hammering in the coin, the visitor makes a silent wish. The oldest coins examined date from the 1870s, and the custom may have become engrained after Queen Victoria's visit in the 1880s. The oak, a slender tree, was already dead by 1927, studded with pennies, a victim of copper poisoning. The tree is now in three three-yard pieces stacked upright like a teepee. It still bears its coat of pennies, and more coins have been hammered into two nearby oaks; still others are strewn among the leaf litter ... The site is of ancient holiness: the tree lies within a circular stone dyke, interpreted as a Druid's circle ... A curse lies on the person who removes anything from the isle. Local people still attach significance to these beliefs, and coin-wishing is obviously still current.'[7]

'When I went to Collyers School in Horsham, I used to cycle past the tree [the Sun Oak at Coolhurst] every day and it was amazing to look at in the summer. I imagine it got its name because of its enormous rounded crown.'[8]

'The Bound Oak is a hollow oak tree in Farley Hill, Berks. Its name refers to the fact that it marks the boundary between the parishes of Arborfield and Swallowfield. A fine wild cherry is growing very close to the oak, giving the impression that it is the oak that is in flower each March.'[9]

'The Milking Oak [Salcey Forest, Northamptonshire], in its heyday, had a spread so huge that the cows grazing on the "lawns" were milked under it in very hot or rainy weather.'[10]

'The Baginton Oak, on the southern edge of the village of Baginton, Warwickshire, sits at the southern tip of a triangular patch of grassland which may be the remnants of an old green ... It is an old hollow pollard, last pollarded probably over a century ago. Concrete was poured into it at one time to prevent fires being lit but that was removed and now wire mesh stretched over the top of the bolling tries to do the same job ... The pub opposite is named after it.'[11]

'Here at Great Barrington [Northamptonshire], we have an avenue of oak trees, planted by a previous Earl Spencer, which leads across an arable field, with crops through the middle.'[12]

'The Nine Brethren', a landmark coppiced oak in Nocton Wood, Lincolnshire.

Aristocratic oaks besieged by EC corn and folk oaks beseeched to death. Landmark trees seem to rise above class and politics:

'The village [Hartley Wintney, Hampshire] is well-known for its glades of English oak, and must be unique in the UK in having these trees planted in rows on the commons surrounding the centre of the village. For the villagers therefore, the Oak is king, and we cherish and are proud of our historic trees, planted by Lady Mildmay of Dogmersfield Park. They are known as the Mildmay or Trafalgar Oaks.'[13]

'Our most noteworthy tree is an oak (now much dwarfed by a massive electricity pylon), which has carved on its trunk a coffin and the letters SC, 1849 – which commemorates the death of Stan Crumpler (a Lytchett name) having been gored near the very same tree.'[14]

'Kett's Oak, Wymondham. Robert Kett led a rebellion against the enclosure of common land in 1549 and rallied support under the oak which now bears his name. He was subsequently hanged from Norwich Castle walls for his part in the revolt. Last year children from Hethersett Middle School planted 12 sapling oaks, grown from acorns from the original tree, and planted these alongside the old tree, which still flourishes, but is supported. Hitler's Oak. Chris Boardman, yachtsman and gold medallist in the 1936 Berlin Olympic Games, was presented with an oak tree by Hitler which was planted at How Hill.'[15]

The diversity which these trees show is echoed by oaks growing in hedgerows and woods. At Castle Malwood near the Rufus Stone in the New Forest is an ancient tree which produces new leaves by Christmas and a second crop in spring. It was first noticed by Dr William How in the 1650s.[16] At Staverton Park in Suffolk, a uniquely well-preserved medieval wood-pasture, there are some 4,000 oak pollards, between 200 and 500 years old.[17] They show every conceivable variation that can occur in natural self-sown oak populations. There are broad and tapering trees, barrel and fluted trunks, smooth barks and dense burrings, trees which hold their leaves all winter and others leafing all the way up the trunk. In Birklands, Sherwood Forest, there are 500-year-old oaks of a highly distinctive shape, and all with dead, stag-headed tops (though the rest of the crowns are thriving). In Wistman's Wood on Dartmoor, the oaks are elfinesque. John Fowles has written an incomparable portrait of them in the closing chapter of *The Tree*:

'The normal full-grown height of the common oak is 30 to 40 metres. Here the very largest, and even though they are centuries old, rarely top five metres. They are just coming into leaf, long after their lowland kin, in every shade from yellow-green to bronze. Their dark branches grow to an extraordinary extent laterally; they are endlessly angled, twisted, raked, interlocked, and reach quite as much downward as upward. These trees are inconceivably different from the normal habit of their species, far more like specimens from a natural bonsai nursery. They seem, even though the day is windless, to be writhing, convulsed, each its own Laocoon, caught and frozen in some fantastically private struggle for existence.'[18]

Extraordinarily, despite world-wide agreement about the importance of maintaining biological diversity, recent European Union 'harmonising' legislation insists on all commercial sources of oak seed coming from a few approved sources, mostly in eastern Europe. An earlier European grouping, the Celts, had a more respectful attitude towards the oak. Like many of the customs of early religions this was absorbed by (or perhaps, in this case, actively infiltrated) the Christian church in Britain. Carvings of oak leaves, acorns, and even galls are to be found in almost every English cathedral and a great number of the older parish churches. Sometimes they are openly displayed in the decorations of fonts and pew-ends, but often seemingly secreted away – under misericords or on the bosses high up in the roof.[19] In a few churches the most pagan symbol of all – the Green Man, wreathed by the oak leaves foaming from his mouth and ears – is found blatantly carved on Anglican capitals.

Ancient and venerated boundary oaks were also Christianised. The 'Gospel Oaks' that frequently occur as place names, with or without surviving trees, may refer to stopping points on the Beating of the Bounds or Perambulation – itself a sanctified version of early fertility rites. At some time during Rogationtide, just before Ascension Day in May, a procession would tour the bounds of the parish, memorising and passing on the knowledge of its course and extent. At many traditional points, often significant oaks, crops were blessed and passages from the gospels read.

In the village of Great Wishford in Wiltshire an elaborate Rogationtide ceremony is still held to affirm the villagers' ancient common rights to gather firewood in Grovely Forest. Its date has been shifted forward slightly, to 29 May (Oak Apple Day – see below), but it still echoes the mixture of Christian benediction, political demonstration and village party that has characterised Perambulations since as far back as the Middle Ages.

The modern ceremony begins in the early hours of 29 May, when the local youths march through the village banging dustbin lids and blowing trumpets and shouting 'Grovely, Grovely and all Grovely!' at the houses. They then go to Grovely Woods to cut vast oak boughs, up to 'the thickness of a man's arm', which are taken back down the hill to the village. Many are set in front of the doors of houses in the village. Others are carried in processions later in the day. One large branch is decorated with ribbons and hoisted to the top of the church tower. It is called the Marriage Bough and is supposed to bring good luck to all those married in the church in the coming year. Later in the morning, four women carrying sprigs of oak travel to Salisbury, accompanied by many villagers and a tremendous banner bearing the Grovely Shout and the commoners' motto 'Unity is Strength'. They dance for a while on the Cathedral Green. (Previously everybody danced the whole six miles to the city, but this was suppressed by the church in Victorian times, because it had degenerated into a revel.) The whole company then go into the Cathedral to make their claim by crying 'Grovely, Grovely and all Grovely!' The phrase 'all Grovely' is crucial. The only other village bordering the forest and having similar common rights was Barford St Martin. But their rights lapsed and Great Wishford now has 'all Grovely'.[20]

Oak sprigs are worn by the Garland King's horse during the Oak Apple Day celebrations in Castleton, Derbyshire.

Oak Apple Day, 29 May, is the anniversary of the triumphant return of Charles II to London at the Restoration of 1660. Of his many adventures during exile, it is his concealment in the oak at Boscobel which seems to have left the most lasting impression, and Charles declared that the day should be set aside as a public holiday 'for the dressing of trees'. Hence the date of the festival, and the tree species associated with it – though why it is Oak Apple – the spongy, crab-apple-like galls formed on oak twigs by wasp larvae at the end of May – Day, rather than simply 'Oak Day' is not clear. In late Victorian times it was simply called 'Royal Oak Day', though there was a

Field oaks near Wenlock Edge, Shropshire.

custom of covering any oak apples attached to sprigs with gold leaf in honour of the crown.[21] Oak Apple Day is still celebrated in many schools and military establishments by the wearing of sprays of oak, though nothing like as extensively as it once was, and has tended to absorb other May-time festivals (see Spring festivals, p. 174).

'As a celebration of the Royalist village [St Neot, Cornwall] saving its church from destruction by the Puritans, a fresh leafy oak branch is put up in the church tower each year and the old branch taken down. Villagers who take part in the ceremony wear a sprig of oak.'[22]

'The chant on Oak Apple Day was "29th May, Oak Apple Day, if you don't give us a holiday we'll all run away."'[23]

'A local name for oak apples [in this case clearly the spherical 'marble galls'] was "chick-chacks", from the sound they make when used as marbles. In Dolton [Devon] 29th May was Chick-chack Day and it coincided with the parade of the village Friendly Society.'[24]

Alongside this traditional mythology of strength and ancient lineage, there is another more modern group of scientific myths, which stress the oak's weaknesses and irregularities in reproducing itself. Despite the fact that it grows all over Britain in most kinds of soil, seedling oaks are rare inside oakwoods – though abundant on heathland, on railway embankments, and even, since set-aside, on fallow arable land. At the historic 1974 conference on the British Oak,[25] there were even foresters who dared to suggest that, though obviously native, the oak was not really at home in Britain. It was a continental tree on the northern edge of its range, and not suited to our late frosts and often cool summers.

But at the same conference the real reasons why oak was reluctant to regenerate in oakwoods also began to emerge. The acorns and oaklings are under siege at almost every stage in their life: eaten by small mammals; often not having sufficient light to germinate because of the end of coppicing in so many woods; and finally, if they do succeed in developing into seedling trees, they are likely to be defoliated by tortrix moth caterpillars parachuting down from their parent tree. Mature oaks can cope with losing almost all their leaves to caterpillars in the spring, simply by growing another set; but oaklings with tiny root systems cannot, and usually die. This is why oaks are so much more successful at regenerating in the open (where their acorns are planted by jays and squirrels), away from the shade and insect rain of their own kind.

Sessile oak, *Q. petraea*. The sessile oak, so called because its acorns (unlike *Q. robur*'s) are not carried on stalks (peduncles), but directly on the outer twigs, has an intriguing distribution. It is commonest in the north and west of Britain, but chiefly in semi-natural woodland, in which it also occurs more sparingly elsewhere in Britain. In hedgerows, plantations and scrub, even in its heartlands, it is largely replaced by the pedunculate oak. The two species seem in fact to regenerate in quite different patterns, the sessile growing well under its own shade and often forming quite dense single-species woodland, but not readily colonising land beyond woods; the pedunculate behaving in exactly the opposite way. Superimposed on this have been the effects of human preferences. Pedunculate oak usually produces a much greater crop of acorns, which were valued as food for pigs and cattle (the Domesday Book, 1086, actually measures the area of woodland in terms of the number of swine it can support), and its stocky trunks and naturally angled branches were prized as more desirable and adaptable timbers. As a consequence it has been widely planted throughout Britain, and has the look of an invasive species. Sessile oak, by contrast, seems largely confined to the areas it must have occupied in the wildwood, and has a relict distribution, analogous to that of small-leaved lime (see p. 116).[26] In the north and west, sessile oak-woods were largely perpetuated by coppicing. The wood was used in iron-smelting, and the bark in the tanning industry.

The culture surrounding sessile oak echoes the tree's rather inferior economic role and is very sparse. Recognisable carvings of the leaves in churches are few. But they are clearly distinguishable on the shrine of St Frideswide in Christ Church Cathedral, Oxford, and carved on bosses at Claydon church, East Suffolk.[27]

The undersides of sessile oak's leaves are downy, which sometimes gives them a silvery sheen. A remarkable colony of such oaks (still thriving) is commemorated in the place name 'Whiteleaved Oak' near Bromsberrow in the southern Malverns.

Hybrids between the sessile and pedunculate oak are often commoner than either of the parents where both species are present, and occur throughout Britain.

Evergreen oak or **Holm oak**, *Q. ilex*, is an evergreen species from the Mediterranean which is widely planted. It is often killed (or defoliated) by severe frosts, but has self-seeded quite widely in southern and central England and Wales, especially on the Cotswold limestone.

A well-known grove are the Bale Oaks in north Norfolk, planted outside the church in about 1716 by Thomas Bullen. But the ancient English oak whose company they joined was a much more interesting tree. Its trailing leaves and acorns are represented in a fifteenth-century stained-glass window in the church, and in the early eighteenth century it was 'so large that ten or twelve men may stand within it. A cobbler had his shop and lodge there of late, and it is or was used for a swinestry.'

The former rector described its end and last rites:

'In 1795 the Bale Oak was severely pollarded, and the Hardys of the village of Letheringsett purchased the wood and the bark (for tanning). It never recovered from this drastic treatment, and a poem was written about the tree in this state, and learnt and passed down to villagers, who can even quote it to this day:

> *Here stand I all in disgrace,*
> *Once the wonder of this place;*
> *My head knocked off, my body dead*
> *And all the virtues of my limbs is fled ...*

By 1860 the Oak had become dangerous, and as the Parish Officers would take no responsibility for anyone getting injured, the Lord of the Manor – Sir Willoughby Jones, Bart., had the tree taken down and carted off to Cranmer Hall, Fakenham. The waggons employed were decked with flags, and all Fakenham turned out to see them pass through.'[28]

Turkey oak, *Q. cerris*, was introduced from southern Europe in 1735. It is widely planted in parks and by roadsides, and well naturalised in sandy soils in southern Britain. **Red oak**, *Q. rubra*, is a North American species with sharply angled leaves, turning deep red in autumn, planted for timber as well as ornament and self-sowing in England and Wales.

Beech, *Fagus sylvatica*. The beech, alone among our large forest trees, has something of a feminine image. When I was a boy growing up in the Chilterns, the tree was always offered as an example of elegance and classicism, the foil to the rugged masculinity of the oak. Fine specimens were often called 'Queen beeches'. Even today, the gracious but solemn plantations of the Chilterns and Cotswolds are referred to reverently as 'nature's cathedrals'.

Yet there have always been other kinds of beech, rowdier, more accessible, less obscured by idealisation: the stumpy workhorse pollards of Burnham Beeches and the gnarled giants of the New Forest; scraps of bristling coppice in the Chilterns; wind-pruned hedges on Exmoor; and, increasingly, these days, wind-*thrown* beeches, with root-plates like small cliffs.

Contrasting kinds – and contrasting images – of the beech have a long history. Even its arrival in this country has been a contentious matter, and it is often claimed to be a Roman introduction (another echo of its 'classical' aura). But beech pollen remains have been found in the Hampshire basin that date from 6000 BC – about 2,000 years after the oaks returned to post-glacial Britain and 500 years before the Channel opened. So the beech just passes the key test of botanical nativeness: it was here when Britain became an island. It advanced under its own steam up to a line between the Bristol Channel and the Wash, and, as a planted tree, much further. In parts of south-east England it became one of the commonest woodland trees.

Beech high forest in winter sunlight, Chilterns.

But early beeches would have been less gainly and upright than those we are familiar with today. In medieval woods, beech was rarely grown as timber and almost never used for buildings. Even in the Chilterns houses were framed in oak. The beech was valued historically as a more basic kind of workhorse. It was an energy source, providing firewood for humans and nuts ('mast') for grazing cattle. As far back as the late Roman period mixed beechwoods supplied fuel for ironworks. In the Weald, beech was the fuel of choice for the glass industry. In the Chilterns, wood cut from pollard (and coppiced) beeches was shipped up to London by barge, for the city's hearths and ovens.[29] Between cuts, on wood-pastures and commons throughout southern England, grazing animals would feed on the mast.

In the eighteenth century, the 'Age of Improvement', beech began to be valued for ornament and timber, and tall, smooth-barked, straight-trunked trees became more frequent both in woods and parkland. Gilbert White thought beech 'the most lovely of all forest trees, whether we consider its smooth rind or bark, its glossy foliage, or graceful pendulous boughs'.[30] William Gilpin, another late eighteenth-century Hampshire cleric and one of the arbiters of Picturesque taste, preferred the old vernacular beech, 'studded with bold knobs and projections', to the tall timber trees, whose branches ran 'often into long unvaried lines, without any of that strength and firmness, which we admire in the oak'.[31] But by the end of the century the timber beech had won the day. The rapid development of the Windsor chair industry (which used turned beech for chair legs) led to the conversion of vast areas of beech coppice and pollards to timber beech lots, especially in the Chilterns.

The situation remains much the same today. The beech's role in popular culture is almost exclusively as a landscape tree – elegant in mature plantations, gothically rugged on old commons, and spectacular in its autumn coloration wherever its grows.

Planted beech landmarks are typically avenues and groves on hilltops:

'At Elveden Hall near Thetford, there is a beech avenue running due south of the main house. Many of these trees have plaques commemorating the date and the planter. Most, if not all these trees were planted by members of various European Royal Families, including Queen Victoria, King Juan Carlos, and the Tsars of Russia.'[32]

One of Paul Nash's many paintings of Wittenham Clumps, a landmark beech grove on the site of an Iron Age fort near Wallingford, Oxfordshire. Nash drew the Clumps obsessively and saw them as a symbol of the repossession of human works by nature. But, as is so often the case with beeches, 'It was the look of them that told most. They were the Pyramids of my small world.'

'There are beech trees on the top of Howe Hill, just outside Ingleby Greenhow. They are used as a logo for the local primary school, and over 45 years ago my husband recalls going up there with the children of the local Sunday school where a service was held.'[33]

Famous hilltop groves include the Seven Sisters at Cothelstone Hill, which can be seen from much of Somerset and South Wales, and Wittenham Clumps on an Iron Age hill-fort in Oxfordshire, which was such an inspiration for Paul Nash's paintings. Wittenham also has the 'Poem Tree', now dead, alas, but still bearing the verse carved by Joseph Tubb in 1844 and celebrating the landscape around the hill-fort.

Graffiti are more usually associated with old pollards, the self-sown, 'vernacular' beeches. It is a venerable tradition, going back to the Romans, who had a proverb for it: *Crescunt illae; crescant amores* – 'As these letters grow, so may our love.' (There is even a possibility that slabs of beech or *bok* wood or bark, etched with ancient Teutonic graffiti, were the first *books*.) In Frithsden Beeches, one of the finest surviving groups of ancient pollards, there are graffiti going back a hundred years, through messages carved by American airmen, to prim Victorian initials. Stretching as the trunks expand, they seem entirely in keeping with the other ways old beech trees register their lifetime's experiences: the lightning scars, woodpecker holes, fungal rots, squirrel-browsed elbows in branches and aerial ponds in crucks, as well as the rounded bosses that mark where the last branches were cropped, a century and a half ago.

The Frithsden pollards also include several trees (known locally as 'the Praying Beeches') with fused branch-stumps, as if the last loppers had deliberately tied the young regrowing twigs together. Young beech shoots will graft very readily, even from close natural contact, as one of the most touching twentieth-century landmark trees demonstrates: 'My father lived in Garforth and Mother worked as a live-in dairy maid in Aberford, so he used to walk down the Fly-line [a disused railway track] to visit her. He found three beech saplings growing by the path and – as he was courting Nellie – he grafte the middle trunk across the left-hand sapling to form the letter N. He gave it the name Nellie's Tree. This would be about 1920, and the tree is still there.'[34]

Another beech oddity in the Chilterns are patches of relict coppice, one at Maidensgrove Scrubs, the other at Low Scrubs, near Ellesborough in Buckinghamshire, which was a parish fuel-allotment apparently last cut in the 1930s.[35] These are particulary intriguing, as the coppicing of beech, once a widespread and successful practice in the area, has proved notoriously tricky to reintroduce. Cut trees, even when protected from browsing, usually die after a few years. The wood at Low Scrubs may provide some clues as to how beech coppicing was carried out.

It is an extraordinary collection of hunched, low-growing trees, surrounded at their bases by masses of twiggy growth. These have the look of 'witches'-brooms', but in fact are the dense network of shoots that beeches throw out when they are cut or browsed, snagged up with fallen leaves and bits of branch. Behind this are two generations of poles, typically two or three stout trunks six to ten inches in diameter, surrounded by a number of thinner poles about two inches wide. Deer-browsing might explain this, but I wonder if the secret of beech coppicing

Beech pollards acquire a remarkable individuality over the centuries.

was to leave a leading pole or two uncut, as was the practice sometimes with hazel. (This was the custom amongst older coppice-workers in nineteenth-century Sussex, where full coppicing was believed to 'bleed the spirit of the trees away'.)

Aside from the uses of its wood and its contribution to the landscape, beech has little in the way of associated custom or folklore. The three-sided nuts make a pleasant nibble in the years when they form, and during the two World Wars were collected in German villages for pressing into oil. The leaves have been made into a potent alcoholic drink – beech-leaf noyau. This is a recipe remembered by a 70-year-old man in the southern Chilterns: 'Wash and dry enough beech leaves to fill your stone jar – cover them with gin. Leave for a week, then strain off the liquid and measure. To each pint add a pound of sugar which is dissolved in half a pint of boiling water. Add a good quantity of brandy and stir together, then leave to go cold before bottling.'[36]

Southern beeches, *Nothofagus obliqua* and *N. nervosa*, are species from Chile and Argentina, extensively planted for forestry since the 1970s, because of their comparative immunity from grey squirrel damage. Self-sown saplings are already beginning to appear in the older plantations.

Sweet chestnut, *Castanea sativa*. The sweet or Spanish chestnut was almost certainly one of the few species that was introduced to Britain by the Romans. Its nuts, roasted over winter braziers or ground more frugally into flour, are satisfying and savoury, and the trees may have been brought over to provide a home-grown supply of chestnut-flour for the legionaries.[37] But it is now an 'honorary native' and in south-east England behaves like a native tree. It is well-established in many ancient woods and propagates itself by seed – though not in the invasive manner of the more recently-arrived sycamore.

Ancient chestnuts are spectacular trees. They develop exceptionally broad trunks for their height, which with age become deeply fissured and covered with burrs and

Frithsden Beeches, Hertfordshire, a beech wood-pasture now grazed by deer.

Sweet chestnut fruits, edible after the first frosts.

bosses. Quite often the fissures run in a left- or right-hand spiral round the tree. (Phil Gates has suggested, not too seriously, that this may be the origin of the name 'Spanish chestnut' for a tree which has no particular affinities with Spain: 'I favour an implausible theory that it is because the twist in the bark is like the swirl of a flamenco dancer's skirt.')[38]

Trees with girths of over 25 feet have been reported from Cranford, Middlesex; Holmbrook, Cumbria; Felbrigg Hall, Norfolk; Wiveliscombe, Somerset; between Bigsweir Bridge and Hudnalls in the Wye Valley; at Studley Royal near Ripon; and around the old observatory at Herstmonceux, Sussex.[39] There are even bigger specimens in Clwyd and Gloucestershire:

'In a field by the side of the Ruthin to Denbigh road, there were three enormous sweet chestnuts. One is quite dead and gone, one is still there as a pile of dead wood, and the third is still just holding its own. The locals say they are 2,000 years old – I doubt this, but they are marked on Ordnance Survey maps.'[40]

'Perhaps the oldest living individual in Longhope [Gloucestershire] is a chestnut tree. This tree stands in the field just east of the old railway line halfway between the Post Office and the Church. The circumference of the tree this October [1993] is 10.5 metres and 10.6 metres at two places ... It is a hollow tree with an inside 'room' measuring roughly seven feet across with about one foot of living and dead tree making an outer ring.'[41]

But the biggest of all is the Tortworth Chestnut in Gloucestershire, which stands at one side of St Leonard's Church. It is unprepossessing at a distance and looks much like a grove of young chestnuts. But inside (it is surrounded by an iron fence) is something quite remarkable: an immense mass of contorted trunk and branch, like wooden lava or a wood-slip. The tree is still very much alive, though, and the collapsed side branches have all taken root and are sending up new shoots continously. It is now virtually a wood itself, more than 30 yards across and with bluebells, dog's mercury and ramsons growing in its shade. It is impossible to date, but a plaque on the fence, dated 1800, reads:

THIS TREE SUPPOSED TO BE
Six Hundred Years Old 1st Jan
1800
May Man still Guard thy Venerable form
From the Rude Blasts and Tempestuous Storm.
Still mayest thou Flourish through Succeeding time,
And Last, long Last, the Wonder of the Clime.

Elsewhere there are also ancient chestnut coppice stools, including one in Viceroy's Wood, Penshurst, Kent, which is known locally as 'The Seven Sisters' from its seven stems. These are reckoned to represent 250 years of regrowth, but the stool must be very much older, as its overall circumference is more than 50 feet.[42] Kent and Sussex are the major areas for chestnut coppice, and thousands of acres are managed commercially to produce chestnut fence-paling.

Many of these single-species Wealden coppices were planted in the mid-nineteenth century. Elsewhere, especially in deer parks, sweet chestnut was often planted for its nuts, which are popularly assumed to be attractive to deer. They grow inside rounded, spiny seed-cases, very similar to conkers, and are often blown from the trees early in October, before they are ripe. Perhaps this is why home-grown chestnuts have a poor reputation as winter nibbles beside the large nuts from Italy and the Balkans. In fact when they are ripe they are crisper and sweeter than imported varieties, and quite palatable raw. (I can remember three gardeners in a Chiltern pub debating the chestnuts of Ashridge Park as passionately as French countrymen discussing grapes. The best were along the Queen's Ride. No, they were in the woods below Lord Bridgewater's monument. The best of all, wherever they

The Tortworth Chestnut, probably not far short of 1,000 years old and now resembling a wooden cave-system more than a tree.

were, were those that hung on until they were brought down by frost.) Modern recipes for cooked wild chestnuts include soup, vanilla-flavoured spread and a stuffing which is then battered and fried as croquettes.[43]

One curious children's ritual with the nuts is 'Philippines': 'There was a custom originally associated with almonds but which became associated with any nut that had twins in one shell. You shared one of these nuts with someone present saying, "Here is a Philippine, share it with me." Next morning the first one to remember greeted the other with "Bon Jour Mon Philippine." The forgetful one had to find a present for the winner. I was astonished to find the next generation still keep up this custom. Perhaps it is a corruption of a German greeting "Vielliebchen", dear little one.'[44]

The long, jagged leaves are also used to make 'fish-bones'. 'The soft leaf tissue would be removed between finger and thumb, great care being taken not to break the ribs of the leaf.'[45]

One final oddity. The long strings of yellow flowers, out in July, smell, as one contributor noticed, 'unmistakably of semen'.[46]

Birch family *Betulaceae*

Silver birch, *Betula pendula*. Birch was one of the first trees to recolonise Britain after the retreat of the glaciers, and it remains an opportunist, pioneering species today. Its seeds are produced in huge numbers and are blown about like dust in the wind. On areas of open woodland or heath, the frizzy seedlings can carpet the ground in a matter of months. In this respect birch plays the same role on slightly acidic ground as ash does on calcareous soils, and where there is not much competition from other species – as in parts of the Scottish Highlands – it can form pure woods, even though the individual trees rarely live more than 80 years.

Its powers of regeneration and rapid growth have inevitably made it unpopular with foresters, who mostly regard it as a worthless competitor with spruces and other commercial trees. So they slash and spray the seedlings and ring-bark mature trees. Nature conservationists do the same to birch that invades heathland and fens.

But silver birches – white-trunked, airily leafed, rich in bird-life – are exquisite trees, and the official hostility that is shown towards any that appear in the 'wrong place' is

Young silver birches, rapid colonisers of open areas on acid soils.

not always popular with the public. In Ashdown Forest, 'the clearing of invading birch on the heathland areas caused such untoward comment by visitors and local dwellers that it has been decided to mount an exhibition portraying the birch's life from seedling to use – covering distribution, management, natural history, poetry, folklore and a big section on woodworking. Birch wine, which is still made as a home brew on a small scale, will be exhibited. A small trade exists making besom brooms with birch and hazel handles.'[1]

These vestigial uses in Sussex reflect the enormous versatility of this prolific raw material. Writing of the great Scottish birchwoods in the 1840s (birch – usually as the prefix birk – is the commonest place-name prefix in Scotland),[2] J. C. Loudon pointed out that the local inhabitants hadn't many other trees to turn to:

'The Highlanders of Scotland make everything of it; they build their houses, make their beds and chairs, tables, dishes and spoons; construct their mills; make their carts, ploughs, harrows, gates and fences, and even manufacture ropes of it. The branches are employed as fuel in the distillation of whisky, the spray is used for smoking hams and herrings, for which last purpose it is preferred to every other kind of wood. The bark is used for tanning leather, and sometimes, when dried and twisted into a rope, instead of candles. The spray is used for thatching houses; and, dried in summer, with the leaves on, makes a good bed when heath is scarce.'[3]

Subsistence is hardly an issue in the Highlands today, and birch is now more often employed as the favoured fuel for smoking haddocks, and as a source of sap for fermenting into wine (an increasingly popular speciality of Deeside).[4] Besoms made from birch spray are still quite popular with gardeners, but the Forestry Commission has stopped using them for beating out fires – not surprisingly perhaps, since young birches are amongst the few deciduous trees that will burn standing.

South of the Border, omelette whisks are fashionable: 'We make whisks by cutting a bunch of birch twigs in spring around bud-burst. Hold each twig at the cut end with a piece of rag and strip off the bark in one pull. Bundle the stripped sticks together and bind with another long stripped birch twig.'[5]

(The adjective 'silver' for birch, from the satin lustre of the papery white bark, is also a comparatively recent Sassenach invention. It is not in any of the editions of Evelyn's *Sylva*, and seemingly first appeared in a poem by Tennyson.)[6]

Recently, birch has begun to be looked at more respectfully by commercial foresters, as an alternative to exotic conifers on poor moorland soils. It is just as good as a source of pulpwood and has the virtue of actually improving the soil rather than acidifying it, as conifers do.

This would have been thought a quite unnecessary and utilitarian defence of the tree by medieval Celtic poets, who loved the birch for its spring greenery and dappled summer light, more golden than under any other native tree. The fourteenth-century Welsh poet Gruffydd ap

Ancient birch trunk in Glen Strathfarrar, Inverness.

Dafydd wrote an elegy 'To a Birch-tree Cut Down, and Set Up in Llanidloes for a Maypole':

'Long are you exiled from the wooded slope, birch-tree, with your green hair in a wretched state; you who were the majestic sceptre of the wood where you were reared, a green veil, and now turned traitress to your grove ... You were made, it seems, for huckstering, as you stand there like a market-woman; and in the cheerful babble at the fair all will point their fingers at your suffering, in your one grey shirt and your old fur, amid the petty merchandise. No more will the bracken hide your urgent seedlings, where your sister stays; no more will there be mysteries and secrets shared, and shade, under your dear eaves; you will not conceal the April primroses, with their gaze directed upwards; you will not think now to inquire, fair poet tree, after the birds of the glen.'[7]

But there is still a vestige of magic around upland birches: 'Near Balmaclellan [Kircudbrightshire] lies the lonely village of Dunscore. One of the houses there is called Letterick, and some years ago the owner died. His last instructions were that his body should be placed in a basket and he should be buried upright near a birch tree. He asserted that he would return regularly in the guise of a crow to keep an eye on things. Needless to say a lone crow does come to sit in the tree at intervals.'[8]

Downy birch, *B. pubescens*, is a similar species, but with a darker trunk and favouring wetter sites, especially in the uplands. The subspecies *tortuosa* (previously ssp. *odorata*) is virtually confined to upland areas of the north. The leaves have a thin resinous coating that has an invigorating balsamic scent, especially after spring rains.

Birches on a steep slope in Glen Strathfarrar. Birchwoods are one of Scotland's 'signature' plant communities.

Alder, *Alnus glutinosa* (VN: Aller, Water aller, Waller, Woller, Orle). Alder is a common tree of riversides, fens and wet woods. It is also, these days, a depressing indicator of the impact that drainage and so-called 'reclamation' can have on wetlands. Alders have deep tap-roots and will survive for many years in dried-out ground, but, as the soil around them shrinks, the fluted upper rootstocks begin to emerge and give the trees something of the look of land-bound mangroves.

Clumps of alder look best in late winter, with their dark twiggy silhouettes hung with the first of the crimson male catkins and the last of the toy female cones. The wood is white when first cut, but gradually darkens to a fresh orange or chestnut colour on exposure to the air. As it does not rot under water, it has sometimes been used for piles for shoring up canal- and river-banks. But its chief commercial uses were in the making of clogs and of charcoal for gunpowder. (Coppices were sometimes deliberately planted near gunpowder works.)[9]

These days it is chiefly used for brush-backs and tool-handles. It is also a traditional lure for woodworm: 'Alder branches are cut and placed in cupboards to prevent woodworm – the beetle will lay in the alder in preference to other wood.'[10]

Grey alder, *A. incana*, from continental Europe, is increasingly planted as an ornamental and roadside tree and on old mine-spoil heaps. It naturalises by sucker and occasionally self-seeds.

Hornbeam, *Carpinus betulus* (VN: Hardbeam, Ay beech). The hornbeam is an abundant tree in parts of southern and eastern England, but is probably the least known of our common woodland tree species. This may be because it is now more confined to woodland – and ancient woodland at that – than oak, beech and ash. It is certainly not because of shyness or inconspicuousness. Mature hornbeams are handsome, sinewy trees, with grey, fluted trunks and crowns of dense, toothed leaves, which on younger trees are often retained through the winter – perhaps the origin of the rare vernacular name 'ay [everlasting] beech'. The fruits are held in stacks of papery wings, like small pagodas.

Hornbeam is a very hard wood (and still known as 'hardbeam' in East Anglia), but no use as timber. Small pieces have been used wherever toughness and resilience are needed: 'In Sussex a main use was for windmill and watermill cogs. Carpenters disliked working it as it quickly blunted their tools. Having myself worked it for implements I can bear testimony to its hardness and difficulty in working. For this reason it was called 'ironwood' and 'lanthorn', because it burnt with a bright flame. Historically it was much valued for its charcoal, which burnt hot enough to smelt iron.'[11]

Fuelwood was certainly the most extensive use for hornbeam, and most trees were managed as coppice or pollards. Big hornbeam areas such as Epping Forest and the woods of east Hertfordshire and the central Chilterns were major suppliers of firewood to London before the days of cheap coal. (Today some are finding a new role as suppliers of pulpwood.) Elsewhere in the Chilterns hornbeam is most often found in long rows on wood-banks and was probably planted as hedging.[12]

Hornbeam survives well in the company of browsing animals, and solitary trees also occur in parks and pastureland hedges. Rather like lime, it has a layer of bast-like fibre directly under the bark which seems to protect the greenwood, and I have seen many trees ring-barked by deer or squirrels with no apparent consequences to their growth.

Hornbeam's natural distribution is probably roughly south and east of the line joining Weymouth, Hereford and Norwich, though it has been quite frequently planted

Very old hornbeam pollard in Hatfield Forest, Essex.

Hazel catkins, widely known as 'lamb's-tails'. (The red female flower is also just visible.)

(often in coppices) throughout the rest of England. Frome in Somerset is very close to its natural boundary, and hornbeams growing there, whether self-sprung or planted, would have been noticed. So it is fitting that an old tree at Hemington School in Frome has become the school's emblem: 'In 1977 the school celebrated its centenary and I researched its history … It seems probable that the tree was in the field before the school was built, as the oldest ex-pupil at the time of the celebration remembered climbing it to escape the wrath of the then headmaster, so it must have been fully grown in 1897, when Mr Riddick climbed it! He planted a new tree in our conservation area to mark the occasion and we also had a seat made to fit around the tree. We chose it as our emblem when we made a flag for the school.'[13]

In Suffolk, in the heart of the hornbeam's natural territory, children can get to know the tree even more intimately. In some of the county's enlightened environmental education courses they can listen to the pulse of rising spring sap through stethoscopes. It is most thunderous in the hardbeam perhaps because of the fabled toughness of the wood.[14]

Hazel, *Corylus avellana* (VN: Halse, Hezzel, Ranger; Lamb's-tails [catkins]). These days hazel is known for its catkins, universally called 'lamb's-tails', its late summer nuts, and occasionally for providing water diviners with their forked twigs. But for much of the past 6,000 years it was a more utilitarian resource. Its foliage was used for cattle food, and its whippy shoots for making the framework for houses and fences.

Hazel was one of the first trees to recolonise Britain after the last Ice Age, coming soon after birch, and for a while it was the most abundant shrubby species. But it is a bush more than a tree, rarely growing above 30 feet in height, and does not tolerate deep shade. When the bigger forest trees began to cover Britain, hazel retreated to more open areas, on cliffs, unstable rocks, river-banks and the like. It was the opening out of the wildwood by early settlers that gave it the chance to spread again.

In the wild, hazel will occasionally grow into a single-trunked small tree. But it is, so to speak, 'self-coppicing' and, if its branches are broken off or debarked by animals, hazel will send up straight new shoots from the base. No doubt these were used by early people before deliberate coppicing was begun some 4,000 years ago.

However they are encouraged, hazel poles have two invaluable properties. They can be split lengthways, and twisted and bent at sharp angles without breaking. This enabled them to be woven, bent back on themselves and even tied in knots. (Thin strips of knotted hazel are used to bind up bundles – faggots – of cut hazel poles.) From Neolithic times the basic product of cut hazel was wattle – split canes woven into a simple warp-and-weft lattice-work. Wattle made hurdles, fencing and the foundation on which wattle-and-daub walls were built. Hazel is also still used to peg down thatch (in which the hazel broaches have to be bent through 180 degrees).

Recently the use of coppiced hazel wood has been going through something of a revival, encouraged by (and encouraging in its turn) the nationwide revival of coppicing itself. There have been some ingenious new uses for wattle, in, for example, motorway sound screens. And the National Rivers Authority has revived the traditional Dutch practice of sinking 'mattresses' of hazel faggots and reed to help fortify the banks of the River Ouse near the Wash. (The mattresses work by catching sediment from the natural tidal flow, which builds up and strengthens the river bed and banks.)

In the Isle of Man, 'above the Great Laxey Wheel, hazel was planted to provide shock-absorbing brushwood for the mine machinery. The woodland is now preserved by Manx National Heritage as part of the mines complex associated with the Wheel.'[15]

A coppiced hazel by the River Teign, near Dartmoor, with wild daffodils beneath.

Small-scale use continues, too. Hazel rods are still widely used for pea and bean sticks,[16] and remain popular with the carvers and whittlers of decorative walking-sticks, particularly because of the contrast between the white wood and the naturally flecked, almost notched bark. One south-country walking-stick maker bends and pegs down young hazel shoots in hedges to 'grow' the curve in the handle (cf. ash, p. 326).[17]

A modern coppice-worker has noticed great variation in the rods from different bushes:

'The hazel rods from different stools seemed to have different splitting qualities and varying suppleness and toughness. Moreover these varying properties of the wood seemed to be associated with particular textures of the bark, so it became possible to predict how a rod would split or how brittle the split rod would be simply by looking at the bark. Much of the hazel at West Wood [North Bedfordshire] has a lovely golden, almost metallic sheen and a fine-grained flaky texture (which could perhaps be likened to an even scatter of tiny flakes of bran). This type of hazel has a strong grain which is easy to split evenly (except when it gets very big and old and tends to become stringy). Another type of hazel rod has a very smooth ground texture to the bark, almost like dark olive-green lacquer-work, and has large, conspicuous widely-spaced lenticels. These rods tend to be very brittle and relatively difficult to split because the split has a tendency to shoot off to the side almost like a conchoidal glossy fracture. "Like splitting a stick of rock", one thatcher described it. Genetic variation is the cause of much of this individuality: adjacent stools of same-age regrowth and in apparently identical conditions can have markedly different features and each rod on a stool has the characteristic of that stool. The individuality is not confined to bark and splitting qualities. On some stools the stems of the new regrowth are a deep maroon-purple, on others they are crimson, and yet others are a washed-out green colour. Some stools have leaves with a purple blotch. Some grow new shoots that are tall and erect, others grow shorter and more prostrate. Some stools have nuts that are long and bullet-shaped, others are nearly spherical or even snub-nosed.'[18]

Hazel-nuts are the other great harvest from the tree. They were one of the staples of prehistoric peoples, especially the Celts, and were highly esteemed: 'An early Irish topographical treatise describes a beautiful fountain called Connla's Well, near Tipperary, over which hung nine hazels of poetic art which produced flowers and fruit (Beauty and Wisdom) simultaneously. As the nuts fell into the well, salmon began feeding off them; whatever number of nuts any of these salmon swallowed, a corresponding number of bright spots appeared on their bodies ... In Celtic legend they [hazel-nuts] are always an emblem of concentrated wisdom, something sweet, compact and sustaining, enclosed in a small hard shell: in a nutshell, so to speak.'[19]

The selection and breeding of cultivated forms dates back to classical times. Tudor farmers and fruit-growers favoured a variety called the white filbert, very similar to the wild filbert, *Corylus maxima*, of Asia Minor. ('Filbert' is named from St Philibert's Day, 20 August, when the nuts are recorded as being ripe.) The name 'cobnut' was not applied to cultivated hazels until later. It derived from a game called 'cob-nut' – 'cob' meant to throw gently – which involved pitching a large nut at a pile of smaller ones. Those knocked off the pile became the property of the thrower.[20]

(The heart of cobnut-growing in England is Kent, and in the village of Ightham there is a pub called 'The Cob-tree'. Until 1995 the pub sign showed a Welsh cob horse standing under a big tree. But Meg Game, grower of and enthusiast for cobnuts, persuaded the brewery to repaint the sign showing the catkins, leaves and nuts of a real cob tree.)[21]

But even wild, native hazel-nuts (*C. avellana*) were regarded as worth collecting in quantity. In the late seventeenth century, John Aubrey (echoing the experience of the coppice-worker above) praised the harvest from the great hazel woods of Wiltshire:

'Wee have two sorts of them. In the south part, and particularly Cranbourn Chase, the hazells are white and tough; with which there are made the best hurdles of England. The nutts of the chase are of great note, and are sold yearly beyond sea. They sell them at Woodbery Hill Faire, &c.; and the price of them is the price of a buschell of wheate. The hazell-trees in North Wilts are red, and not so tough, more brittle.'[22]

In 1826 the owner of Hatfield Forest, Essex, complained that: 'as soon as the Nuts begin to get ripe ... the idle and disorderly Men and Women of bad Character from [Bishop's] Stortford ... come ... in large parties to gather the Nuts or under pretence of gathering Nuts to loiter about in Crowds ... and in the Evening ... take Beer and Spirits and Drink in the Forest which affords them an opportunity for all sorts of Debauchery.'[23]

It is a depressingly familiar complaint by landowners. All fruit harvests, wild or cultivated, are quite properly occasions for socialising and celebration.

Nutting also generated several ingenious devices, including gathering-bags looped over the wrist and hazel-nut-crackers fashioned from hazel-wood. I have seen a pair of these made by a Sussex hurdle-maker in the 1930s, which he used to carry when working in the coppices in autumn. After shaping a piece of straight wood with his knife, he soaked it, doubled it over, and then bound it tightly with a strip of split hazel until it dried out.

In Great Houghton, Northamptonshire, a more rough-and-ready implement was used: 'In the village, until 20 years ago there was a lane with a hazel hedge boundary, leading to a farm gate. The gate was used to crack the hazel nuts, and the lane had been known for as long as anyone can remember as "Crack-nuts". Development swept away lane, hedge and name. One house built

Illustrations of some of the cultivated varieties of cobnut and filbert, from Pomona Britannica, 1812. The more robust, oblong filberts are native to south-east Europe.

on the land was the new rectory, and the then rector's wife thought Crack-nuts to be an inappropriate address for a rector, so with great imagination the area was renamed Rectory Close.'[24]

Perhaps this was retribution by the clergy for the ancient indignities of 'Nutcrack Night'. This was the evening when nuts, stored away to ripen, were first opened. (In Cleveland it was 15 November.)[25] In some parishes there was a custom for the nuts to be taken into church the following Sunday and cracked noisily during the sermon.

In fact most venerable customs involving hazel are far from straightforward. In many, the use of hazel-wood seems neither here nor there; it is simply the most convenient wood for the job. But behind this there are sometimes hints of white magic (hazel was lucky as well as bountiful), sly humour, commonplace economic custom – echoes of the character of the shrub itself. In Abbots Ann, Hampshire, the parish still keeps up the medieval custom of awarding 'Virgin's Crowns' made of hazel. A plaque in the church explains:

'The ceremony of this ancient burial rite takes place at the funeral of an unmarried person who was born, baptised, confirmed and died in the Parish of Abbots Ann, and was a regular Communicant. Such persons must also be of unblemished reputation.

The Virgin's Crown is made of hazelwood and is ornamented with paper rosettes, with five white gauntlets attached to it. The gauntlets represent a challenge thrown down to anyone to asperse the character of the deceased.

The Crown suspended from a rod is borne by two young girls habited in white with white hoods, at the head of the funeral procession. After the funeral the Crown is carried to the Church and is suspended from the gallery near the West Door, so that all who enter the Church on the following Sunday will pass under it. There it remains for three weeks. If during that time no one has challenged or disputed the right of the deceased to the Crown, it is hung in the roof of the Church with a small scutcheon bearing the name and age of the person concerned, and the date of the funeral, and there the Crown remains until it decays and falls with age.

Most of the Crowns are awarded to women, but men are not excluded, provided they fulfil the same conditions.

The present Church was built in 1716, and the oldest Virgin's Crown still in existence approaches that date.'[26]

Hazel is also used in the Corporation of London's Quit Rents Ceremony:

'The Ceremony ... is said by distinguished antiquarians to be the oldest surviving Ceremony next to that of the Coronation itself [over 750 years]. It is feudal in origin and character, since it represents the rendering of rents and services in respect of tenure of two pieces of land, one being a piece of waste land called "The Moors" in Shropshire and the other being a Tenement called "The Forge" in the Parish of St Clement Dane in the County of Middlesex. The services being rendered by the original tenants of these pieces of land having been commuted in kind by the Sovereign, the rents are only a token payment in kind. This is why they are called Quit Rents and Services, since thereby the tenant goes "quit" and free of all other services.

In respect of "The Moors", the Quit rent consists in the presentation of a blunt knife and a sharp knife. The qualities of these instruments are demonstrated by the Senior Alderman or the Comptroller and Solicitor of the City of London, who will bend a hazel rod of a cubit's length [taken from Shropshire] over the blunt knife and break it over the blade of the sharp knife. Hazel rods of this length were used as tallies to record payments made to the Court of Exchequer by notches made with a sharp knife along their length and after the last payment split lengthways with a blunt and pliable bladed knife, one half being given to the payer and the other half being retained by the Court to vouch its written records. On behalf of the Corporation of London, he will then render them to the Queen's Remembrancer on behalf of Her Majesty.'[27]

(Hazel tally-sticks, incidentally, were used in one Chiltern pub into the 1980s as 'the slate' for a local joiner.)[28]

'Mrs Griffiths had a Welsh slate sink that she used for separating the milk, and there was a hole at the bottom that could be plugged and unplugged. During the summer months when the milk had to be left for a whole day to separate, she rubbed hazel leaves on the slate bottom before she poured in the milk, and this helped stop the milk turning sour.'[29]

Hedges

Hedges are widely regarded as being uniquely British landscape features, deliberately created by farmers during the last two or three centuries. Not a single part of this assumption is true. There are recognisable hedges in northern France, in the Austrian Alps, across much of the United States, and even in the Peruvian Andes. In Britain there is documentary proof that many are over a thousand years old, and biological and archaeological evidence puts the origins of some back in the Bronze Age. Even hedges that consist of tidy lines of trees and bushes are not always the result of planting. Something very close to what we call a hedge can be an entirely spontaneous feature, forming for example in the debris at the edges of rivers prone to flooding and by the side of strips of scree and landslip. (I have seen similar, if more ephemeral, wisps of self-sown scrub colonising the bulldozed earth-banks during the building of new roads.) More permanent hedges arise when shrubs colonise an already established human boundary. All across the Texas prairies, nineteenth-century barbed-wire fences have been turned into dense hedges by seedling trees and bushes taking root at the base of the fences and being protected by them from browsing animals.[1] Precisely the same process can be seen along any temporarily neglected British fence or ditch – though it is rarely permitted to proceed beyond the bramble, scrub willow and ashling phase.

Natural, unplanted hedges were also created when fields were cleared directly out of the wildwood, leaving a row of wild trees as a boundary. This process has been continuous since the early Middle Ages, and many rich hedges (evocatively christened 'woodland ghosts') are all that remain of woods destroyed comparatively recently. A celebrated example is at Shelley in Suffolk (see p. 117); another is 'the very outgrown hedge, with its 40-yard wall of small-leaved lime stools' which backs onto the Massey Ferguson factory in industrial Coventry, whose origins David Morfitt has traced to a wood grubbed out in the eighteenth century.[2]

Yet even planted hedges continually aspire to the condition of linear woodland. Max Hooper's well-known formula, that the number of woody species in a 30-yard stretch equals the age of the hedge in centuries, is a rule-of-thumb, not an exact equation.[3] But it does make the general point that with age hedges become progressively more complicated and richer in species. What generally happens is that bird-ferried or wind-blown seeds of new species take root in the shelter of the hedge and in their early years are helped to compete against the established shrubs by the process of hedge-cutting. And the average rate of establishment seems remarkably constant: one species per century per 30-yard stretch.

But the exceptions are as interesting as the ones that follow the rule. 'Reed hedges' in Fenland ditches, for example; the tall, moss-clad beech windbreaks on the Blackdown Hills in Somerset; the stone and turf banks of Devon and Cornwall, which are always called 'hedges' locally, though they barely carry any shrubs, and which, near Land's End, are sometimes as much as 3,000 years old.

There are modern exceptions, too, which, if Hooper's rule were applied too literally, would also appear to be prehistoric. Round the Stiperstones in Shropshire, eighteenth- and nineteenth-century squatters and free miners planted hedges around their smallholdings that were full of domestically useful shrubs, such as damson, gooseberry, laburnum, spindle and blackthorn. In some hedges the 30-yard count is in excess of 20 species.[4] Near Hargate, Lincolnshire, there are farm-hedges which echo the similarly resourceful medieval practice of planting orchard trees in hedgerows and headlands: 'There is a beautiful hedgerow of apples, plums, pears and rhubarb here, all the more beautiful for being one of the hedges left by greedy farmers. Nearer home along the same stretch are plums on one side and almonds and walnuts on the other.' In other parts of the country, often in deep countryside, there are hedges of laburnum, spiraea, fuchsia and flowering currant.

Alas, hedgerows of all kinds have continued to be destroyed at a rate that has scarcely diminished since the black days of the 1960s and 70s. Figures released by the Institute of Terrestrial Ecology in 1994 showed that total hedgerow length fell from 341,000 miles in 1984 to 266,000 in 1990, accounting in six years for a third of the entire post-war loss. Between 1990 and 1993, a further 6,750 miles disappeared. These figures are even more graphically confirmed at a local level. Vikki Forbes's survey of hedge loss in the parish of Ardleigh, Essex, shows that the period between 1980 and 1990 was the worst since records began. Between 1960 and 1980, 19 miles of hedgerow were destroyed; in the ten years between 1980 and 1990, 10 miles.[5]

The finding that many surviving hedges have in the same period become 'derelict' (i.e. tall, gappy and unmanaged) is more contentious. Although regular cutting and layering is appropriate where stock containment is required and is a good way of encouraging farmers to retain hedges, hedgerow management is now being zealously advocated almost for its own sake. As a result, even the tall Exmoor beech windbreaks are being hacked down to the level of Midland quicksets.[6] So-called derelict hedges – the wide, meandering rows of shrubs allowed to flower and fruit, studded with old pollards, festooned with creepers and supporting a vast range of ferns and flowers in their shade – are what most of us understand by ancient hedges and are the locations of many of the plant stories featured in this book.

Hedges along an old green lane, showing the rich mixture of habitats that results from not being too over-tidy: standard trees, pollards, bushy thickets and dense hedgerow bottom vegetation.

Dew-plant family *Aizoaceae*

Hottentot-fig, *Carpobrotus edulis* (VN: Sally-my-handsome). A South African succulent, with brilliant, silky flowers in pink or yellow, Hottentot-fig is widely naturalised on cliffs and rocks near the sea in the south-west, and more rarely elsewhere. The brown fruits, the 'figs', are edible, though not much used in Britain. Sally-my-handsome is a corruption of the old scientific name for the genus, *Mesembryanthemum*.

Goosefoot family *Chenopodiaceae*

Fat-hen, *Chenopodium album* (VN: Muckweed, John O'the Nile). This common weed of cultivated ground was once a valued crop – a prehistoric staple, probably. It is not difficult to see how it rose to this status. It forms pale green, mealy-leaved swarms close to human settlements, particularly where rubbish is thrown out – in middens and stackyards in the Iron Age; today, in muck-heaps at the edges of fields and between the rows of well-dunged crops (ironically, often its domesticated relatives such as spinach and sugar-beet). Early people would not have been slow to try such an accessible and abundant vegetable (especially as it has large, fatty seeds) and would not have been disappointed. The whole plant, eaten raw or cooked, is as pleasantly tangy as kale or young broccoli.

Fat-hen is one of those plants whose remains have been found all over Europe in prehistoric settlements. It was part of the last, possibly ritual meal of Tollund Man, the 2000-year-old corpse found in a peat-bog in Jutland in 1950. He had been hanged, either as a punishment or as a sacrifice, and thrown into the bog, where the acid peat in effect pickled him. His last meal was a gruel that included the seeds of fat-hen, gold-of-pleasure, black-bindweed, wild pansy, barley and linseed.

Fat-hen seems to have been important or plentiful enough in some areas to have whole settlements named after it. Its Old English name was *melde*, and the place-name specialist Eilert Ekwall believed that Melbourn in Cambridgeshire, for instance – Meldeburna in 970 – was the stream on whose banks *melde* grew.[1] Later place-name experts would not be so confident. But the association with dung has persisted in surviving local names. In Shropshire it is still known as 'muckweed', but also, more obscurely, as 'Jack (or John) O'the Nile' – or 'nail', which would be pronounced 'nile' in the Midlands.

'I came into the area some 50 years ago when hand weeding behind a horse scuffle was paramount. My first job was weeding sugar-beet (at two shilling and sixpence a week, plus keep) and John O'Nile was the name used by the wagoner, who had no doubt it was "Nile" not "nail".'[2] (But 'nail' does fit with another cryptic Shropshire vernacular name, 'Johnny O'Needle'.)

Deciphering the origins of plant and place names is notoriously full of pit-falls. But in the small farming village of Milden in Suffolk – Meldinges in *c.* 1130 – some villagers have no doubt about the *melde* root of their settlement's name. In the 1970s they commissioned a local blacksmith to make a six-foot-tall cast-iron statue of a fat-hen plant and placed it on a fieldside road-verge on the

Left: purple dew-plant (a relative of Hottentot-fig from South Africa) naturalised at the Lizard, Cornwall. *
Right: fat-hen growing as a weed in Norfolk, ironically amongst its cultivated relative sugar-beet.

* Also in Guernsey on a steep cliff; with 3 distinct shades of pink and also on roadsides.

boundary of the parish. It must be one of the most bizarre – and ambivalent – village signs in the country: a memorial to a dung-hill weed that quite likely gave the settlement its name, sited just yards away from drifts of the real plant, which modern Milden farmers are forever trying to drive out of the fields.

Good-King-Henry, *Chenopodium bonus-henricus* (VN: Mercury, Mercree). For a plant often classed as a weed, Good-King-Henry is surprisingly handsome – upright, shapely and with triangular, sometimes red-tinged, leaves. It is almost always found close to habitation in similar positions to fat-hen, in the rich soils of stackyards, hedge-banks, cultivated ground and rubbish-tips, and it may well have been first introduced as a crop plant by Bronze Age settlers from southern Europe. It is a rather bland-tasting but pleasantly textured green vegetable, undergoing a revival in modern herb gardens.

The name is an anglicised version of the German *Guter Heinrich*, 'Good Henry' (Henry being a Teutonic elf rather like our Robin Goodfellow). The 'Bad Henry' with which the name is meant to contrast is dog's mercury (see p. 256), a poisonous species whose form is vaguely similar to a young *Chenopodium*.[3]

Other goosefoots can be used as pot-herbs, as can most of the closely related oraches. **Garden orache**, *Atriplex hortensis*, probably from Asia, is increasingly grown as a leaf vegetable and occasionally escapes to the wild. **Common orache**, *A. patula*, and **spear-leaved orache**, *A. prostrata*, are native species of waste and disturbed ground. **Grass-leaved orache**, *A. littoralis*, and **frosted orache**, *A. laciniata*, are quite common along sandy coasts throughout Britain.

Sea-purslane, *A. portulacoides*, is another coastal species, but a much more distinctive plant. In summer, its mealy, silvery leaves provide one of the key tones in the chequer of pastel greys, greens and purples that cover the

*Milden's village sign commemorating its name plant, fat-hen (*melde *in Old English), which was once a staple food-plant.*

Another edible goosefoot, sea-beet (an ancestor of cultivated beets), on dunes in north Cornwall.

upper reaches of saltmarshes on the south and east coasts. It has a particular liking for the edges of small patches of salt water, and 'a bird's-eye view of an east-coast marsh would show nearly every little creek bordered with a light grey band produced by these plants'.[4] Sea-purslane prefers well-drained soils, and the development of these characteristic fringes comes about when the tide overflows the edges of pools and creeks, depositing silt amongst the plants growing on their edges. As a result, the level of the bank is gradually raised above that of the marsh. And as the ground becomes higher and better drained, so the sea-purslane grows more luxuriantly, and eventually blankets out other species.

The leaves make an excellent, crisp ingredient for salads.

Sea-beet, *Beta vulgaris* ssp. *maritima* (VN: Sea spinach, Wild spinach). A large-leaved, straggly perennial which grows on sea-walls, shingle and waste ground on most parts of the coast, except in northern Scotland. It is obviously a close relative of cultivated beets such as Swiss chard, mangel-wurzel, sugar-beet and beetroot, and specimens with red-veined leaves, of the kind developed into beetroots, are quite common in the wild. The leaves are tangy and substantial cooked as a spinach, and one of the most popular of wild vegetables.[5]

Common glasswort or **Marsh samphire,** *Salicornia europaea*, and related species (VN: Samphire, Sanfer, Sandforth). Marsh samphire is *not* the wild vegetable whose gathering Shakespeare described in *King Lear* as a 'dreadful trade'. (That is the cliff-growing rock samphire, *Crithmum maritimum*, see p. 288.) It is a plant of the muddiest zones of saltmarshes, and collecting it is more a draggled than a dreadful business. But it is a special plant, and many people's introduction to the lesser-known reaches of wild-food gathering. It has an aura not shared by many other edible wildings, which comes home most strongly when you see that it is sold (e.g. in Sussex)[6] in fishmongers, alongside the cod and cockles, rather than in greengrocers with the 'land' vegetables. In King's Lynn, Norfolk, it is hawked around the streets by a local picker with a horse and cart, who already has a 'traditional' street-cry: 'Any samphire, you ladies?'

It occurs right around the coast of Britain, but East Anglia is where it is best known and most widely used, and the area which has accumulated the richest lore about the species. North Norfolk is where I first made its acquaintance in the 1960s, as a shiny, succulent plant, rather like a plump, jointed pipe-cleaner, which appeared on areas of bare mud and the edges of creeks from late May onwards. I soon learned that picking shouldn't really begin before the longest day, and that the healthiest specimens were those 'washed by every tide'. There were other, more dubious traditions. The samphire was once pickled by filling jars with the chopped shoots, covering with spiced vinegar and leaving in the local baker's oven as it cooled off over the weekend.[7] What its condition was like afterwards one can only guess.

In those days samphire was gathered when it was

Marsh samphire, an early coloniser of bare mudflats, and one of the best wild vegetables.

about six to nine inches high by being pulled up by the roots. This is now illegal, under the Wildlife and Countryside Act, except with the landowner's permission, and it is more responsible (and time-saving in the long term) to cut the tender tops of the plants with scissors. The stems can be eaten raw as a crisp and salty salad plant, or boiled briefly like thin asparagus and dipped in molten butter or warm oil. They are eaten traditionally by holding the root-end and drawing the stems between the teeth, to strip the flesh off the central spine.

Samphire will keep for a few days, provided it is dry. Left damp after washing, it rapidly wilts – as it does if the roots or cut ends are stood in fresh water, which sucks the sap out of the plant. Samphire's succulence, which is the source of its tangy savour, redolent of iodine and sea breezes, is a biological adaptation to enable it to survive in a salt-water environment. The plant contains sodium salts in solution, to balance the 'sucking' (osmotic) pressure of the sea and prevent the plants being dehydrated. The concentration of salts is so high that they were once used in the making of glass and soap (hence the name 'glasswort'). The plants were dried, and then burnt in large heaps. The ash was heated with sand until it fused into a crude glass, or leached with limewater to make a solution of caustic soda. This was evaporated and the resulting crystals of caustic soda (sodium hydroxide) were used to make better-quality glass or heated with animal fats to make soap.

Lancashire is another samphire centre: 'Pickled samphire is popular in Wigan. Special journeys used to be made by Wigan people to collect samphire from the Ribble marshes in September.'[8]

In other parts of Lancashire, samphire was known as 'sandforth' and gathering the plant as 'sandforthing'.[9] (A friend of mine, accustomed to the Norfolk name and pronunciation and never having seen it written down, used to spell it 'sandfire', which is a wonderful description of the plant in autumn as the tips turn a tawny-red over the flats.)

Samphire is moving up-market now. It had an honoured place in the wedding breakfast of Prince Charles and Diana in 1981, delivered fresh from the Sandringham Estate in Norfolk;[10] and it appears increasingly on the menus of smart restaurants, though sometimes just as a garnish for fish, like a maritime parsley (and often commercially imported from Brittany).

Its distribution and abundance are also constantly changing. The annual species (which are highly variable, with 20 to 30 'sorts' distinguishable in south-east England)[11] are amongst the earliest colonisers of fresh, bare estuarine mud. They can grow as thick as grass in the first few years, forming samphire 'lawns', but decline as the mud stabilises and perennial saltmarsh species move in. In East Anglia, samphire's *locus classicus*, it may actually be increasing. The Eastern coastline is slowly sinking as a result of natural earth-movements and the warming of the sea, flooding more drained land each year and opening up new samphire sites all the time.

There is a story that in 1953, after the terrible east coast

Shrubby sea-blite, a local shrub of east- and south-coast shorelines.

floods, a monster samphire plant, six feet tall, was found in a creek near Blakeney in north Norfolk. It was strapped to a bicycle crossbar, taken to the local pub and hung above the bar like a prize fish. Forty years later I saw a more credible development for myself, in a creek in much the same place. A crumbling plastic dinghy, moored a few yards off-shore, was carrying the remains of a presumably self-sown crop of samphire. I had a vision of the owners using it like an outsize maritime equivalent of one of those mustard-and-cress cartons, towing it in for the occasional snip in season, and then setting it adrift again.

Shrubby sea-blite, *Suaeda vera*, is a scarce evergreen shrub which forms dense, scrubby colonies on some beaches and the upper reaches of saltmarshes on the south and east coasts of England. It is a species with a distinctly Mediterranean distribution in Europe, yet it seems perfectly adapted to the often turbulent conditions of the North Sea coast. In shingle that is tossed about by winter storms it grows with a low, creeping habit, which allows the stems to become buried under the shifting stones. New shoots then sprout from the submerged stems, even when they are two or three feet below the surface.

Shrubby sea-blite can grow up to four feet tall, and its thickets form distinctive landmarks on, for instance, Blakeney Point, Norfolk, and Chesil Beach, Dorset. They are also landmarks for migrating small birds, which often use them as refuges.

Annual sea-blite, *S. maritima*, is a more widespread relative, a small, succulent plant of the middle and lower reaches of saltmarshes. It is an unprepossessing pale green in summer, but makes a useful, if salty, salad plant, and, as a bonus, is splashed with vivid red in autumn.

Spineless saltwort or **Russian thistle**, *Salsola kali* ssp. *ruthenica*. This is one of a disparate group of plants (mostly of dry habitats) popularly known as 'tumbleweeds', from the way in which they spread their seeds. The parent plants dry off after flowering, are uprooted by the wind and then blow about the landscape, scattering seeds as they go.

Spineless saltwort is a native of eastern Europe and Asia, but there is a celebrated colony on the ash-tips at the back of the Ford Motor Works in Dagenham, Essex, which was first established in 1934 and which may have arrived here via North America. *Salsola ruthenica* is also naturalised in dry plains country in the USA, and the popular explanation is that seeds first arrived at the Essex factory in packing material from Ford America. This was dumped close to the area where the warm ash from the foundries was tipped, and the seeds, finding the conditions similar to those in their arid native habitats, germinated and prospered.

The Dagenham colony has declined since the development of much of the tip for a car park. But on windy days after a warm summer it is still possible to see a few of these prairie plants bowling about behind the security fences.

Purslane family *Portulacaceae*

Springbeauty, *Claytonia perfoliata*, is a lax annual widely naturalised from the North American west coast, especially on sandy soils. The fleshy upper leaves are joined into a kind of saucer under the small white flowers, and the plant is edible, either as a salad or as a cooked vegetable. In America it is known as 'winter purslane'. **Pink purslane**, *C. sibirica*, also from North America, is naturalised in damp shady places, especially in the north and west.

Pink family *Caryophyllaceae*

Chickweed, *Stellaria media*. An abundant weed of gardens, roadsides and waste and cultivated ground, staying green – and quite often in flower – throughout the winter. Chickweed has always been valued as food for poultry and cage-birds, and, in small quantities, as a vegetable, in salads or stir-fries. **Three-veined sandwort**, *Moehringia trinervia*, is a pleasing, if modest, plant of bare areas in old deciduous woods and hedge-banks. The clumps resemble chickweed at a distance, but each leaf is marked with three clear veins. **Sea sandwort**, *Honckenya peploides*, is one of the earliest colonisers of sand-dunes and shingle, and remarkable for its sprawling concertinas of geometrically stacked leaves. These, as is the case with many seashore plants, are succulent and edible, and in parts of northern England were pickled. **Spring sandwort**, *Minuartia verna*, has a taste for lead-rich soil and is frequently found on old lead-mine spoil-tips in northern England and North Wales, and by the sides of streams flowing near them.

Greater stitchwort, *Stellaria holostea* (VN: Headaches, Stinkwort, Wedding cakes, Milkmaids, Star-of-Bethlehem, Brassy buttons, Shirt buttons, Poor-man's-buttonhole, Daddy's-shirt-buttons, Snapdragon, Poppers). This is a familiar spring flower of hedge-banks and wood-rides, much loved for its modestly beautiful white flowers, which have been likened to – and no doubt used as – buttonholes. Other local names, some of which have survived, remark on the ease with which the stalks break (snapdragons, snapcrackers) and its habit of noisily firing off its seeds (poppers, pop-guns). A woman from Kent recalled to a contributor how this became a childhood game while her family picked hops: 'When her mother and family went hop-tying in May she was pushed to the hop garden in a wooden baby cart. This was then filled round her with tangled stems of greater stitchwort from the hedgerows, with all the round seed capsules already ripening, the popping of which would keep her absorbed and occupied while her mother tied the bines.'[1]

Snow-in-summer, *Cerastium tomentosum*, is aggressive in gardens and widely naturalised in dry places. **Procumbent pearlwort**, *Sagina procumbens*, is a common plant of paths, brickwork and lawns that forms moss-like cushions. In the Scottish Highlands and Islands it was once believed to be the first plant Christ stepped on when he came to earth after the resurrection, and it was revered as a magically protective, softening species. It would soothe a woman in labour if placed under her right knee, and protect cows, calves and milk if put in the forehooves of the bull.[2] It still has its friends: 'In cracks between paving stones and at the edges of pavements. Diminutive, but when examined closely has an attractive foliage and colour. And it is always nice to see a wild plant growing in such apparently inhospitable surroundings.'[3]

Corncockle, *Agrostemma githago* (VN: Kiss-me-quick). One of the most attractive of cornfield annuals, with purple flowers which are folded or furled like a flag before they open. Once it was abundant enough to be regarded as a menace, because it made bread bitter-tasting and possibly even poisonous. Cleaner seed-corn and modern herbicides have virtually eliminated it from the fields, and these days it is seen only occasionally, when old pastureland is ploughed, or when it has been deliberately sown in 'wildflower gardens' (e.g. in Earlham churchyard, Norwich).[4]

Ragged-Robin, *Lychnis flos-cuculi*, is a declining plant of damp places, especially marshes, meadows and woodland rides. 'Ragged' is from the much-divided, thread-like petals. These are normally pale to deep pink, though white flowers are very occasionally found and have become popular with gardeners in recent years.

Rose campion, *L. coronaria*, is a popular garden plant from south-east Europe that frequently escapes and is occasionally naturalised. 'Campion' (see the several species below) is a fourteenth-century word, a doublet of 'champion', and it may be that the flowers of this family were seen as champion's flowers – 'fit for a garland'. But the epithet did not begin to be attached to plants in English until the sixteenth century, and an alternative explanation is a later meaning of 'champion' or 'champaign', 'level and open country'.[5]

Greater stitchwort – 'poor-man's-buttonhole' – decorating a Hampshire hedge-bank in late April.

Corncockle painted by one of the Clifford family of Frampton. The picture catches the way the petals are furled like flags before opening.

Bladder campion, *Silene vulgaris*, is a widespread roadside plant, whose bladder-shaped calyx can be 'popped' before the flowers have opened. It is one of the favourite food-plants of the little insects known as frog-hoppers, notable for surrounding themselves with protective froth whilst feeding. John Gerard, ever felicitous at naming, called it 'Spatling Poppie', 'in respect of that kinde of frothie spattle, or spume, which we call Cuckow spittle, that more aboundeth in the bosomes of the leaues of these plants, then in any other'.[6] **Plymouth campion**, *S. vulgaris* ssp. *macrocarpa*. 'A pink flowered bladder campion growing on two sites on Plymouth Hoe. A native of the Mediterranean, it has been in the city since 1921. It is now considered a subspecies, with no particular claim to protection, although the local Parks dept responsibly direct their mowers to leave it undisturbed during May and June.'[7]

Red campion, *S. dioica* (VN: Adder's flower, Robin Hood, Cuckoo flower). One of a dozen or so spring flowers that share a name, and a season, with the cuckoo, red campion is at its best where it grows with bluebells and white stitchwort and ramsons, in hedgerows, in woods, and on northern and western sea-cliffs. But it has an oddly patchy distribution, abundant, sometimes, in one parish and absent from the next, even though there are no obvious habitat differences. Although the flowers are most usually rose-red, pink- and white-bloomed varieties are common. Red campion also hybridises with its close relative white campion to produce another range of intermediate-coloured flowers.

White campion, *S. latifolia*, is an annual or biennial herb of roadsides, waste places and cultivated fields. Its white flowers (sometimes up to an inch across) give off a slight scent of clove-pink at night, which makes them very attractive to moths.

Sea campion, *S. uniflora*, is a tufted perennial with grey-green leaves and an abundance of white flowers, and part of the spectacular display of spring flowers – including thrift, early-purple orchid and spring squill – which adorn many sea-cliffs, especially along the western seaboard of Britain. Sea campion also thrives on shingle

Ragged-Robin, a declining species of marshes, meadows and damp woodland.

and established sand-dunes, and in a few inland sites in mountainous areas of Scotland, Wales and the north of England. **Nottingham catchfly**, *S. nutans*, no longer occurs on the site from which it was named, Nottingham Castle Rock, but can still be found in dry grassy places in the county (and at a few sites elsewhere). It resembles a rather deflated bladder campion, and opens its flowers only at night. **Sand catchfly**, *S. conica*, by contrast, has an egg-shaped calyx vastly bigger than its diminutive pink flower. It is a scarce plant of two quite distinct habitats, which have only their sandiness in common: the East Anglian Breckland and a sandy common in Worcestershire, and seashore dunes in eastern and southern England.

Soapwort, *Saponaria officinalis* (VN: Bouncing Bett – especially applied to the double-flowered form). There is no wishful thinking in this name. Soapwort is a detergent-herb, plain and proven. Simply rubbing a leaf between the fingers will produce a slight, slippery froth. Boiled in water, the plant produces a green lather with the power to lift grease and dirt, especially from fabrics. The detergent effect is due to the presence of saponins – chemicals which, like inorganic soaps, appear to 'lubricate' and absorb dirt particles.

Soapwort's properties have been exploited across Europe and the Middle East, where it is native. It has been cultivated for laundering woollens in Syria, used as a sheep-wash prior to shearing in the Swiss Alps, and in Britain employed as a soaping agent by medieval fullers, who beat the finished cloth to clean and thicken it (one medieval name was 'foam dock'). And because vegetable saponins are so much gentler than soaps, *Saponaria* has been used much more recently for washing ancient tapestries. 'The natural silk produced by the Hart-Dyke family [and destined for royal wedding dresses] was washed with this vegetable detergent.'[8] In the Victoria & Albert Museum it was last used for cleaning fragile fabrics in the 1970s. The National Trust have also used it, for bringing up the colours in antique curtains.[9]

Although soapwort may be native along rivers in parts of south-west England and North Wales, some colonies are probably relics of this ancient use in laundering, especially when they grow by the sites of old mills: 'Just outside the churchyard at St Winnow, beautifully situated on

Sea campion on the cliffs of the Lizard Peninsula, Cornwall.

a tidal creek of the River Fowey, there was a mass of soapwort until the site was developed. The soapwort has gone for ever, but I cannot believe that it was not originally planted for one of its important uses, the washing of church vestments, and with the tidal river nearby.'[10]

But most soapwort, on road- and rail-sides and waste ground, originates from more mundane garden throw-outs. This is especially true of the decorative double variety, *flore pleno*, which – in honour of the species' long association with exuberant washerwomen – is known on both sides of the Atlantic by the splendid name of Bouncing Bett.

Childing pink, *Petrorhagia nanteuilii*, is a delicate annual, not recognised as growing in Britain until 1962. It is now known only from two sandy places near the sea in Sussex, having disappeared from two in Hampshire since 1965.[11] The name derives from the cluster of tiny pink flowers (the childings or children) wrapped in the papery calyx.

Cheddar pink, *Dianthus gratianopolitanus*, is in Britain confined to limestone ledges in and near the Cheddar Gorge. The Cheddar pink is an exceptionally attractive plant – tufted, grey-leaved, topped with rose-coloured, clove-scented flowers – and, after its discovery in the early eighteenth century, it became as famous as Cheddar cheese. It was dug up by tourists and locals alike, transplanted to rockeries, sold to alpine plant merchants. By the late nineteenth century, some guidebooks to the Mendips were declaring it extinct. But the plant clung on in the more inaccessible corners and crevices in the Gorge, and it is now thoroughly protected. It can normally be seen through field-glasses, and, for those who strike up a fancy for it in their alpine gardens, there is a range of cultivated varieties on the market.

Clove pink, *D. caryophyllus*, is the Tudor 'gillyflower' and the chief ancestor of clove-scented pinks and carnations. Clove pink was probably introduced from southern Europe by the Normans, since when it has become naturalised on a few old walls. Some traditional stations include the walls of the gents' toilet at Rochester Castle, Kent, and the remains of Beaulieu Abbey in Hampshire. **Pink**, *D. plumarius*, is also naturalised on banks and old walls, as at Fountains Abbey in Yorkshire and Ludlow Castle, Shropshire. It is the ancestor of modern pinks.

Cheddar pink, safe from pickers, on a limestone outcrop in the Cheddar Gorge.

Maiden pink, *D. deltoides*, is a small, native species with unscented deep-pink flowers now found in a few scattered areas of sandy grassland and rocky ground, including the East Anglian Breckland. **Sweet-William**, *D. barbatus*, is a popular garden plant from southern Europe, occasionally escaping to waste places. **Deptford pink**, *D. armeria*, is a native species, not unlike Sweet-William in form, now very rare and decreasing in dry grassy places in the south. The one place it may never have grown is Deptford, east London. The name 'Deptford pink' was attached to this species by the seventeenth-century herbalist and botanist Thomas Johnson, who provided an excellent and unambiguous portrait of it for his edition of Gerard's *Herball*, published in 1633. But the plant that Gerard found at Deptford, and which prompted the name, was almost certainly maiden pink, to judge from his own evocative description: 'There is a Wilde creeping Pinke, which groweth in our pastures neere about London, and other places, but especially in the great field next to Detford, by the path side as you go from Redriffe to Greenewich, which hath many small tender leaues, shorter than any of the other wilde Pinkes set vpon little tender stalks, which lie flat vpon the ground, taking holde of the same in sundrie places, whereby it greatly encreaseth; wherevpon doth growe little reddish flowers.'[12]

As the first name which is given to a plant generally has priority in botanical protocol, the misleading label stuck, and the East End 'acquired' a plant which may not have grown there since before the city of London was built.

Common bistort. The leaves are used in a spring pudding in the north.

Knotweed family
Polygonaceae

Common bistort, *Persicaria bistorta* (VN: Dock; Easter ledges, Easter ledger; Easterman giants; Easter May-giants; Water ledges; Pudding grass, Pudding dock; Snake-weed; Pink pokers). Only botanists and southerners use the name 'bistort' for *Persicaria bistorta*. It is an awkward piece of anglicised Latin, probably meaning 'twice-twisted' and referring to the contorted root (hence 'snake-weed'). To those who live in its heartland, up in the north Pennines, and especially in the fellside villages between Halifax and Carlisle, it is known simply as 'dock' or 'pudding dock', or occasionally by one of the other names that refer to its central role in a traditional springtime pudding made from the cooked leaves and various combinations of oatmeal, egg and other green herbs. Easter May-giants, Easterman giants (or sometimes Easter mangiants) are derivations from the French *manger*, to eat. (A nineteenth-century dictionary of Cumbrian dialect has a wonderful phonetic rendering of the name Easterman giants: 'Easter-mun-jiands (EE,STTHUR'R'-MU'-JAAI'NTS)'.)

Bistort was also once called 'passion dock', and the pudding almost certainly originated as a cleansing, bitter dish for Lent, traditionally eaten in the two weeks before Easter. It was an obvious choice for such a recipe. The dock grows in dense patches in damp meadows and pastures, on river-banks and roadsides, and occasionally in wet woodland clearings. (One aficionado from the Calder Valley in West Yorkshire swears the best crop came from his grandfather's grave on Sowerby Top.) The long-stalked, heart-shaped leaves are well developed by Easter, and are unmistakable amongst the grass because of their ribbed tops and silvery undersides, which flash in the March winds.

Varieties of dock pudding have been eaten on a local basis in the north Pennines for centuries, but began to attract national attention in 1971, when the 'World Championship Dock Pudding Contest' was inaugurated in the Calder Valley, around the villages of Hebden Bridge and Mytholmroyd. The competition, which is still kept up, brought many memories and recipes to the surface: 'As kids we used to go out with carrier bags and collect the first sprouts of dock leaves. We would sell them to the housewives in the village, who would inspect the quality and give us two shillings or half a crown depending on how good they were.'[1]

The competitors are required to follow a basic recipe including dock leaves, chopped nettles, onions, oatmeal and seasoning, all fried in bacon fat. But dock pudding is a dish for which every village, and maybe every family, has its own recipe. Cumbria, still the dock-pudding centre, despite the Calder Valley's higher profile, shows the range of local flourishes and variations which can be added to this essentially simple dish. The following recipe is from Raughtonhead Women's Institute, just south of Carlisle: 'Take a small quantity of nettles, cabbage or young leaves and shoots from Brussels sprouts, kale or curly greens, three or four dandelions or [common] dock leaves, three good leeks and a good handful of the herb known in Cumberland as Easterman Giants, plus two or three gooseberry or blackcurrant leaves may be added. Cook in a little water and add some cooked barley. Chop and mix greens with barley. Serve as it is or with eggs or have an egg and oatmeal beaten in.'[2]

From Carlisle itself: 'We used to gather dandelion leaves, young ones, and nettles, also young ones, Easter ledges from the churchyard, all chopped up with two new sticks of rhubarb, bound with beaten egg and pearl barley, put in a cloth and boiled.'[3]

And from Beetham, down below Kendal: 'The leaves are washed and chopped up like mint and mixed with a dumpling mixture, either vegetable or suet, rolled and chopped into slices and fried with bacon fat. Serve with a grilled dish.'[4]

A more sophisticated Cumbrian recipe, involving rolling and skewering the leaves and then simmering them in milk and butter, was noted by a visitor from Epping Forest in Essex.[5] In fact, Epping, and the river valleys and wet commons surrounding it, are one of the southern strongholds of the species, so perhaps the emergence of Essex dock kebab is not an impossibility.

Amphibious bistort, *P. amphibia*, is a surprising plant of ponds, canals and ditches, with spikes of deep pink flowers above floating, oblong leaves, like an aquatic orchid. A shorter form occurs as a land-weed. **Redshank**, *P. maculosa*, a miniaturised bistort, is an abundant annual

Amphibious bistort, which has two forms, adapted to water and land habitats respectively.

Ray's knotgrass, a local species of sand and shingle beaches, mainly in the west.

of open and cultivated ground. **Water-pepper**, *P. hydropiper*, grows chiefly in damp meadows and waterlogged woodland rides. Its leaves have a burning peppery taste.

Buckwheat, *Fagopyrum esculentum*. Originally from Asia, buckwheat was once widely cultivated in Britain to provide flour as well as food for cattle and poultry, and was taken by early settlers to North America. There it prospered, and remains a favourite base for pancakes and biscuits. (Brittany crêpes are made from it too.) The British seem to have lost their liking for the nutty-tasting flour, and here it has virtually vanished as a crop. But rogue and relic patches, tinged with tan and orange in the leaves, can colour the edges of fields, especially near woods where grain has been scattered for pheasants, and give a hint of what a buckwheat crop must have looked like.

Knotgrass, *Polygonum aviculare*, is still sometimes known as iron-grass because of its tenaciously wiry stalks and roots. It is an abundant weed of bare places, gardens and cultivated ground.

Japanese knotweed, *Fallopia japonica* (VN: Japweed, Sally rhubarb, German sausage). Japweed is now officially regarded as the most pernicious weed in Britain, and it is illegal to plant it deliberately in the wild. Its rampaging spread across Britain in the late 1970s and 80s is regarded as a parable of the dangers of casually introducing alien plants into the countryside – which goes to show how quickly perceptions of plants can change when their behaviour, or gardening fashion, begins to shift.

Japanese knotweed was introduced to Britain from Japan sometime between 1825 and the 1840s, and was an immediate hit with gardeners. Its dense sheaves of canes (up to six feet tall), heavy, heart-shaped leaves and spires of tiny white flowers suited the Victorians' austere taste. It was recommended for naturalising in the shrubbery by none other than William Robinson (see p. 43), who described it as 'most effective in flower in the autumn' and advocated planting in groups of two or three – though he did warn that neither it nor its larger cousin, giant knotweed, can 'be put in the garden without fear of their overrunning other things, while outside in the pleasure

Japanese knotweed, which the Victorian gardening writer William Robinson thought 'most effective in flower in the autumn'.

ground or plantation, or by the waterside where there is enough soil, they may be very handsome indeed'.[6]

But even the pleasure ground and plantation proved too restricting for them, and when their formidable powers of colonisation were realised they were thrown over the garden wall onto railway embankments and rubbish-tips. From these strongholds they advanced even further, able to sprout from the smallest fragments of root as well as by the remorseless extension of their whole root systems.

In the footnotes of local naturalists' societies' newsletters and county floras, it is possible to map Japanese knotweed's inexorable spread across Britain. It was first noticed in the wild in London in 1900. It had reached a rubbish-tip in Langley, Middlesex, two years later, and was in the smart adjacent village of Denham by 1918. The sharp-eyed botanist George Claridge Druce found it near Exeter in 1908. It was in Suffolk by 1924, West Yorkshire in the 1940s and Northumberland in the 1950s. By the early 1960s its colonies stretched across Britain from Land's End to the northern tip of the Isle of Lewis. In many places they seemed so dense as to exclude all other species, and to be virtually ineradicable.

A species advancing as aggressively as this was bound to generate myths in its wake. In part of south Hampshire the arrival of the Japanese weed was blamed on the military: 'Our garden suffers from the invasion of knotweed from the adjoining Telecom exchange ... I have seen a photograph taken by the previous occupant, which seems to show an old army hut approximately where the infestation is now centred. Maybe freshly billeted troops from overseas brushed some seeds from the seams of their uniforms or corners of kit bags?'[7]

In one corner of Cornwall in the 1930s, it earned the nickname of Hancock's curse, having spread from the garden of someone with that name; and there is a story that a house in the same area was reduced in price by £100 because its garden was overrun by knotweed. These days it is regarded as a serious nuisance across much of the West Country: 'The Devon Community Council is embarked upon an eradication campaign beginning with the parish of Buckland Monachorum, with the aim of making the parish a knotweed-free zone.'[8]

Not everyone in Devon is so hostile though. In Chulmleigh, a father makes crude pan-pipes for his children by using dry knotweed stems of different lengths and diameters.[9] It has been successfully used as cattle fodder at the Cardiff City Farm (a practice once widespread in central Europe);[10] and it produces moderate-quality handmade paper.[11]

Most encouragingly, the habit of eating the young shoots (which are used as a vegetable in Japan) has begun to spread. The knotweed is well naturalised along the eastern states of Canada and the United States, and is reportedly eaten by Japanese immigrants there. It was first introduced to a wider public by the pioneer writer on American wild foods, Euell Gibbons, in the 1960s. In his classic *Stalking the Wild Asparagus*, he calls it 'a combination fruit-vegetable' and stresses that, although it may *look* like an asparagus when it first appears, it has a tart taste more like sorrel or rhubarb (members of the same family), and recommends making jam or pies from the shoots when they are no more than a foot tall.[12]

Quite independently, people in Wales have been discovering knotweed's culinary potential. In parts of Dyfed, the young shoots and leaves are cooked like spinach. In

The other face of Japanese knotweed: dense and almost ineradicable thickets.

Russian-vine, a rampant climber from the Far East that is sometimes known as 'mile-a-minute-plant'.

Swansea children 'suck the sharp-tasting juicy stems during high summer' and know the plant as 'Sally rhubarb'. (It is an exact and evocative name, 'sally' being an old variety of 'sallow', traditionally used to describe low scrubby willows – of which a thicket of knotweeds can be reminiscent.) In Clwyd, the plant is known as 'German sausage' – though I suspect this is because of the speckled-brown appearance of the cylindrical stems, rather than anything to do with their taste.[13]

In Bristol in the 1960s it was precisely the jungle-like profusion of knotweed thickets that attracted one group of youngsters:

'The area "our gang" played in would now be regarded as inner-city dereliction; a few acres of waste land, partly the result of Hitler's bombs, and partly due to the demolition of houses, the occupants of which were removed to the city's periphery ... Its crowning glory was the Japanese knotweed, which covered at least half of this wilderness of ours. Not knowing its name we christened it "Bambarb", because it showed characteristics of both bamboo and rhubarb. The Bambarb grew much taller than the average fourteen-year-old, so consequently became a great hiding place from rival gangs and the local Bobby ... and for enacting boyish fantasies of jungle warfare and tales of exploration. When the land was finally redeveloped in the early 70s, our beloved wilderness was taken from us for ever.'[14]

It is worth adding that, where Japanese knotweed grows under some constraint and one can put aside the prejudices and real worries that have come to focus on it, William Robinson's admiration is well justified. One such site is the industrial area of Sheffield. Here the River Don throws a corridor of wildness right through the heart of the city, wreathing distant views of the cathedral in an almost Amazonian luxuriance (see p. 66). Crack willows sprout from the rootings of broken water-borne branches and in places grow into fantastic multi-stemmed shapes, like mangrove clumps. The banks and riverine islands are thick with showy immigrant flowers, Himalayan balsam, Michaelmas daisy, soapwort – and Japanese knotweed. Every so often the river floods, with water pouring in from Pennine feeder streams, and it is consequently banked with high stone walls on either side. This has several consequences for the Japanese knotweed. It cannot spread outwards, for a start. And whenever the river floods, the accumulation of dead stems that usually carpets the ground is swept away, and in its place comes silt full of seeds and bulbs from plants in the high Pennine woods – bluebells, wood anemones, ramsons and celandine. They readily root in the rich mud underneath the knotweed stands, and come into flower as the new season's shoots begin to grow. And so in late April, before the Japweed leaves are fully open and the canopy closed, the banks of the Don look like nothing so much as some strange urban hazel coppice.

Giant knotweed, *F. sachalinensis*, similar to *F. japonica* but often taller, is also spreading, though as yet it is a much scarcer plant. Hybrids between the two species are also cropping up.

Russian-vine, *F. baldschuanica* (VN: Mile-a-minute-plant). A vigorous white-flowered scrambler from China that can cover walls, hedges and sometimes whole aban-

Common sorrel. The leaves are refreshingly sharp, like young plum-skins.

doned buildings in urban areas. It grows at a prodigious rate – hence its modern nickname, mile-a-minute-plant – and has been known to extend 50 yards from its roots along a hedge or fence.[15] It is always a garden escape originally, and, never setting seed in this country, can spring only from stray pieces of root or stem, though it is increasingly found some distance from houses. A pink-flowered variety has been found on a rubbish-tip in Dartford.[16]

An interesting hybrid between Japanese knotweed and Russian-vine (*F. japonica* × *F. baldschuanica*) appeared in 1987 in a nature reserve on one-time railway waste ground in Harringay, London. It is more elegant and less aggressive than either of its parents and has leaves shapely enough to make it a serious contender as a garden scrambler or ground-cover plant in the future. In the warm summer of 1993 it was being visited by many different species of native insect.[17]

Black-bindweed, *F. convolvulus*, is a common annual weed of arable fields and waste places, a little like a straggling, triangular-leaved sorrel. Impressions of the seeds have been found on pots in several Bronze Age excavations, and it is almost certain that they were used as a source of food in prehistoric times.[18] Black-bindweed may even have been cultivated, in a rough-and-ready way, allowed to grow amongst the main crop and then harvested with it.

The seeds also formed one of the ingredients of the ritual meal eaten by Tollund Man (see fat-hen, p. 95).

Common sorrel, *Rumex acetosa* (VN: Sour docks, Sour dogs, Sour dots, Sour ducks, Sour grabs, Sour sabs, Sour sap, Sour sops, Soorocks; Vinegar plant, Vinegar leaves; Rain; Green sauce, Bread and sauce; Sugar stick – stems only). Sorrel's reputation as a sour-tasting plant is echoed in almost every one of its common names. Yet sour isn't really the right description at all, suggesting something as painfully, tartly dry as a sloe. Sorrel leaves are more like the skins of young plums, sharp, astringent and refreshing. Nearly two centuries ago John Clare described how parched field-workers would chew them raw to slake their thirst.[19]

Sorrel leaves are still nibbled by children across Britain, and increasingly used in salads, soups and sauces for fish. A woman from the Baltic serves sorrel soup with hard-boiled egg, just as spinach soup is served in eastern Europe.[20] But the taste of cooked sorrel is closer to

Broad-leaved dock, a common target for weedkiller sprays, but still widely used for rubbing on nettle stings.

rhubarb than spinach. A fascinating recipe from the 1930s exploits this fruitiness by using the leaves in a turnover: 'In Lancashire they still use the fresh young leaves of wild sorrel as a substitute for apple in turnovers. I have made it myself, and very good it is, with plenty of brown sugar and a little moisture on the leaves. Sorrel is in season between apple and gooseberries, i.e. in April and May.'[21]

Sorrel is still common on road-verges and river-banks and in grassland of all kinds (except 'improved', reseeded meadowland). Where it grows in quantity, its flower-spikes float like a red haze amongst the grasses in May and June. The leaves are arrow-shaped and grow both from the stem and at the base of the plant. In seashore sites, they can become almost succulent.

Sheep's sorrel, *R. acetosella*, is a shorter species which can give its own rusty tinge to sandy, well-drained and usually acid soils. **French sorrel**, *R. scutatus*, is a European species with shield-shaped leaves, much used in cooking. It is naturalised here and there on banks and walls, chiefly in the north, including one long-established and improbably-sited colony in North Yorkshire: 'At Settle, French sorrel has established itself on the bridge that carried the former A65 Leeds–Kendal trunk road across the River Ribble. It has been there since 1928.'[22]

Broad-leaved dock, *R. obtusifolius* (VN: Docken, Dockan). An abundant perennial of fields, gardens and waste places, still universally used, by children especially, to rub on nettle stings.

'A new name invented by my son when he was three is "doctor leaf" for dock, which is what he thought we were saying because it makes nettle stings better. The name has been used by us ever since.'[23]

'When my friends and I were out exploring the woods and fields we'd often take a bottle with "potion" in it to put on any nettle or insect stings. This we made by chopping up dock leaves and lemon-balm leaves and shaking in water.'[24]

It has also been used as a more serious salve by adults: 'My grandmother's practice was to collect young dock leaves before breakfast. These were still wet with dew. They were thoroughly washed before being added to pure melted pigs' lard in a stout saucepan. The mixture was allowed to reduce on a low heat until the residue was a pale green colour, after which it was strained into clean jars and, when set, sealed. This ointment was used for the treatment of piles.'[25]

'Still the best antidote to nettle stings. Works perfectly on dogs' feet, too (one of our dogs is sensitive to nettle stings).'[26]

In Lancashire it has also been used 'for cleaning dogs' backsides', and in Sussex for protecting dairy products from the heat, like butterbur (see p. 377), an ancient practice which gave broad-leaved dock the obsolete name of 'butter dock': 'My mother (who made Cheddar cheese and butter in her youth) used dock leaves to keep butter cool in the summer. They were draped over the dish, which was kept in cool water.'[27]

Monk's-rhubarb, *R. pseudoalpinus*, also had its large, roundish leaves employed as a 'butter dock'. Like **patience dock**, *R. patientia*, it is a native of continental Europe. Both species were once grown as pot-herbs and are naturalised in a few waste and grassy places.

Distinctive native docks (most of which have also been used at times to relieve nettle stings) include **clustered dock**, *R. conglomeratus*, preferring damp places, **wood dock**, *R. sanguineus*, which has a preference for old woodland, and **curled dock**, *R. crispus*, which is probably the commonest species. A variety of the last species that grows on sand-dunes and shingle (var. *littoreus*) is remarkable because the dying stems of one year's growth form a kind of tent over the new shoots to protect them from winter sea-spray. **Water dock**, *R. hydrolapathum*, has long, palm-like leaves and is a stately plant of the edges of canals, ponds and rivers, often growing in shallow water.

Thrift family *Plumbaginaceae*

Common sea-lavender, *Limonium vulgare*. Sheets of sea-lavender, often growing as densely as bluebells, turn whole tracts of marshland on the south and east coasts of England into glorious, rippling sheets of pale mauve and lilac. In August heat-waves at Stiffkey and Blakeney in Norfolk I have seen mirages of distant lavender plains floating above the flat calm of the estuaries.

Common sea-lavender grows chiefly in the middle level of the saltmarsh and is a very variable species, growing taller and with larger trusses of flowers on the drier reaches of the marsh. Here it is often joined by the smaller, paler and earlier flowering species, **matted sea-lavender**, *L. bellidifolium*, and the diminutive **rock sea-lavender**, *L. binervosum*, which also grows on shingle beyond the edges of the marsh (and on maritime rocks and cliffs elsewhere). On the north Norfolk coast these

Thrift – which may have acquired this common name from its tight and economic tufts.

three intermingled species form a subtly changing tapestry of mauves, not just through the summer flowering season, but well into autumn. Like their cultivated relatives, the statices, they are everlasting blooms, harvested as a wild crop (not always legally) by coastal dwellers, and sold in bunches to tourists, as a popular radiator adornment after a coastal holiday. The trade has even reached the main shopping precinct in Bletchley, Buckinghamshire, where rather old and pallid sea-lavender is hawked from baskets as 'lucky white heather'.[1]

There are nearly thirty other species and subspecies of sea-lavender growing in Britain, most of them rare and highly local. One village near Brighton on the Sussex coast has, remarkably, two. **Rottingdean sea-lavender**, *L. hyblaeum*, from the rough shores of Sicily, has escaped from the rockeries of this elegant seaside village to become naturalised on nearby cliffs, close to *Limonium procerum* ssp. *procerum*, which otherwise grows as a native only in western Britain.[2]

Thrift, *Armeria maritima* (VN: Sea-pink, Cliff clover, Ladies' cushions, Heugh daisy). An accommodating perennial found in almost every kind of seashore location. The compact pink-flowered cushions grow on sand-dunes, shingle, marshland edges, cliffs, even stone walls near the sea. The origin of the most commonly used name, thrift, is obscure; it may derive from 'thriving', i.e. evergreen, or, more likely I suspect, from the tight and economic tufting of the leaves, which serves as a way of conserving the plant's fresh water in salt winds. Geoffrey Grigson memorably described ambling across acres of these padded cushions on Annet in the Isles of Scilly as like 'a dream of walking on soft rubber which has squirted into flower'.[3]

At any rate, 'thrift' was once well enough known to be used as a punning emblem on the back of the old twelve-sided threepenny bit, the coin whose awkwardness made it the most frequently consigned to money-boxes. It is still the emblem of the Fourth City Building Society.

'Heugh daisy' is a very local Scots and northern name, from 'heugh', meaning a cliff or ravine.

Thrift's blooms can vary from deep pink to white, and they have been a favourite for garden edging at least as far back as the sixteenth century, when John Gerard recommended them 'for the bordering vp of beds and bankes'.[4]

Sea-lavender in a saltmarsh, north Norfolk. One of the great shoreline landscape plants in high summer.

Charles Rennie Mackintosh's portrait of thrift or sea-pink, painted on Holy Island, 1901.

This, of course, is exactly how the hazy pink flowers and neat green cushions seem to arrange themselves in the wild, fringing the tops of Devon drystone walls, lining rock crevices in Pembrokeshire, or clustering around the high-tide line on Lancashire sand-dunes.

Other varieties of sea-pink, some of them quite large, occur a long way from the sea, on mountains and river shingles – relics of the post-glacial era, when the species was probably widespread in rocky open country. They also have a taste for lead-rich soils and are often found on mine-spoil tips. The thrift family's scientific name, *Plumbaginaceae*, derives from the ancient belief that they could cure lead poisoning: 'In a picnic place near Woodhall in Wensleydale [40 miles from the sea], where a stream running from the old lead-mines often floods the area, there is often a wonderful display of thrift. Some say that this is a relic of the wooden bungalows that were erected there by a philanthropist for city dwellers. But with thrift being a Plumbago, I feel it may have been there first.'[5]

Peony family *Paeoniaceae*

Peony, *Paeonia mascula*. Britain is a thousand miles from the wild peony's native home in the Mediterranean scrublands. Yet this handsome shrub with its luscious crimson flowers has had at least two newsworthy 'wild' appearances on these shores.

The first was possibly a seventeenth-century 'urban myth', or a simple confidence trick. In the original edition of his *Herball* (1597), John Gerard claims to have found peony growing 'wilde vpon a conie berrie [rabbit warren] in Betsome, being in the parish of Southfleete in Kent'.[1] But the editor of the second edition, the scrupulous Thomas Johnson, adds a sceptical, reproving footnote to Gerard's entry: 'I haue beene told that our Author himselfe planted that Peionie there, and afterwards seemed to finde it there by accident: and I doe beleeue it was so, because none before or since haue euer seene or hard of it growing wilde since in any part of this kingdome.'[2] If this was a 'plant', so to speak, it is one of a long history of botanical hoaxes, though none of them has ever achieved the notoriety of the fake skull of Piltdown Man.

But in 1803 a small clump of naturalised peonies was found growing on the uninhabited island of Steepholm, in the Bristol Channel. The discovery of this exotic beauty in such a desolate place caught the imagination of the Romantics, and one of Coleridge's mentors, William Lisle Bowles, composed a poem to this 'one flower, which smiles in sunshine or in storm, there sits companionless, but yet not sad'.[3] The lateness of its discovery is curious, given that the island had been a magnet to botanists since the middle of the sixteenth century. William Turner, author of the first serious study of British plants, *A New Herball* (1551), may have visited it in 1562, followed by de l'Obel (1581), Sir Joseph Banks and John Lightfoot (1770s). They recorded many of the southern European herbs that still crowd the island's craggy slopes – henbane (a hypnotic), caper spurge (the 'katapuce' in Chaucer's 'Nun's Priest's Tale', an ingredient in a potent laxative), coriander, wild leek, greater celandine and alexanders. Such a suite of foreign medicinals growing together is a strong hint that there was once a physic garden on the island. A community of Augustine monks, an order noted for their medicinal and gardening skills, lived there between 1166 and 1260, and John Fowles believes they may have deliberately exploited Steepholm's equable

Wild peony, in Britain known only from the island of Steepholm.

oceanic climate for the bulk growing of Mediterranean herbs.[4] Yet despite the fact that peony was one of the most prized and frequently grown wild herbs, and has flamboyant red flowers up to five inches across, not one of Steepholm's early botanisers – a distinguished and observant company – mentions so much as a glimpse of it. This has tempted some sceptics (remembering Gerard's sleight of hand, perhaps) to speculate that it might have been slipped in to the island by a visitor later – and less reverent – than the Augustinians.

Yet the wild peony is in flower for only seven days, and could easily have been missed; and a twelfth-century monkish herb-farm is still the most plausible explanation for its presence, given its importance in the medieval *materia medica.* Gerard provides a prodigious list of ailments for which it was effective – jaundice, 'torments of the belly', falling sickness, and 'the disease of the minde'. Perhaps there was just a tinge of guilt in his long description of its most effective use, for those afflicted by '*Ephialtes* or nightMare, which is as though a heauie burthen were laid vpon them, and they opressed therewith'.[5]

St John's-wort family
Clusiaceae

Perforate St John's-wort, *Hypericum perforatum.* A bright perennial of open woods, rough grass and road-verges, up to two feet tall and topped with golden-yellow flowers from June to September. Its sun-like blooms, usually at their best around late June, and the blood-red juice which exudes from its stems made it one of the key plants burned in the Midsummer Day Fires which flourished across Europe until the end of the last century (and still persist in a few remote corners). The ritual fires, intended symbolically to 'purify' communities and crops, took all manner of forms. There were torchlight processions through the streets, blazing gorse bushes dragged around cattle pens, wheels bound with straw, which were set alight and rolled down steep hills, and huge bonfires on village greens or local hilltops, onto which were thrown magical 'sun-herbs' such as corn marigold and St John's-wort.[1]

Marsh St John's-wort in a New Forest bog.

The festival was to some extent absorbed by Christianity, since Midsummer's Day conveniently coincided with the Feast of St John the Baptist on 24 June – hence the saintly name for what was originally a pagan plant. (The Revd Hilderic Friend quotes another striking blend of paganism and Christianity associated with the Midsummer rites: 'About Hanover ... I have often observed devout Roman Catholics going on the morning of St John's Day to neighbouring sandhills, gathering on the roots of herbs a certain insect (*Coccus polonica*) looking like drops of blood, and thought by them to be created on purpose to keep alive the remembrance of the foul murder of St John the Baptist, and only to be met with on the morning of the day set apart for him by the church.')[2]

But the fires were prehistoric in origin, probably based on sympathetic magic, and were intended both to mimic (and thus strengthen) the power of the sun, and, by allowing the smoke of the burning herbs to waft over the fields, to protect crops and cattle from other, more malevolent summer 'heats': lightning, drought, field-fires.

The 'perforate' in this species' name refers to the translucent dots which speckle the leaves, and which look like tiny holes against the light. They, too, were interpreted 'sympathetically', as a sign that the plant was a remedy for wounds. (Modern herbal medicine, interestingly, unites both beliefs, and employs *Hypericum* poultices and salves for both wounds and burns.) In fact the 'perforations' are tiny resin glands, and are responsible for the aromatic, foxy, not always pleasant smell of the foliage of most members of the family.

Imperforate St John's-wort, *H. maculatum*, is a similar species but lacks the translucent dots, and prefers damp sites in hilly areas. **Wavy St John's-wort,** *H. undulatum*, is an intriguing and attractive species with distinctive crimped leaves, confined to a few marshy places in Devon, Cornwall and west Wales, which is sadly declining. **Trailing St John's-wort,** *H. humifusum*, is a small-leaved scrambler of acid woods and heaths, which smells of lemon as much as fox. **Slender St John's-wort,** *H. pulchrum*, lives up to its Latin name and is the family's most elegant member, delicate in habit and with the pointed flower-buds beautifully tipped with deep red. It is quite widespread on dry soils in open woods and heaths. **Marsh St John's-wort,** *H. elodes*, not only haunts damp heaths and marshy fields, but will often grow in shallow ponds and cattle wallows. Its round, greyish leaves are covered with a dense layer of soft hairs, which catch dew and rainwater. It will grow in large colonies in favoured places, such as the damp, grazed 'lawns' in the New Forest. After showers these St John's-wort swards can look as if they have been covered with glistening gossamer.

Tutsan, *H. androsaemum*, is a bushy plant, with a procession of bright yellow flowers from June till September, which give way to shiny berries, green at first, then tinged with red, and finally purplish-black. The leaves are heart-shaped, up to four inches long, and when dry have an evocative, fugitive scent, reminiscent of cigar boxes and candied fruit. For this reason, as well as their

The delicately bronzed leaves of tutsan, here growing in a limestone pavement, Gait Barrows, Lancashire.

Rose-of-Sharon, a garden plant from the Near East, frequently naturalised.

convenient shape, they were used as bookmarks – especially in Bibles ('Bible leaf' is an obsolete West Country name), perhaps because of the plant's reputation as a benign and healing herb. ('Tutsan' is itself a corruption of the French *toute-saine*, meaning, roughly, 'all-heal'.) Tutsan leaves were still being used as Bible markers in parts of Somerset up to the Second World War.[3]

Tutsan is very much a plant of the west, of damp woods and shady banks. In Devon and Cornwall it is one of the most characteristic plants of hedge-banks, and even grows on some sea-cliffs. But it also occurs scarcely in moist and sheltered places throughout Britain, and is always an intriguing plant to find, lurking deep in a crevice in the limestone pavements below Ingleborough in Yorkshire, or in a sunken lane in the Hampshire Weald. Rather surprisingly, given the popularity of the berries with birds, tutsan cultivated in gardens rarely seems to become established beyond them.

Stinking tutsan, *H. hircinum*, is a garden plant from the Mediterranean naturalised in a few shady places. The 'stink' appears to some people to resemble goats, to others overripe apples. **Rose-of-Sharon**, *H. calycinum*, an evergreen creeping shrub from the Near East with flowers up to three inches across, is widely planted in gardens and in municipal displays in parks and by roadsides. Its aggressive root system has enabled it to naturalise freely and persistently. One feral colony in Mickleham, south London, has been known since 1836. Another, in Wiltshire, stretches for hundreds of yards along the side of the railway line at Norton Bavant.[4]

Lime family *Tiliaceae*

Small-leaved lime, *Tilia cordata* (VN: Pry, Linden tree). The ubiquity of planted common lime (*Tilia × vulgaris* – see below) on streets, avenues and village greens has obscured the fact that its two parents – small-leaved and large-leaved lime – live here in the wild and are amongst the most beautiful and historic of all British native trees.

Small-leaved lime is the more widespread of the two, but is strongly and mysteriously local. It grows in widely separated clusters, chiefly in the South Hams region of Devon, the Mendips and Wye Valley, north Essex around Earl's Colne, west Lincolnshire, the Derbyshire Dales and the Lake District. In most of these areas it survives as coppice – perhaps the most striking underwood there is, with straight, steel-grey poles and small, heart-shaped leaves.

But 6,000 years ago, in the warm 'Atlantic' period, it would have grown as an uncut, rugged, maiden tree, draped with fragrant flowers in July. And, to judge from pollen deposits from prehistoric lime-blossom, it was

The fragrant flowers of small-leaved lime.

once the commonest tree throughout lowland England. But it is a southerly species and below a mean summer temperature of about 20°C ceases to produce fertile fruit. When the climate began to cool around 3000 BC, lime could no longer rely on seed to perpetuate itself and was more or less frozen in the sites it already occupied. As the wildwood was progressively cleared or grazed, lime retreated even further and survived only where soil conditions suited it and it was maintained by coppicing.

Yet it must still have been a comparatively well-known tree as late as the Anglo-Saxon period, to judge by the number of place names which feature it. 'Linde' was Anglo-Saxon for lime, and many parish and wood names preceded by lynd- or lin- refer to the presence, past or current, of *Tilia cordata*. At Lyndhurst and Linwood, both in the New Forest, lime has gone. But it survives in Linwood, Lincolnshire, and in individual woods, for instance, Linsty Hall Wood, Grizedale, and Lynderswood at Black Notley, Essex.[1]

Since then the number of lime woods has shrunk, but the territories of the 'lime province' have remained tenaciously constant. This has much to do with lime's longevity and its remarkable powers of regenerating from broken roots, toppled trees and buried branches. It will even layer itself, rooting along ground-level shoots much like a bramble. In the severe winters of the early 1980s, coppice shoots in Lady Park Wood in the Wye Valley were welded to the ground by the ice, took root and sent up new stems. Lime sometimes survives in hedges which are the 'ghosts' of cleared woods. In the parish of Shelley in Suffolk, a hedge along 600 yards of lane consists almost entirely of overgrown small-leaved lime coppice. A search of early maps showed that the lime strip coincided exactly with the northern edge of Withers Wood, which was cleared in the early nineteenth century, leaving, it seems, just enough to form a hedge.[2]

Coppicing itself can prolong the life of small-leaved lime almost indefinitely. In Tiddesley Wood, Worcester, there is a still-living stool 'about 10 metres across. It is represented as a circle of ... 15-year-old stems, and I regularly take groups of people into it, when walking in the wood. It is rather like being in a small cathedral.'[3]

The distinctive grey-tinged, straight poles of lime coppice.

In Swanton Novers Great Wood, a similar hollow stool measures 27 feet across.[4] But by far the most impressive stool is at Silk Wood, next to the Westonbirt Arboretum in Gloucestershire. It consists of 60 sizeable trees growing in a circle 48 feet in diameter. It used to be assumed that they were unconnected individuals, planted or accidentally grown as a circular grove. But DNA 'fingerprinting' of the trees has shown that they are part of a genetically identical clone, whose parent tree – originally at the centre of the circle – no longer exists. Oliver Rackham and Donald Pigott, one of the world's leading authorities on limes, have estimated the clone's age as at least 2,000 years, by assuming that coppice growth was harvested at 25-year intervals as it grew outwards from the mother tree. But there is now private speculation that it may be more than 6,000 years old – which, if it were to be regarded as a 'single' organism, would make it probably the oldest living thing in Britain, born in the summertime of the limes and outstripping any of the ancient yews.

There are younger but no less remarkable landmark limes in the Lake District. In a scatter of woods – for instance, Linsty Hall Wood, Wash Dub Wood and Hartsop Low Wood – there are huge small-leaved lime trees that look as if they are ancient pollards. They also have something of the look of massive wooden stalagmites, and Dr Pigott, who has investigated them, realised that they were old coppice stools whose roots had been exposed by soil erosion. At some of the sites, pieces of rock more than eight inches in diameter are embedded in the exposed roots up to four feet above present ground level. Comparisons with coppiced limes of known age and calculations of the average rate of soil erosion around the roots has led Dr Pigott to estimate that the limes are probably 1,000 years old and possibly much older.[5]

Lime coppice was used in much the same way as other coppice-wood, as fuel, hop-poles, etc. (An unusual modern use for part of the crop from Collyweston Great Woods National Nature Reserve in Lincolnshire has been to make sticks for the Rutland morris men. Lime wood is well suited for morris-stick thwacking, having a tight grain and not splintering even when hit hard.)

Larger poles (and timber trunks) were stripped for what is called 'bast' – the fibrous layer between bark and greenwood, which was twisted into ropes (and generated another group of lime place names, e.g. Bastwick in Norfolk).

Grinling Gibbons's exuberant carving in lime wood: a detail from the fireplace in the state dining-room at Chatsworth.

Lime wood itself is pale, soft and cuts very cleanly. It has been a favourite with wood-carvers since at least the Middle Ages and was the wood used by Grinling Gibbons (1648–1721) in his elaborate decorative work. The effects that can be achieved with it can be seen, for example, in a Gibbons frieze in St Paul's Church, Covent Garden, where there is a wreath of flowers and fruit – some with stalks only a fraction of an inch wide. They have both a breathtaking intricacy and a kind of Shaker simplicity, with the still-visible chisel marks giving a freshness and depth to the carvings. Horace Walpole, fifty years after Gibbons's death, wrote that 'there is no instance of a man before Gibbons who gave to wood the loose and airy lightness of flowers, and chained together the productions of the elements with a free disorder natural to each species'.

David Esterley, who restored the Gibbons carvings damaged in the 1986 fire at Hampton Court Palace, echoes Walpole and says: 'Fresh limewood foliage carving seems to float, with leaves and petals as light as air.' He also discovered that the carvings were not left exactly as they had been carved by Gibbons, but were roughly finished-off with Dutch rush (see p. 13): 'The moment I rubbed a dried section of this tubular stalk against a piece of limewood it became clear that the key to Gibbons's long-lost technique had been found. The rush left behind on the wood exactly those curious striations which are discernible on Gibbons's carvings.'[6]

In Hallwood Green, Gloucestershire, there are two ancient pollard small-leaved limes that mark the boundary between Dymock and Much Marcle and which were the site of open-air church services up to the 1920s: 'The services were held every Sunday afternoon during the summer months. The vicar ... gave his sermon while standing on an old pair of blacksmith's bellows which at that time stood under the trees on the green.'[7]

Another remarkable group of landmark limes is the avenue at Turville Heath in Buckinghamshire. There are many eighteenth- and nineteenth-century avenues, but these almost invariably consist of common limes. The trees at Turville are unique in being small-leaved limes. They were planted in the 1740s, by William Perry, along what was then the coach road to his house at Turville Park. He had been Lord Lieutenant of Radnorshire, and Donald Pigott believes he may have brought the trees from local woods in Wales. Despite losses to gales, there are still 35 statuesque trees with massive gnarled trunks and sheaves of small leaves. Alas, the gaps have been filled partly with common limes, and more recently with nursery grown *Tilia*, rather than by cuttings from the original trees, which would have kept intact the unique character of this avenue.[8]

Large-leaved lime, *T. platyphyllos*, is a tall and elegant tree, and one of our rarest native hardwoods. Up till a few years ago it was known in any quantities only from a few woods in the Pennines, Wye Valley and Cotswolds (though widely planted elsewhere). Then, in the early 1990s, Francis Rose and colleagues discovered a string of some 16 new sites along the foot of the South Downs, between Hampshire and East Sussex. Most of the trees are in ancient copses or on boundary banks.

The most impressive site, Rook Clift near Reyford, is a remarkable place and one where it is hard to believe that the trees could have gone unnoticed for so long. Even from a lane a quarter of a mile away, the limes look stupendous, their domed, billowing crowns standing clear of all the other trees. There are more than 50 in the wood, a mixture of huge coppice stools, some maybe 1,000 years old with poles that have something of the sinewy grey sheen of tropical hardwoods, a few pollards and some handsome maiden trees, sprung from seed or old stools and now maybe a century or so old, and all bearing bright, large, viridian leaves.

T. platyphyllos is now cropping up in other, similar sites (I have found coppiced stools at the foot of a Chiltern escarpment at Fawley Bottom, Buckinghamshire) and it looks as if large-leaved lime was a component of the original wildwood on the lower levels of the chalk, just as it was on limestone.

Lime or **Common lime**, *T.* × *vulgaris*, is the fertile hybrid between the two native species above and occurs naturally in the wild where both parents grow together, as in woods in the Wye Valley and Derbyshire Dales. But these trees tend to differ slightly from the variety that has been so abundantly planted in town streets, parks and churchyards throughout Britain, which almost certainly originated with stock imported from the continent in the mid-seventeenth century. It is hard to see why this latter form became so popular, apart from its fast growth and tolerance of grotesque lopping and mutilation at the hands of municipal authorities. It sprouts great nests of

'Preaching limes' at Hallwood Green, Gloucestershire.

side shoots at the base and around burrs on the trunk, and the leaves are notorious for the aphid honeydew which in summer rains down on anything beneath them. It has its friends, nonetheless:

'A lime tree was planted to replace a willow tree in the centre of Stratton St Margaret, near Swindon. It was a lucky tree and survived a Second World War aeroplane crash which exploded near the site.'[9]

'Five lime trees were planted on the slope of Corton Denham Hill [Somerset] behind the church when the five bells were rehung after the church was rebuilt in 1869–1870.'[10]

'In Ewell village, Surrey, the "Grove" consists of lime trees planted to commemorate the accession of William III to the throne in 1689, and originally contained 38 trees – one for each year of William's age. There are still some of the originals remaining.'[11]

'We have two lime trees outside our old house in Enfield, side by side, only three yards apart. In 1981 we took a holiday in Sweden, to an old country rectory in Gotland. In the garden our cousin showed us two large lime trees growing side by side and close together and in full view of the windows of the house. When his parents took up residence there (in the early twentieth century), they were newly married, and following an old Swedish custom, they planted two lime trees close together to symbolise their marriage, and gave the trees the same names as themselves – Gustav and Lydia. Does the same tradition exist here, or were the earlier occupants of our house Swedish?'[12]

Perhaps the most romantic lime avenue is at Kentwell Hall in Long Melford, Suffolk, which is recorded as having been planted in the late 1670s. The limes are hybrids, but combine the features of their parents in a more attractive way than usual, and Oliver Rackham believes they may be of English and perhaps even local origin.[13] They seem originally to have been pleached (that is, had their lower side branches joined) as a formal avenue, but they have now grown exceptionally tall and wide-spreading. They also have curious crooks and swellings in their upper branches produced by generations of mistletoe (see p. 243).

All groups of lime trees, of whatever species, are wonderfully fragrant when in full blossom in July. They are also the noisiest of trees at this time, and the roar of bees in them can often be heard 50 yards away. The blossom makes a rich tea, *tilleul*, which was recommended as a mild sedative during the last war. Other bits of lime are used domestically, too: 'We rolled the inner bark or bast of common lime into a tightly packed cigarillo, perhaps two to three mm thick, and smoked it in the ordinary way. The effect was rapid and heady.'[14]

The young leaves make refreshing sandwich fillings. After the aphids have been at work they have been described as tasting like 'honey-coated lettuce leaves'.[15] Even the small round fruits are just about edible and have a curious cocoa-like taste. In the late nineteenth century the French chemist Missa tried to patent a chocolate substitute made from a mixture of ground-up lime flowers and fruit.

Left: statuesque small-leaved limes in an avenue at Turville Heath, Buckinghamshire.
Right: the billowing crowns of large-leaved lime at a recently discovered site: Rook Clift, West Sussex.

Plants, places and names

Plants often derive their names from close association with a particular place. It may be the place where the species was first discovered, or first named, or to which it is exclusively confined. Oxford ragwort grows all over Britain now, but it began its spread from the Oxford Botanic Garden. Bath asparagus was sold in Bath, as well as growing chiefly in woods within a 20-mile radius of the city. Plymouth pear has been known in hedges near Plymouth since 1870 but was found near Truro in 1989. Tunbridge filmy-fern still grows near Tunbridge Wells, though its distribution in the British Isles is mainly western. The endemic Arran service-tree and Lundy cabbage are still confined to the islands commemorated in their names. Cheddar pink, though it is quite common further south in Europe, in Britain grows wild only in the Cheddar Gorge. There is even a 'Rottingdean sea-lavender', a plant from Sicily naturalised on the cliffs of the Sussex seaside town.

Conversely, places are frequently named after the plants that grow there: odd and noticeable plants (such as box); abundant plants (such as beech and fern); economically useful plants (such as cress, as in Kersey, Suffolk); and old, isolated and conspicuous oaks, thorns and pear-trees.

But it is dangerous to assume that a place-name component which resembles the name of a plant necessarily derives from one. Buckhurst, Essex, is a 'beech-wooded hill', but Buckingham is 'river-bend land held by Bucca's people'. Holmstone Beach, Kent, is named from the Old English *holm*, meaning holly; but Holme in Huntingdonshire is from the Old Norse *holm*, meaning 'island' or 'raised ground in a marsh'. It is always essential to trace the name back step by step to its first use and spelling to be sure of its origins and meaning.[1]

Plants also appear frequently in field and street names. The latter are only occasionally named after genuine botanical residents; the former almost invariably were. Some straightforward Oxfordshire field names are Primrose Shaw, Thistle Field, Broom Hill, Hazeley Mead, Blewbottle Coppice (after an old name for the cornflower) and Gorsty Mead; more obscure are Chesscroft ('chess' is an old Oxfordshire name for the grass *Bromus secalinus*, which grows amongst wheat) and Guldfurlong, named after another arable weed – the corn marigold.[2]

Caroline Giddens, of the Exmoor Natural History Society, drawing in part on N. V. Allen's work, has compiled a list of botanically influenced Exmoor place names. Here is her selection of names based on shrubs and smaller plants:

*'Brompton Regis in the Domesday Book is Brunetone and Gerard said "the country hereabout is strewn with Broom". I should think this meant gorse (*Ulex *spp.) rather than broom (*Cytisus scoparius*). We also have*

Above: Tunbridge filmy-fern – first named at Tunbridge Wells, Kent, but also found in several areas of western Britain.
Top right: Plymouth pear, confined to the outskirts of the city and to a few hedges over the Cornish border.
Bottom right: Cheddar pink. In Britain it is found only on a few cliffs in the Cheddar Gorge, Somerset.

Broomstreet. Gorse here is locally known as furze, which gave rise to Furzebury Brake and Furzehill Common, which was Furshulle in 1270. Most hill farms have their furze brake, i.e. gorse-covered hill.

The blackberries give us Bramblecombe and Brimblecombe, and Brendon comes from Bramble Hill.

Ferny Ball is a bracken-covered hill to this day, and Ivystone Rock is an ivy-covered promontory into the Bristol Channel.

Billbrook, a village with the longest ford in England, derives its name from bilders, an old name for watercress.

It is thought that an area named Cowlings derives from ling fields for cattle – the area was heather-covered until enclosed in the nineteenth century.

Cuckolds Combe is interesting, cuckold being an old name for burdock, and Riscombe means rushy combe, from the Old English rysc, meaning rushes.

Nettlecombe is obvious, as is Snowdrop Valley [see p. 421], which is still visited in spring for its carpets of white flowers.'[3]

Mallow family *Malvaceae*

Common mallow, *Malva sylvestris* (VN: Pick-cheeses (for the seeds); Bread and cheese). This is a common and widespread perennial of road-verges, footpaths, farmyards, beaches and waste places of all kinds. It can become a rather straggling bush up to four feet tall (and with a habit of picking up dust on its hairy leaves), but the flowers are beautiful – deep pink with darker stripes, as sculptured as bone china, and often staying in bloom right through the autumn. (In 1993 I saw a plant whose petals had an almost pure white ground colour on a road-verge in Eriswell, Suffolk.)

The leaves are edible, though rather mucilaginous, and quantities of the pollen have been found in excavations at the Roman fort at Bearsden, just north of Glasgow. The leaves, flowers and seeds were all eaten by the Romans, both for food and as a kind of preventative medicine. (Pliny said that a daily dose would make you immune to all diseases.) As the pollen has been found only in the Roman levels of the excavation, it is possible that it was deliberately cultivated by the legion.[1]

British children still nibble the small, round seeds, which are widely known as 'cheeses' because of their shape. They have a bland, slightly nutty taste. In Norfolk they are called 'pick-cheeses'.[2]

Musk mallow, *M. moschata*, is a fairly widespread perennial of dry places. Chalk pastures, roadsides, churchyards and old quarries are typical sites. The poppy-like flowers are usually pink, though white ones occur, and have become popular in modern cottage gardens. By contrast with the crisp flowers of common mallow, musk mallow blooms have a soft-edged, accommodating form that suits these informal groupings. The flowers have a musky scent (and the leaves, too, to a certain extent), especially when brought indoors.

Tree-mallow, *Lavatera arborea*, is a tall, shrub-sized mallow of western shorelines, growing at the base of cliffs and on the rough ground just inland from the beach. In Devon and Cornwall especially, specimens are often transplanted into gardens – from which they sometimes escape again.

Marsh-mallow, *Althaea officinalis*, is the plant whose roots were originally used to make the famous sweet. They were dug up from the marshes of the Thames estuary, and contain enough starch, sugars, oils and gelatinous matter to be turned into jelly simply by infusing them in water.

In France the dried roots (*hochets de guimauves*) are still sold in chemists as teethers. They have the advantage of growing ready-shaped for the job – they resemble thin pale carrots – and can be sucked like a dummy. They are hard and fibrous enough for a baby to chew on, but slowly soften as their mucilage is released (which, as a bonus, is calming to the stomach).

Now, in Britain at least, this is a declining species, its velvety, grey leaves and soft pink flowers a surprise to find along the brackish creeks and muddy paths of the south and east coast marshlands.

Common mallow, frequent on roadsides.

Marsh-mallow, whose roots were once used to make the well-known sweet.

Rough marsh-mallow, *A. hirsuta*, is an attractive but exceptionally rare annual, now known only from four sites in southern England, all warm, south-facing banks on chalky soils. 'Hairy' mallow is not an arable weed, and almost certainly not an 'alien' (though its European distribution is definitely biased towards the south), but seems to need a habitat which is rather scarce this far north: warm, open and disturbed semi-natural grassland. English Nature's Species Recovery Programme is helping to provide it with just such conditions.[3]

Hollyhock, *Alcea rosea*, is a garden plant of uncertain parentage (though probably west Asian) which was introduced to this country by the Crusaders. Now even the most blowsy and flamboyant varieties can be seen growing ostentatiously in the wild, either as garden throw-outs or as self-seeders. They crop up along railway banks and roadsides, at the edges of commons and car parks, and even on sunken boats, providing one observer with a small landscape masque rather suitable for such a statuesque plant: 'At Pulborough, Sussex, I have spent a fascinating lunchtime at a curved creek of the local, tidal river, watching an old grounded boat full of blooming hollyhocks rise with the tide, only to sink out of sight again in front of my eyes, and then a while later, rise out of the water again – hollyhocks and all.'[4]

'When my mum was little she used to make hollyhock fairies by turning the flower upside down, cutting the stem off and pushing a dry poppy seed head in where the sepals meet the stem, using a pin to prick a face into the poppy head and pushing short pieces of dried grass up between the sepals and the petals.'[5]

Sundew family *Droseraceae*

Round-leaved sundew, *Drosera rotundifolia*. The sundews are our main family of insectivorous plants. Their fleshy leaves are covered with red-tinged hairs, each tipped with a drop of sticky, translucent 'dew'. Small insects are sometimes trapped by these, and slowly digested by enzymes secreted by glands in the centre of the leaf. Sundews don't rely wholly on insect protein, but they are plants confined to acid peat-bogs and damp moorland, where nutrients in the ground can be in very short supply.

Yet, where conditions suit them, they can grow in quite large colonies and a sundew 'meadow' against the setting sun is a dramatic spectacle, the rufous leaves and beads of dew catching and sparkling in the light. Early herbalists were struck by the fact that the dew persisted even through the hottest sun (though they did not seem aware of its insect-catching properties), and believed that by sipping 'the distilled water thereof … the naturall and liuely heate in mens bodies is preserued and cherished'.[1] This was the origin of a whole range of potions based on sundew's syrupy, seductive but highly functional secretions. In the Lancashire peat-digging areas the plant was called 'youth grass' and was harvested as a kind of catch-crop. Here and throughout Europe, it was mixed with a variety of spices to make a liquor called *Ros Solis*, which was regarded as a source of youthful looks and strength, virility and longevity.

Inevitably, sundew was also believed to be a love-charm, a reputation enhanced when its mysterious power to lure and entrap other creatures was eventually realised.

One of our native insect-eating plants: oblong-leaved sundew, Cranesmoor, New Forest.

This belief persisted in the Isle of Man, even amongst teenagers who may have been unaware of the roots of what they were doing. In the 'year of love and peace', appropriately, they added another twist to a tradition that had been evolving for at least four centuries: 'In 1968, a labelled dish of the plant in the Manx Museum was gradually pilfered by Douglas secondary school children. The idea is that you slip the plant into the pocket or clothing of the person you want to attract.'[2]

Great sundew, *D. longifolia*, and **oblong-leaved sundew**, *D. intermedia*, are scarcer and more local species. The spectacular **pitcherplant**, *Sarracenia purpurea*, is from a related family of insectivorous species and is native to eastern North America. It is well naturalised in central Ireland (outside the scope of this Flora) after being deliberately planted in Roscommon in 1906. But a few more transient patches have been found here and there in boggy sites in England.

Rock-rose family *Cistaceae*

Common rock-rose, *Helianthemum nummularium*. This low, often creeping, evergreen shrublet with cheerful sulphur-yellow flowers from June to September, is virtually confined to calcareous sites – downs, cliffs and rocky hillsides, especially where the grass is short. On old chalk grassland in Wiltshire it has a special liking for anthills and fairy rings.[1] In some places, on limestone pastures in Wharfedale, Yorkshire, for example, rock-rose can grow so densely that on warm, breezy days, its pollen scents the air for hundreds of yards around.

White rock-rose, *H. appeninum*. This elegant species, so characteristic of the limestone districts of southern Europe, occurs in a couple of areas in south-west England. **Hoary rock-rose**, *H. canum*, is plentiful in a few places on Carboniferous limestone in the north and west, with its major British stronghold on the Great Orme.[2]

Violet family *Violaceae*

The violets are a variable and promiscuous family, apt to throw up all kinds of sports and hybrids. For the first fifty years of this century botanists – fonder in those days of 'splitting' than 'lumping' species – responded in kind, and dutifully logged every slight variation. In the Revd Keble Martin's loving and meticulous account of the flora of Devon (1939) – a county abundant with violets – there are some 40 species, crosses, forms, colour oddities, eccentric shapes and local varieties, including one delectable type found only in the villages of Marldon, Berry Pomeroy and Dartington, *Viola odorata* var. *variegata*: 'This colour-variety may be distinguished from var. *dumetorum* in having its white petals irregularly splashed and streaked with violet, whereas in *dumetorum* the reverse of the upper petals is coloured to a greater or less degree with reddish purple.'[1] By 1948, when the Revd H. J. Riddelsdell's copious Gloucestershire Flora appeared, the roll-call of violet types had risen to 50, including no fewer than 18 varieties of the hairy violet, *V. hirta*, and an intriguing form *leucantha* of the early dog-violet, *V. reichenbachiana*, from Pighole near Tidenham: 'A considerable colony of pure white flowers, many spurless; and then looking very much like the flowers of *Oxalis acetosella* [wood-sorrel]!'[2]

Clive Stace's *New Flora of the British Isles* (1993) prunes this wildly proliferating catalogue (and maybe a little of its glamour too) down to a mere 28 species, sub-species and hybrids. The following are some of the best known and most interesting:

Common dog-violet, *V. riviniana*, is the commonest and most widespread species, flowering in deciduous woods, hedge-banks and old pastures from April to June. The flowers are unscented (hence 'dog' violet, to distinguish it from the scented 'sweet' species), variable in colour and often rather stumpy. It was almost certainly this species that John Clare wrote of in his little-known poem 'Holywell'. Most violet poetry is little more than purple poesy, and it is a relief to read the sympathetic and unornamented clarity of Clare's vision:

> *And just to say that spring was come,*
> *The violet left its woodland home,*
> *And, hermit-like, from storms and wind*
> *Sought the best shelter it could find,*
> *'Neath long grass banks, with feeble flowers*
> *Peeping faintly purple flowers.*[3]

Early dog-violet, *V. reichenbachiana*, is a slightly more petite species, with a preference for ancient woodlands, not occurring much outside England and normally in bloom by March. Both these two species of woodland violet can respond rampantly when light is allowed into a wood. In Hayley Wood in Cambridgeshire a forty-fold increase in the number of violet flowers has been recorded after coppicing.[4]

Sweet violet, *V. odorata*, prefers rather more open habitats, in hedges and scrub as well as woods. It is a

Common rock-rose haunts sunny chalk and limestone sites. Cheddar cliffs, Somerset.

Common dog-violet, Lincolnshire.

native species, although many colonies are obvious escapes from cultivation, especially those in churchyards and on village greens and banks. It is easily recognised by its large, hairy, pale green, heart-shaped leaves and tufted habit. The flowers may be deep bluish, purple, white, even cream; but the rich-red form (Riddelsdell's var. *rubropurpurea*) is rare, and was understandably the favourite plant of one Bedfordshire woman: 'My mother's most specific botanical memory is of a red violet that used to grow inside a gateway of a grass field between Rushden and Newton Bromswold. I suspect it was a colour variant of the sweet violet.'[5]

The fragrance of the *V. odorata* flowers can be very strong and they have been used in the making of perfume as far back as Classical Greece. (The scent can seem to be curiously fleeting, though this is a phenomenon of our sense of smell, not the flower. One of the chemicals contributing to the scent of violets is ionine, which has the ability temporarily to deaden the smell receptors that detect it.) In medieval Britain sweet violets were one of the strewing herbs used as early household deodorants. They also had a role in herbal medicine, especially for insomnia, headache and depression. And they led John Gerard to say some wise words about the more subtle, psychosomatic healing effects of plants. He wrote that

'the blacke or purple Violets, or March Violets of the Garden, ... haue a great prerogatiue aboue others, not onely bicause the minde conceiueth a certaine pleasure and recreation by smelling and handling of these most odoriferous flowers, but also for that very many by these Violets receiue ornament and comely grace: for there be made of them Garlands for the heade, nosegaies and poesies, which are delightfull to looke on and pleasant to smell to, speaking nothing of their appropriate vertues; yea Gardens themselues receiue by these the greatest ornament of all, chiefest beautie, and most gallant grace; and the recreation of the minde which is taken heereby, cannot be but verie good and honest: for they admonish & stir vp a man to that which is comely & honest.'[6]

Heath dog-violet, *V. canina*, is normally pale blue, with a cream-coloured spur, and quite common on grassy heaths and fens on sandy, acid soils. **Hairy violet**, *V. hirta*, is rather similar to the early dog-violet, but covered with

Teesdale violet, a rarity restricted to short-turfed limestone pastures in the north.

short hairs, and preferring open habitats on calcareous soils. The flowers are normally a conventional violet colour, but can vary from white to slate-grey.

Teesdale violet, *V. rupestris*, is a famous local rarity from short-turfed limestone pastures in the northern Pennines and Cumbria. Much of the population that grows on the crumbly 'sugar limestone' of Upper Teesdale in Durham was destroyed when the Cow Green reservoir was built on the site at the end of the 1960s. It is a dwarf, tufted plant, with roundish flowers in shades of white, pale blue or reddish violet. **Fen violet**, *V. persicifolia*, is an even rarer species, confined recently to just two fenland sites in Cambridgeshire, Woodwalton Fen and Wicken Fen, both now National Nature Reserves. To judge from its previous much wider distribution and its favoured haunts in Ireland, the fen violet is a short-lived perennial with very fussy habitat needs: short, damp calcareous turf which is subject to winter flooding and periodically grazed or disturbed in some way to help distribute its seed. The blooms are perhaps the most beautiful of all the native violets, the mottled bluish-white flowers often suffused with a mother-of-pearl sheen.

Field and wild pansies, drawn by Caroline May in 1834 and 1842, showing the variation that was exploited in cultivation.

Pansies are as muddled – and muddling – a group as the violets proper. The most frequent species, the little **field pansy**, *V. arvensis*, is becoming increasingly common as an annual weed of arable and disturbed ground. On set-aside land and newly made road-verges, its long-stalked flowers can be found winding amongst the grasses in almost any month of the year. They are variable in size and usually pale yellow in colour, but are often suffused or blotched with purple, especially on the upper 'ears' or in streaks in the 'eye'. (Pansies, more than any other native species, have blooms which are irresistibly reminiscent of 'faces' – as many of the obsolete names testify, e.g. three-faces-under-a-hood, cat's face.) Occasionally the whole flower is violet. These multicoloured forms can be difficult to tell from the **wild pansy**, *V. tricolor* (VN: Heartsease) – especially as this hybridises with *V. arvensis*, and throws up plenty of colour variants of its own. As the Latin name indicates, there are often three different shades on each bloom – most commonly violet, yellow and blue, but also reddish-purple, rusty-red, white or an almost blackish, velvety blue. They can also be pure yellow, or pure purple. One mid-May, on Barnham Cross Common in Norfolk, I found sheets of heartsease of both colours – and of most possible combinations between – growing on a patch of sandy soil that had been scorched by a fire the year before. These may have been the subspecies *curtisii*, which normally grows on coastal sand-dunes but also on the inland sands of East Anglia's Breckland (especially on firebreaks). A few miles to the east, a householder found other forms taking more unconventional advantage of environmental change: 'In Harleston, south Norfolk, during the recession, wherever there were For Sale notices and run-down lawns, there were purple heartsease in hundreds. They love the sandy soils here.'[7] *V. tricolor* is a species which can grow both as an annual and as a short-lived perennial, and colonise grassland on acid soils as well as cultivated ground. In Derbyshire and central Wales it begins to overlap with the southern outposts of the mountain pansy, and yet more intriguing liaisons begin ...

Mountain pansies from Perthshire, showing the extremes of colour variation. Intermediates and blooms with blotched 'eyes' and 'ears' are common.

The **mountain pansy**, *V. lutea*, is a squat, vivacious plant, and its exquisite flowers, normally pale yellow, are held on stiff stalks that seem to spring straight from the turf. On hill pastures you can sometimes see acres of them shivering in the wind. In Shropshire, for instance, 'in the 1940s, it was still possible to walk from Ratlinghope via Squilver and Shelve to Bromlow Callow through field after field washed with mountain pansies'.[8] Here, as in many other places, they have declined because of agricultural pressure. But, in general, the mountain pansy is a resilient plant. It will tolerate quite heavy grazing, and enjoys soils with a high mineral content – even lead spoil-tips. Where it occurs on limestone soils (as in the Derbyshire and Yorkshire Dales) it is usually in areas where rainfall has leached out the calcium carbonate and left other minerals behind.[9]

Like other *Viola* species it is highly variable in colour and form. Many of the Scottish colonies are var. *amoena*, which is entirely purple save for some yellow streaking around the 'eye'. A magnificent form of this – deep purple and an inch across – grows on Ben Lawers on Tayside. Forms with purplish ears grow in high Teesdale hay-meadows, often mixed with *V. tricolor*. In grasslands on the banks of the River Tyne in Northumberland, the two species unquestionably cross (the seeds of the *V. lutea* travelling down-river to meet the more lowland *V. tricolor*) and form spectacularly variegated hybrid swarms: 'These are our special glory, growing in sheets in fields and clearings near the river. Colours vary, in every possible combination of blue, purple, white and yellow.'[10]

It is the tendency of the pansy tribe to be naturally 'sportive' that spurred the early cultivators into action in the mid-nineteenth century. The first garden varieties were raised simply by progressively selecting the most interesting chance seedlings from the various forms and crosses of *V. tricolor* and *V. arvensis*. Then William Thompson, Lord Gambier's gardener at Iver in Buckinghamshire, added strains of *V. tricolor* from Holland to the breeding stock, and produced the first all-blue variety. There followed the famous 'Beauty of Iver', which had a broad 'face' of pure yellow, encircled by an edge of sky-blue.

Meanwhile James Grieve of Edinburgh was breeding violas, using as his starting point our native mountain pansies and the pert, long-stalked, blue- or white-flowered **horned pansy**, *V. cornuta*, from the Pyrenees (which is naturalised in some places in Britain). The road to the vast array of modern garden pansies had been opened.[11] Conventionally these are now lumped together under the scientific tag of *Viola × wittrockiana*, which is believed to contain genes from all our native species. It frequently self-seeds or escapes into open and waste places, and, it hardly needs to be added, is already back-crossing with its parents.

Tamarisk family *Tamaricaceae*

Tamarisk, *Tamarix gallica*. A small tree, with sprays of soft pink flowers from June to October. It has been much planted in seaside gardens, as its feathery, scale-like foliage is almost immune to the dehydrating effects of salt sea winds. It was originally introduced from the Mediterranean before the sixteenth century, and it has more recently been planted out to stabilise shingle banks, notably at Westward Ho, Devon, Chesil Beach in Dorset, and Shoreham, Sussex.[1] So much has been planted around Hastings that the whole town is sometimes called Tamarisk Town, and boasts an alleyway by the name of Tamarisk Steps.[2]

In many of these areas it has become naturalised, looking very much at home as its whippy branches bend and gnarl with the prevailing wind. In Cornwall these are sometimes used by fishermen in making lobster-pots.

White bryony family *Cucurbitaceae*

White bryony, *Bryonia dioica* (VN: Mandrake, Wild vine). White bryony is the only native British member of the gourd or cucumber family, and in summer its greenish-white flowers, five-lobed leaves and coiled tendrils are very reminiscent of the greenhouse vegetable's. But there, any similarities end. In winter, bryony bears strings of green berries which turn a brilliant orange-red, and which – like the rest of the plant – are dangerously poisonous.

The most toxic part are the roots, which are white, succulent and often as much as six inches thick, with an acrid and bitter taste. Despite this, cattle are sometimes fatally attracted to them. A notorious case occurred in the 1960s, during the laying of a pipeline across pasture land. The operation entailed removing some 60 yards of hedgerow in which bryony had grown luxuriantly – and innocuously – for many years. Neither the farmer nor the

Tamarisk, from the Mediterranean, is frequently planted as hedging by the south coast.

workmen recognised the roots, nor realised they were poisonous. Forty milking cows were turned into the field the morning after the work had been finished. By the afternoon four were dead, and post-mortems showed that each had eaten more than 4 lb of root. Curiously, as has been reported with other cases of bryony poisoning, two other animals which had eaten roots but survived developed a craving for the plant, 'and during the following summer, searched the hedgerows for it and ate leaves, stem and flowers, whenever they were available. This caused attacks of acute indigestion and diarrhoea with almost complete, but temporary cessation of milk secretion.'[1]

In rural France, there was a popular belief that bryony root would stop or slow down milk production in humans, too, and there used to be frequent cases of poisoning amongst weaning mothers. In this country the uses of the plant in folk-medicine were generally more cautious. Very small portions of the root were taken as a purgative or, distilled, as an external treatment for sunburn, boils, whitlows and other skin eruptions.[2]

White bryony painted in an almost medieval arrangement by one of the Cliffords of Frampton in 1846.

White bryony climbing through a churchyard yew, Oxfordshire.

The most bizarre use, which probably continued until the start of the eighteenth century, was in the construction of fake mandrake roots. Mandrake (*Mandragora autumnalis*) is a nightshade, and, like other members of the family (see p. 300), was anciently employed as a pain-killer and narcotic. The roots were also believed to have powerful magical and aphrodisiac properties (and, incidentally, to shriek when they were pulled out of the ground) because they occasionally grew in the rough shape of a human figure. Women would wear them round their necks or waists to help them conceive.

Since true mandrake is a native of the Mediterranean and hard to cultivate in Britain, the market was wide open for counterfeiters, who would carve out humanoid figures from any convenient root (bryony was the favourite), sometimes planting grass seeds in the root to grow into 'hair' (echoing the hair-roots of the true mandrake). John Baptista Porta described exactly how it was done by 'couzeners and conycatchers': 'You must get a great root of bryony, or wild nep, and with a sharp instrument engrave in it a man or a woman, giving either of them their genitories: then make holes with a puncheon into those places where the hairs are wont to grow, and put into those holes millet, or some other such thing which may shoot out his roots like the hairs of one's head. And when you have digged a little pit for it in the ground, you must let it lie there, until such time as it shall be covered with a bark, and the roots also be shot forth.'[3]

For all its dangerous and dubious history, white bryony is one of the most attractive ornaments of hedgerows and wood margins in the south and east of England, especially in deep winter, when the berries have been glazed by frost: 'I lately indulged a long held desire: to drape a vine of bryony berries round my neck. The flaming beads had always held this temptation for me, but, of course, as a child I was forbidden to touch poisonous berries.'[4]

Other members of the gourd family – melons, cucumbers, marrows, pumpkins and ornamental gourds – can spring up as casuals where vegetable refuse is tipped.

Willow family *Salicaceae*

Black-poplar, *Populus nigra* (VN: Water poplar). On late afternoons in March, especially when there is a patchy sun glinting from the west, parts of the Vale of Aylesbury in Buckinghamshire are suffused with an exotic orange glow. All over the flood-plain, by dykes and lanes and thin streams, rows of craggy pollards begin to shine, as if they have been coated in amber. Closer to, the sprays of twigs seem kaleidoscopic. They have ochre bark, ginger-shellacked buds, and the germs of what will soon be voluptuous crimson catkins. This spectacular display is the largest concentration of our grandest native tree, the black-poplar, in all its spring finery, and there is not another treescape like it in Britain.

It is astonishing that such a conspicuous species should have passed so thoroughly out of common knowledge. But up till the mid-1970s, when the distinguished botanist Edgar Milne-Redhead began to study it, the black-poplar was overlooked by the public and regarded

Black-poplars in March, Vale of Aylesbury, Buckinghamshire.

by most botanists as indistinguishable from nursery-bred hybrid poplars. Even the definitive 1962 *Atlas of the British Flora* lumped *P. nigra* var. *betulifolia*, as our native race is properly styled, along with all the various 'Italian blacks' so beloved of municipal authorities. In the early stages of Milne-Redhead's survey it looked as if there might be fewer than 1,000 trees surviving. Twenty years later some estimates put the national population at between 2,000 and 3,000. But my own estimate would be double that figure, especially as the Aylesbury Vale population has never been properly censused and almost certainly exceeds that of all the rest of Britain put together. The tree's contribution to local landscapes is also beginning to be appreciated, as is what is proving to be a fascinatingly complicated social and ecological history.

Once you have an eye for them, it is hard to believe that mature black-poplars could ever have been mistaken for any of their characterless hybrids. They are distinctive, not just in their spring flush, but at all times of the year. They have thick fissured trunks, covered with massive bosses and burrs, grow to over 100 feet if uncut, and often develop a pronounced lean in middle age. The branches turn downwards at their ends, often touching the ground, then sweep up again into sheaves of twigs, as if they have been caught by a gust of wind. The catkins are followed by masses of shiny, tremulous, beech-shaped leaves. Black-poplars appear in several of John Constable's landscapes of the Stour Valley (still a good place for the tree), though art historians have repeatedly misidentified them as willows or elms. One with a typically deep fork and a lost limb is in the several views that the painter made of the river by Flatford Mill; another vast tree lours in the background of *The Hay Wain* (1821) – which was itself probably planked with black-poplar wood. The lightweight, springy timber was used for brake-blocks, clogs, and even a clutch of arrows which was discovered aboard the Elizabethan galleon, the *Mary Rose*. It is a very heat- and fire-resistant wood, and was consequently also popular for floor-boards, especially in oast-houses. The large forked trunks grow naturally into the shape required for cruck-framed buildings. In Frampton on Severn, Gloucestershire, what looks like a black-poplar cruck survives over the hearth of an Elizabethan house not far from an exceptionally tall and gracious tree on the village green.[1] A Shropshire family told Edgar Milne-Redhead that the top floor of their country house was made of black-poplar wood so that the staff who lived there would be less likely to set fire to the house with their candles and oil lamps!

Native black-poplars develop pronounced burrs and bosses on their fissured trunks.

The highly distinctive appearance of black-poplars meant that they were also employed as landmark trees. One ancient, weatherbeaten tree (*c.* 200 years old), in the Bourne Gutter near Berkhamsted, marks the intersection of parish, manor and county boundaries.

The decline of the black-poplar, both on the ground and in our culture, says much about changes in the landscape over the past few centuries. In prehistoric times, it was a tree of winter-flooded riverine woods (a virtually extinct habitat itself in Britain now), growing with willow, downy birch, alder, ash and elm. It would have survived the clearing of these forests as a tree of riversides and wet meadows, but not the wholesale drainage of the landscape. Like holly and yew, black-poplar is dioecious, and the red male and green female catkins grow on separate trees. To regenerate, male and female trees need to grow moderately close to each other, and the fertilised seed needs to fall on mud which is still damp in June, and which remains damp and bare during the seedlings' first

Black-poplars figure frequently in Constable's paintings, as in this study of the river near Flatford Mill. The Stour Valley in Suffolk is still a good area for the tree.

critical months into the autumn. Such conditions virtually vanished once serious agricultural drainage works began in the seventeenth century. The surviving trees became marooned, ghosts of a wilder and wetter landscape, and from then on could regenerate only when they were deliberately struck from cuttings or 'truncheons' (which were mostly taken from male trees, because of the vast quantities of fluff produced by female catkins). The new trees were planted out as boundary markers on low-lying and often flooded grazing-land, and were usually pollarded to make them more stable and to provide a crop of wood for bean-sticks, thatching spars and fruit baskets. In the last hundred years many of the populations have become even more embattled. Increasingly they are being replaced by more upright and faster-growing hybrids. Elsewhere the policy of regenerating the tree by cuttings is producing increasingly vulnerable cloned colonies.

In the first 20 years of his still ongoing survey, Edgar Milne-Redhead discovered only one natural seedling, in a ditch in Gedgrave, Suffolk; and for a while only one place was known where male and female trees grew close enough together – and far enough away from hybrid trees – to ensure 'pure' pollination, at Hallwood Farm Marl Pit in Cheshire. In 1979, seedlings from these trees were planted out at Sturminster Newton in Dorset. 'They are well spaced on the banks of the River Stour and the mill stream and should grow into fine specimen trees. [Unlike in cloned colonies] considerable variation is already noticeable in the angle of branching ... the shape of the leaves and the colour of the leaf stalks, some being green and others red.'[2]

Since 1992 what look like relict wild populations have been discovered on an island in the Great Ouse at Fenlake near Bedford and at a site by the River Exe within the city bounds of Exeter.[3] And along the Fairham Brook at Widmerpool in Nottinghamshire there is a collection of unpollarded trees, including six large females, which 'have every appearance of being a native grouping'.[4]

In all these places – and others where there are no females, but the trees are unmanaged – the black-poplar is proving that it has more than one stratagem for reproduction. Fallen trunks will often strike roots, if they land on moist ground, and send up new vertical shoots. So, less certainly, will green boughs and twigs which are broken off and washed into muddy river-banks. In Dorset 'there are two black poplars growing in my home village of Hazelbury Bryan, which started life as stakes for a chicken run and which accidentally grew'.[5] And in a few places (e.g. at Binham in Norfolk) I have seen new trees rising from what are in effect suckers, where a tree has been destroyed but its root system survives. These tangles of regenerating root plates and branches (plus occasional clumps of seedlings) must have been what young black-poplars looked like in their ancestral flood-plain forests.

The survey has also revealed the great variety that still exists in the species and in the range of places where it survives. One of the most northerly individuals, the Sunderland Point Tree in Lancashire, is said to have come from America on a ship trading with Lancaster and to have been planted by the skipper. (This may be an example of reintroduction, since European black-poplars were introduced to New England by early settlers and became so well established that American botanists described them as native early in the nineteenth century.)[6] The genuine North American black-poplars are known as 'cottonwoods' – a name that is echoed for black-poplars of an unspecified origin further inland in Lancashire: 'The black poplar is known locally as the Cottontree on account of the way its seed pods look like cotton seeds on the ground. The village is named Cottontree and the inn, which is much older than the village and stands on an old drove road by the tree, is called the Cottontree Inn.'[7]

Further south, the tree is frequent in river valleys

all along the Wales–England border.[8] In Cheshire, near Alderley Edge, I have seen tall, narrow roadside trees whose bark, for once, merits the description 'black'. (No really adequate explanation of this epithet has yet been given.)

In Aston on Clun in Shropshire is probably the best-known black-poplar in Britain, which every 29 May is dressed with multicoloured flags and bunting. The flags remain on the tree until the following year, when they are taken down and replaced with new ones. The earliest records of the ceremony date from 1786, and it is believed to commemorate the wedding of squire John Marston to Mary Carter. Mary Carter is said to have so liked the flags that adorned the tree during her wedding that she gave a sum of money for the festivities to be repeated annually. (The wedding is now portrayed in a pageant which forms part of the ceremonials surrounding the replacement of the flags.)[9] But the date of the flag-dressing – Oak Apple Day, which in 1660 Charles II declared a public holiday 'for the dressing of trees' (see p. 76) – suggests that this 'Arbor Day' ceremony may have become attached to an already existing festival. Perhaps the wedding of John Marston and Mary Carter was deliberately arranged to coincide with the late May fertility rites – which long preceded King Charles's Restoration. The ancient tree was blown down in a gale in September 1995, but a cutting taken from it some years before was quickly re-established on the same site.

On the Welsh side of the border there are several fine trees, including one 20 feet in girth and 80 feet high a mile north of Ruthin, Clwyd, and the landmark tree near the bridge in Newtown, Powys.[10] In Gloucestershire (one of the few counties where the tree was distinguished in an early Flora)[11] the trees were meticulously surveyed under the direction of Sonia Holland, and in 1992 numbered 355 (264 males and 91 females).[12] Herefordshire almost certainly has as many, including the remarkable cluster of 80 pollards on Castlemorton Common, which are still harvested for use in the village. There are also vast and majestic trees close to the churches at Blakemere and Hollybush.

Norfolk's trees were surveyed in 1992, and 54 mature specimens were found, mostly in river valleys and farm hedgerows, though there is a fine female on the village green at Old Buckenham.[13] Suffolk has some hundreds of surviving trees, and (perhaps because Edgar Milne-Redhead had his roots in the county) seems already to be taking the tree back into parish life:

'A female tree stands by the River Lark in "The Ebney Gardens", Bury St Edmunds. Another notable female grows in The Street at Dalham ... A facet of the black poplar in Suffolk is its alleged ability to foretell approaching rain. It is possible that this originated from the highly mobile leaves rustling in the breeze [cf. aspen] and sounding very much like the downpour that might be on the way.'[14]

'We have several large old black poplars in the parish of Marlesford. There are two huge specimens by the Ford. We have successfully transplanted cuttings which now seem to be well established. The two trees just to the east of Milestone Farm have grown side by side from the fallen trunk of an earlier tree.'[15]

'There is one on the village green at Bardwell. It is growing on a slight mound at one end of the Green known locally as the Stocks. Cuttings have been taken of it, and one planted next to it, and one in the churchyard.'[16]

There is a hollow black-poplar outside Honeypot Hall, Wattisfield. ('When we lived there our cat had four kittens in it.')[17] And a stately, unusually straight tree by Butley Church has become the archetypal black-poplar silhouette.

But the Vale of Aylesbury remains the tree's classic location. In the 2-km square (tetrad) containing the villages of Long Marston and Astrope there are more than 270 black-poplars visible from roads and footpaths alone. Amongst them there are specimens in just about every conceivable form: maidens, short and tall pollards, even a couple of two-storey pollards, cut once at about 10 feet, and then the leading shoot cropped again 10 feet above that. Trees are regenerating from fallen trunks and from the tips of low branches which have become buried in mud. I have even found one windthrown pollard sending out a new shoot from the *underside* of the root plate. The colour, fissuring and degree of burring on the trunks of the Vale's colonies are enormously varied, suggesting that they have originated from many sources. These days the trees are pollarded chiefly to ensure their stability. But historically the lopped wood was used for cattle food, and for making matches, bean-poles and fruit baskets,[18] and a

The Aston on Clun Flag Tree, before it was toppled by a gale in September 1995.

A puzzle: this photograph found its way into Edgar Milne-Redhead's black-poplar files unlabelled. It looks as if it may be one of the largest black-poplars in England, yet no one has yet been able to identify the site.

peculiarly local form of wattle: 'The wood was used with willow to make sheep hurdles, for pens to confine sheep at night when they were brought down from the Chiltern downland. The wood was also used for making rifle butts in the 1914 war.'[19]

Slowly the tree is coming back into visibility. A survey organised by the *Daily Telegraph* in 1994 located dozens of previously unknown trees. And the appearance of mis-namings, deliberate or accidental ('dark popular' has been heard in Kent), is a sure sign that it is also winning back its place in popular imagination. Edgar Milne-Redhead's eightieth-birthday cake was decorated with black-poplar leaves and had a representation of the Butley tree on top. But the best tribute to his virtual rediscovery of the tree would be the re-establishment of at least one example of the flood-plain wildwood that was the black-poplar's aboriginal home. Plans have already been put to the Forest Authority.[20]

Hybrid black-poplar, *P.* × *canadensis*. Many cultivars are grown for ornamental and timber plantings. 'Serotina' is the commonest, and one of the most widespread road-side and parkland trees in Britain. It arose in France in the early 1700s. 'Regenerata' has become popularly known as the 'railway poplar' because of its frequent planting beside railways in southern England. It is an unmistakable vari-ety, with a very cluttered habit, the trunk sprouting shoots and snags in all directions.

Lombardy-poplar, *P. nigra* 'Italica'. The narrow, up-swept outline of the Lombardy-poplar never looks really at home in the British countryside, and the tree is always planted, never naturalised. But its shape has made it a personal landmark for some:

'My mother had a 70-foot-high Lombardy poplar in her garden at Riseley [Bedfordshire] until earlier this year [1992] when we were worried about its stability and I had it lopped half-way up. A couple of people in our road, while acknowledging this was a sensible thing to do, said they were a little disappointed, because the first thing they did each morning was to look at the tree to gauge the direction of the wind.'[21]

'I have some Lombardy poplars in my garden that were used as guides by pilots flying into RAF Halton during the war.'[22]

Aspen, *P. tremula*, is known by, and named in Latin from, the way its leaves tremble in the breeze. Other poplars have the same habit, but aspen's quivering stands out because of the serrated edges of its leaves and the sound they make – a rustling whisper, as if they were being spattered by rain. Gerard Manley Hopkins caught their rhythm in his lament on the aspens cut down in 1879 at Binsey near Oxford:

My aspens dear, whose airy cages quelled,
Quelled or quenched in leaves the leaping sun,
All felled, felled, are all felled;
Of a fresh and following folded rank
Not spared, not one ...[23]

Aspens are colonial trees, spreading by sucker to form what are sometimes quite extensive groves, though indi-vidual trees rarely live more than 50 years. In the lowlands they prefer ancient woods or open heathy woodland on clay plateaux which become thoroughly wet in winter. But they flourish best in cool climates (they were one of the first trees to return to Britain after the retreat of the Ice Age), and in the uplands they grow in a wider range of habitats, including rocky gorges and stream valleys. I have seen them growing with hazel on steep-sided limestone slopes in the Derbyshire Dales, with an understorey that included globeflower, wood anemone, salad burnet and bloody crane's-bill – meadow, woodland and downland species flourishing together in the way they probably did across England in the post-glacial period, 9,000 years ago.

White poplar, *P. alba*, is a large broad-crowned tree, with a grey bark, pitted by black diamonds, which is not uncommon in parks and on roadsides. It is dramatic in spring, when the young shoots and leaves are coated with thick white down and from a distance the tree appears to be covered with snow. Later, the upper sides of the leaves turn a glossy green. White poplar was previously believed to have been introduced from southern Europe in the seventeenth century. But Oliver Rackham has found many references to it under its old name of 'abel' in thirteenth-century documents, so it is either a much earlier introduction or a native (though it is very rarely found inside woods).[24] Whatever its origins, it propagates itself successfully by suckers.

Grey poplar, *P.* × *canescens*, is a similar but often larger tree with more greyish down and foliage. It is believed to be a hybrid between the white poplar and the aspen, but such crosses do not seem to occur naturally in this country, and most specimens are of planted origin. Nevertheless it suckers even more vigorously than either of its parents, and colonies are often found in damp woodland.

Balm-of-Gilead, *P. candicans*, and **western balsam-poplar**, *P. trichocarpa* (and various hybrid balsam-poplars) are widely planted, especially in damp places, and sometimes naturalise by suckers. In mid-April the buds and young leaves exude a resin that fills the air around with the aroma of incense, which in a warm spring breeze can be nosed out from at least 100 yards away, and which has been immortalised in Alan Bennett's diaries:

An aspen wood on the shore of Loch Torridon.

'For years I didn't know of its existence, only around this time of year one occasionally caught a whiff of something so intoxicating that it seemed to promise opportunity, fresh beginnings, the turning over of new leaves; it seemed the very breath of spring. Once I thought it was daffodils (the daffodils were under the tree); in Rome I thought it was a waiter's aftershave (we were dining outside); once I even thought it was sweat and that the person concerned (we were walking in the Parks at Oxford) must be like a saint and have the odour of sanctity. It was only a few years ago that I came upon a line of these poplars in the grounds of Alec Guinness's house and was told by Alec that it was a tree on the site that had decided them to live there ... I once took a branch from the tree into Penhaligon's, hoping they manufactured something similar. They didn't, but a dreamy look came into the assistant's eye as she sniffed, and she called in her colleagues, who were similarly entranced.'[25]

In the 1980s, Oliver Gilbert discovered a colony of self-sown poplar hybrids on some waste ground in central Leeds. The parents were found on a nearby traffic roundabout and proved to be 20 female balsam-poplars (*P. candicans*) and two male black-poplars (*P. nigra*). The large quantities of seed which are shed in June pile up as white fluff wherever there is shelter, and germinate within 24 hours after rain. Most eventually die from drought, but increasing numbers are surviving, producing dense, uneven-aged stands, often hundreds strong, on patches of waste ground nearby. The oldest seedlings were at least 15 years old in 1995. This site and two sewage works in London are the only recorded instances of alien poplars – most of which grow as single-sex clones – regenerating by seed in Britain.[26]

Crack-willow, *Salix fragilis* (VN for willow in general: Withy, Sallies, Wullies, Saugh, Sauchan). Crack-willow is well named. The trunk grows fast but is apt to split open under its own weight. Its tendency to collapse is encouraged by its favoured habitats, which are damp, flood-prone fens and river valleys. But by riversides (and roadsides, where the roads run above dykes and damp ground, as in east Norfolk), crack-willows are often planted to stabilise the banks and are usually pollarded to reduce the chances of splitting and allow light on to the water. These rows of tufted trees – sheaves of narrow leaves above a wizened, knobbly, leaning trunk, with half its roots in the water – are a quintessential part of the landscape of lowland rivers, especially the Thames and Cam. At the beginning of the century they helped create the drowsy atmosphere of Kenneth Grahame's *The Wind in the Willows*, and inspired Arthur Rackham's illustrations, and no slow river seems right without them now. (The two unrelated Rackhams – Arthur and Oliver – have between them quite transformed the image of the pollard in the twentieth century.)

Crack-willows make the most contorted of all pollards. Adventitious roots creep earthwards from inside their hollow centres. The crowns are often so full of holes, crevices and rotting leaves that they nurture second-storey woodlets of their own: ash, holly, gooseberry, elder, ferns, honeysuckle and brambles. Anita Jo Dunn, who looked at 400 willow pollards on the banks of the Evenlode and Windrush in west Oxfordshire and elsewhere, found 74 species growing in their crowns.[27]

The wands that are cut from pollard willows, on a five- to ten-year rotation usually, will themselves root immediately to make new willow trees (as will a collapsed portion of crack-willow trunk – cf. black-poplar, p. 135).

In Middleton Cheney, Northamptonshire, 'about thirty years ago, some fencing needed to be erected, and the farmer charged with this task, Johnny Pollard [*sic*], hammered 36 willow stakes into the ground as fence posts. A line of crack-willows has grown from the stakes – a distinctive landscape feature along this stretch of the Farthinghoe brook. Mr Pollard's grandson still lives locally and the story is well known.'[28]

I have seen the same thing happen by gravel-pits, where anglers have cut V-shaped willow rod-rests and left them in the ground.

Crack-willow poles, if cut young and thin enough, can be woven, though this is usually done with osier (see below). Rough-and-ready open-air baskets for cattle food were made from pollard cuttings until a few years ago. In the Lake District they were known as 'swills': 'Swills are

Ancient pollard willows. Crack-willow can regenerate from the rooting of split and fallen branches.

large woven baskets made of willow wood. The branches for the body of the swill and the thin hazel branches for the rims were soaked for some days in a dammed-up stream. These were widely in use for holding cattle food and washing, both wet and dry.'[29]

In the Severn valley they were 'cribs': 'Mr Riddle [Thornbury, Gloucestershire] is aged 65 and last made a crib in this way 10 years ago. He said it took about two days for a man to cut the branches off a pollarded crack willow tree and weave the crib. The crib was 7 feet in diameter and the uprights were willow branches, 2 to 3 inches in diameter, around which were woven branches around 2 inches in diameter up to a height of two and a half feet. Several of the local farmers and farmworkers used to make these kind of cribs which lasted outside for about three years.'[30]

White willow, *S. alba*, is a large and spreading tree, with grey bark and silver-felted leaves, which stream dramatically in the wind. Native by streams and ponds and in marshes, it is also frequently planted. **Cricket-bat willow**, a tall, straight-trunked form, grown for cricket-bats in damp ground, in Suffolk and Essex especially, is var. *caerulea*. **Golden willow**, var. *vitellina*, has a cultivar 'Britzensis', with bright orange twigs, which are cut for basketry. It is often grown as low pollards along the rhines in the Somerset Levels.

A memorial seat at Rougham, Norfolk, made out of living willow wands.

Osier, *S. viminalis*, is a fast-growing shrub, especially when coppiced, with long leaves and straight shoots, which turn a shiny yellow-brown when mature. It grows in fens, ditches and damp places throughout lowland Britain, but has also been widely planted in osier-beds to be harvested for basket-making. More than 60 different osier species, hybrids and cultivated varieties are grown in Britain, with different strengths, growth rates and flexibility, and in colours ranging from pale yellow to deep purple. (There is a naturally purple-barked willow, *S. purpurea*.) They are traditionally used for basket-making, but 'wicker-work', as stripped and woven willow is called, is going through a renaissance at present, and osiers are appearing in novel situations. Some are being woven whilst still alive, with their cut ends stuck into the ground, to form living fences or pergolas. The Department of Transport is using a variant of these for sound screens along the edges of motorways. Wicker 'mattresses', 10 feet high, are filled with earth, into which the willow roots are planted. They are irrigated where necessary to prevent drying out, and other plants such as ivy and honeysuckle are added at intervals. And, using more conventionally dried willow, the sculptor Serena de la Hay makes life-size willow figures for gardens, the strips of willow seeming like sinews of muscle.

In Chediston, Suffolk, there has been a revival of a 'willow-stripping' ceremony at a willow-grower and basket-maker's farm: 'For several years we have held a willow stripping, usually at the first full moon in May. We construct a Green George figure which is dressed with the willow strippings, and at the end of the festivities he is danced up to the pond and ceremoniously cast in.'[31]

Many fast-growing willow (and poplar) varieties are currently under trials for large-scale cultivation as fuel-wood: 'Over the next five years we shall plant another six hectares of land [by the River Blackwater, Essex] and the crop will be coppiced every three or four years for about thirty years. The willows absorb and sequester carbon dioxide from the atmosphere and require very little spraying or fertiliser, so that they are environmentally friendly. The crop is cut by machines, chopped and then burnt in specially designed boilers. In Sweden small towns are heated by this method and it is hoped that in time the same thing will happen here.'[32]

Goat willow, *S. caprea*, and **grey willow**, *S. cinerea* (VN: Sallow, Pussy willow). These are the commonest and most widespread willows in Britain, growing in ditches, reedbeds, scrub, wet woodland, hedges and urban wasteland. They are best known for the silky, silver-grey male catkin buds – the 'pussy willows' – which appear in late January and become brilliant yellow in March. (They used to be called 'goslings' at this stage, because their texture and colour were like newly hatched geese.) Because so little else was in leaf or flower at this early season, sprays of sallow have frequently been used as 'palm' to decorate churches at Eastertide.

'I remember attending Sunday School on Palm Sunday and wearing a sprig of pussy willow, and my father was always told "You get your hair pulled if you don't wear a piece of palm on Palm Sunday." '[33]

Osiers, cut and tied with osier 'rope', Somerset.

White willow, gleaming in autumn sunshine by the River Severn, near Welshpool, Powys.

'As a child I used to make puppets out of pussy willows. To do this, take about 10 pussy willows, while they are still silvery, before the pollen appears, and thread them on white cotton. This takes patience and some of them will break, so you need some spares. Put knots on either end of the cotton to stop them falling off. Attach more cotton for the strings and tie these to a twig. Move the twig and the puppet wriggles like a snake.'[34]

Bay willow, *S. pentandra*, is chiefly a northern species, with wide, dark leaves which have the scent of bay. **Almond willow**, *S. triandra*, is a species of southern and central England whose twigs and leaves smell (and taste) slightly of almonds. It is sometimes called snake-skin willow from the way it sloughs reddish layers of bark.

The willows have one other important claim to economic fame. Bitter infusions of willow bark were anciently employed by country people as a remedy for chills, rheumatism and 'the ague'. (Herbalists rationalised this tradition by appealing to sympathetic magic, arguing that, as willows tended to grow in wet places, they would be good for diseases engendered by the damp.) The remedy worked, and early in the nineteenth century the active ingredient, salicylic acid, was isolated, both from willow bark and from meadowsweet (another plant of damp places). This led in 1899 to the synthesis of what was to become the world's most widely used – and useful – synthetic drug, acetylsalicylic acid, which the pharmaceutical company Bayer called Aspirin, after the old botanical name for meadowsweet, *Spiraea ulmaria*.

'Pussy willow', or 'goslings' – the male catkins of goat willow.

Cabbage family
Brassicaceae (or *Cruciferae*)

London-rocket, *Sisymbrium irio*. London-rocket shot up profusely and mysteriously from the ruins of east London in the spring following the Great Fire of 1666. It was a source of wonder and speculation at the time, not because of its appearance (it is an ordinary enough mustard-like plant, a foot or two tall with small yellow flowers) but because of the vast quantities that sprang from the piles of rubble and burned-out timber houses. One contemporary writer concluded that 'these hot bitter plants with four petals and pods were produced spontaneously without seed by the ashes of the fire mixed with salt and lime'. (The name 'rocket', from the Latin *eruca*, is, incidentally, simply a generic name applied to many hot-tasting species.)

But seeds, of course, there must have been. London-rocket is not a British native, but a common annual of waste places in the Mediterranean, and stowaway specimens already established in London's commercial quarters were the most likely origin. The great swathes of warm, disturbed ground left in the wake of the Fire would have created ideal conditions for a population explosion.[1]

The numbers dwindled over the years, and the species was probably extinct in London by the early nineteenth century. Then a colony was discovered at the Tower of London in 1945. There were other reported sightings from bombed areas during and after the Second World War, but most of these turned out to be the similar but taller **eastern rocket**, *S. orientale*. This is now one of the most abundant and distinctive plants of remnant waste-land in the City of London, especially around building and development sites, and suffuses whole areas with yellow in late summer. **Hedge mustard**, *S. officinale*, is a common annual or biennial of hedgerows, arable fields and wasteland. It grows about two feet tall with clusters of small yellow flowers from May to September. The stems can be very stiff and tangled, which has given rise to the modern vernacular name of 'barbed-wire plant'. (In Australia, where it is also naturalised, it is known as 'Wiry Jack'.) The leaves have been used as a green vegetable. **Flixweed**, *Descurainia sophia*, is an attractive and unusual mustard, up to three feet tall, with feathery leaves and sheaves of long, erect seed-pods. It grows in waste places and on roadsides, and is especially common on the sandy soils of East Anglia.

Garlic mustard or **Jack-by-the-hedge**, *Alliaria petiolata*, is an abundant herb of hedge-banks and woods, smelling mildly of garlic. It has long been used as a flavouring: in sauces for fish and lamb in the seventeenth and eighteenth centuries, and as an ingredient for spring salads today.[2] In 1993 it was being sold for a pound a bunch in a smart Italian delicatessen in London's Covent Garden.[3] Jack-by-the-hedge is a biennial, and the soft, nettle-shaped leaves can be picked from September, when they first begin to show, until late spring, when the brilliant white flowers appear.

Garlic mustard, whose leaves are a good addition to spring salads.

Woad, *Isatis tinctoria*, may or may not have been the dark blue dye with which the Ancient Britons daubed their skins. This is one of the stories from Caesar's *De Bello Gallico*, which is notoriously prone to generalisations based on Caesar's one short visit to Britain. But there is no doubt that it is a dye plant of great antiquity, probably brought to Britain by Celtic immigrants from western and southern Europe. Glastonbury, in Somerset, derives its name from the Old Celtic *glasto*, and means 'a place where woad grows'. Names from the Anglo-Saxon 'wad' are more frequent: e.g. Wadborough in Worcestershire ('woad hills'), Wadden Hall, Kent ('nook where woad grows'), Waddicar, Lancashire ('newly cultivated land where woad grows'), Waddon, Dorset and Surrey ('woad hill').[4] In a sparkling essay on the plant written in the early 1950s, Geoffrey Grigson describes a farm close to his home in Wiltshire 'called Woodhill Park. Woodhill did not mean the "hill near the wood". In 1086, in the Domesday Book, the farm was called "Wadhille", the hill where the "wad" or woad was cultivated.'[5] Wiltshire, Dorset, Somerset and Gloucestershire seem to have been the main centres of woad cultivation, though there were others in East Anglia. It was cultivated in the Fens up until the beginning of this century, and reputedly last used in the dyeing of policemen's uniforms. In Essex a 'Blueman Farm' is believed to be named from woad-growing.[6]

The process of manufacturing the dye was quite elaborate. The woad leaves were crushed to a pulp in a mill, and then moulded into balls, which were allowed to dry in the sun (but protected from the rain) until the pulp began to ferment. A crust formed over the balls, and care was taken to ensure that this did not split. When fermentation was complete the balls were pulped again in the mill and again formed into cakes. The whole cycle was repeated a third time before the fully-fermented balls were thoroughly dried and sent off to the dyer. It took a hundredweight of leaves to produce 10 lb of the final dye, and the ammoniacal stench of the fermentation process was so notoriously disgusting that Elizabeth I issued a proclamation that woad production had to cease in any town through which she was passing.[7]

Woad was made virtually obsolete by the growing popularity of indigo, which gave a stronger, faster blue, and was cheaper, even though it had to be imported from the tropics. But woad has become naturalised on a precipitous cliff close to one of the ancient cultivation areas in Gloucestershire, and in 1818, when the colony was first discovered, 'the cliff was quite golden with it about the end of May'.[8] It has proved remarkably persistent in spite of – or perhaps because of – repeated disturbances to the site:

'On the red marl cliff called the Mythe Tute, north of Tewkesbury, woad grows in a small colony. Early writers on botany considered it to be indigenous in this area in the nineteenth century, but in the nineteenth century it was cultivated around Wotton-under-Edge. We have watched the colony since the mid-1960s, and the number of plants show great variation, from the twenties down to two plants in the early 1970s. Then cliff-falls and the removal of dead elms provided open soil and the biennial woad again flourished, so that in the late 1970s there were over a hundred flowering stems. The colony is again declining as the surface becomes overgrown. In the mid-1980s we were surprised to see three plants growing on the verge of the M5 nearby.'[9]

With the growth of interest in natural dyes, woad is experiencing something of a revival. It is, for example, being grown by craftspeople in Essex, Somerset, and in West Sussex, where the seed stock has been bulked up at the reconstruction of an Iron Age farm at Butser.[10]

But woad is an unusually attractive plant, not simply what Ruskin would have scorned as a 'chemical factory'; and its aesthetic virtues have also been rediscovered – or perhaps noticed for the first time. Its yard-high stems, its long, almost succulent leaves, which seem to shine like stained glass with an inner, immanent blue, and its foamy clusters of brilliant yellow flowers – all put together, as Grigson wrote, rather like a 'wireless mast' – have made it a grail for flower-arrangers, who have also begun to grow it, providing another source of seed. In one Suffolk churchyard, woad's new role and ancient resilience came fortuitously together in the summer of 1992:

'Earlier this year, I thought I had found something special in Framsden churchyard. Woad has only two established colonies in England ... From its position I guessed that the seed had been dropped by a bird sitting on the church gutter. But where had the bird got the seed? Surely not from the nearest wild colony, in Surrey [a chalk-pit near Guildford]. Perhaps somebody had been experimenting with growing it for its dye? I soon met two or three people who had done just that in years gone by. Then, on a return visit to Framsden, I met a lady in the churchyard who told me that somebody in the village grew a variety of plants for their natural dyes, and that she had passed woad to several of her friends. It looked well in flower arrangements, and was often used in the church. The position where the plant was growing was not only under the church gutter, it was also outside the door where the flower ladies removed their dead flowers.'[11]

Wallflower, *Erysimum cheiri* (VN: Gilliflower, Milkmaids). A species native only in rocky places in southern

Woad, the ancient dye plant, is undergoing a revival. It is also becoming popular with flower arrangers.

Greece and the Aegean. The wild forms have very fragrant yellow or brownish-orange flowers and it was for these that the plant was introduced to Britain at an early date. There is a story that it was often planted on the walls of castles and manor houses so that the scent could waft through the windows of the bedchambers. Whatever their origins, yellow-flowered plants indistinguishable from wild specimens have clung on to many ancient buildings and walls – for instance, Bury St Edmunds Abbey, Suffolk; Holy Island Priory, Northumberland; Waltham Abbey and the castle and Roman walls at Colchester, Essex; and many of the Oxford colleges. Plants closer to garden varieties appear more widely, on stonework in churchyards and railway cuttings, for example. Having very lightweight seeds which can be carried upwards by the wind, they sometimes take root even on chimneys.

A different species, the violet-flowered *Erysimum linifolium*, has been christened 'the Manx wallflower' on the Isle of Man and 'has been known on the ruins of Rushen Abbey and nearby quarries for more than a hundred years'.[12]

Dame's-violet, *Hesperis matronalis*, is a cottage-garden favourite introduced from Europe in the sixteenth century, and now increasingly naturalised in waste places and on roadside verges, but rarely far from houses. The flowers can range from purple to white, and have a sweet violet scent (though this is less evident in the purple varieties).

Hoary stock, *Matthiola incana*, is possibly a rare native of the southern chalk cliffs, now probably confined to Sussex and the Isle of Wight, and is the parent of the garden stocks. In the wild it has short sprays of large, deep violet flowers with a clove-like fragrance, which produce seed-pods often more than four inches long. It was developed into the strain of 'Brompton stocks' in the eighteenth century at the Brompton Road nursery in London.[13]

Winter-cress, *Barbarea vulgaris*, is a medium-sized 'rocket' widespread in dampish, disturbed habitats, on road-verges, river-banks and rough grassland. Its clusters of small yellow flowers appear from April. The leaves have been eaten as a vegetable but are exceedingly bitter.

Dame's-violet, a cottage garden favourite now widely naturalised.

Its cousin, **American winter-cress**, *B. verna*, has proved much more popular and successful as a vegetable. Originally from south-west Europe, it has been introduced as a cultivated vegetable not just to the United States and northern Europe, but to South Africa and Australasia. It has become widely naturalised in all these places.

In country areas of North America winter-cress is gathered from the wild as well as being cultivated. In Virginia the over-wintering rosettes are sold in supermarkets as 'creesy greens', and the young, unopened flower-spears are cooked like broccoli.[14]

Water-cress, *Rorippa nasturtium-aquaticum*, is the only British native plant which has passed into large-scale commercial cultivation scarcely altered from its wild state (or perhaps I should say 'states': there are 10 other closely related species and hybrids, of which three are the chief ones used in commerce).

Water-cress was traditionally picked wild from the edges of fast-flowing streams, where it can grow in thick drifts. It was important enough for settlements to be named after it – e.g. Kersey, Suffolk ('cress island'); Kesgrave, also in Suffolk ('ditch or grove where cress grew'); Kersal, Lancashire ('the haugh [flat alluvial land] where cress grew'); and Kershope, Cumbria ('cress valley').[15] But John Evelyn, in *Acetaria: A Discourse of Sallets*, refers to it rather disparagingly as 'the vulgar *Water-Cress*', one of two salad vegetables which are 'best for raw and cold Stomacks, but nourish little'.[16] It became more fashionable in the eighteenth century, when its anti-scorbutic properties were realised, and by the nineteenth century it was certainly under small-scale cultivation, especially in areas where there were clear chalk streams, such as Wiltshire and the north Chilterns. Henry Mayhew made a fascinating first-hand record of the water-cress trade in Victorian London:

'The first coster-cry heard of a morning in the London streets is of "Fresh wo-orter-creases". Those that sell them have to be on their rounds in time for the mechanics' breakfast, or the day's gains are lost ... At the principal entrance to Farringdon market there is an open space, running the entire length of the railings in front

Hoary stock, a rare native ancestor of garden stocks.

Winter-cress, an early-flowering species.

and extending from the iron gates at the entrance to the sheds down the centre of the large paved court before the shops. In this open space the cresses are sold, by the salesmen or saleswomen to whom they are consigned, in the hampers they are brought in from the country ... The market – by the time we reach it – has just begun; one dealer has taken his seat, and sits motionless with cold – for it wants but a month to Christmas – with his hands thrust deep into the pockets of his grey driving coat. Before him is an opened hamper with a candle fixed in the centre of the bright green cresses, and as it shines through the wicker sides of the basket, it casts curious patterns on the ground – as a night shade does.'[17]

Water-cress, common but often contaminated in the wild by liver-fluke larvae. Cooking kills them.

Special railway tracks – 'Watercress lines' – were established to run the crop up to London and are still referred to by this name in north Hampshire and near Stamford in Lincolnshire.[18]

Water-cress grown commercially in beds has the advantage over the wild form of growing in water drawn directly from underground springs or bore-holes and isolated from the surrounding land by concrete or chalk channels. It thus avoids the danger – always, unfortunately, a threat with wild cresses – of picking up larvae of the fluke, *Fasciola hepatica*, which attacks and severely damages the livers of sheep and cattle, and can do the same to humans. The fluke is transmitted through a complicated cycle involving a pond snail. The larvae hatch from eggs in the sheep's droppings and 'swim' through the damp grass until they reach an area of water where the snails live. They burrow into the snails' soft tissues, multiply and metamorphose, before escaping and swimming to the water's edge. Here they crawl onto vegetation, including water-cress, and wait for the plant to be eaten by grazing animals (or foraging humans, for that matter) so that the whole cycle can begin again.

One picker, to be on the safe side, follows the same rule as is applied to shellfish: 'Water-cress growing wild should never be eaten when there is an "r" in the month'[19] – though I am afraid this is a rather contracted view of the fluke's ingenious breeding cycle. The only way to ensure you are eating fluke-free cresses is to pick plants growing *in* fast-flowing, clean water (preferably over chalk) and not from the banks; and, to be doubly sure, turn them into soups: cooking kills all stages of the fluke.

This is not to say that cultivated cresses are utterly without hazards. Although they are unmistakable when in flower, with four-petalled white flowers on top of the swaying hummocks of green, at other times they can be mistaken for a few other species with similar leaves that cohabit with them in running chalk-rich water. I have twice found sprigs of fool's water-cress (see p. 293) – not very pleasant eating, but fortunately not poisonous – tangled up with bunches of supermarket water-cress.

An up-and-coming cousin of water-cress is **Walthamstow yellow-cress**, *R.* × *armoracioides*, a speciality of the damp wasteland round Walthamstow Reservoirs in north London. It is a hybrid between **creeping yellow-cress**, *R. sylvestris*, and **Austrian yellow-cress**, *R. austriaca*, an introduced species whose nearest locality is some miles from Walthamstow. Just how it got together with its cousin is still a mystery.

Horse-radish, *Armoracia rusticana*, is a common perennial of waste ground and roadsides in England, though very scarce in Scotland. No one is sure when it arrived in this country from its native western Asia, but it was certainly well before 1548. That year William Turner remarked that it 'groweth in Morpeth in Northumberland and there it is called Redco'. Fifty years later, Gerard was recommending it as being preferable to mustard as a condiment for meat.[20] Even then, its deep, spreading tap-roots must have been building up large and tenacious colonies in untended ground – a process which can be seen most strikingly today in the extensive troops on railway embankments.

Horse-radish does not produce its sprays of white flowers every year in this country. But its long, crinkled leaves are unmistakable, and smell slightly of horse-radish root if crushed between the fingers. Digging up the roots

Horse-radish, introduced from Asia by the sixteenth century, is now common on roadsides.

of wild plants is illegal on private land except where you have the landowner's permission. But horse-radish is frequent enough in the rough corners of gardens and allotments for some to be likely to be within legal access. In the Fens it was regarded as a gleaner's perk when farmland was being ploughed: 'Horseradish grew on the roadsides. Men came on bikes with sacks and dug it to sell. When Grassfield was ploughed, Jimmy Greenwood (a well-known hawker) followed the plough and picked up horseradish to sell it.'[21]

The labour of excavating and extracting its intricate, woody and pungent roots is followed by the ordeal of peeling and grating them. This can end in tears worse than result from the most blinding onions: 'A Russian friend suggested to us that we should deep-freeze the root first and then grate it after peeling. In this way the effect on the eyes can be avoided.'[22]

There are all manner of recipes for turning the grated root into sauces and condiments, but none to match the formula of the flamboyant and reforming nineteenth-century chef Alexis Soyer (author of *Shilling Cookery for the People*). It is called, with some justification, 'The Universal Devil's Mixture':

'Put in a bowl a good tablespoonful of Durham mustard ... mix with four tablespoonfuls of Chilli vinegar. Add to it a tablespoonful of grated horseradish, two bruised shallots, a teaspoonful of salt, half ditto of Cayenne, ditto of black pepper, and one of pounded sugar, two teaspoonfuls of chopped chillies, if handy. Add the yolks of two raw eggs. Take a paste brush, and after having slightly seasoned each piece [of meat] with salt, rub over each piece with the same, probing some into the incisions. First broil slowly and then the last few minutes as near as possible to the Pandemonium fire.'[23]

Coralroot, *Cardamine bulbifera*, is a scarce and very local species of old deciduous woodland with a strangely disjointed distribution. Genuinely wild colonies occur in two quite distinct and widely separated regions: the wet, acid woodlands of the Weald in Kent and Sussex, and the drier, usually more calcareous Chiltern beechwoods. (There is a third, possibly native cluster, around Needwood Forest in Staffordshire.)

Coralroot is one of the most subtly beautiful woodland plants, its soft, pale lilac flowers held on stalks a foot or 18 inches high and coming into bloom at the same time as, and often amongst, the bluebells. It has been widely introduced into woodland gardens, where it often becomes naturalised. There are, for instance, established colonies in the well-known gardens at Knightshayes Court and Dartington Hall, both in Devon, at Cliveden in Buckinghamshire, and in Ellen Willmott's old garden at Warley Place in Essex.

Coralroot, a very local species of old woodland in the Chilterns and the Weald.

Coralroot spreads both by the extension of its root network and by means of the little purple-brown bulbils that lie between leaves and stem. These are easily dislodged from the plant by birds or strong winds, from June onwards. Given these twin means of reproduction it is a little surprising that the plant hasn't escaped more often from its garden strongholds. But there are feral colonies in, for example, Smallcombe Wood, Bath; a small hornbeam copse near Wellingore Hall Park, Lincolnshire; woodland at Silverdale, Lancashire (naturalised from an adjoining nursery); and Scalby Churchyard, Yorkshire, where it has been since at least 1900, and 'since it reached the wall of the churchyard it has not looked back'.[24]

Cuckooflower or **Lady's-smock**, *C. pratensis* (VN: Our Lady's smock, Milkmaids, Fairy flower, May flower, Coco plant). Cuckooflower is the 'approved' English name for this common and widespread species of damp grassland, roadsides, ditches and river-banks. But 'lady's-smock' is both more current and more expressive. The flowers vary in colour from very pale pink to mauve, and are slightly cupped or 'frocked' (though 'smock' was once less than complimentary slang for a woman, on a par with our 'bit of skirt', and there may be allusions in the name to what went on in springtime meadows).

Caroline May's painting of cuckooflowers. The double variety (left) and hose-in-hose (right) were painted from specimens in the vicarage orchard at South Petherwyn, Cornwall, in 1852.

In some places the flowers can be so dense as to colour the ground: 'In Norfolk I always knew it by its other common name of Ladies Smock and also as Milkmaids. I remember they grew in such abundance on my local common (South Wootton, just outside King's Lynn) that I could pick armfuls every day. There are pictures of me [*c.* 1970] with my arms full posing, ironically, beside the massive pipes which were soon to drain the common and denude it of its glorious pinky-mauve display.'[25]

In parts of Devon the double-flowered form *flore pleno*, and even hose-in-hose forms (where one normal bloom grows through the centre of another – cf. primrose, p. 168) are not uncommon.[26]

As for 'cuckooflower', it is a name which this species has shared, at a local level, with at least a dozen or so other spring flowers, including wood anemone, red campion, greater stitchwort, wood-sorrel, bluebell and early-purple orchid, and the coincidence of its flowering with the arrival of the cuckoo is probably no better than for any of these other species.[27] In fact, in the south of England it is usually in flower by March, and in the extreme south-east often as early as February after a mild winter: 'My birthday falls on 14 April, and when I was a child my father told me that on that date the old woman of the woods let the cuckoo out of her basket (locally there used to be Cuckoo Fairs). For the last twenty years or so my children have given me cuckoo flowers on my birthday and it is a very rare year when there are none around by 14 April. This year I got a bunch on Mother's Day, 13 March!' (East Sussex, 1994)[28]

But in 1994 at least, correspondents showed that in many other parts of Britain, the first full blooming of the cuckooflower was a fairly accurate predictor of the first hearing of the bird itself:

	Cuckooflower	Cuckoo
Farnham, Surrey	10 April	21 April
Dymock, Glos.	15 April	24 April
Lugwardine, Hereford	21 April	25 April
Cutnall Green, Worcs.	23 April	24 April
Upton-by-Chester, Ches.	24 April	29 April
Kidlington, Oxon.	27 April	28 April
Gt Houghton, Northants.	21 April	1 May
Shrewsbury, Shrops.	29 April	1 May
Forest of Ae, Dumfries	30 April	4 May
Gourock, Renfrewshire	6 May	8 May[29]

The semi-transparent seed-pods of honesty, which was probably named from their 'see-through' quality.

The young leaves have a peppery taste, and make a useful substitute for cress in salads and sandwiches.[30]

Hairy bitter-cress, *C. hirsuta*, is an abundant annual or biennial in gardens and other cultivated ground, and in flower for much of the year. It has the look of a miniature water-cress, and, like lady's-smock, makes a pleasantly tangy addition to salads. **Wavy bitter-cress**, *C. flexuosa*, is a similar, but taller and often perennial species of ditches, stream-banks and marshes.

Tower cress, *Arabis turrita*, is one of the most precisely and idiosyncratically placed plants in Britain. Its only surviving colony (it was introduced from southern Europe in the early nineteenth century) is on old walls at St John's and Trinity Colleges, Cambridge. It also grew at Magdalen College, Oxford, till the colony was destroyed during the rebuilding of the walls in the 1850s. **Bristol rock-cress**, *A. scabra* (a native), is almost as local, growing only on limestone rubble in the Avon Gorge. It is an attractive white-flowered plant, with a tufted, alpine habit, and has suffered from the predations of rock-gardeners.

Aubretia, *Aubrieta deltoidea*, is a mat-forming species introduced from south-eastern Europe as a garden plant, and widely naturalised on walls and stony banks close to habitation.

Honesty, *Lunaria annua* (VN: Moonpennies, Bread-and-cheese). A purple- or occasionally white-flowered biennial from south-east Europe, commonly grown in gardens, and often escaping on to waste ground and road- and rail-sides. Its English name, which appeared in the sixteenth century, derives from the round, semi-transparent seed-pods, which have the lustre of fine vellum and can hang on the dry skeleton of the plant through the winter. Perhaps they suggested silver coins to some gardeners, but I am more persuaded by the theory that honesty 'derives its name from the "see-through" quality of its seed-pods'.[31] In Alison Uttley's 'Little Grey Rabbit' stories the pods are called 'windows'.

Common whitlowgrass, *Erophila verna*, is a modestly cheerful and many-branched annual, which will grow in the smallest cracks in calcareous walls, rocks and dunes. The small white flowers are amongst the first to appear in March, and are followed by seed-pods a little like miniaturised versions of honesty's. It was used by medieval herbalists in the treatment of whitlows.

Common scurvygrass, *Cochlearia officinalis*, is a rather straggly perennial with white flowers up to one centimetre across, and glossy, fleshy, kidney-shaped leaves. The leaves have a high Vitamin C content, and before the availability of citrus fruit were widely used in the prevention of scurvy on board ship, as recommended by, for example, Captain Cook. Even inland, they were sometimes included in spring tonics, mixed up into a bitter infusion with fruits and other herbs. Scurvygrass occurs on saltmarshes, dunes and sea-cliffs right around the coasts of Britain, and on tidal river-banks and wet mountainous areas inland.

Danish scurvygrass, *C. danica*, used to be confined, like most of its relatives, to coastal habitats, and to a few inland railway-line sites, where its seeds are believed to have been introduced with ballast from the seashore. But in the late 1980s it began to appear along the edges of major roads, almost always on the central reservation. By 1993 there were colonies on stretches of motorway and trunk roads in 320 10-kilometre squares. There are now concentrations on, for example, the M4, M5 (especially around Cardiff and Cheltenham), M6 and M56, along many reaches of the A1, the A5 in Anglesey, the A11 in Suffolk and the A30 in Devon, and on dual carriageways around Christchurch and Poole in Dorset. It has even crossed the Scottish border and appeared on the A74 in Dumfriesshire.

On many roads it seems to be spreading at the rate of some 10 to 15 miles per year, and rapidly builds up continuous colonies many miles long. In March and April, its profusion of low-growing, small white flowers can look like a layer of hoar-frost on the edge of the central reservation.

There can be little doubt that the turbulence caused by

Danish scurvygrass, rapidly colonising the edges of motorways and trunk roads.

A seed-pod of shepherd's-purse, showing the resemblance to the little leather pouches carried by medieval peasants.

fast-moving traffic is what is wafting its seeds along at such a remarkable rate, and that the bare, stony edges of trunk-road verges (often liberally doused with spray from de-icing salt) are a congenial habitat for this native of maritime shingle banks.

But its preference for central reservations is still a puzzle (though it seems to be becoming less pronounced the more the species spreads around the road system). It may be partly a consequence of the greater speed of traffic along the lanes nearest the central reservation, the more powerful slipstream and spray they create, and the better drainage of the raised (usually) central reservation.[32]

Gold-of-pleasure, *Camelina sativa*, is a native in south-east Europe, and cultivated in Britain as a source of fibre and for its oil-bearing seeds since Neolithic times. It is now a rare casual (usually sprung from birdseed), but quite striking for its yellow elliptical seed-pods.

Shepherd's-purse, *Capsella bursa-pastoris* (VN: Mother's heart). Shepherd's-purse gets its English name from its heart-shaped seed-cases, which resemble the little purses or pouches which were worn by medieval peasants, hung by draw-strings from the belt. The similarities with a purse don't end there: when they are fully ripe, the pods will break in two, and spill out pale coppery seeds. There was a rather cruel children's game, played in Germany as well as Britain, where one child would persuade another to pick a ripe shepherd's-purse seed-case, and, when it broke, say the child had broken his or her mother's heart. Fortunately, perhaps, this is one of the few ancient plant games that seems to have become obsolete. Shepherd's-purse is a very common annual of cultivated ground and waste places throughout the British Isles. **Pink shepherd's-purse**, *C. rubella*, is an introduced weed from the Mediterranean, with red-tinged petals and sepals. It is normally rare and sporadic, but in the early 1980s so much appeared in the rough grassland of Boxmoor Common, Hertfordshire, that whole patches of the Common appeared quite russet-coloured, even from a distance.

Alpine penny-cress, *Thlaspi caerulescens*, is a neat, rosette-forming biennial or perennial, topped with small bunches of white or purplish flowers. It grows chiefly in the Pennines, on limestone grassland and rocks which are naturally – or artificially – contaminated with heavy metals. The cress (like some of its relatives) is a 'hyper-accumulator' of metals such as nickel, lead and zinc, and this has led to some promising experiments in using it as a natural cleanser of toxic waste sites. The cresses are planted on the contaminated soil, allowed to take up the metals, and then simply pulled up.[33]

Wild candytuft, *Iberis amara*, is an attractive white- or mauve-flowered annual, sometimes reaching 12 inches but usually shorter, confined as a native to the chalk, especially in the Chilterns.

Dittander, *Lepidium latifolium*, is a tall perennial native of damp ground near the sea on the east and south coasts, and introduced or naturalised in a few places inland, for instance by the Grand Union Canal in London, and along a considerable stretch of the main road east out of Baldock, Hertfordshire, where it has been since at least 1929.[34] In flower dittander resembles a small-leaved horse-radish, and its roots have a similar hot and pungent taste. Before the introduced horse-radish (see above) became fashionable, dittander was picked, and occasionally cultivated, for use as a condiment.

It was also once a standard herbal treatment for leprous sores, and in 1990 a Sussex botanist found a colony growing on ground once occupied by Chichester's oldest hospital – the Hospital of St James and Mary Magdalen, which was established in the twelfth century (safely outside the medieval city walls) specifically to care for lepers.[35] This discovery prompted John Palmer, from north Kent, to notice that the only three established colonies of dittander that he had found in this part of the county were *all* close to the sites of turn-of-the-century hospitals: Bexley Hospital, near Dartford Heath (opened in 1898); Joyce Green Hospital, on Dartford Marshes (opened 1902); and the Old G.L.C. Southern Hospital at Darenth (opened 1890 but now demolished).[36] It stretches credibility that dittander was still being cultivated as a leprosy herb at the end of the nineteenth century, but so does the possibility that all three sites are near hospitals purely by coincidence.

Hoary cress, *L. draba* (VN: Thanet cress, Thanet weed). The white-flowered umbels of this pepperwort are increasingly common on waste ground, roadsides and field borders in England and Wales. It arrived in this country from mainland Europe at Thanet, in Kent, in a manner that has become part of local mythology: 'This plant was first recorded here in the early nineteenth century, introduced after the battle on the Island of Walcheren in the Napoleonic wars. The soldiers, stricken with fever, were brought to Ramsgate on palliasses stuffed with hay which was later given to a Thanet farmer, who ploughed it into his fields as manure. The cress then appeared in great quantity and spread over Thanet and now grows along the south coast. One of its English names is Thanet weed.'[37]

Cabbage, *Brassica oleracea*. Our wild, or sea, cabbage, is a variety of the same species as the cultivated cabbages, and occurs on cliffs in a few scattered places in Wales and southern England, and around western European coasts as far as the Mediterranean (where it appears as ssp. *robertiana*). But whether our British colonies are natives, part of the ancestry of one of Europe's most widely domesticated and diversified vegetables, or garden throw-outs gone wild is an open question. N. D. Mitchell of Newcastle University has studied our surviving *B. oleracea* sites

Garden candytuft naturalised at Breakheart Quarry, Gloucestershire.

Wild or sea cabbage, a possible ancestor of cultivated varieties, is confined in Britain to sea-cliffs.

(some 30 in number) and has found that almost all of them are close to or within towns and villages, and that individual 'clumps' are often short-lived. (Though they can be strongly rooted: one saved a climber who lost his footing on a south-coast cliff in the early 1980s.) The population on the Cliffs of Dover is unique in that it has been recorded as growing wild for over 400 years: 'This population has perhaps the strongest claim to native status, yet Dover has had extensive garrisons for centuries and has been a major route for invaders. It was also one of the main areas of Saxon settlement during the early post-Roman period. Thus cabbages may have been grown here for food since AD 500 or earlier, and it is still a major area for cabbage cultivation.'[38]

At Whitby, North Yorkshire, wild cabbages grow particularly near old allotments, and at Tynemouth, Northumberland, close to the site of an old priory. (There is a surviving local name of 'Monk's cabbage'.) Mitchell believes that the evidence points to the wild cabbage being a Roman or Saxon introduction which escaped from cultivation to form naturalised colonies in suitable habitats, which have then been repeatedly 'topped up' by bird-sown specimens from cultivated patches nearby. Certainly cultivars such as red cabbage, savoys and kale (though these may contain genes from other *Brassica* species) nearly always revert to the form of the wild species after a few years of self-seeding.

Oil-seed rape, *B. napus* ssp. *oleifera*, is an increasingly common arable plant, grown for its oil-bearing seeds and occasionally for fodder. Its flowers can turn whole areas of the countryside a dazzling yellow, and have a penetrating, kale-like smell – qualities which have combined to make it a controversial crop, especially among beekeepers. (It is a favourite, and reliable, food plant for bees in spring, but gives a tang to the honey that is not to every human's taste.) Understandably oil-seed rape is now the commonest casual member of the cabbage family, self-seeding on roadsides and field edges.

Black mustard, *B. nigra*, is a fairly common yellow-flowered cabbage of cultivated ground, waste places and river-banks. The seeds of cultivated varieties are ground to make mustard flour – often mixed with those of **white mustard**, *Sinapis alba* ssp. *alba*, from southern Europe, which is quite widely naturalised and is also the mustard of 'mustard and cress'. **Charlock**, *S. arvensis* (VN: Runch, Runches), is a similar yellow crucifer and a common native annual of arable fields and waysides. The rough and bitter leaves have been used as a green vegetable in times of food shortages.

Hairy rocket, *Erucastrum gallicum*, is a yellow rocket from Europe, naturalised in a few places, especially in Wiltshire: 'This rather aptly named plant occurs frequently on Ministry of Defence land on Salisbury Plain, on the disturbed ground of tank tracks, shell holes, etc.'[39]

Lundy cabbage, *Coincya wrightii*, is a substantial, perennial cabbage, often as much as a yard tall, which grows only on cliffs on the south-east of Lundy Island in the Bristol Channel. The whole plant is covered in fine, silvery hairs. It is the sole host plant of two beetles, also endemic to the island; and, given the time for such a specific adaptation to occur, the Devon botanist Dr Elliston Wright considered that the Lundy cabbage may have been isolated on its island redoubt since the early Pleistocene era – perhaps half a million years ago.[40]

Isle of Man cabbage, *C. monensis*, is similar but smaller, and hairless or nearly so. It grows on sandy ground, especially on the Isle of Man, though also in a score of other seashore sites from South Wales to north-west Scotland.

(A rather more popularly known Isle of Man 'cabbage' is the **cabbage-palm**, *Cordyline australis*, from New Zealand: 'It is widely exploited by the tourist industry as a sign of the island's mild climate, and features largely in the literature advertising the resort [Peel]. It seeds itself rapidly, and its spread through the island owes a lot to a charity sale held at Bishopscourt in 1905. It is known by all as the Manx Palm.'[41])

Oil-seed rape, a widespread arable crop and now an increasingly common casual in the wild.

Sea-kale, *Crambe maritima*, is one of the largest and most striking of all maritime plants, and in the few places where it occurs in quantity – now very few, sadly – it dominates the shorescape. From a distance the tightly clumped plants look like enormous sea-urchins, or a rotund desert cactus that has taken to the shingle. Closer to, they have something of the character of outsize cauliflowers. The domed sprays of white flowers are held above, and partly wrapped by, fleshy, glaucous blue-green leaves, which at the base of the plant can be 12 to 18 inches long. The single, small colony that survives on the north Norfolk coast at Cley (deliberately planted there in 1912)[42] is a well-known landmark on the walk out to Blakeney Point.

Children in Anglesey have had ingenious fun with the outsize leaves and round green seeds: 'In Cemlyn Bay [they] used it when playing shop – the leaves representing cabbage and the fruit, peas.'[43]

But sea-kale's main attraction has always been as a vegetable delicacy. For centuries, on south-coast beaches and cliffs, coastal dwellers would cut the young shoots, especially where they had been naturally blanched while growing up through the shingle or sand. In some places, locals would watch for the shoots to appear in the early spring and heap seaweed or sand over them to make the blanched shoots extend even further. At Dungeness in Kent (which still has one of the biggest colonies in Britain) this custom continued into living memory: 'Many years before the nuclear power station I can remember locals kicking shingle over the plants in early spring to blanch the shoots.'[44]

To judge from contemporary records sea-kale was abundant along the south coast in the eighteenth century. There was some trade in the shoots at local markets, and it was grown by a few specialist gardening enthusiasts. Gilbert White sowed some in his garden at Selborne on 6 April 1751, with seed gathered on a visit to Devon the previous August.[45] But its status began to change when White's near neighbour, the distinguished Hampshire botanist William Curtis, popularised the vegetable with his pamphlet *Directions for the Culture of the Crambe Maritima or Sea Kale, for the Use of the Table* in 1799. As a result the plant was taken up by London society, and the demand at Covent Garden increased enormously. Fresh shoots – and, for that matter, roots and seeds – were now gathered for export, not just for home use.

Above: sea-kale, on the shingle at Cemlyn Bay, Anglesey.
Left: wild radish is a widespread and variable annual, with flowers principally yellow or white, often thinly veined with lilac. The garden radish is a separate species of unknown and ancient origins.

The impact on some populations did not seem to be immediate. On the Isle of Man at the end of the nineteenth century, it was collected from the shore in cartloads: 'It was fashionable to collect the winter roots for forcing, and junior civil servants had to affect pleasure when their seniors made them presents of the fresh shoots.'[46]

According to Devon naturalists, it was still 'very frequent' in the county in 1915. But by 1939, the Revd Keble Martin, editor of the great *Flora of Devon*, described it as 'much less common than it used to be', and blamed erosion of the cliffs as well as the gathering of the roots for cultivation.

That is a fair statement of the sea-kale's contemporary status; and it is probable that its slow decline is due now more to the instability and human disturbance of many coastal habitats than to uprooting.

Wild radish, *Raphanus raphanistrum*, is a common annual of cultivated and waste ground, with rather delicate cruciform flowers of white, yellow or lilac, and rough, bristly leaves. It is probably a southern European native, but was in Britain by prehistoric times. The yellow-petalled forms are apparently commoner in the north and west. (The cultivated radish, which occasionally escapes, is a different species, *R. sativus*.)

Mignonette family
Resedaceae

Weld, *Reseda luteola* (VN: Dyer's rocket). A tall biennial with lax leaves and spires of feathery, greenish-yellow flowers, which is quite common throughout much of Britain on dry disturbed ground, from arable field edges to railway sidings. It is probably native on chalky and sandy soils in southern and eastern England. But it gives one of the best yellow dyes, and its distribution has almost certainly been extended by cultivation, which has been going on since Neolithic times. The principal growing areas were Yorkshire, Lincolnshire, Essex and Kent, and relics of weld catch-crops may still persist in these areas amongst genuinely wild lineages.

'A native plant grown as a crop in north-east Essex often in with the corn. All parts of the plant were harvested and used in the process of dyeing locally woven cloth yellow. Evidence of this tall, majestic plant is still to be seen on field margins and roadside verges.'[1]

'In Yarnton, Oxon, weld was grown in the Manor Park, and the dye was used for colouring yarn, and mixing a paint known as "Dutch pink". As this needed expert cultivation, a family known as Dunsley was brought down from Lincolnshire where the plant was well-known. They tended the plants, and passed on their knowledge.'[2]

The use of weld, like other ancient dye-plants, is undergoing a revival – as are the fortunes of the plant itself in some parts of the country.[3] On the sandy soils of East Anglia's Breckland, where the Forestry Commission have recently clear-felled large areas of conifers, weld and viper's-bugloss have sprung up in prodigious quantities, the separate yellow and blue flowers making one of the most dazzling colour combinations in eastern England.

Wild mignonette, *R. lutea*, is like a miniature and paler-flowered version of weld, with a similar distribution and taste in habitats, so that the two species often grow side by side. The flowers have a musky scent, but nothing

Weld, East Suffolk. A traditional dye plant and quite common as an arable weed on light soils.

Wild mignonette, a native of bare places on sand and chalk in the south and east.

like as strong or sweet as that of **garden mignonette**, *R. odorata*, an annual from the Mediterranean, which occasionally escapes and is found as a casual on tips and in waste places near gardens.

Heather family *Ericaceae*

Rhododendron, *Rhododendron ponticum*. This, the commonest of the pink-flowered rhododendrons, is regarded as a menace in many parts of western Britain, where it has escaped from gardens and built up impenetrable, obstinately rooted, evergreen thickets at the expense of almost all other species. But it may just be reclaiming old territory. It was a species whose natural distribution included much of Europe before the last glaciation, and pollen and other remains have been found in Ireland. Now it is not regarded as native west of the Black Sea (except in Spain and Portugal), and all British feral colonies have originated from garden escapes. Provided the soils are moderately acidic, rhododendron can spread prolifically by both seed and sucker.

It is not universally hated. In the drier climate of East Anglia it is less inclined to invade, and the thickets on the RSPB's reserve at Minsmere, Suffolk, for instance, are tolerated as cover for birds. (They were the nesting site of the first pair of Cetti's warblers to breed in the county.) In conifer plantations they provide some bright colour in late spring. And even where rhododendron is at its most imperialistic, in Wales, it does have some historic interest, as James Robertson argues: 'The big Welsh estates (the retreats of the wealthy who were attracted west by the lure of the picturesque) have left a substantial legacy in the contemporary Welsh landscape and plant life of Wales. There are many species that are encountered again and again, primarily in what are now taken to be woodlands. They include the remnants of box knot-gardens, lime avenues, snowberry and flowering nutmeg plantings, yew hedges, stately bay trees, laurels and *Rhododendron ponticum*, which has established a particularly renowned place in the landscape of Snowdonia.'[1]

Marooned bushes and trees of many other *Rhododendron* species and varieties (especially **yellow azalea**, *R. luteum*) occur in parks and plantation woods throughout western and upland Britain.

Trailing azalea, *Loiseleuria procumbens*, is our only truly native 'azalea' – a domed shrub with small pink flowers, found in rocky places above 1,300 feet in the Scottish Highlands. **Blue heath**, *Phyllodoce caerulea*, is one of the rarest and most elusive plants in Britain, confined to a handful of high moorland sites in Scotland, with flowers (mauve more than blue) which appear only irregularly.

Heather, *Calluna vulgaris* (VN: Ling). In an entirely natural situation, heather is a less showy and gregarious plant than the one that colours whole areas of upland moor and southern heath in August. It straggles about the understorey of northern pine and birch woods, and some open beech and oak woods in the lowlands, but elsewhere flourishes only where there is no natural tree cover – on gale-blown western cliffs and sand-dunes, for example, and above the tree-line on Scottish mountains.[2]

The great sweeps that are called 'heaths' because of the abundance of heather species are almost always the result of human activity. Heathland develops when trees are cleared on poor, acidic soil and grazing animals (or fire) prevent them becoming re-established.

There are no precise historical figures for the amount of heather-clad land. At the end of the seventeenth century Gregory King estimated that there were 10 million acres of 'heaths, moors, mountains and barren land' – about a quarter of the land surface.[3] For those who shared its frugal habitats, heather was one of the basic raw materials of domestic life, like the bracken and gorse with which it often grows. It has been used as fuel, fodder and building material where wood was in short supply (sometimes standing in as the framework of wattle-and-daub), and its springy stems have been bundled up into thatch and brooms and woven into ropes. Its roots were carved into knife handles, particularly for the ceremonial Scottish dirk. The spikes of honey-scented lilac flowers make an orange dye, a sweet tisane and a spectacular beer. It is so softly supporting and fragrant that it was used for human as well as animal bedding, and Scottish settlers took it to America with them, naturalising one of their national symbols thousands of miles beyond its natural range.

Rhododendron in the New Forest – an obstinately naturalised escape these days, but once a north European native.

The widespread abolition of common rights and subsequent loss of grazing meant that in the nineteenth century and early twentieth century a great deal of heath returned to woodland. More, trapped in its simplistic image of 'wasteland', has succumbed to development. Today, there are probably not much more than two million acres across all of mainland Britain, and most of this is in the uplands, where heather moorland is conserved for sheep and grouse. In England, heath now covers less than one-third of one per cent of the land area. The losses have been greatest on the southern heaths, with many areas such as south Dorset and the Breckland losing three-quarters of their heathland this century.[4]

But in the surviving heaths some of the traditional uses of heather have been kept up, albeit on a reduced scale. In the New Forest it has been used to repair potholes in the rides and tracks: 'The local estate sends a tractor round tipping gravel where needed, but our sage says, "T'aint no good unless they puts down 'eather first." He must be right, too – the holes are still there!'[5]

Like bracken, it was also used as a packing material, as in Clwyd: 'Bricks used to be taken from Buckley by tramroad to Connah's Quay, wrapped in straw to avoid breakages. When foot and mouth threatened exports to Ireland, because of a ban on straw, a trade was created for heather as a substitute. This was gathered from the hills around Buckley and heaped up at the brickyards ready for use. There are still patches of heather beside the former tramroad and on old coal banks around Knowl Hill.'[6]

On the Isle of Man, 'heather was wound into ropes called Gadd. This was strong enough to be used for mooring boats.'[7]

Around Ashdown Forest in Sussex: 'While it isn't permitted to take hives on the commonland of Ashdown, beekeepers still move their hives into the gardens and land adjacent to the Forest to take advantage of the heather bloom.'[8]

On other Sussex heaths, dead growth is cut and used as firelighters.[9] In Scotland, people still sometimes put a sprig of heather under their beds,[10] echoing its old use as a mattress plant – or maybe just for luck: 'White heather is considered very lucky, especially in Scotland, and many if not all bridal bouquets have a sprig of it; also wedding, birthday and celebration cakes.'[11]

Heather ale holds a special place in the mythology of the Highlands. It is a brew of ancient legend, lost recipes and a recent dramatic revival. Enthusiasts for the beer believe it to have been first developed by the Picts 4,000 years ago and have as their evidence some drinking vessels which were found during the excavation of a Neolithic settlement on the island of Rhum, in 1985. A crust on the inside of the cups contained pollen spores of oats, barley, heather and meadowsweet. The archaeologists believed this to be the remains of a fermented beverage and were able to produce a moderately acceptable ale using the same ingredients.

Heather ale was once one of the staple drinks of the Highlands. It was made by clan 'yill wives' and drunk from cattle horns. It is even possible that the first 'uisge-beatha' (Gaelic for whisky) was produced by condensing the alcohol from hot heather ale against stone walls.[12] But across the Highlands it went into decline during the wars with the English and the Highland Clearances, though the brewing tradition was kept up in the Hebrides and Orkneys. When Thomas Pennant visited Islay in 1774, he found that the islanders made ale from 'the tops of young heath, mixed with a third part of malt and a few hops', and in some isolated spots such brewing may have survived into the nineteenth century.

But its reintroduction on a commercial scale on the mainland had to wait until the 1990s and the enthusiasm of a Glasgow home-brewer, Bruce Williams. The story of his enterprise reads like a Scots legend in its own right. In 1993 a woman gave him a recipe in Gaelic, which had reputedly been handed down through ten generations. After many false starts, Williams found a way of making the process work on a commercial scale, and in 1994 he launched his remarkable, honey-scented, bitter-sweet 'leann fraoch' (Gaelic for heather) ale. His heather comes chiefly from Argyllshire: 'It is picked by local pickers, July to August, in the Oban to Connel area. North Connel hills' south-facing crops are earliest. Only flowering tips are removed. In August, the majority is picked between Tyndrum and Loch Awe, also on the "Rest and be Thankful" Glen (which many pickers did!), mostly pale purple and white ling heather. Heather is abundant in Argyll. It is a wasted resource. The sheep farmers burn it ... The plants do not appear to be damaged by cropping; the previously cut plants in my own garden flowered again the next year.'[13]

The brewing process begins conventionally enough, with the heather flowers added as an aromatic flavouring to the malted barley, as if they were hops. (They also provide, through their nectar, a small amount of additional, fermentable sugar.) The hot brew is subsequently filtered through a bed of heather tips.

Bruce Williams's enterprise has given a boost to the revival of interest in herb beers, and ales bittered with wormwood, bog-myrtle, ginger, lemon and even the original 'ale-hoof' (ground-ivy, see p. 317) are increasingly appearing in commercial brewers' catalogues.

There are various other heather species – some highly local. **Dorset heath**, *Erica ciliaris*, is large-flowered, and confined to damp heaths in Dorset and the West Country. **Cross-leaved heath**, *E. tetralix*, is common on wet heaths and moors throughout Britain. **Bell heather**, *E. cinerea*, usually grows on dry heaths. **Darley Dale heath**, *E.* × *darleyensis*, is a cross between Irish heath (*E. erigena*) and the central European spring heath (*E. carnea* or *herbacea*)

Indigenous heathers: ling on the North Yorkshire moors (left) and Dorset heath (above).

which was planted in the 1930s on a bank beside the A12 near Woodbridge in Suffolk and is now well established. It flowers between November and June. **Cornish heath**, *E. vagans*, is more or less confined to the magnesium-rich serpentine rocks of the Lizard Peninsula in Cornwall, where it grows in at least six colour variants.[14]

Bilberry, *Vaccinium myrtillus* (VN: Blaeberry, Whortleberry, Whinberry, Wimberry; for fruits: Whorts, Hurts, Urts). Bilberry-picking transports one more thoroughly to the role of hunter-gatherer than even blackberrying. The fruit is virtually unknown in cultivation – a source of constant bafflement to anyone who has sampled the dark, winy berries. (Commercial blueberries are a larger and less flavoursome American species: see below.) Bilberry's favoured habitats are in wild places, too, on acid heaths and moors in the northern and western uplands and as an understorey in Scottish pine-woods. And, because the shrub is low-growing and the berries often hidden under the leaves, much of the collecting must be done on hands and knees. In a BBC documentary in the mid-1970s, a group of volunteers were marooned on Exmoor as an experiment in survival. They spent much of their time browsing for bilberries on all fours, and an anthropologist, who was one of the party, remarked that they resembled nothing so much as a troop of foraging chimps.

But bilberry-picking has always been a great social and family occasion, with slightly differing customs in different parts of Britain. (And different names: the fruits are 'hurts' in Surrey, for instance, and 'whorts' in Somerset.)[15] Bilberries in Devon are picked mainly on Exmoor and Dartmoor and cooked in pies. Picking them used to be a local cottage industry with its own 'Whort Sunday' celebrations in August.[16]

'When I was living at Nether Stowey [Somerset] in the 1970s, whort picking had just ceased to be a commercial activity, giving a seasonal income to the villagers. However it had not long gone, and the locals always recommended the wearing of wellingtons whilst picking because of the danger of adders.'[17] (This may have been an example of the frequent practice by local pickers of exaggerating dangers, to deter 'outsiders'.)

Cornish heath on cliffs above Mullion Cove on the Lizard Peninsula.

Bilberries, once an important wild harvest in upland Britain.

Other threats, ancient and modern, have sadly curtailed picking in parts of the north-west, including the Isle of Man:

'Bilberries occupy a place in the Manx way of life beyond the simple attraction of flavour. A number of families reserve the first Sunday in August as the right day for gathering them without being aware that they are participating in a custom from much earlier times. Laa Luanya, Quarter Day, was formerly observed on 1st August by climbing a nearby hill and indulging in riotous games and lovemaking. There are many accounts of local clergy preaching against the practice, and as late as 1820 a preacher named Glick went up South Barrule in order to deter its observance. There is evidence from one of the last native speakers of Manx that there was still a handful of young people keeping up the custom in the early part of the present century. For a long time previously the activity had been disguised under the euphemistic excuses of "looking for a well", or "gathering bilberries". [Now] the accident at Chernobyl has deterred many people from picking them for fear of contamination. How long it will be before people's confidence returns remains to be seen.'[18]

The heart of mass bilberry-picking is the Welsh border country:

'Considerable numbers of people still pick whinberries, some of them [in Condover, Shropshire] using the traditional coarse-toothed metal comb [known as a *peigne* in France] with which the berries can be relatively quickly stripped from the plants.'[19]

'Bilberries are always known as "wimberries" locally, and there is an almost ritual annual return to traditional picking places. They were very important as a crop up till the Second War. Picking wimberries by whole families (taking the kettle up to make tea) on the Stiperstones and Long Mynd and selling them to dealers to be sent away (for dye, mainly) would keep children in boots for the next winter. There are school log-book references over the whole area for days taken off for wimberry picking.'[20]

In Gwent: 'The most worthwhile crop is the bilberry, in these parts known as wimberries. Although they grow all over the mountains, there are only a few favoured areas on north-facing slopes where they grow in such numbers that they can be picked commercially. Around here this is a traditional summer job which in the past was an important help to the family finances. One lady told me that when she was a child she and her brothers and sisters had to pick enough wimberries to buy their winter shoes. Today, most people pick enough just for their freezers, but a few dedicated pickers pick for the market and supply local shops, or sell to others for their freezers.'[21]

Wimberry tarts and pies are sold in pubs around Shropshire and the Marches. In Yorkshire they are known as 'mucky-mouth' pies and served as part of funeral teas. In a hotel in Hawes, North Yorkshire, I have eaten a traditional bilberry dessert, with the fruits (flavoured with a few sprigs of mint) baked in a Yorkshire pudding.

Hybrid bilberry, *V. vitis-idaea* × *V. myrtillus*, a cross between the bilberry and its just-about-edible cousin the cowberry (*V. vitis-idaea*), was first discovered on the Maer Hills in North Staffordshire in 1870. It also grows now on Cannock Chase. It is fertile and sets edible fruits.[22] **Blueberry**, *V. corymbosum*, the larger-fruited American bilberry, now grown commercially in many parts of Britain, is naturalised from bird-sown seed on heathland in south Hampshire and Dorset. **Cranberry**, *V. oxycoccos*, is a low shrub of bogs and very wet heaths, with bright pink flowers, but not always producing the round to pear-shaped, edible orange-red fruits. Those used in cranberry sauce are usually from the **American cranberry**, *V. macrocarpon*, which is grown on a small scale commercially in Britain and naturalised in a few peaty places. **Bearberry**, *Arctostaphylos uva-ursi*, is a locally common shrub in rocky moorland in northern Britain. The shiny red berries are sharp but edible – as are the black fruits of the rarer **mountain bearberry**, *A. alpina*.

Primrose family *Primulaceae*

Primrose, *Primula vulgaris* (VN: Spinkie). The primrose is the *prima rosa*, first flower of the year. Despite blooming almost throughout the year in sheltered Cornish hedge-banks and Sussex copses, its pure yellow flowers and tufted habit – arranged naturally into the form of a posy – have made it a universal token of spring, and especially of Easter. For generations bunches were picked as presents for parents and decoration for churches. They are also still used occasionally in the making of 'Pace' or 'Pasche' eggs (see Spring festivals, p. 174): 'On Good Friday we wrapped ivy leaves, onion peelings, primrose and celandine and gorse flowers around an egg, then newspaper and string, before hard-boiling a panful of eggs to go Pasche egging on Easter Monday, with bread and butter and a packet of crisps for a picnic.'[1]

'In Chesterton Wood, Warwickshire, old Mr Tulley,

Cranberry 'hummocks' on Abbots Moss, Cheshire. The closely related American species is used for cranberry sauce.

who owned the wood, always opened it on Good Friday, so that local people could go and pick primroses to decorate the Churches, and his son has continued the tradition.'[2]

A woman who went to boarding school in Sussex remembers her annual Easter Botany Walk: 'We went in crocodile to Ditchling Beacon, and then broke ranks and foraged in coppiced woodland thick with primroses, which we were allowed to pick and bunch up with wool ties and send to our parents for Easter.'[3]

'Bunches were tied with wool, and then attached to a twig, which was carried horizontally, so that the flowers were not crushed.'[4]

'We used to pick very large amounts of primroses every springtime. Our lady at the Big House used to take them up to a London hospital. My mother packed the primroses in forest moss, having made the flowers into neat bunches.'[5]

'A 90-year-old friend of mine, whose family were land and mill owners in the Minchinhampton area of Gloucestershire for centuries, remembers her mother's spring wedding, when wild primroses and cowslips were strewn all along the church path for the bride to walk on.'[6]

'Primroses were an annual cash crop for the family when I was young, and my mother, sister and I daily walked miles to pick them in season. The laden baskets would be gently emptied onto the kitchen table, and we would sit for hours making up small bunches of flowers, with two leaves per bunch, tied with cotton from a sewing reel, stems trimmed to uniform length, and then placed in large, shallow bowls of water overnight. In the morning they would be carefully wrapped in tissue paper, inside cardboard flower boxes, and then one member of the family had the two-mile precariously stacked bicycle-push to Gwinear Road station, from where the boxes would steam their way to the London and Birmingham markets ... Oh, to see again such abundance of primroses growing everywhere.'[7]

A more formal celebration is Primrose Day on 19 April, when primrose flowers are placed on Disraeli's statue in front of Westminster Abbey (and also on his grave at Hughenden in Buckinghamshire).[8] They were the politician's favourite flower, and Queen Victoria regularly sent him bunches from Windsor and Osborne. After his death in 1881, the botanist Sir George Birdwood suggested inaugurating a 'Primrose Day' and the custom has been kept up ever since.

Primroses and other 'first flowers' of the spring.

The picking of primroses, especially *en masse* ('Gipsies used to pick pillowcases full to sell in Devon towns in 1930s and 40s'),[9] began to get a bad name in the more conservation-minded 1970s and 80s. Yet there has never been any real evidence that picking primroses, as distinct from digging them up by the root, has any effect on their numbers. Oliver Rackham can find no correlation between public access and primrose abundance in East Anglia, and puts three of the most publicly used rural woods in the region amongst his top twenty primrose woods.[10] And in Devon, a local and little-known commercial enterprise, involving extensive primrose-picking, led to an investigation which for the most part confirmed that little harm resulted.

In the early part of this century, the owner of Hele paper-mill in South Devon decided to bring 'a breath of Devon air' to the buyers of his paper. He arranged for bunches of primroses to be picked and sent off to valued customers. The primroses were picked from woods and hedgerows by the wives and children of the mill-workers.

'Local boys and girls picked the primroses in March and April – 25 primroses and 5 leaves in each bunch tied with string, and took them to the Mill, where they were boxed up and sent. My son and daughter partook of this "pocket-money" exercise which had gone on for years before we moved here.'[11]

The custom soon spread to other Devon paper-mills and continued when most of the smaller firms were bought up by the Wiggins Teape Group. But in the mid-1970s, the company received a good deal of adverse publicity about what they were doing, and in 1977 they invited a team of ecologists from Plymouth Polytechnic to make an independent assessment of the effects of the annual harvest.

Although the team uncovered a certain amount of *ad hoc* picking by children, most was done on a contract basis on a few farms in the South Hams district. The numbers picked seem, on the surface, to have been prodigious. In 1978, some 13,000 boxes of primroses, containing a total of 1,300,000 blooms, were mailed out. But the operation was carefully organised, with no evidence of mass picking. 'Only a few blooms were picked from any one plant at any one time; for the packaging and posting process, only immature blooms can be used, so that the recipient receives fresh lasting material (which can last from seven to 14 days). Thus the pickers selected young blooms, and open flowers remained on the plants. Clearly, although the picking reduces the total number of flowers able to set seed, no plant has all of its flowers removed.' The Plymouth team also noted that 'all involved in the operation, farmers, pickers, packers and organisers at the hall, enjoyed being involved, and the annual event, lasting for up to three weeks, was regarded as something of an occasion.'[12]

A number of comparative plots were set up, and the team's preliminary conclusion was that 'the level of picking carried out is not a serious biological threat to the survival of *Primula vulgaris* in the South Hams', especially as individual plants seemed to have a life of about 15 to 25 years.

But the custom was wound up a few years later, in the light of hardening public attitudes towards wild-flower picking and a perceived reduction in primrose populations nationwide. Whether this had anything to do with picking is doubtful. Primroses have always had a rather odd and scattered distribution, both locally and nationally. Where they have declined or disappeared, it is usually as a result of their habitat becoming unsuitable – drained, sprayed, or shaded out perhaps. Where they do not occur, it is usually because the local climate or soil does not suit them.

In 1944, Professor Ronald Good published a classic study of the distribution of primroses in Dorset. He travelled 'every road and major track in Dorset' and discovered a distinct pattern as to where primroses did, and did not, grow. In the west they occurred in woods and hedge-banks. In the east they were largely confined to woods. Nowhere did they grow in hedge-banks but *not* in adjacent woods; and there were two conspicuous areas in which there were virtually no primroses at all. Professor Good explained the pattern in terms of the distribution of soils and rainfall in the county. Primroses seem to prefer damp conditions, and the rainfall is noticeably higher in the west. Although hedge-banks dry out more quickly than woods, in the west they are always moist enough to support primroses. In the east only woods on clays and loams are usually sufficiently moist. The two primrose 'gaps' corresponded roughly with the chalk and sand areas, which have poor water-holding capacity.[13]

It is an explanation which at first sight seems to hold true for the country as a whole. In the extreme west, primroses will grow anywhere – on sea-cliffs, stone walls, even along the middle of country roads. The further east you move, especially into chalky areas, the more they are confined to woods and shady banks. But problems arise when you consider an area such as central Suffolk, where the rainfall is on average less than half that of east Dorset but where primroses still grow abundantly in many hedge-banks. Some of these are ancient sites, perhaps once the bank-and-ditch system of ancient woods, so perhaps continuity plays a part. But so, clearly, does disturbance. Primroses have not only colonised new motorway banks on unsuitably dry soils, but have shown a great willingness to spread where paths are opened up or broadened in woods. This is partly because of the increased light. But trampling and traffic help shift the primroses' rather immobile seeds that otherwise have to rely on rain-splash or the packhorse labours of ants to move even short distances.

Oliver Rackham has suggested other factors in the equation and believes that primroses will really prosper only where soils are rich and have a higher than average level of mineral nutrients. And, though they can tolerate deep shade, they need regular bursts of light to flower and set seed. In Buff Wood, Cambridgeshire, an area of coppice was cut for the first time in over 60 years. The primroses, which had been inconspicuous, began to flower in the following November and produced at least a hundred times their usual output of flowers for the next eighteen months.[14]

In any of these large populations, and especially in the great linear colonies along West Country hedge-banks, you will notice variations in the colour, texture and size of primrose flowers. Any with pure white flowers may be escaped specimens of one of the Mediterranean subspecies (sspp. *balearica* and *sibthorpii*).[15] But the native species can produce flowers in every shade from deep yellow to palest cream, including a delightful variety with white flowers round a pale yellow 'eye'. There is also a pink – or more accurately rhubarb-and-custard – form. It is most frequent in churchyards and on banks close to villages, so there is some doubt about its origins. But it also occurs in much wilder sites, especially in west Wales, and is so constant in its coloration that it is almost certainly a genetically different native form: 'Pink primroses grow in a little wooded hollow, near a small pool on a steep bank, close to the water and shaded by trees, and also alongside a small stream flowing into the pool. They have grown in this location for many years. The site is at least ¾ mile from the nearest house, and there is no path or track near to it.'[16]

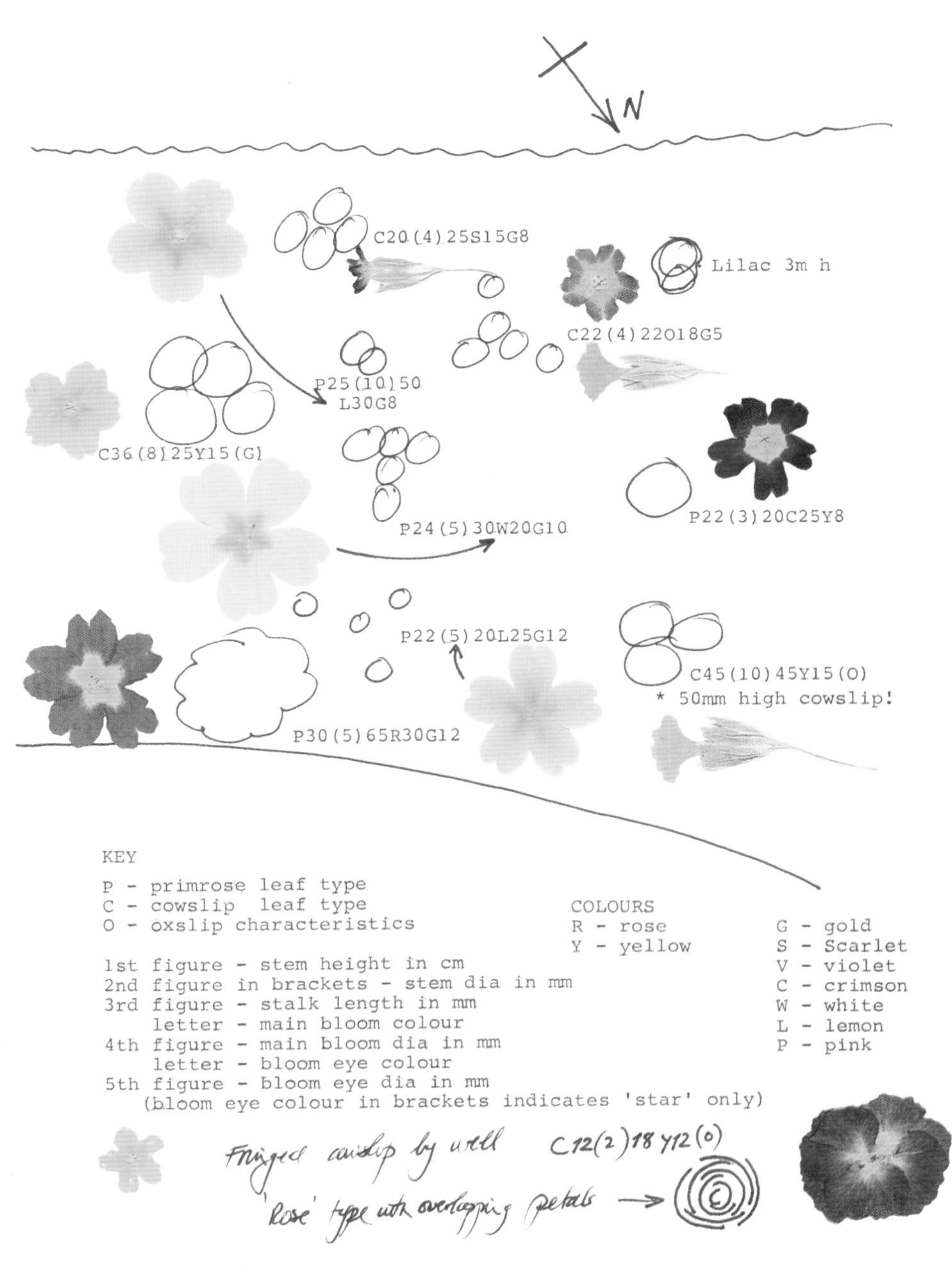

Ian Hickling's plan of his wild Primula corner, with pressed specimens of each variety and a map of their spatial relations (see p. 169).

As well as colour varieties, primrose flowers can adopt unusual, even bizarre forms. The commonest is the umbellate form (var. *caulescens*), in which the flowers form a spray on top of a longer, cowslip-like stalk. Doubles also occasionally spring up as chance sports, and one with soft shell-pink flowers has recently been taken into commercial cultivation as 'Sue Jervis'. Much rarer, though greatly prized by early cottage gardeners, are Jack-in-the-greens, where the normal calyx is replaced by a ruff of miniature primrose leaves, and hose-in-hose, in which a second complete flower grows through the centre of the first. And Gerard described an extraordinary 'amiable and pleasant kind' found in a wood at Clapdale near Settle: 'it bringeth forth among the leaues a naked stalke of a grayish or ouerworne greenish colour: at the top whereof doth growe in the winter time one flower and no more, like vnto that single one of the fielde; but in the sommer time it bringeth foorth a soft russet huske or hose, wherin are conteined many smal flowers, sometimes fower or fiue, and oftentimes more, very thicke thrust togither, which maketh one entire flower, seeming to be one of the common double Primroses, whereas indeed it is one double flower made of a number of smal single flowers, neuer ceassing to beare flowers winter nor sommer, as before is specified.'[17] (I have seen something similar in 'fasciated' primulas, where the stalks of several flowers fuse together. In one garden in Wigginton, Hertfordshire, a specimen appeared in which the stems had virtually disappeared. It was like an alpine primrose, with the flowers forming a tight, tufted dome amidst a wreath of leaves.)

There are, finally, less festive strains amongst the primrose's cultural associations. Shakespeare's metaphor of the 'primrose path of dalliance' in *Hamlet* is a long way from the Victorian custom of planting primroses on the graves of small children. In parts of Buckinghamshire these are often on the west side of the churchyard, producing sadder paths of primrose yellow in April.[18]

Oxlip, *P. elatior* (VN: Paigle). The 'oxlip' which Shakespeare features on his 'bank where the wild thyme blows' and again, in *The Winter's Tale*, where Perdita distinguishes it by the adjective 'bold', is not *Primula elatior*, but what is now more usually known as the false oxlip, a hybrid between the primrose and the cowslip (see below). The true oxlip had not even been officially discovered when Shakespeare was writing and was presumably taken, by ordinary country people and botanists alike, for one of those aberrant forms that primroses are apt to throw up. Its blooms are a similar shape and colour to primrose's, but smaller and more bell-like, and are held in a graceful, drooping, one-sided cluster at the top of the stalk.

Yet, despite the fact that they have a very limited distribution in eastern England, oxlips must have grown in some woods, as now, in colonies millions strong; and the comparative lateness of their recognition is nearly as puzzling as that of the fritillary (see p. 403). John Ray made the first definite record of 'Primula veris elatior pallido flore' in 'Kingston and Madingley woods [Cambridgeshire] abundantly and elsewhere' in 1660.[19] But the standard botanical work of the late eighteenth and early

Oxlips in their heartland on the East Anglian boulder clay. Bradfield Woods, Suffolk.

nineteenth centuries, Sowerby's *English Botany*, though carrying a picture of the true oxlip, rejects it as a separate species: 'In describing the Primrose, we expressed a suspicion that the Oxlip might be a variety of that rather than the Cowslip, or possibly a hybrid between the two. We are still much inclined to the latter opinion and that it has originated from a Primrose impregnated by a Cowslip; its external habit, the contraction towards the middle of the leaf, and the umbellate flower-stalk indicating (according to Linnaeus's ingenious idea) the father, while the blossom itself, in form, colour and scent, most resembles its mother.'[20]

It was not until the 1840s that two Essex botanists, George Gibson and Henry Doubleday, finally clinched the matter. It was Doubleday's observations of the plants around the village of Great Bardfield in Essex that was crucial: 'They cannot be hybrids, for the primrose does not exist in the parish and these oxlips grow by the thousand in the meadow and in the moist woody places adjoining: in one instance a meadow of about two acres is entirely covered by them, being a mass of yellow bloom.'[21]

The true oxlip has drooping, bell-shaped pale yellow flowers held chiefly on one side of the stem.

The species was subsequently named 'the Bardfield Oxlip' to distinguish it from the primrose–cowslip hybrid. Alas, it no longer grows in Great Bardfield, or in many of its previous haunts (including old meadowland – of which there is precious little left in East Anglia). But it has survived in a triangle of land on the boulder clay between Stansted in Essex and Bury St Edmunds and Stowmarket, Suffolk, and in a smaller area west of Cambridge. In these areas it grows in most of the ancient woods, sometimes in great quantities and occasionally replacing the primrose altogether. It is even hanging on in some non-woodland habitats, such as damp field-edges and hedgerows. A hedge-bank near Dovedenhall Wood, Suffolk, is typical in being a relic of a wing of the wood which was grubbed out around 1800.

The few oxlip 'outliers' may be relics on a much grander scale, isolated remnants of a time when oxlips were more widespread in southern and eastern England. One is at Dickleburgh Fen in Norfolk. Another is between Berkhamsted and Bovingdon, on the Hertfordshire–Buckinghamshire border, in a group of four woods on a chalk–clay soil not unlike boulder clay. They came to light in the early 1970s, and there is a slight possibility that they were planted by a local clergyman, the Revd Moule, who was noted for such botanical evangelism. But if so he was very thorough, as some of the colonies are in remote, barely visited spots. It is more feasible that they are a genuinely wild, relict population, especially as the four woods were once part of a large, continuous stretch of ancient woodland (shown on a map of 1766).[22]

All the oxlip populations are probably in a roughly stable state at present. Like primroses, they have benefited greatly where coppicing has been reintroduced. But this has been offset by the increased number of deer and pheasants in East Anglia, which have a taste for the blossoms. Oxlips hybridise with both the cowslip and primrose, producing an almost complete range of intermediate types. There are even occasional pink oxlips, though these are usually the result of crosses with garden primulas. In his garden in Berkshire, Ian Hickling found an oxlip hybrid with the characteristic one-sided bell-flower habit, but bearing blooms of a 'vivid red-terracotta' with oxlip-yellow 'eyes': 'This plant is a natural seedling from the many primulas I have growing here, and a descendant of an oxlip plant brought from East Anglia in 1963, planted in Moulton, Northamptonshire, and then brought here in 1973, from which hybrids with primrose and cowslip have appeared.'[23]

Given the attractiveness of the oxlip's habit, it is surprising that such little use of it has been made in breeding garden primulas.

Cowslip, *P. veris* (VN: Hey-flower, Peggles, Paigles). Cowslip wine, cowslip balls ('tisty-tosties'), cowslips strewn on bridal paths and worn as chaplets on May Day – the cowslip's cultural history suggests a flower that was once as abundant and accessible as the buttercup. No wonder that its dramatic decline between the 1950s and 1980s was felt so keenly.

'One bank holiday the family went to Chingford Plain for the day. There were a dense mass of holidaymakers like ourselves escaping the dirt and grime of our home streets in East London. We found a solitary cowslip where no doubt all its companions had been picked. We encamped and my grandmother sat by the flower all day with it covered by a paper bag to prevent anyone else noticing it. We left that evening with that one flower still intact hoping it would survive at least until the next weekend.'[24]

As with the primrose, it was not over-picking that put paid to cowslips in so many parishes, but the relentless advance of modern farming, especially the ploughing of old grassland and a massive extension in the use of chemical herbicides (which extended to wayside management up to the mid-1980s). And banished along with the flower went a host of vernacular names, for instance 'culverkeys' and 'bunch of keys', from the jiggling egg-yolk flowers that John Clare delightfully called 'cowslip-peeps'; 'freckled face', from the orange spots that lie at the base of the petals, and which Shakespeare thought were the source of the flower's faintly fruity, dill-like scent. (In *A Midsummer Night's Dream*, the fairy chants: 'In their gold coats spots you see/ Those be rubies, fairy favours,/ In those freckles live their savours ...') In George Meredith's novel *The Ordeal of Richard Feverel* (1859), there is even a cowslip malapropism on a par with 'wooden enemies' for 'wood anemones': on forages with his unrequited lover, Richard used to call the flowers 'coals-sleeps'. Now we are reduced to 'paigle', which today is used rather indiscriminately for any wild primula, and 'cowslip' itself, which, though it hasn't the most pretty of origins (it is a euphemism for 'cow-slop' or cow-pat), does at least hint at the kind of company cowslips like to keep.

In Lambley, Nottinghamshire, a dearth of wild specimens has meant that the ceremony of 'Cowslip Sunday' has had to resort to garden-grown flowers: 'Cowslip Sunday is celebrated in Lambley on the first Sunday in May. Nowadays the occasion is marked by having a basket of cowslips on the altar at the morning service in the Parish Church ... Formerly, when cowslips grew more profusely in the wild, parties of people travelled out from nearby Nottingham on Cowslip Sunday to buy bunches of cowslips picked by local children. Some local residents, now in their eighties, remember selling the flowers to day trippers. There is said to have been a special Cowslip Sunday hymn but this is now forgotten.'[25]

But in the 1990s, the cowslip is showing signs of recovery. On the chalky and light-soiled areas of England and Wales that were its stronghold, it has begun to return to unsprayed verges and village greens and to colonise the banks of new roads – no doubt assisted here and there by the scattering of wild-flower seed-mixtures. On downland where grazing pressures have eased, for instance at Tring Park, Hertfordshire, vast masses have reappeared. (Cowslips seem to flower more profusely amongst rank grasses and scrub than amongst large numbers of spring-grazing sheep.) And they are flourishing in the increasing number of churchyards that are being looked after with an eye to their wildlife. Lambley may yet be able to use wild flowers again in its ancient ceremony.

Cowslips share in their family's tendency to produce a profusion of sports and variations, and churchyards are probably the most likely place to find orange-flowered forms and the variety known to gardeners as 'Devon Red'. These are hard to differentiate from conventional cowslips except by their colour and are almost certainly natural, native forms. But, as they are seen most often in mixed populations of primulas, close to habitations, there is always the possibility that some are back-crosses with red polyanthuses (themselves originally crosses between the cowslip and various primroses).[26] Another unusual form is recorded in some handwritten notes in the back of my copy of Druce's *Flora of Oxfordshire* (1886), made by a previous owner, J. M. Albright: 'May 1913. Found cowslip in Henley Nap, 15 inches high with 87 flowerets on the stem. Two other cowslips on same root about 10 or 12 inches high, had 43 and 49 flowerets respectively.'

Primula hybrids. All the above species (including their various forms) hybridise with each other. The commonest cross is between the primrose and the cowslip, which gives rise to the 'false oxlip', *P. vulgaris* × *P. veris*. This occurs wherever both species are frequent and differs from the true oxlip in holding its primrose-like flowers upright and splayed at the top of the flower-stalk. It is the chief ancestor of garden polyanthuses.

But wherever cultivated primulas and polyanthuses (especially red-flowered species and varieties such as the purple-flowered *P. juliae* from the Caucasus and *P. hirsuta* from the Alps) grow in the proximity of wild species, extraordinary progeny can result, including what are known as 'hybrid swarms', in which all kinds of intermediates and back-crosses occur. (Churchyards are famous for this kind of miscegenation, e.g. at Barking in Suffolk.)

In 1839, the floriculturalist Charles M'Intosh reported in the magazine *The Flower Garden*, that a Mr Herbert had 'raised from the natural seed of one umbel of a highly manured red cowslip, a primrose, a cowslip, oxlips of the usual and other colours, a black polyanthus, a hose-in-hose cowslip and a natural primrose bearing its flowers on a polyanthus stalk. From the seed of that hose-in-hose cowslip I have since raised a hose-in-hose primrose.'

A comparable swarm arose in the wild primula corner of Ian Hickling's garden in 1994 (see above), including one particularly striking throwback: 'a cowslip hybrid which has evolved a petal fringe colour rather after the

Cowslips saved from the mower in a Cambridgeshire churchyard.

Plants from a hybrid swarm of Primulas in Barking churchyard, Suffolk, showing colour variants of the primrose and of the crosses between primrose and cowslip ('false oxlips' – the chief ancestor of the polyanthus).

Bird's-eye primrose, a Yorkshire speciality.

fashion of the old "gold-laced" polyanthus, despite there being no flower of that type around to my knowledge last year when hybridisation took place.'[27]

The story of how the great range of garden primulas and polyanthus was developed from these natural crosses is told at length in Roy Genders' *Collecting Antique Plants* (1971).

Bird's-eye primrose, *P. farinosa*. Geoffrey Grigson once suggested that this exquisite upland species, with its posy of coral-pink flowers carried over a rosette of dusty, greying leaves, should replace the White Rose as West Riding's county flower.[28] Reginald Farrer, the great nineteenth-century Yorkshire explorer and plantsman, would probably have agreed. He had seen alpine species all over the world, yet still rated this little primula from his native hills above all the rest.

Bird's-eye primrose needs rather specialised conditions – calcareous 'flushes', where lime-rich water runs or seeps through peat – and as a consequence the plant is virtually restricted to the limestone country of the northern Pennines and Lake District. Round Pen-y-Ghent and Ingleborough in Yorkshire it flowers in its millions in

May and June. But it is the *particularity* of the places in which bird's-eye grows that helps make it such a beguiling flower: on the edges of limestone pavements where the slabs vanish under the rising turf; in pink ripples that trace the route of water leaching down hills (often in company with butterwort, the 'starfish plant');[29] on boggy verges under drystone walls, especially in dips in the road. I have even seen tufts perched precariously on tiny islands in the middle of rushing brooks.

Scottish primrose, *P. scotica*, is like a smaller bird's-eye primrose. One of Britain's few endemic species, it is confined to damp grassy places near the sea in northern Scotland.

Water-violet, *Hottonia palustris*, is a beautiful water-plant found in ditches and shallow ponds, with fronds of ferny, underwater leaves and whorls of delicate lilac flowers, each with a yellow eye. It is rather scarce and confined pretty much to eastern England. **Cyclamen**, *Cyclamen hederifolium*, is the familiar autumn-flowering garden plant from southern Europe. It is quite widely naturalised in hedgerows, churchyards (e.g. Ibstone, Buckinghamshire, and Shillingford St George, Devon)[30] and a few woods, chiefly in southern and south-western England.

Yellow pimpernel, *Lysimachia nemorum*, is a creeping, bright yellow-flowered perennial, quite common in old deciduous woods and by shady streamsides in most of Britain. **Creeping-Jenny**, *L. nummularia*, is a more southerly species, distinguished from yellow pimpernel by its nearly circular leaves and cup-like flowers. A much more floriferous cultivar is sometimes found naturalised close to gardens. **Yellow loosestrife**, *L. vulgaris*, is a tall and handsome perennial of fens, reedbeds, ditches and riversides, bearing its yellow flowers in pyramidal spikes. It often grows with its unrelated namesake, purple-loosestrife (see p. 234), and flowers at the same time in midsummer. **Dotted loosestrife**, *L. punctata*, is a rather similar species to the above, from south-east Europe, increasingly naturalised from gardens in damp places and waste ground.

Bog pimpernel, *Anagallis tenella*, is a distinctive mat-forming plant of bogs, brooksides and damp flushes in peaty ground. Its delicate pink flowers, borne on long stalks, often grow in great quantity, and their effect has been well caught by a contributor who saw them in Shropshire: 'Whole eiderdowns of them in a boggy stream near Offa's Dyke.'[31]

Scarlet pimpernel, *A. arvensis* (VN: Old man's weathervane, Poor man's weatherglass). A common and widespread creeping annual of cultivated and bare waste ground, this has long been known as a combined sundial and 'weather-glass', opening its red petals at about 8 a.m. and shutting them at two in the afternoon or if the weather becomes dull or wet. The subspecies **blue pimpernel**, ssp. *caerulea*, crops up very rarely in arable land, mostly in southern England.

Top: water-violet, Barton Fen, Norfolk.
Above: scarlet pimpernel.

Spring festivals

To say that most spring ceremonies and traditions involve plants would be true, but would be to miss the point: the encouragement and celebration of new growth – both wild and cultivated – is what these ceremonies are for. They are, to use that much misused phrase, fertility rites.

A surprisingly large number survive in modern Britain, yet because of religious, political and commercial pressures they have tended to coalesce around a few key dates, pagan quarter days and Christian festivals merging for convenience with twentieth-century bank holidays. The sacred and secular elements become blurred in a similar way. Only one occasion has no ceremonials attached to it and that, ironically, is the most 'natural' of all – the spring equinox of 21 March, sometimes optimistically called 'the first day of spring'.

One key historical factor must be taken into account in considering the match between the dates of various festivals and the 'natural' calendar. Up until the mid-eighteenth century two different calendars had been operating simultaneously in Britain, the 'Old Style' Julian calendar and the 'New Style' Gregorian calendar. In 1751, Lord Chesterfield's Act provided that the Gregorian calendar should become the norm throughout Great Britain and its dominions. By this time the discrepancy between the Old and New Styles had reached 11 days and, to normalise affairs, Parliament decreed that the days between 2 and 14 September 1752 should be omitted. From then on, natural events were tagged with a calendar date of 11 days later. So, if primroses traditionally flowered on 21 March in a village, they now bloomed on April Fool's Day.

The various species associated with spring festivals are discussed under their individual entries, but the following are some of the chief festivals that involve plants.

The Christian festival of Eastertide begins with Palm Sunday, when sprays of pussy willow or yew are sometimes used as substitutes for true palm. Primroses have become the flower of Easter itself and are often used to decorate churches.

May Day is the occasion of the old Celtic festival of Beltane, which is echoed in dozens of ceremonies across Britain: in Padstow, Cornwall, cowslips are worn in the Obby Oss procession; in Oxford, a Jack-in-the-Green cloaked in hawthorn leaves careers through the city. 'May birching' is largely obsolete, but involved fixing sprigs of plants to people's doors. The plants were chosen either because of their symbolic associations or because their names rhymed with the epithet regarded as most apt for the householder. So, plum, holly or briar meant, respectively, glum, folly or liar.[1] May garlands are still made on May Day in many country schools

Above: 'The Vuz Dance of Flowers', a spring 'trade dance' revived in West Torrington, Devon in 1994 (see p. 233). Right: the Garland King, covered in flowers and foliage, like a Green Man, is carried on horseback throughout the Garland Day celebration in Castleton, Derbyshire, on 29 May.

(and, more traditionally, in a few villages). At Charlton-on-Otmoor, there is a belief that the local May Day garland ceremony is an almost thoroughly Christianised relic of an old pagan festival. The Rector writes:

'With the coming of Christianity the missionaries had two choices with this, as with other customs – they could suppress it or adapt it. It would seem that they adopted the second course. It was clearly impossible to continue a pagan spring festival, so that ended; instead a Christian festival was held in honour of the Blessed Virgin Mary. This was, or became, associated with the figure on the rood in the church, representing the Lord's Mother. With the coming of Christianity, therefore, the pagan mother-goddess was no longer worshipped ... It is a traditional custom, from time immemorial in the village, that children make little crosses covered with flowers ... Since 1963 they bring them in procession to the church, where a service takes place, followed by dancing in the village street. The verse makes it clear that [the carol they sing] relates not to the May garlands carried by the children, but to the decorated "garland" on the north end of the screen, which indeed stood "at the Lord's right hand".'[2]

The more secular garland ceremony at the Oxfordshire village of Bampton has, ironically, migrated to the more overtly Christian festival of Whitsuntide. The flowers used in what is a partly competitive ceremony must be wild:

'The fields belonging to the old Busby brothers were filled with every flower you could think of – Moondaisies, Harebells, Goozie Ganders, Pots and Pans, Clovers, Ragged Robins and Quaker Grass. We picked yellow flags from the brook, because these went on the top of the garland. The flowers were usually kept in a tin bath until Sunday evening. To make the garlands, two willow sticks were tied in circles and placed one inside the other, tied at the top. The grown-ups, mainly Mums, would then tie the flowers (which by this time we had bunched in small bunches) in identical order up each side of the hoops. When all the sides were covered, the garland would be hung on the line, splashed with water and left till morning.'[3]

Other festivals which doubtless began as May Day rites for encouraging growth in fields and woods have also moved towards the end of the month, often joining the civic commemoration of the Restoration of Charles II on 29 May (e.g. Oak Apple Day and Grovely, p. 75, and Arbor Day at Aston on Clun, p. 137). Rather more have clustered around the movable feast of Rogationtide (the fifth week after Easter, leading up to Ascension Day). Rogation Sunday became officially sanctioned by the Church for the blessing of crops, which was combined with the social business of reaffirming land boundaries and common rights in the ceremony known as Beating the Bounds or Perambulation. Plants were invariably involved in this, being amongst the most frequent natural features marking boundaries, as well as instruments (in the form of elm or willow wands) for beating them. The seventeenth-century poet and populist preacher George Herbert, Rector of Bemerton in Somerset, listed the benefits of the ceremony, including 'a blessing of God for the fruits of the field; Justice in the preservation of bounds; Charitie in living walking and neighbourly accompanying one another'.[4]

Mock-orange family
Hydrangeaceae

Mock-orange, *Philadelphus coronarius* and hybrids (VN: Syringa). More popularly, but confusingly, known by the scientific name of shrubs in an entirely different family (the lilacs), 'syringa' is a common garden shrub with heavily fragrant white flowers, introduced from south-eastern Europe. It usually naturalises on waste ground and railway embankments by means of suckers or discarded roots. In some areas of north London, though, there are uneven-aged colonies which seem to have sprung from seed.

Gooseberry family
Grossulariaceae

There is a popular belief (shared by some early twentieth-century botanists) that all currant bushes found growing in the wild are bird-sown naturalisations from gardens. Although this is undoubtedly the origin of many specimens, the three commonest edible species – gooseberry, red currant and black currant – are found growing most happily amongst native vegetation in ancient, undisturbed habitats, and I personally have no doubts that they are all British natives.

Gooseberry, *Ribes uva-crispa* (VN: Goosegogs). A spiny bush, with small hairy berries, gooseberry occurs throughout Britain – in deciduous woodland (where it often fails to fruit), hedges, scrub and occasionally rooted in drystone walls. The fruit appears in late May and is usually ripe enough to eat by mid-July (though it is rather sharp and dry without cooking). A wild bush with wine-coloured berries was found by a contributor in Banffshire.[1]

This humble wild fruit was the ancestor of all domestic gooseberries, including the most sumptuous dessert varieties, and the story of its transformation is a remarkable example of grass-roots plant-breeding. There is no record in Britain of the introduction of any cultivated varieties, nor of cultivation itself beginning before the sixteenth century. English gardeners domesticated the plant by taking in promising specimens from the wild, and by the early nineteenth century gooseberry-growing had become a cult amongst cottagers in the industrial north and Midlands. At the end of the century 2,000 named varieties were in circulation.[2]

Traditional Oldbury tarts, from Oldbury-on-Severn in Gloucestershire, were reputedly made from wild gooseberries – though, being cooked for the Whitsuntide Fairs at the end of May, they must have been made from preserved, not fresh fruit. They were small raised pies, teacup-sized, filled with gooseberries and brown sugar. They are still occasionally served (made from cultivated fruit) in restaurants in the Severn Vale.

Red currant, *R. rubrum*, is a small shrub which occurs chiefly in the southern half of England. Truly wild specimens are more or less confined to ancient deciduous woodland, river-banks and wet alder and willow scrub (carr). The berries are small and pale – often a translucent white. Cultivated red currants are believed to have originated from a cross between this species and **downy currant,** *R. spicatum*, a similar but much rarer native species of limestone hills in the north.

One arcane children's game uses red currant leaves: 'When I was a child in West Oxfordshire my grandfather showed me how to make a "Secret Gang" sign with the leaf of a redcurrant bush. Press the leaf on the inside of your wrist and hold it there until it becomes limp and

Gooseberries painted by Sir William Hooker, one of the foremost botanical illustrators of the nineteenth century.

moist. Take it off and rub the leaf pattern imprinted on the wrist with dust or oil. The result is a very clear imprint which will last for several days, if not washed.'[3]

Black currant, *R. nigrum*, is scarcer and more local than the red currant, but found in similar habitats. The leaves have a strong smell of black currant, and on damp days are detectable some distance away. The densest and most convincingly wild populations of black currant are probably in the fens and broads of East Anglia, where they are sometimes plentiful enough to be worth picking, as the Norfolk naturalist Ted Ellis noted: 'In the summer of 1947, following a notoriously hard winter which decimated the populations of woodland and garden birds in East Anglia, it was found possible to gather large quantities of excellent black, red and white currants from the wild bushes round some of the broads.'[4] Black currant fruits were used in the making of hot drinks or 'robs' for sore throats.

Flowering currant, a familiar garden plant from North America, occasionally found in hedgerows.

Flowering currant, *R. sanguineum*, from western North America, is a popular garden plant which has occasionally become naturalised. There is also a 200-yard farmland hedge made purely of this species at Lexham, on the Swaffham–Fakenham road in Norfolk, which is 'absolutely unmissable in flower'.[5]

Stonecrop family *Crassulaceae*

Navelwort, *Umbilicus rupestris* (VN: Wall pennywort, Coolers). Navelwort is conspicuous in rocks, walls and stony banks in the west of Britain, from its fleshy, circular leaves and spires of straw-coloured flowers. It is 'navelwort' because the stalk grows from the centre of the leaf, forming a dimple on top. The leaves have long been used by children as imitation coins or plates: 'As a child I used to make tiny Victorian posies in the leaves of pennywort, by pricking a hole through the "navel" into the stalk, and filling it with small sprays of flowers such as cow parsley, forget-me-not, violets and vetches.'[1]

Navelwort is one of those species that defines the damp south-west of Britain. Yet it does occur occasionally further east, for instance in the humid valleys of the Sussex and Kentish Weald, especially on old churches. In the sixteenth century, perhaps because of a damper climate or cleaner atmosphere, its distribution was wider, and John Gerard writes of it growing at Westminster Abbey, 'ouer the doore that leadeth from *Chaucer* his tombe to the olde palace'.[2] (Thirty-five years later, by the time of Thomas Johnson's second edition of Gerard's *Herball*, it was gone.)[3]

The name 'coolers' probably derives from the use of the sappy leaves as an ointment for burns, like a poor-man's house-leek (see below).

Orpine, *Sedum telephium*, is a relative of the showy garden 'ice-plant' or butterfly stonecrop, *S. spectabile*, with its succulent, silvery leaves and late-summer pink flowers so beloved of butterflies. Orpine is also succulent and usually pink-flowered, but its leaves are grey-green and flaccid, its blooms sparse, and the whole plant something of a recluse in shady hedge-banks and woodland edges. It occurs throughout Britain, but is nowhere common, and colonies near houses may well be naturalised from cultivated forms. Despite this, it has a rich folklore, particularly as a divinatory plant. One obsolete name, midsummer men, referred to a custom widespread on

Navelwort, named from the dimple formed where the flower-stem joins the centre of the fleshy leaf.

Midsummer's Eve, in which young men and women would stick slips of orpine in cracks in house-beams and joists. The cuttings were placed in pairs, to represent two sweethearts; and the way in which they inclined towards or away from each other was believed to predict the likely progress of the romance. If either of them withered, one of the pair was thought likely to die.

The name 'vazey flower' from Surrey is almost certainly a neighbourhood coining and seems to have occurred nowhere else: 'Orpine, whose name we didn't know, we called Vazey Flower, I think because of the squeaky noise the leaves made if you rubbed them together.'[4]

Roseroot, *S. rosea*, is a yellow-flowered stonecrop, growing on mountain rocks and sea-cliffs in north-west Britain. The root, when cut, develops a strong scent of roses, and it was grown in cottage gardens to make scented waters.

Biting stonecrop or **Wall-pepper**, *S. acre*, is a prostrate perennial with bright yellow flowers, which is quite common on sand-dunes, shingle, open grassland, walls and pavements throughout Britain. It will even grow on roofs, where its presence may explain the longest and most cryptic vernacular name of any British plant, 'welcome-home-husband-though-never-so-drunk'. This name is shared with another denizen of roofs, the **house-leek** or **sengreen**, *Sempervivum tectorum*. This is a species native to mountainous areas of Europe, which has been widely planted in rockeries and on all kinds of domestic stonework. It cannot be said to be naturalised, but can be very persistent. House-leeks were originally grown on roofs as magical protection against lightning, especially on thatched houses, which were very prone to fire. (They are still planted on pigsty roofs in the Forest of Dean.)[5] The thick sap from the leaves was also used as a cooling ointment for burns and scalds, the consequences of fire. But which of these was the primary use, and which the derivative, it is hard to say.

Wall-pepper covering an old roof in Dorset. It has the most cryptic vernacular name of any British flower: 'welcome-home-husband-though-never-so-drunk'.

Saxifrage family *Saxifragaceae*

Meadow saxifrage, *Saxifraga granulata*. This is a declining species of old grassland, especially chalk downland, churchyards and damp hay meadows. Three centuries ago its buttery-white flowers grew, John Parkinson reported, in central London, close to Grays Inn Road.[1]

Rue-leaved saxifrage, *S. tridactylites*, is a small annual, widespread but declining, usually occurring on lime-rich walls, and readily identified by its three-lobed, often red-tinged, leaves.

Mossy saxifrage, *S. hypnoides*, is probably the commonest of the upland saxifrages, forming fine-leaved mats on damp rock-ledges and by streams as far south as Exmoor. It was first discovered in Lancashire before 1640,[2] and is sometimes known as Dovedale moss.

Purple saxifrage, *S. oppositifolia*, is a dramatic plant of damp mountain ledges and scree, chiefly in Scotland, but coming as far south as the limestone uplands of South Wales and Yorkshire. It is tufted in habit, with quite large pale-pink to purple flowers, often blooming against the snow. The great walker Alfred Wainwright has given an evocative description of it in the Dales: 'April visitors will ever afterwards remember Penyghent as the mountain of the purple saxifrage, for in April this beautiful plant decorates the white limestone cliffs on the 1,900-ft contour with vivid splashes of colour, especially being rampant along the western cliff (overlooking the descent to Hunt Pot), which it drapes like aubretia on a garden wall.'[3]

It is no surprise that the species was much gathered from the wild for sale in Covent Garden.[4]

London pride, *S.* × *urbium*, is a popular garden plant, produced by crossing St Patrick's-cabbage (*S. spathularis*), from Ireland, and Pyrenean saxifrage (*S. umbrosa*). It is usually a sterile hybrid and has no particular connection with London. But escaped plants are widely naturalised on walls and rocks and by streams throughout Britain.

Celandine saxifrage, *S. cymbalaria*, is a yellow-flowered species from the eastern Mediterranean, escaping from gardens in shady places. Although it has been recorded in the wild only during the last hundred years, it occurs at a number of historic sites, e.g. Rievaulx Abbey, Yorkshire (founded 1131), Beaulieu Abbey, Hampshire

Meadow saxifrage is now confined to old grassland, including churchyards. Rockbourne, Hampshire.

(1204), and on Roman masonry at Chesters in Northumberland, hinting at the possibility that it may be a much earlier, unnoticed introduction.[5]

Fringecups, *Tellima grandiflora*, is a familiar garden plant from North America, widely naturalised in damp woodland, especially in the West Country. **Opposite-leaved golden-saxifrage**, *Chrysosplenium oppositifolium*, is a creeping, mat-forming perennial whose bright yellow flowers, cupped in green, leafy bracts, form trickles of gold on shady stream-banks and in woodland flushes as early as mid-March.

Grass-of-Parnassus, *Parnassia palustris*, is a handsome species, with heart-shaped leaves and ivory-white chalice-shaped flowers, now more or less confined to damp pastures, moors and marshes in northern Britain. It once occurred more widely and can still be found in the Norfolk Broads and fens. The Flemish botanist de l'Obel called the species 'Gramen Parnasi' in 1576, and the English translation, Grass of Parnassus, is attributed to Lyte in 1578.[6] It seems a rather extravagant tag until you examine the flowers closely. The oval petals are not in fact pure white, but veined with translucent green stripes, and are cupped around a nest of glistening yellow stamens.

Left: purple saxifrage – 'snow purple' – at 2,000 feet on Penyghent, Yorkshire. Above: Grass-of-Parnassus, a beautiful relative of the saxifrages which grows in damp upland sites.

Rose family *Rosaceae*

Brideworts, *Spiraea* species, are shrubby perennials from Europe, North America and Asia, with white or pink plume-like flowers, widely grown in gardens and naturalised on waysides and rough ground throughout Britain. *S. douglasii* (steeplebush) and *S. alba* and hybrids are probably the commonest in the wild. In Wales they are occasionally used for hedging on smallholdings.[1] **Sorbaria**, *Sorbaria sorbifolia*, from north Asia, is rather similar to the brideworts in its foamy panicles of flowers. But this is very much an opportunist urban plant, spreading rapidly by suckers and increasingly self-seeding into walls and pavement cracks in south-east England.

Meadowsweet, *Filipendula ulmaria*. In July, the frothy white flowers of meadowsweet spill out of damp ditches and across riverside meadows throughout Britain: 'Sweet Green Tavern is a pub in the heart of Bolton, on an island between two arterial roads near the railway station. In the eighteenth century the area was the site of "Sweet Green House", so called because of the meadowsweet that grew prolifically all around.'[2] Meadowsweet may have been named initially because it was used to flavour mead, the drink, not because of its adornment of meads and meadows. But vernacular names will always broaden if there is an opportunity for them to absorb new layers of meaning. So, 'mead-wort' or 'mead-sweet' growing in a meadow would become more memorable – and durable – by evolving into 'meadowsweet'.

Some of the obsolete local names – e.g. bittersweet, new mown hay – refer to the mixture of scents in different parts of the plant. There is a basic dill-like aroma running through meadowsweet that one Somerset contributor nailed exactly as 'marzipan';[3] but it is tinged with musk and honey in the flowers and with the sharpness of pickled cucumber (or carbolic) in the leaves – the difference between 'courtship and matrimony', as one unknown cynical namer believed.

Dropwort, *F. vulgaris*, is like a small meadowsweet and is more or less confined to calcareous grassland in England.

Cloudberry, *Rubus chamaemorus* (VN: Noop, Nowtberry). A subalpine shrub, confined to mountain bogs and moors north from Derbyshire and North Wales. The fruit – like an orange dewberry – rarely forms in Britain (unlike Scandinavia), because of an overwhelming preponderance of male plants. Cloudberries were so hard to come by in the Berwyn Mountains that in the parish of Llanrhaiadr anyone bringing a quart of the berries to the parson on the morning of St Dogfan's (the parish saint) Day would have his tithes remitted for the year.[4]

It is not easy to see another reason why anyone should scour the mountains for them. They make a thin marmalade, but are indifferent eating. As one North Yorkshire contributor puts it: 'They are known as "nowtberries" here because they taste of nowt! I tried to get up Noughtberry Hill one day in July, but was driven back by a heath fire. When I did get there two years later, the cloudberry plants had completely smothered the whole area.'[5] (Grigson finds a less entertaining but probably academically more correct explanation of the Northern 'knout'/'nowt' prefixes in the Middle English word 'knot', meaning a hill. And 'cloud-' originates from Old English *clud*, also meaning a hill, rather than from evocative moorland mists.[6] Names, like fruit, ripen with age.)

Cloudberry in flower high in the Cairngorms.

Cloudberries rarely fruit in Britain.

Chinese bramble, *R. tricolor*, the far-eastern equivalent, is much grown as ground-cover in gardens and is beginning to naturalise in hedgerows and shrubberies. It fruits rather more readily than *R. chamaemorus*.

Raspberry, *R. idaeus*, is locally common on heaths and in open woods throughout Britain, often appearing in large colonies after clear-felling or coppicing. It is a native shrub, though in waste ground and near habitation it may be naturalised from garden varieties. The canes grow to about six feet, spread by suckering, and tend to die back after they have fruited in their second year. The familiar orange-red fruits are usually ripe by July, and at this stage they come away very easily from their pithy core, known as the hull. (I have picked them as late as October, though, from canes sheltered by, but growing through, a bramble patch, so that a single bush appeared to be bearing two different kinds of fruit.)

It is not hard to see why the raspberry has been the *Rubus* species taken most extensively into cultivation, given that its canes are virtually spine-free and spread with more restraint than brambles. The most successful cultivation and hybridisation is carried on in Scotland, whose climate seems to suit the plant; and some of the new varieties – together with naturally occurring hybrids and sports – are well established in the wild north of the border. A yellow-fruited form grows on Speyside.[7]

Loganberry, *R. loganobaccus*, is naturalised on railway embankments, commons, old allotments and waste places. It is one of a number of crosses between raspberry and blackberry species (including, more recently, the tayberry and boysenberry) which may occur spontaneously in the wild. **Salmonberry**, *R. spectabilis*, from North America, is grown chiefly for the ornamental value of its pink flowers and is naturalised in a scatter of woods and hedgerows. A promising sport of this – very tall, with mauve flowers – is the most likely identity of a *Rubus* found growing wild by a contributor near Dunbar, East Lothian: 'I was surprised when the berries ripened on the 15th June, about three weeks before the raspberries. The berry was almost identical, except in size and colour. It was bright orange and bigger than a wild raspberry, but smaller than the cultivated ones in a local farm. I would say that the biggest of the berries would be about the same size as the biggest of the wild brambles. It has a pleasant taste, although the jam that I made from 3 lb of them is not as good as the better-known ones. Some of the plants are much taller than me, indeed one is about 12 feet tall. The berry comes off the hull cleanly.'[8]

Bramble or **Blackberry** species, *R. fruticosus* agg. (VN: Black heg, Blegs). Blackberrying is the one almost universal act of foraging to survive in our industrialised island and has a special role in the relationship between townspeople and the countryside. It is not just that blackberries are delicious, ubiquitous and unmistakable. Blackberrying, I suspect, carries with it a little of the urban dweller's myth of country life: harvest, a sense of season, and just enough discomfort to quicken the senses. Maybe the scuffling and scratches are an essential part of the attraction, the proof of satisfying outdoor toil against unruly nature.[9] It is a tradition going back thousands of years (blackberry seeds have been found in the stomach of a Neolithic man dug up at Walton-on-the Naze, Essex), but in these times of increased fencing-off of whole areas of countryside, it is looked on with hostility by some landowners: 'One West Midlands farmer didn't like "townies" coming and "stealing" his blackberries. He wrote to the *Birmingham Post* advising other farmers to follow his example and put up signs saying the berries had been sprayed with poison.'[10]

But it can be a connoisseur's pastime, as well. Over 400 microspecies have been recognised in Britain, each one differing subtly in fruiting time, size, texture and taste. In some varieties you may detect hints of plum, grape, apple or lemon. Dave Earl of Southport, Lancashire, has drawn up his own gourmet's list:

'Perhaps the commonest bramble in the North-west is *Rubus tuberculatus*; whilst the fruits can be pleasant, they resemble those of *R. caesius* [the dewberry, see below] in consisting of a few large drupelets, and it is quicker to gather well-formed fruits. Some of the smaller fruiters can be very seedy and sour ... One of the best is *R. dasyphyllus*, a common species of the hills, the flavour being pleasantly sweet. Of course flavour is very variable and I find that this can vary from sour to sweet to watery on the same panicle. I have often been put off eating the berries of the cultivated *R. procerus* for this reason.

Blackberries, the one fruit still widely gathered from the wild.

Nevertheless, with a good few ounces of sugar and some stewing they are ideal for jam and pie fillings. I much prefer the berries of *R. nemoralis* to those of *R. procerus*. The fruits are of good size and flavour, and the species is common on acid soil, particularly on the site of former mosslands. If you are particularly hungry, *R. gratus* has large berries ... Members of the Wild Flower Society recently enjoyed feasting on the berries of bushes on Carrington Moss, Greater Manchester. This was despite the fact that the bushes grew along the perimeter fence of the gas works. The taste topper for me, however, was *R. bertramii*, encountered on Ashton Moss, Ashton-under-Lyme, in August 1993. There are only a few bushes there, but a good stand can be found at the north end of Knutsford Heath in Cheshire. Some bushes are very local in distribution. The "Bollin Blackberries", known to my grandparents and picked by myself as a child to make Auntie's crumble, were not described until 1971. *R. distractiformis* – "from the north banks of the River Bollin, Hale" – was named by Alan Newton, the expert who identified most of the brambles in the Liverpool Museum Herbarium. Even though this bramble had not been described as a distinct species, my Aunt preferred *R. distractiformis*.'[11]

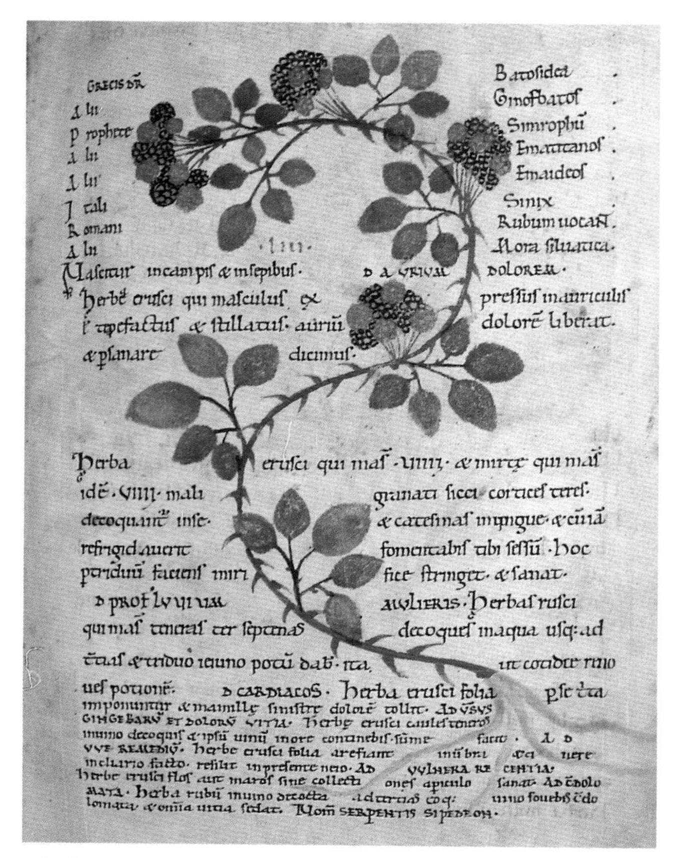

The bramble, from a herbal made probably at Bury St Edmunds Abbey in the eleventh century and which contains some of the earliest naturalistic plant drawings.

Blackberries form in clusters at the end of mature shoots, which die back after two or three years' fruiting. The berry at the very tip of the stalk is the first to ripen, and the sweetest and fattest of all. A few weeks later, the other berries near the end ripen. These are less juicy, but fine for pies and jams. By the end of October, the remaining berries have often picked up mildews and bacteria and turned sour or cloying. There is an old saw (still widely known) to discourage picking these late, inferior specimens which says that the Devil pisses, or spits, on the blackberries on Michaelmas Night.

Earlier this century, wild blackberries were sometimes picked commercially, as in Somerset and Oxfordshire:

'They were collected by local children for sixpence a pound. They were used for dye. The money usually financed trips to the fair.'[12]

'In Eynsham, Oxfordshire, fruit was picked and sold to Coopers Marmalade of Oxford, who had the contract to make jam for the troops in the First War. Lots of local people did this to earn a few pennies. The fruit was taken into Oxford in an old pram.'[13]

Recipes are legion: old – such as blackberry crowdie, made with toasted oats, cream and rum; modern – blackberry vinegar, with the berries marinated with sugar in white vinegar and served with grilled goat's cheese; and instant – apple and blackberry sandwiches.[14] The most delicate is a junket, invented I imagine for the very young or very old, made simply by squeezing the juice from some very ripe berries and allowing it to set overnight in a bowl in a warm room. The result is the pure, ethereal essence of blackberry, hanging between liquid and solid.[15]

Bramble is also, of course, both loathed and respected for its thorns and powers of entanglement. Bushes were once planted on graves, to cover less sightly weeds and deter grazing sheep, but probably also as an echo of more ancient and magical hopes of keeping the dead in and the Devil out: 'The Old Man of Braughing can sleep in peace. For once more the villagers have completed the traditional ceremony of sweeping Fleece Lane and putting brambles on the old man's grave. The "old man" died at a ripe old age and the bearers were carrying his coffin down Fleece Lane to the churchyard when they stumbled on some stones and the coffin lid fell open and broke. The Old Man sat up and soon afterwards was married again. When he eventually died he left cash for Fleece Lane to be swept clean of stones each year and for brambles to be placed on his grave to keep the sheep off.' (Hertfordshire, 1957)[16]

In a churchyard in Cheshire the grave of a nineteenth-century bramble expert is adorned, deliberately or by happy accident, with a bush of *R. laciniatus*, a cut-leaved species of unknown provenance: 'In the village of Lower Peover there is in the churchyard an impressive tomb-stone of the Warren family. It has an unusual bramble planted in the grave and a stylised version carved up the tall stone. J. B. L. Warren [Lord de Tabley] wrote *The Flora of Cheshire* in 1899. He took a special interest in the *Rubus* family. It is good to see such a potent memorial to his great work.'[17]

The long, arching stems of bramble (and wild roses) were once known as 'lawyers' because of the trouble you have escaping if you happen to fall into their clutches.[18] An especially vicious form in Honeypot Wood, Norfolk, has been nicknamed '*Rubus Boadicea*'.[19] But they were apparently looked on more kindly by Victorian men of a romantic bent, because of the excuse they gave for paying attention to ladies' long skirts. The Pre-Raphaelite poet Thomas Woolner introduces them as welcome obstructions in his epic 'My Beautiful Lady' (1863):

We thread a copse where frequent bramble spray
With loose obtrusion from the side roots stray,
And force sweet pauses on our walk.
I lift one with my feet and talk
About its leaves and stalk.

Or maybe that some thorn or prickly stem
Will take prisoner her long garment's hem:
To disentangle it I kneel,
Oft wounding more than I can heal,
It makes her laugh, my zeal.

The ambivalent image of the bramble patch, simultaneously grasping and protective, permeates even the technical and supposedly rational business of woodland management. Despite bramble being an entirely natural component of woodland vegetation, waxing and waning according to the light allowed it by the tree canopy, commercial foresters attack it ruthlessly, in the belief that it 'smothers' young trees. Traditional woodland managers can view it rather differently: 'Bramble has been the saviour of our working coppices in Suffolk, as it has given vital protection to young coppice shoots (especially small-leaved lime) from the attentions of deer. As a consequence it is encouraged where browsing pressure is high.'[20]

The succulent, white-bloomed fruit of the dewberry.

Dewberry, *R. caesius*. Dewberries are quite common in hedge-banks, woodland edges and scrub, especially in damp places on calcareous soils. They have a low arching growth and can be recognised immediately by their berries, which have a few large drupes covered with white bloom, like miniature grapes. They are so juicy that they are difficult to pick. Snipping off a few berries with scissors or secateurs is the answer, after which they can be eaten like cocktail cherries, on their own sticks.

Shrubby cinquefoil, *Potentilla fruticosa*, is a relic of the last phases of the Ice Age. Twelve thousand years ago this handsome yellow-flowered shrub, the ancestor of the many colour forms now popular in gardens, grew in the

Shrubby cinquefoil, Upper Teesdale.

glacial meltwaters right across England. Now it has retreated to just two northern sites: the riverside boulder-rubble close to High Force waterfall in Upper Teesdale, and ravines of damp limestone scree above Wastwater in the Lake District. It seems to need these unstable, often winter-flooded habitats to survive competition from other shrubby species.

Silverweed, *P. anserina* (VN: Traveller's ease, Traveller's joy). The ground-hugging leaves of silverweed – silky green above, pale matt grey beneath – are common on waysides, in waste places and in the compacted soil of footpaths and field entrances, where they can form downy mats spangled with five-petalled yellow flowers. No wonder that they were believed to give some relief to travellers if stuffed inside their shoes. Deering wrote in 1738 that 'it is certain that your Carriers wear the Leaves in their Shoes which keep them cool and prevent a too immoderate sweating of the Feet which causes a Soreness in them'.[21]

A silverweed trail. The downy grey leaves were believed to ease aching feet.

The roots were once eaten (and possibly cultivated) in some upland areas of Britain. They were baked and boiled, or dried and ground into a rough flour. John Ray saw children at Settle in Yorkshire eat them raw. They are a meagre vegetable, but taste a little like parsnips, and in the Scottish Highlands were eaten in times of famine right up to the end of the nineteenth century.

Tormentil, *P. erecta*, is a common, low-growing perennial of short acid grassland on heaths, moors and roadsides, with small four-petalled yellow flowers. **Creeping cinquefoil,** *P. reptans*, is a larger and more aggressive species with reddish runners and larger five-petalled flowers, common in gardens and waysides and on cultivated and waste ground throughout England and Wales. It is included, with its five-fingered leaves clearly picked out, amongst the stone carvings in Southwell Minster, Nottinghamshire.[22] **Barren strawberry,** *P. sterilis*, is an early flowerer (in March sometimes) of hedge-banks, scrub and woodland rides and clearings. It is distinguishable from the true wild strawberry by its smaller white flowers and pale, blue-green leaves with the end tooth shorter than the adjacent two.

Wild strawberry, *Fragaria vesca.*

'As a child in the Yorkshire Dales I collected lots of wild strawberries on our evening walks before going to bed. In the morning we had wild strawberries with our breakfast cereals. When I returned to school we were asked to talk about our holidays. I spoke about my camping holiday and described all the plants and flowers and the delicious wild strawberries. My class teacher listened in disbelief and said, "There is no such thing as a wild strawberry. It must have been from a garden." I was ridiculed by him for exaggerating and felt humiliated. But I wondered how you could get a cultivated strawberry if there were no such thing as wild strawberries!

I still find lots around the country on our walks. My three children know the story of the wild strawberry that doesn't exist. The first one we find each year gets special treatment as we joke about its not existing and then eat it!'[23]

This nine-year-old's logic is still lost on many people, who believe that any edible fruit growing wild is a garden escapee, or some kind of degeneration from an original cultivated variety. Our native strawberry is not a direct

Wild strawberries, often found lurking under leaves and in flavour often superior to cultivars.

ancestor of modern commercial varieties, but is a species in its own right, and in flavour often superior to cultivars. In limestone areas such as the Derbyshire and Yorkshire Dales, the fruits are often warmed by heat reflected from the rock and become soft, fragrant beads of sweet juice. (I have seen a sixteenth-century recipe for a form of strawberry shortcake, made with almond flour and cooked simply by leaving the paste in the hot sun for a couple of hours.) Wild strawberries are very delicate when ripe and are easily bruised if gathered directly into a basket or jar. In Sweden, they thread them individually onto long grass stalks.

They occur quite commonly in open woodland and scrub throughout Britain, especially on calcareous soils. In parts of Gwent they are sufficiently common to have places named after them, for instance Cwmsyfiog – 'Strawberry Vale' – near Tredegar.[24] The plants often appear in large quantities after felling in the southern beechwoods, especially where the soil has been disturbed. In Sussex and Hampshire, 'strawberry slidders' were features of steep woods where the felled trunks had been dragged downhill. These eruptions are partly from buried seed, suddenly exposed to the light. But the species spreads chiefly by runners, which can extend very quickly among its glossy, trefoil leaves.

Wild strawberry also finds sunny railway embankments a congenial habitat: 'We gathered wild strawberries into jam-jars from the "batters" – the banks of the railway lines between Kirkby Lonsdale and Kirkby Stephen. Strawberries flourished on the stony, well-drained slopes.'[25]

I have been shown white-berried strawberries growing on the rubble of disused railway lines. But the fruit is almost tasteless, and after a couple of years reverts to its normal colour. Altogether more interesting is the robust form known as the alpine strawberry, which will go on fruiting until the first frosts. It may have been one of these that Gilbert White picked 'ripe ... on a bank' at Selborne on 10 January 1790, after an exceptionally mild winter.[26]

Strawberries are represented in several churches, notably in St Mary's, at Whalley in Lancashire, where both leaves and berries are figured on a misericord, their respective shapes echoed by hop leaves and fruit on the supporters.[27] They also inspired William Morris: 'He was sitting in Kelmscott Manor courtyard, and noticed a thrush swoop down and take a wild strawberry in its beak. He rushed indoors and designed the famous "Strawberry thief" pattern. The wild strawberry patch is still there.'[28]

Garden strawberry, *F.* × *ananassa*, is the cultivated strawberry, bred chiefly from American wild species and occasionally naturalised on rubbish-tips, waysides and (like its wild cousins) railway embankments: 'I have found garden strawberries growing on the edge of railway lines near Patney and Etchilhampton, presumably from the pips of fruit thrown from passing trains.'[29]

A detail from William Morris's 'Strawberry Thief' design, inspired by the wild fruit.

Wood avens or **Herb Bennet**, *Geum urbanum*, is a common yellow-flowered perennial of woods and hedgebanks throughout Britain. Its root has a spicy clove smell and was widely used in herbal medicine. Herb Bennet is a corruption of the medieval Latin 'herba benedicta', 'the blessed herb'. **Water avens**, *G. rivale*, is a more glamorous and secretive species, with cup-shaped flowers flushed with purple, pink and dull orange. It haunts riversides and damp woods, but in southern England has a curious and very local distribution.

Mountain avens, *Dryas octopetala*, is an exquisite mat-forming arctic-alpine that has probably been continuously present in Britain since before the Ice Age. It must have been widespread immediately after the retreat of the glaciers, but is now confined to calcium-rich rock-ledges and crevices in mountainous areas in the north and west, especially in Snowdonia, the Lake District and the western Highlands. The flower is pure white, but seems to have a sheen, reflected from the dense central cluster of golden stamens. The seed-head that follows is a feathery sheaf, twisted at the top as if it were a sweet-wrapper.

D. octopetala, as its name suggests, usually has eight petals. But doubles are quite frequent, and in the Burren in Ireland, where the limestone pavements are covered by sheets of the flower in May, the botanist Mary-Angela Keane has seen an individual with 17 petals.

Agrimony, *Agrimonia eupatoria*, is a common wayside perennial, sometimes more conspicuous for its rust-coloured hooked fruits than its spikes of small yellow flowers. It was an important plant in herbal medicine, recommended for snake-bite and 'elf-shot' in the Middle Ages, and later, and more practically, as a mild astringent

Mountain avens was probably widespread after the last glaciation, but it is now confined to mountainous areas in the north and west. The white flowers are followed by feathery, twisted seed-heads (right).

Agrimony, common on waysides, and in flower from June.

for unsettled digestive systems and catarrh. **Fragrant agrimony**, *A. procera*, is an altogether larger plant, whose leaves and flowers both have a pleasant balsamic scent reminiscent of walnut leaves. Although it is scarcer and more local than common agrimony, and pretty much confined to heavy clay soils, it has been used for pot-pourris and scented pillows in some areas.

Salad burnet, *Sanguisorba minor*, is a short herb of chalk and limestone grassland throughout Britain, sometimes abundant enough for the leaflets to scent the air with cucumber when you walk over them. They are slightly bitter to the taste, but have long been used as a salad green,[30] and as a cooling addition to summer drinks.

The round, rust-speckled flower-heads, a little like scabs or blood clots, made it a signaturist's favourite, for staunching wounds – hence the Latin name '*sanguisorba*'. But for one budding eighteenth-century botanist – the Revd Gilbert White – it was burnet's more palpable strengths that made the deepest impression. On 6 July 1765, he was riding back to his brother's house across the Hampshire hills and was struck by how, despite the privations of a long drought, the downs were still kept green by this diminutive plant:

'The downs between Alresford & Andover are full of Burnet: so full in many places that it is almost the only herb that covers the Ground: & is eaten down very close by the sheep, who are fond of it ... It is worth observation that this herb seems to abound most in the poorest, & shallowest chalkey soil ... Near Waller's Ash I rode thro a piece of Ground of about 400 acres, which had been lately pared by a breast plough for burning: here the burnet was coming-up very quick on the bare ground, tho' the crown of the root must have been cut off of course along with the turf: this shews that it is a plant tenacious of life, since it springs from the severed root like plantain.'[31]

Fodder burnet, *S. minor* ssp. *muricata*, a leafier subspecies from southern Europe, occurs occasionally as an escape or relic in southern England. **Great burnet**, *S. officinalis*, a taller herb, with a much more pronounced 'blood-clot' flower-head, is confined to old damp grassland in hay meadows and by riversides, chiefly in the Midlands, South Wales and northern England.

'The flower-heads of great burnet were made into wine in Westmorland until at least 1950, their name being curiously corrupted to "Burnip".'[32]

Pirri-pirri-bur, *Acaena novae-zelandiae*, is a mat-forming Australasian species with very spiny fruits that found its way to this country caught up in imported sheep-fleeces. It is now well naturalised in bare and sandy areas of southern and eastern Britain, for instance on Kelling Heath on the Norfolk coast and on Lindisfarne. Here its burs stick to children's clothes as effectively as to sheep and 'defy the washing machine'.[33]

Lady's-mantle, *Alchemilla vulgaris* agg., is a group of grassland species differing in only the smallest characteristics. They grow mostly in unimproved grassland and woodland rides and on rock-ledges. Most of them do not occur at all in southern England, though the large garden species, *A. mollis*, from the Carpathian mountains, is naturalised here and there. The often nine-lobed leaves of lady's-mantles, like cloaks or umbrellas, fold up overnight

Lady's-mantle. The dew collected by the leaves was prized by alchemists and herbalists.

and catch the dew on their soft hairs. Plant-dew was highly valued by early herbalists (see sundew, p. 125) and this made *Alchemilla* prized as a simple, prescribed for wounds, infertility and impotence. The alchemists also required the purest dew for turning base metals into gold – hence the name *Alchemilla*, 'little alchemist'. Such a powerful and magical herb was bound to be christianised, and some time in the late Middle Ages it was named Our Lady's Mantle, and eventually lady's-mantle.

Parsley-piert, *Aphanes arvensis*, is a small, inconspicuous greenish-flowered annual, found on cultivated and bare ground throughout Britain. The name comes from *perce-pierre*, a plant that could push its way through stony ground. So, by sympathetic magic, it was believed to be similarly capable of breaking up stones of the kidney and bladder. The frayed, parsley-like leaves made anglicisation of the name easy. Parsley-piert has also been used more rationally as a salad ingredient.

Wild roses, *Rosa* species. The rose is England's national flower, but none of the likely candidates for this honourable position is a native wild species. The Red Rose of Lancaster is the Mediterranean *Rosa gallica*, the White Rose of York almost certainly a hybrid between the native field-rose, *R. arvensis*, and the damask rose *R.* × *damascena*. Even the heraldic roses carved on churches and memorials throughout the land are modelled more on the Holy Rose of Abyssinia (*R. richardii*), the oldest of all cultivated roses and figured in paintings on the murals at Knossos in Crete, than on any indigenous briar.[34]

But our native roses are amongst our best-loved and most familiar flowers, even though the barbarous schedules of modern hedge-cutting rarely allow them to flower. There are 14 species currently accepted as native to Britain. Four of these – dog-rose, *R. canina*, field-rose, *R. arvensis*, harsh downy-rose, *R. tomentosa*, and sweet-briar, *R. rubiginosa* – are common hedgerow climbers. Numerous subspecies and hybrids occur naturally, and three of these species have played a modest part in the evolution of various groups of cultivated rose. (Wild roses

The dog-rose, the most abundant of our native rose species.

have also frequently been been used as grafting stock for cultivars. A Buckinghamshire contributor recalls how cottagers would take wild roses from the hedges to graft garden roses, 'if they got given a bit from one of the wealthier houses'.[35])

Dog-rose, *R. canina* (VN: Ewemack), is the most abundant and widespread species, and also the most variable. It can be a low-growing scrambler or climb 30 feet up a woodland tree to flower in the crown, like a rain-forest vine. The sweet-scented blooms vary in colour from deep pink to white. *R. canina* has produced a small number of cultivated varieties, the best known of which is 'Abbotswood', a chance hybrid with an unknown garden variety, which has scented, double, pink flowers. (It originally appeared in the garden of Harry Ferguson, of tractor fame.) It is also in the breeding line of the 'Alba' group.

An old riddle, 'The Five Brethren of the Rose', gives an effective way of identifying roses of the *canina* group. It is a folk-riddle that has been passed on orally since medieval times. This is a version transmitted through a line of distinguished gardeners, from Canon Ellacombe to Edward Bowles to William T. Stearn:

> *On a summer's day, in sultry weather,*
> *Five brethren were born together.*
> *Two had beards and two had none*
> *And the other had but half a one.*[36]

(The 'brethren' are the five sepals of the dog-rose, two of which are whiskered on both sides, two quite smooth and the fifth whiskered on one side only.)

Field-rose, *R. arvensis*, is a scrambler, whose arching stems rarely reach more than three or four feet above ground level. It always has pure white flowers, with conspicuous golden anthers, a pin-like column of styles and a musky, honey scent. It is the ancestor of the Ayrshire roses, now very rare in cultivation. **Sherard's downy-rose**, *R. sherardii*, is commonest in northern Britain. It is a beautiful and subtle rose with downy leaves and deep pink, velvety petals, which surprisingly has not been taken into cultivation at all.

Sherard's downy-rose, common in Scotland but very rare in southern England.

Sweet-briar, *R. rubiginosa*, the 'eglantine' of Shakespeare and early writers, is noticeable for its sticky, apple-scented leaves, which become especially fragrant after rain. It is scattered throughout Britain, but most frequent in chalk scrub in southern England. A semi-double sport (reputedly discovered in a Cheshire lane) was named 'Janet's Pride' and was the precursor of a group of spectacular and vigorous sweet-briars bred by Lord Penzance in the 1890s, including 'Amy Robsart', 'Meg Merrilies' and 'Lady Penzance'. Most of the earlier cultivars are probably extinct. They included var. *concava*, in which both the deep pink flowers and the leaves were concave, like little spoons; various doubles, including the very fragrant 'William's Sweetbriar' (var. *multiplex*); and var. *marmorea*, with marbled flowers.[37]

'A local farmer's wife [Whitby] recalls, on the way to school, picking the tender young shoots of wild roses, to suck in the place of sweets which were not available to country children. I think this could have arisen in more calcium-rich areas, where the sweet-briar rose grows. This has glands which secrete an apple-smelling liquid leaving a pleasant taste.'[38]

Rose-hips (VN: Heps, Itchy-coos). The oval red fruits of wild roses have long been used as food. Gerard speaks of them making 'most pleasant meates and banketting dishes, as Tartes and such like'.[39] In the eighteenth century they were made into a purée, by the laborious process of slitting the fruits in half, cleaning out the pith and seeds, leaving the shells to soften (without cooking) and then rubbing them through a sieve.

During the Second World War, rose-hips came into their own in the form of rose-hip syrup, whose taste all wartime children can recall as vividly as dried egg. Nutritional scientists had known since the 1930s that wild hips had a higher proportion of vitamin C than any other common fruit or vegetable. (A cup of rose-hip pulp provides more vitamin C than 40 fresh oranges.) But it wasn't until the war began to disrupt our usual sources of the vitamin (especially citrus fruits) that the Government began seriously to consider the use of rose-hips. In 1941 the Ministry of Health initiated a scheme for voluntary collection. One hundred and twenty tons were gathered that year for processing into syrup and distribution to small children. By the end of the war, by which time the collection was being co-ordinated by the County Herb Committees, the annual harvest was averaging over 450 tons.

Schools, scout troops, guide companies and children's gangs all took part, and many contributors remember the expeditions in late summer (the hips had to be collected when they had just turned red, to maximise the vitamin C content) and hauling the sackfuls back to the local school or village hall. With collectors being paid 3d a pound, some entrepreneurial youngsters were able to earn more than just pocket money.

The picking was encouraged most in the north of England, where the hips reputedly had a higher vitamin content, but storage sometimes became a problem: 'One year, on the children's first morning back from the holidays, the teacher sat down at the piano to play the opening hymn of the morning assembly. Her hands fell with a flourish on the keys, but not a sound came forth. Inspection revealed that in the holiday-quiet schoolroom mice had busied themselves eating fleshy bits of rose-hip and stuffing the seeds beneath the piano keys until they were packed solid!' (Cumbria)[40]

Collecting hips for National Rose Hip Syrup went on

Rose-hips, known locally as heps and itchy-coos, and used in the making of that wartime staple, rose-hip syrup.

until the early 1950s. It was sold at a controlled price of 1s 9d per 6-oz bottle, though mothers and children were able to obtain larger quantities, at reduced prices, from welfare clinics. (It is still made commercially today, though from farmed 'wild' roses.)[41]

But many country-dwellers made (and still make) their own syrup, often following the precise instructions given by the Ministry of Food in their booklet *Hedgerow Harvest*.[42] The process involves mincing, stewing and then, crucially, straining through a jelly-bag to remove the prickly seeds, which can be a dangerous internal irritant. Boiled again with sugar and reduced, the hips make a bright red syrup which has been used for drinks, mousses and summer puddings.[43]

As for the seeds, they weren't an entirely useless by-product, as generations of children have known of their potential as an itching powder:

'I was a schoolmaster till 1987, and rose-hips were still being used then. It is not in fact the seeds that are used but the hairs attached to the seeds and the lining of the hip.'[44]

'During the late 60s, children in Dundee continued the tradition of using rose-hip seeds from the dog-rose and also ornamental roses planted in urban landscaping as itching powder. This activity is well recorded (although I suspect the effects are largely psychosomatic). However, I have never seen recorded anywhere the name applied by children to the seeds, the hips, and even the whole plant – "Itchy-coos". The second half of the name (translated as "cows") is a complete mystery but the term seemed to be quite widely used.'[45] (In the 1960s, the English rock group the Small Faces, for instance, sang about 'Itchy-Coo Park'.)

Burnet rose, *R. pimpinellifolia*, is a low-growing species, largely confined to dry sandy places near the sea and to calcareous areas inland. It has white flowers, frequently tinged with cream (or more rarely with pink), prominent golden stamens, and the sweetest smell of any native rose – a mixture of honey and jasmine. The hips are almost round, blackish-purple in colour, and the leaves are small and oval, very like those of salad burnet (hence the name).

A colour variety with slightly yellower flowers is known to rosarians as the Dunwich Rose, from its discovery on the sandy cliffs of Dunwich in Suffolk in 1956. It has gone from the cliffs, but still survives in heathland a mile or two inland – and, it hardly needs to be added, in several local gardens.[46]

R. pimpinellifolia has been used in rose-breeding as far back as the early 1800s, when the double varieties known as 'Scotch roses' were fashionable. The best known is 'Stanwell's Perpetual', with very double, soft pink flowers.

The hybrid between the burnet rose and the dog-rose, *R.* × *hibernica*, occurs occasionally in the wild. (I have seen it in the Yorkshire Dales.) It has bright pink single flowers, but most of its other characteristics are midway between those of its parents.

Japanese rose, *R. rugosa*, is a low-growing, viciously spiny Asian species with single pink flowers, often planted in masses on urban roadsides and roundabouts. It can expand into very large clumps by suckers and is increasingly naturalised on banks, dunes, waste ground and the sites of old gardens. Thickets conspicuous enough to have become local landmarks include those along the railway embankments at Bletchley, Buckinghamshire; in the sandy fen behind the coast road at Aldeburgh, Suffolk;

Burnet rose, whose flowers have the finest scent of any of our wild species – a mix of honey, clotted cream and jasmine.

Japanese rose, widely naturalised – as here, in the steelworks area of Sheffield.

covering the cliffs between Coldingham Bay and St Abb's Head in Berwickshire; and on beaches at Flimby, Cumbria, and Carylon Bay, Cornwall, not far from the St Austell china-clay works.[47] An exceptionally large colony is along the York ring road, though this is probably an extension of an ornamental planting: 'They are growing on rough ground adjacent to and overlooking the roadway, in some cases with other bushes, to a depth of 10 yards, and extend along the road for almost 800 yards.'[48]

Japanese rose was one of the spiny species chosen by the Essex police in 1993 for planting as anti-thief barricades. These also included *Berberis* spp., hawthorn and gorse, and were collectively nicknamed, in a fine example of Estuary Latin, '*Burglaris disembowelis*'.

'These plants make an attractive alternative to conventional fencing or the use of barbed wire on walls.'[49]

Red-leaved rose, *R. glauca*, better known to gardeners as *Rosa rubrifolia*, is a species rose from central Europe, quite widely naturalised from bird-sown seed. It is striking for its greyish, red-tinged leaves and diminutive pink flowers, more like a bramble's than a rose's.

There are many other foreign rose species and cultivars which have naturalised from gardens – or lived on there after the house itself has vanished. A few have been thought lost, but have then been rediscovered in abandoned corners and brought back into cultivation. The most celebrated is the deep-coloured Victorian rose that was found by the late Humphrey Brooke, scrambling up the wall of Woolverstone Church in Suffolk. The bush was over a hundred years old but still flowering. Brooke, an eccentric but distinguished and knowledgeable rosarian, thought it had the strongest scent of any rose he knew and once recounted to me how a blind friend had 'put his nose in a bloom and said that if this scent was available in a bottle it would put every tart in Europe out of business'. The Woolverstone Church rose is now back on the market as 'Surpassing Beauty of Woolverstone'.

The wild-rose screen at Great Warley, Essex. The white roses at the top of the screen represent purity; the lower, earthbound red roses, blood and passion.

Wild cherry blossom, April.

Wild cherry, *Prunus avium* (VN: Gean, Mazzard, Murry). The wild cherry is arguably the most seasonally ornamental of our native woodland trees. The drifts of delicate white blossom are often out in early April, just before the leaves, while in autumn its leaves turn a fiery mix of yellow and crimson. Even the bark – peeling to reveal dark, shiny-red patches – is extravagantly colourful for a British tree.

Cherry is a tree of the southern half of Britain and prefers chalky soils, even though it will grow on acid plateaux, as in the Chilterns. This is one of its favourite areas, and it can form large colonies by its vigorous and prolific suckers. When these are at the edges of woods, as they often are (cherry needs light to regenerate), they can make the entire wood seem to be ringed with white at blossom-time. A couple of weeks later, when the flowers have fallen, the woods are ringed again, on the ground. After the great storms of October 1987 and January 1990, there was another cherry delight the following spring: windblown trees blooming horizontally in the woods, like flowering hedges. In Fingest, in the south Chilterns, gean blossom is used to decorate the church at Eastertide.[50] (In the former Czechoslovakia boughs of cherry were cut on the feast of St Barbara, 4 December, and kept indoors so that they would flower on Christmas Eve. The girls would then take them to midnight mass under their cloaks, and boys would try to steal them.)[51]

Cherry blossom at dusk in the Chilterns.

Cherry is short-lived, but the timber is valuable, being reddish-brown in colour, tough and capable of being polished to a finish resembling mahogany. The trunks often produce large burrs with elaborate grain patterns, which are much sought after by turners.

Another asset is the fruits, which are produced in great quantities in some years. They are small, yellow, red or black, thin-fleshed, and popular with birds and small mammals. (You will sometimes find piles of cherry stones on tree-stumps where mice or squirrels have been feeding.) They can be bitter, but are perfectly edible to humans from late July and are, of course, the ancestor of cultivated cherries. They are also the best fruits to use for cherry brandy, following a recipe similar to that for sloe gin: a small bottle full of wild cherries, and a couple of tablespoons of sugar, topped up with brandy and left for three or four months.

'On the farm track adjoining two estates in the village [Colinsburgh, Fife] there is a Cherry Lane. Children (and adults) will spend hours up the trees picking "geans" when in season. An arrangement with the landowner is such that anyone can climb to pick geans so long as the trees incur no damage.'[52]

'One lady in our village of Llys-y-fran said that 70 years ago she would collect wild cherries for her mother to make a pie. Now the hedges are cut by machine, and there are none.'[53]

A final cherry product is the sticky resin which exudes from the trunks when they are damaged. This has been used by children and forest workers as a bitter-sweet chewing-gum.

St Lucie cherry, *P. mahaleb*, is a southern European species quite well naturalised by railways and in plantations in the south. **Japanese cherry**, *P. serrulata*, is widely planted in parks and by roads, and often persists as a relic. (The pink-flowered, multi-petalled cultivar is one of the commonest ornamental trees in Britain, but it does not naturalise.) **Fuji cherry**, *P. incisa*, is a small Japanese tree, naturalised in oakwoods at Chinnor Hill in the Chilterns. **Bird cherry**, *P. padus*, is a native shrub or small tree, found chiefly in woods and by streams in northern Britain but coming as far south-west as the Derbyshire Dales and

The flower-spikes of bird cherry, a species found chiefly in the north.

the gorges of the Welsh Marches. It also has a curious outlying population in ancient woods in Norfolk and is a speciality of Wayland Wood, near Watton, the real-life site of the 'Babes in the Wood' legend.[54] It carries drooping spikes of white flowers in May (giving it the name of 'wild lilac' in parts of North Yorkshire)[55] followed by black, bitter cherries. But it is often most conspicuous in summer, when it is likely to be almost leafless and festooned with the silken tents of the defoliating small ermine moth, *Yponomeuta padella*.

Blackthorn, *P. spinosa* (VN for fruit: Sloes, Slones, Bullums). The definitive account of the popular culture of the blackthorn – and one of the most vivid and concise biographies ever written of a native tree – was penned by William Cobbett more than 150 years ago:

'Everyone knows that this is a Thorn of the Plum kind; that it bears very small black plums which are called Sloes, which have served love-song poets, in all ages, with a simile whereby to describe the eyes of their beauties, just as the snow has constantly served them with the means of attempting to do something like justice to the colour of their skins and the purity of their minds ... These beauty-describing sloes have a little plum-like pulp which covers a little roundish stone, pretty nearly as hard as iron, with a small kernel inside of it. This pulp, which I have eaten many times when I was a boy until my tongue clove to the roof of my mouth and my lips were pretty near glued together, is astringent beyond the powers of alum. The juice expressed from this pulp is of a greenish black, and mixed with water, in which a due proportion of logwood has been steeped, receiving, in addition, a sufficient proportion of cheap French brandy, makes the finest Port wine in the world ... It is not, however, as a fruit-tree that I am here about to speak seriously to sensible people; it is of a *bush* excellent for the making of *hedges*, and not less excellent for the making of walking sticks and swingles of flails. The Black Thorn blows very early in the spring. It is a Plum and it blows at the same time, or a very little earlier, than the Plums. It is a remarkable fact that there is always, that is every year of our lives, a spell of cold and angry

Flower-frosted branches in a 'blackthorn winter'. Sloes follow in the autumn.

weather just at the time this hardy little tree is in bloom. The country people call it the *Black Thorn winter* and thus it has been called, I dare say, by all the inhabitants of this island, from generation to generation, for a thousand years.

This Thorn is as hardy as the White Thorn; its thorns are sharper and longer; it grows as fast; its wood is a great deal harder and more tough; it throws out a great deal more in side-shoots ... The knots produced by these side-shoots are so thickly set, that, when the shoot is cut, whether it be little or big, it makes the most beautiful of all walking or riding sticks. The bark, which is precisely the colour of the Horse Chestnut fruit and as smooth and as bright, needs no polish; and, ornamented by the numerous knots, the stick is the very prettiest that can be conceived. Little do the bucks, when they are drinking Port wine, ... reflect that, by possibility, for the "fine old Port" which has caused them so much pleasure, they are indebted to the very stick with which they are caressing their admired Wellington boots.'[56]

Blackthorn is an abundant shrub of hedges, scrub and open woodland throughout Britain, and its reputation and uses have scarcely changed in the past century and a half:

'I firmly believe in the "Blackthorn winter" and have observed for over twenty years that the blackthorn has flowered during a bitterly cold spell of weather, usually after a "false spring".'[57]

The fertile cross between blackthorn and the cultivated plum, *P.* × *fruticans*, can produce thorns more than two inches long and tough enough to penetrate a tractor tyre.[58]

'Blackthorn thorns have a bad reputation for puncturing tractor tyres ... They have a habit of producing patches of sucker growth. For example, many neglected hedges have spread out to become several yards wide. When these patches are cleared it leaves many sharp stubs and stumps, many between pencil and thumb thickness, and these can be sharp and hard enough to puncture a tyre. Blackthorn is a hard wood, so much so that "corkscrew" walking sticks of blackthorn are highly prized because they are so rarely found. (In contrast soft-stemmed shrubs and trees like sallow, aspen and birch are more readily deformed by twining honeysuckles.)'[59]

Blackthorn was also the traditional wood for Irish

The spines of blackthorn make it a common hedging species.

shillelaghs, once definitively described by the Chairman of the Pharmacology Department at the University College of Los Angeles as 'an ancient Hibernian tranquilliser'.

Sloe gin is now a more popular drink than imitation port and is traditionally made by gathering sloes in late September or October, pricking them with a skewer, half filling a bottle with them, adding a few spoonfuls of sugar and covering with gin. (Commercial sloe gin, made by Gordons in Britain, disappointingly uses east European fruit.) The purple, almond-flavoured liqueur is ready to drink by Christmas, but improves with age. When the liquid is finished the sloes can be eaten, neat or processed: 'As a by-product, the pitted, gin-soaked sloes can be dipped into melted chocolate, which is then allowed to set.'[60]

As one Sussex woman reported, 'My uncle had a saying: "He likes his women fast and his gin sloe." '[61]

Wild plums, *P. domestica* agg. (VN: Bully tree, Crixies, Winter crack). It is possible that the tree known as the bullace (*P. domestica* ssp. *insititia*), with large, sloe-like fruits, is native in woodlands in Britain. But its fruits are barely distinguishable from naturalised, dark-fruited damsons. And the many varieties of relict or bird-sown gages, 'prunes' and damsons that are found growing in the wild form such a continuous spectrum that they are best all treated generically as 'wild plums'. Many are the outcome of ancient crosses (accidental in some cases) between the blackthorn and various sweeter-fruited plum species introduced from Asia, in a lineage of Byzantine complexity. With such an ancestry it is no real wonder that feral plums are one of the best wild foods (many being edible straight off the tree, unlike sloes) and represent a huge genetic reservoir.

Most are to be found in hedges. They are sometimes deliberately planted, as windbreaks, pollinators or linear orchards. Some spring from bird-sown stones or discarded human picnics. Whatever their origins, they frequently spread by suckers.

'Plum appears to be one of the more persistent relic species which remain on the sites of long-deserted habitations. In Shropshire the species occurs near the sites of ruined cottages on common land, including The Cliffe and Clee Liberty.'[62]

'They were commonly planted in Kent as windbreaks for orchards. Nowadays supplanted by various kinds of poplar and European grey alder.'[63]

'In Suffolk, many old cottages have disappeared during agricultural development, or under wartime airfields. These old sites are often marked in the remaining hedgerows by cherry plum and bullace.'[64]

Wild plums in the Lake District.

In Essex – round Dunmow, Harlow Common and Catmere End near Duxford, for instance – wild damsons, be they blue-bloomed, green or yellow, are known as 'crixies'.[65] At Edlesborough and Weston Turville, near Aylesbury, Buckinghamshire, two highly local varieties were used both for eating and for dye-making:

' "Edlesborough Prunes" were once prized in our parish for jam-making and for fresh fruit. During the Second World War they were sent to Covent Garden in crateloads. They provided the Parish with prosperity and ran alongside our Straw Plait industry. The Prune provided dye for the Luton hat trade. There are still old trees in the hedges, but they are falling down.'[66]

'The "Aylesbury Prune" is almost unique to Weston Turville, and I am told the fruit was used to produce an indigo dye for the colouring of the "pearl" pattern of straw plaiting which was a cottage industry here until the late nineteenth century.'[67]

'After the opening of the railway large quantities of fruit [Aylesbury Prunes] were sent all over the country and during the Second World War much jam was made for the men at Halton Camp, but the trade virtually ceased after the war.'[68]

Damson colonies in Staffordshire and Cheshire were also harvested to produce dyes for the Manchester cotton-mills and Stoke potteries:

'Damsons seem to occur in every garden and field around this area. This is explained by the local story that the fruit used to be collected by locals and brought by basket to the pottery firms in Stoke-on-Trent for use as a dye.'[69]

'The Burgh of Galashiels has a long association with wild plums, which recalls a successful skirmish against a band of Englishmen in 1337 just outside the town. The English were gathering wild plums when they were surprised and overcome by a group of local men who later called themselves the "sour plums of Galashiels".'[70]

But wild plums are not usually that sour. They can be used instead of cultivated damsons and greengages in pies and jams, or as sweeter substitutes for sloes in alcoholic cordials. An old recipe is for damson cheese, a thick, sugary jelly made from strained damson pulp, which was served as a condiment with cold meats. When the King of Nepal was on a state visit to Britain in the 1980s, he ordered a large quantity of damson cheese to accompany the roast lamb banquet he was throwing at London's Guildhall.[71] And an edible oddity I discovered one autumn in a hedge outside an orchard at Bourne End, Hertfordshire, was naturally sun-dried damsons. The hedge had been cut in late summer with the first plums already formed, and the trimmings lay beneath, covered with dry, wrinkled fruits that tasted exactly like thin-fleshed prunes.

Cherry plum, *P. cerasifera*, is the most readily distinguished of the true species, and the earliest flowering. The white blossom often appears as early as February, with or just a little before the glossy, bright green leaves. Cherry plum is native to south-east Europe and is widely planted and naturalised in hedges. Pure hedges of this species in the Chilterns and Vale of Aylesbury are characteristic of land owned (and settlements built) by the Rothschild family in the nineteenth and early twentieth centuries.

The fruit does not form often, but is red or yellow and crisp-fleshed.

Cherry laurel, *P. laurocerasus*, is an evergreen from south-east Europe, almost ubiquitous in gardens and parks, and widely naturalised in woods from its purplish-black berries. It spreads rapidly by suckers and can shade out most other vegetation. But badgers enjoy the often horizontal trunks as scratching posts; and children, as usual, have discovered improbable properties in the leaves: 'We used to write our names with a twig on a laurel leaf and then put the leaf inside our clothes onto our backs. After a while we would take the leaf out and the writing would have gone brown and very easy to read.'[72] The leaves smell quite strongly of almonds when bruised or scratched, because of the presence of small quantities of prussic acid. They were once used in insect-killing bottles.

Wild pear, *Pyrus pyraster*. The status and distribution of the wild pear in Britain is uncertain. The true wild species is usually a more or less spiny tree with a rounded crown and small, hard, roundish fruit. Feral pears (referred to as *P. communis*), sprung from pips of discarded cores, are more pyramidal in outline and oval in leaf shape. But in practice it can be very hard to distinguish them, especially if they do not bear fruit. Some authorities regard wild pears as an introduction from Europe (though it is hard to see why the tree would be introduced, given that the fruit is inedible). But Oliver Rackham has found abundant early historical references to the wild pear and believes it is probably a native, now dramatically reduced in numbers. It is frequent in place names and is one of the six commonest trees mentioned in Anglo-Saxon charters as boundary features. Most of these landmark pears appear to have been hedgerow or free-standing trees well away from possible cultivation, though the handful of trees Rackham knows in East Anglia are all inside ancient woods. (One in Hayley Wood, Cambridgeshire, had its top blown off in 1974. Its exceptionally dense wood, which only just floats in water, was sold to a harpsichord maker.)[73]

Yet even trees of questionable origin – wildings sprung from discarded cores, or relics of cultivation – are an asset where they grow, with their dark, chequered trunks and mantles of dazzling white blossom in April:

'In the parish of Whatcote, Warwickshire, there are some 20 mature pear trees, not in an orchard, but spread out over several hundred acres of farmland in hedgerows.

The hard, inedible fruits of the wild pear.

Apparently the village was noted for its perry, made from the fruit.'[74]

'There are wild pears in ancient boundary hedgerows, in the Vale of the White Horse, south of Kingston Bagpuize.'[75]

Plymouth pear, *P. cordata*, is probably an English race of a shrubby wild pear which grows chiefly in Brittany and the west of Spain and Portugal. There are about 15 trees left in Britain, mostly in hedge-banks and factory buffer-land around the city of Plymouth, where it was first discovered in 1865. It rarely grows much taller than about fifteen feet in England and has distinctive early-flowering pink and white blossom which has been graphically described as smelling of 'decaying scampi'. The pears are small and hard, but 'blet' like *Sorbus* species late in October, producing a gritty fruit with a recognisable pear flavour.[76]

No one is sure whether it is a native or not. The Plymouth trees – and another small group on old hedge-banks near Truro in west Cornwall – are not far, as the blackbird flies, from the populations in north-west France and could be either ancient local inhabitants, dating back to the days before the English Channel opened, or recent bird-sown immigrants. But, unlike their French counterparts, they grow largely in suburban habitats rather than ancient woodland and could conceivably have been introduced by humans.

Either way they represent a scarce and maybe unique genetic resource and, as a bonus, hybridise well with domestic pears, which may give them a horticultural value in the future. So every effort is being made to conserve the trees. Suckers have been planted out in two Plymouth City Council nature reserves, for instance, and fruit is stored in the Kew Seed Bank. Perhaps (given the success of the campaigns to conserve the fig trees of Sheffield and the Buckinghamshire military orchids) they should be permitted to become more public trees, respected and protected by the people of Plymouth.

Crab-apple, *Malus sylvestris* (VN for fruit: Scrogs). The true wild crab is a comparatively scarce tree of old hedges and woods, nothing like as common as the 'wilding' – the collective name for chance seedlings sprung from the pips of discarded domestic apples. Crab-trees are spinier than cultivated varieties and, when mature, form trees with rounder and denser crowns. In hedgerows they can be conspicuous landscape features, especially when in blossom, and they are the third most mentioned species as boundary features in Anglo-Saxon and Welsh charters, occurring in nearly 10 per cent of 658 charters examined.[77]

Crab-apples are small, hard and sour, but make a good pickle, a pectin-rich base for jams, and the best of all wild fruit jellies, ranging in colour from yellow to deep pink, depending on the colour of the fruit. They can also be roasted and served with meat or added to warmed ale and winter punches. These are the crabs that 'hiss in the bowl' in Shakespeare's *Love's Labour's Lost*, and whose form the mischief-making Puck assumes in *A Midsummer Night's Dream*: 'And sometimes lurk I in a gossip's bowl,/ In very likeness of a roasted crab,/ And when she drinks, against her lips I bob'.[78]

The crab-apple was the most important ancestor of the cultivated apple, *M. domestica*. More than 6,000 named varieties of this have been bred over the centuries, of which probably only a third still survive. Because they

Crab-apple blossom.

must cross-pollinate to produce fruit, cultivated varieties do not come true from seed, and when a pip germinates it is likely to turn into anything, with genetic echoes, perhaps, from centuries-old ancestors. I know a green lane near Bovingdon in the Chilterns, not far from an area of one-time orchard land, in which there are three wilding trees, one with apples like miniature Cox's Pippins, another whose fruit has a bitter-sweet, almost effervescent taste, like sherbet, and a scent of quince, and a third whose long, pear-shaped apples have a warm, smoky flavour behind the tartness, as if they had already been baked. On the shingle beach at Aldeburgh in Suffolk there is a prostrate apple of unknown provenance which bears fully ripe fruit before the end of June.

Wildings represent an extraordinary genetic reservoir in these days of mass-market fruit-growing, both of lost varieties and of potential new ones. Many of the most famous names in the apple dynasty came from such random springings, often in the most unpromising surroundings. Granny Smith reputedly grew from an Australian woman's compost heap; Keswick Codlin from garden rubbish in Ulverston Castle; Shaw's Pippin from the council refuse tip at Wheathampstead in Hertfordshire.

Sometimes cultivated apples are deliberately set in the wild, echoing the frugal medieval practice of planting fruit trees in hedgerows or in the headlands at the end of the common fields:

'In this part of the world [Appleby (OE for 'apple village'), Cumbria] enclosure hedges were usually planted in the ratio of three thorns to one apple (very occasionally pear). There was a tremendous trade in hedging trees from "instant" nurseries all over the place. Any variety would do, and very rarely would the true crab be available, so our relics are the varieties of the time, and of cider pulp – any source of pips. We have not lost so many hedges up here as elsewhere in England, so many old ones survive.

Two forms of 'wilding' apple (sprung from discarded cores) on Cholesbury Common, Hertfordshire. The genes of many old – and possibly 'lost' – apples are preserved in the immense variety of wildings.

Immense variety of fruit is exhibited, by now scraggy and small, but I know of some I pick for culinary use. I work mainly with the MoD Forester for the North of England, on Warcop Army Range, where old trees persist. He is raising new stock from pips from old enclosure hedges.'[79]

'During the 1940s I spent several years in the Evesham district and had the pleasure of meeting an old countryman who looked after hedges. He told me that at the appropriate time of the year he would walk round with a pocket full of apple graftings. When he came across a suitable crab-apple stock, he would graft on a "proper" apple graft. Hence the large number of apple trees in hedges around Harvington, in particular.'[80]

Rowan, *Sorbus aucuparia* (VN: Mountain-ash, Witch wiggin tree, Keirn, Cuirn). Once widely planted by houses as a protection against witches, rowan is now better known as a street tree, a role it performs with civic efficiency. It has airy, ash-like foliage and orange-red berries, which only occasionally drop on pavements to annoy peevish pedestrians and car-owners.

It may have been the vivid colour of the berries that originally credited rowan with protective powers. Yet it was the wood that was regarded as the potent part of the tree. Up to the early years of this century, rowan boughs were hung over stables and byres in the Highlands, used for stirring cream in the Lake District, made into divining rods in Yorkshire, and cut for pocket charms against rheumatism in Cornwall.[81] In parts of Scotland there is still a strong taboo against cutting down a rowan tree, especially when it is close to houses.[82] The tree outside Gavin Maxwell's house at Camusfearna was famously cursed by the poet Kathleen Raine, with what Maxwell believed to have been disastrous consequences.[83]

But it is only on the Isle of Man that traditional rowan customs survive to any extent:

'In the Isle of Man [rowan] is still widely known as Cuirn. Apotropaic crosses of broken-off twigs ... are still hung above the lintel of house doors (inside) on May Eve. Formerly these were used on animals, animal shelters, churns, etc.'[84]

'Two twigs about five inches long are plucked, never cut, and a split made in the centre of one with the thumbnail. The other twig is pushed through and bound round with thread twisted by hand from sheep's wool taken from the hedge. If the wool is from the native Loaghtyn sheep so much the better. The cross is put up over the door of each house, inside, fresh each year, on May Eve.

Rowan (with juniper) at Little Langdale, Cumbria.

On the same night it is as well to have primroses in the house. In more profligate days they would be strewn on the doorstep.'[85]

Magically protective or no, rowan's pale bark and brilliant berries make it a striking tree to find outside an upland farm, or framed against dark pines on an autumn heath. A fourteenth-century Irish poet wrote: 'Glen of rowans with scarlet berries, with fruit fit for every flock of birds; a slumbrous paradise for the badgers in their quiet burrows with their young.' ('Deirdre Remembers a Scottish Glen')[86]

Rowan berries are fit for humans, too, and with a few crab-apples make a sharp, marmaladish jelly traditionally served with game and lamb. The tree grows in woods and scrub throughout Britain, chiefly on acid or light soils, but sometimes on chalk.

Wild service-tree, *S. torminalis* (VN: Chequer tree, Maple cherry; (for fruit) Chequers, Chokers). The wild service is one of the most local and least known of our native trees, yet it has a fascinating ecological and social history, revolving chiefly around its small but exotically flavoured fruits. The story of its rise, decline and subsequent rediscovery is something of a parable of the fortunes of our scarcer domestic plants. The fruits were a Neolithic staple, gained enough popularity at one time for houses, farms and pubs to be named after them, and then passed into obscurity, as the tree's ancient woodland habitats were destroyed and more glamorous fruits became cheaply available.

Wild service, or 'chequer-tree', in autumn foliage. The leaves are often blotched with scarlet and copper.

For much of the year a mature service-tree is an inconspicuous tree, a little like a maple in habit and leaf-shape, and with a dark, red-tinged bark that is sometimes cracked, like mud baked in the sun. But for two brief spells it can be spectacular. In May it is covered with white blossom, almost as thickly as a hawthorn. And from late October, after the brown fruits have set, the leaves on many of the trees turn a brilliant red tinged with copper and ochre. The best description of service-trees in autumn, and of the one-time popularity of their fruit, is given by John Clare. His poem, written about 1830, is entitled 'The Surry Tree', which was a nineteenth-century Northamptonshire name, echoing the Old English root of service, *syfre*.

Tree of tawny berry rich though wild
When mellowed to a pulp yet little known
Though shepherds by its dainty taste beguiled
Swarm with clasped leg the smooth trunk timber grown
& pulls the very topmost branches down
Tis beautiful when all the woods tan brown
To see thee thronged with berrys ripe & fine
For daintier palates fitting then the clown
Where hermits of a day may rove & dine
Luxuriantly amid thy crimson leaves ...[87]

In Bernwood Forest, Oxfordshire, where wild services have been left to grow on amongst newly planted conifers, the splashes of scarlet among the dark evergreens are dramatic.

The unravelling of the history and distribution of this secretive tree owes much to the efforts of Patrick Roper, who lives in the tree's heartland in the Weald. The survey he organised through the Botanical Society of the British Isles, beginning in 1974, showed it to be more widespread than had been believed, but largely confined to ancient woods and hedgerows on clays in eastern and southern England and limestone in the west. Its northernmost station is the southern Lake District.[88]

The loyalty of wild service to ancient woods is proba-

The berries of the wild service-tree are edible once they have been softened, or 'bletted', by frost. Their taste has hints of dried apricot, tamarind and sultana.

bly due to the fact that, in our climate, the tree spreads almost exclusively by suckers. Seedlings are rarely found, and I have only once seen them developing into young trees, in Roydon Wood in the extreme south of Hampshire. Suckering, on the other hand, can be prolific. In Longmans Grove Wood, near Rockhampton, Gloucestershire, there is a patch of close on a thousand wild service-trees, probably all suckers from a single parent and covering an acre of ground. From the air in autumn they show up as a large, irregular crimson circle.[89]

The main concentrations include the Severn and Wye Valleys and much of the Welsh Marches; Bardney Forest in Lincolnshire; many ancient woods and hedges in Essex, Suffolk, and east Hertfordshire; wind-pruned woods on cliffs and in river valleys in Cornwall and Pembrokeshire (where the trees still have the unique local name of 'maple cherry'); a curious suburban cluster south and east of Birmingham, especially round Solihull and Tanworth, perhaps relics of the Forest of Arden[90] (they even occur in hedges in urban parks in the Bournville area); and the old Rockingham Forest area in Northamptonshire, John Clare's countryside, where service boughs used to be carried in May Day processions and Beatings of the Bounds.[91]

But the tree's main stronghold is the clays of the southern Weald, from Kent to Hampshire. Wild service often occurs in parks and pastureland here, and reaches a much greater size. What may be the biggest tree in the country, at Parsonage Farm, Udimore, in East Sussex, has a 13-foot girth and in a good year bears two tons of berries. At Brooks Green in West Sussex, there is a hedgerow that may possibly have originated from seedlings (though it is more likely a woodland 'ghost' – see Hedges, p. 92): 'We have about 25 wild service-trees, mostly in a long line down an old field boundary. We have always assumed that they spread by means of bird droppings from birds which sit on the fence. There is a very large tree in a small wood some 300 yards away. Mistle thrushes love the berries.'[92]

These southern counties are the area where the tradition of eating the fruits persisted longest. The brown berries, which may be round or pear-shaped, are hard and bitter at first, but as the autumn progresses (or once they are picked and taken indoors) they begin to 'blet' and turn soft and very sweet. The taste is not quite like anything else that grows wild in this country, with hints of apricot, sultana, overripe damson and tamarind, and a lightly gritty texture. They were a boon when other sources of sugar were in short supply, and were enjoyed as a kind of natural sweet by children. W. A. Bromfield described how they were sold in markets in the 1850s: 'The fruit is well known in Sussex by the name of Chequers, from its speckled appearance, and is sold both there and in this island [Isle of Wight], in the shops and public markets, tied up in bunches, principally to children.'[93]

The custom was to hang the chequers up in strings in the kitchen or over the hearth: 'I was born in St Michaels near Tenterden, Kent, and well remember gathering the berries from the tree which grew in the hedge bordering our garden. They were picked in clusters and threaded on lengths on string, layer upon layer, which were then hung in the larder to ripen. We were then allowed to pick them singly, as they ripened, leaving the strings.'[94] (Some modern children seem to enjoy service berries nibbled like this just as much as their predecessors.)[95]

The Weald, and Kent especially, is also the site of a still unresolved connection between Chequers pubs and chequer trees which may be the source of this local name for the wild service. There is an unusually high proportion of Chequers Inns in the area (12 in the Canterbury, Kent, Yellow Pages alone),[96] and too many have chequer trees growing in their gardens or close by to be explained away as a coincidence. There are trees, for instance, at Chequers Inns at Rowhook, West Sussex, Smarden, Kent, and (at one stage removed) Four Elms, Kent: 'The wild service-tree grows by our village pond in Four Elms, which coincidentally has a "Chequers Garage" near one corner of it, although the name came with the garage from a "parent" garage of that name in the neighbouring village of Bough Beech. This Chequers Garage originated from a Chequers Inn which gave up as a pub about 30 years ago.'[97]

Patrick Roper believes that the link may lie in an alcoholic drink made from chequer berries and that the tree and its fruit derived their local name from being served in Chequers Inns. (A chequer-board was often hung outside public houses, as brass balls are outside a pawnbrokers.)[98] But it is doubtful if the drink was a beer, and there is certainly no etymological connection, as has sometimes been suggested, between the Latin *cervisia* (or Spanish *cerveza*) and 'service'. (Its root is the OE *syfre*.) Most probably the drink was a ratafia, like sloe gin and cherry brandy, and made by steeping the service berries in spirit.

A modern version from Essex uses whisky and fruit from Halstead Wood: 'Cover ripe service with cheap whisky, and add sugar at about 1 lb per pint. Leave to infuse for a couple of months. Give a good shake every week and then strain. Ours has a lovely golden colour and a very distinctive and agreeable taste.'[99]

The Chequers Inn at Smarden has an undated house recipe for a kind of wild-service wine: 'Pick off bunches in October – Hang on strings like onions (looks like swarms of bees) – hang till ripe. Cut off with scissors close to checkers [*sic*] – do not pull out – Put in stone or glass jars – Put sugar on – 1 lb to 5 lb of checkers – Shake up well. Keep airtight until juice comes to top. The longer kept the better – Can add brandy.'

Service-trees also haunt non-Chequers inns. A gnarled specimen grows out of a Barnack stone wall opposite the Hare and Hounds in Greatford, Lincolnshire.[100] Another is in the garden of the Snooty Fox, Lowick, Northamptonshire.[101] The tree at the Chequers, Gedney Dyke, Lincolnshire, was planted in 1992, and is a practical example of renewed interest in the tree and its fruit.[102] A somewhat improbable wild service memorial tree was planted on the campus at Welwyn Garden City to celebrate Nick Faldo's golfing achievements (as he was brought up in the town).[103] Even the pale, tough, close-grained wood is coming back into use now that more trees are being discovered. (In France, even though it is much commoner, it is the most expensive native wood.) A musical-instrument-maker from Billingshurst, West Sussex, uses it for making the jacks which pluck the strings of harpsichords.[104]

Service-tree, *S. domestica*. What is sometimes called the true service-tree is a native of central and southern Europe, with rowan-like leaves and large pear- or apple-shaped fruits, which is occasionally grown in botanic gardens and parks in Britain. But for more than three centuries there has been a cryptic specimen in the depths of Wyre Forest, Worcestershire, that seemed, from its situation, as if it might have a genuinely wild ancestry.

The Wyre 'Whitty Pear' was first recorded in the Forest in 1678 by Alderman Edmund Pitts, who wrote about it in *Philosophical Transactions* for that year in a note entitled 'An account of the Sorbus pyriformis'. He plainly regarded it even then as a very ancient tree. He described the fruit 'in September, so rough as to be ready to strangle one. But being then gathered, and kept till October, they eat as well as any medlar.' The names 'whitty', 'whitty-tree' and 'wicken' were widely used in the region for the rowan and were no doubt borrowed for this rowan-like tree. In 1842 it was burned down, reputedly by a poacher to spite a local magistrate who was fond of it. Fortunately, two grafts had been taken from the tree some years before by the Earl of Mountmorris and grown on in the grounds of nearby Arley Castle. One of these – already a mature tree – was planted out in 1913 at the whitty pear's ancestral site in Wyre Forest, where it continues to thrive.[105] Subsequent grafts were planted elsewhere in the vicinity of the Forest, and the fine specimen in the Oxford Botanic Garden is believed to have originated from Wyre stock (possibly seed – though the fruits are rarely fertile in our climate).

And that might have been the end of the story of *S. domestica*'s presence in Britain, had it not been for the remarkable discovery between 1983 and 1991 of two indisputably wild populations of the species on cliffs in South Glamorgan. It was an extraordinary find, not least because this area has for centuries been one of the most intensively botanised in Britain.

There are 13 or 14 separate trees, stunted by the wind and growing on inaccessible ledges on crumbling limestone, so that they rarely or never fruit – all reasons, perhaps, why they were not noticed for so long, or mistakenly assumed to be rowans. They are multi-trunked, and microscopic counts of the very narrow rings in dead branches suggest an age of up to 400 years for the individual living trunks, and possibly of over a thousand years (cf. small-leaved lime, p. 118) for each whole tree.

More recently, biochemical tests have verified that these are indeed true service-trees, and more closely related to the Wyre Forest whitty pear (and its Oxford descendant) than to cultivated trees from continental stock – which suggests that there may once have been a more widespread native population of *S. domestica* in Wales, the South-West and the Midlands. (There are very old records from Cornwall, for example, which Patrick Roper suggests should be re-examined, along with any stunted 'rowans' on Cornish cliffs.)[106]

Dr Quentin Kay, of the University College of Swansea, who has co-ordinated much of the research, comments: 'We have not found any records of medieval (or indeed modern) cultivation of *S. domestica* in South Wales, and deliberate introduction in the past seems most unlikely for these inaccessible cliff sites; in fact the most likely explanation is that the Glamorgan populations are fully native and have survived in their cliff refuge sites since their arrival during the spread of the deciduous trees in the early Post-glacial!'[107]

Whitebeam, *S. aria*. In late April, the opening leaves of whitebeam – conical and white-coated – are like magnolia buds just before they bloom. John Evelyn may have been thinking of this species (rather than its relative, the wild service) when he wrote in 1670: 'The *Service* gives the *Husbandman* an early presage of the approching *Spring*, by extending his adorned *Buds* for a peculiar entertainment, and dares peep out in the severest *Winters*.'[108] It did not acquire the name whitebeam, 'white tree', until the eighteenth century, despite retaining the felted white underside to its leaves throughout the year.

Whitebeam is a tree chiefly of tall scrub and the edges of woods on calcareous soils and is seen at its best against the dark foliage of yew in the Wye Valley or on the Hampshire beech hangers. But in the Chilterns it is sometimes found deep inside ancient woods on acid plateaux.

The whitebeam family is remarkable for producing highly local endemic species, often confined to single rocky gorges and growing nowhere else in the world. Stace lists 14 of these neighbourhood *Sorbus* species from England, Scotland and Wales, all subtly different in height, habit and the shape and colour of leaves and fruit.[109] All their fruits are edible when 'bletted', like those of the wild service-tree, and most are in need of local English names: Arran service-tree, *S. pseudofennica*, is confined to steep granite banks in Glen Catacol, Arran. *S. arranensis* is also confined to the Arran Islands, though more widely distributed on steep stream-banks. *S. leyana* is represented by a few small shrubs on limestone crags near Merthyr Tydfil, Brecon. *S. minima* grows on limestone crags near Crickhowell, Brecon. *S. anglica* is local in woods and rocky places on Carboniferous limestone in Wales, Shropshire and south-west England. *S. leptophylla* grows on limestone crags in two areas in Brecon. *S. wilmottiana* is restricted to woodland and scrub in the Avon Gorge. *S. eminens* is local in rocky Carboniferous limestone woodland in the Avon Gorge and Wye Valley; *S. porrigentiformis* in rocky limestone woods from south

Opening whitebeam leaves, like magnolia buds just before they bloom.

Devon to Brecon; *S. lancastriensis* in rocky scrub on limestone in west Lancashire and Westmorland; and *S. vexans* in rocky woods (*not* on limestone) near the Bristol Channel in Devon and Somerset. *S. subcuneata* grows in rocky oakwoods near the north coasts of Devon and Somerset, and *S. bristoliensis* in rocky scrub and woods on limestone in the Avon Gorge. French Ales or Devon Whitebeam, *S. devoniensis*, grows in woods and hedges on well-drained soils and is widespread in Devon and very local in east Cornwall and the Isle of Man. The fruits were reputedly sold for eating in Devon, but certainly not in living memory.[110]

To confuse the picture still further, Swedish service-tree, *S. hybrida*, from Scandinavia, Swedish whitebeam, *S. intermedia*, from the Baltic, and *S. latifolia* from south-west Europe are widely planted and naturalised from bird-sown seed, often in rocky places. And a small naturalised group of the orange-berried whitebeam, *S. croceocarpa* – a species whose origins are still mysterious – has recently been found at Lleiniog, on the north shore of the Menai Strait in Anglesey.[111]

Juneberry, *Amelanchier lamarckii*, is a North American shrub, naturalised in some woods on sandy soils in southern England, most extensively south of Pulborough in Sussex. It has light, well-spaced branches, and its purple-tinged young leaves and large white flowers make a striking show in the understorey in April, before many native trees are in full leaf. The dark berries rarely form in this country, but are edible.

Wild cotoneaster, *Cotoneaster integerrimus* (VN: Great Orme berry). This is a small native shrub, of which less than 20 specimens survive (1994 figures) on the windswept limestone cliffs of Great Ormes Head, Caernarvonshire, where it has been known since 1783. On the cliffs it is greatly outnumbered by a naturalised Himalayan cotoneaster, *C. integrifolius*. Stace lists no fewer than 45 species of this genus, most of them from south-east Asia, which have become naturalised in various sites in Britain as a result of bird-sown seed.[112] As at the Orme, they are as likely to crop up in natural sites (cliffs and chalk scrub, for example) as on urban walls and in waste places. The number is likely to increase, given that there are now some 80 species being cultivated in this country for their ornamental fruits, which are as attractive to birds as to gardeners. **Firethorn**, *Pyracantha coccinea*, is a southern European shrub, very widely planted in gardens and increasingly naturalised in the same manner as the cotoneasters.

Medlar, *Mespilus germanica*, was, with quince, mulberry and walnut, one of the quartet of trees often planted at the corners of herb gardens and orchards. But, unlike its old partners, medlar's fruits have gone out of fashion, and the tree has become one of those intriguing species (along with, for example, asarabacca and true service-tree) whose provenance – and distribution, for that matter – are shadowy issues, prone to throwing up myths.

It is hard to mistake, with its dark, contorted trunk (spiny in the wild forms) and solitary white flowers sitting

Juneberry, at Frensham, Surrey. An early-flowering North American shrub, naturalised in a few woods in the south.

Medlars, like service berries, are edible once they have softened. The taste is a little like baked apple.

on the branches like camellias, and most of the vaguely wild specimens are obviously relics of plantings in orchards or parks. It is, strictly, a native of south-east Europe and south-west Asia, anciently introduced and barely ripening its fruit unaided in this country.

So what is one to make of the individuals which have been found in much wilder situations in woods and hedgerows in the extreme south-east? Gerard reported that it could be found 'oftentimes in hedges among briars and brambles'.[113] In 1831 John Stuart Mill found a spiny specimen in a hedge on Redstone Hill, near Reigate, Surrey (still there in the 1980s),[114] and there are others in apparently semi-natural woods in Surrey and Sussex. Were they bird-sown, or examples of the once widespread peasant practice of planting orchard trees in the wild? Or was the eminent botanist A. H. Wolley-Dod right to believe that medlar might be native in the extreme south-east?[115]

The fruits are like large, brown rose-hips, and in our climate become edible only when they are 'bletted' – made soft and half-rotten by frost, like service-berries (see p. 205). In the Mediterranean region they can be eaten straight off the tree. The flesh tastes a little like baked apple, but with the consistency of chestnut purée. The slightly 'high' flavour and granular texture made them popular for serving with whisky. They also made jellies, preserves and fillings for pies or were baked and eaten directly out of their skins with a spoon.

Hawthorn or **May-tree**, *Crataegus monogyna* (VN: White-may, Whitethorn, Thorn-bush, Quick, Quick-thorn, Mother-die; Bread-and-cheese (usually for leaves only); May (for blossom); Awes, Asogs, Azzies, Aglets, Agags, Agars, Arzy-garzies, Boojuns, Hoppety-haws (for fruit)).

The May-tree is the only British plant to be named after the month in which it blooms and seems to have acquired its eponymous title some time in the sixteenth century. Its blossoming marks the cusp between spring and summer, and the old saw 'Cast ne'er a clout ere May is out' almost certainly refers to the opening of the flowers, not the end of the month. It was the ancestor of the Maypole, the source of May Day garlands and the decoration of Jacks-in-the-Green and Green Georges, and one of the models for the foliage which wreathes the faces of Green Men carved in churches and inns. Superstitions about the flowers – and especially about the dire consequences of bringing them inside the house – persist more widely than for any other species. Isolated hawthorns are treated with respect, too, and, though often little more than bushes, are the most frequent trees mentioned in Anglo-Saxon boundary charters. The thorn is also the tree after which the Anglo-Saxon runic letter þ is named. It is a formidable list of honours for a tree which these days is often written off as 'scrub'.

There are some obvious reasons for the attention which was paid to hawthorn historically – the combination of thorns and red berries, for instance, which suggested a tree associated with protection and sacrifice, perhaps even the source of Christ's crown of thorns. The blossom too – white, heavy, sweet-and-sour, 'the risen cream of all the milkiness of May-time' as H. E. Bates put it[116] – must always have been cheering after a dark winter.

Yet these features hardly explain such an extravagant history of veneration. Why should a small, commonplace and not especially long-lived tree have been so often chosen to mark boundaries and meeting places? How could its blossom be so festively flourished at spring ceremonials, yet be banished from houses? How did a shrub that seems to bloom most typically in middle and late May become the symbol of May Day itself, the beginning of the whole cycle of spring festivals?

This last anomaly is perhaps the easiest to explain. Before the revision of the calendar in 1752, which did away with 11 days (see p. 174), May Day occurred on what, in the modern (Gregorian) calendar, became 12 May.

This is the kind of date on which hawthorn customarily breaks into bloom today, so it is a fair assumption that, for all but the last two centuries, may blossom would have begun opening on May Day – at least in the warmer south and west of Britain. W. G. Hoskins put the current date for the Midlands as 18 May: 'On that day these miles of snowy hedges reach perfection, so dense and far-reaching that the entire atmosphere is saturated with the bitter-sweet smell whichever way the summer wind is blowing.'[117]

This evocative description of the Midlands fieldscape highlights another crucial change in the status of hawthorn. Those endless, single-species quickthorn hedges simply did not exist before the great parliamentary enclosures of the eighteenth and nineteenth centuries, in which something like 200,000 miles of thorn hedge were planted. Before that, hawthorn was a frequent (rather than abundant) component of mixed-species hedges and of chalk scrub, fens and woodland clearings. In blossom-time it was isolated individuals and clumps that would have shone out in the landscape, not the billowing ribbons which we see today.

But hawthorn is notoriously erratic in its flowering, and greatly influenced by late winter and spring temperatures. A contributor who has recorded its first flowering regularly in Minehead, Somerset, found that over the period 1984 to 1994 this swung between the extremes of 19 April (1989) and 26 May (1987).[118] Its flowering is also influenced by altitude, soil, shade, and these days by provenance (many of the earliest leafing and flowering specimens are from a cultivated variety introduced for hedging from the Netherlands). On the first two days of May 1994, for example, the first may blossom was recorded in Minehead, Somerset; Burgess Hill, West Sussex; Rayleigh, Essex; Much Marcle, Herefordshire; Cresswell, Staffordshire; and Tadcaster, North Yorkshire.[119] More revealing was Judith Allinson's meticulous account from Yorkshire, relating the first blooming of the may to altitude. At 60 m near Leeds, the flowers were just appearing on 16 May; at 100 m (Skipton) on 23 May; at 316 m (Cowside, near Settle) on 16 June; but not until 24 June at 345 m on the same hill. At 388 m, near Malham Tarn, the

Hawthorn blossom, 'the risen cream of all the milkiness of May-time' (H. E. Bates).

first bushes did not begin to flower until the beginning of July and one was still in flower on 29th.[120]

The general coincidence that existed formerly between the flowering of the may and May Day itself, the Celtic festival of Beltane, might begin to explain the tree's reputation. Yet it has never generated an unqualified welcome for may flowers. Although the blossom has been used profusely for garlands and open-air decoration, there is still a widespread superstition (respected, if not literally believed – like that of walking under ladders) that may blossom is unlucky inside a house, and likely to presage a death:

'When I was teaching in Warwickshire, a child brought in a bunch and her class teacher seized them between her finger and thumb and flung them out through the window. (However Proust describes the church in Combray, and particularly the altar, being decorated with May for the Virgin.)'[121]

'As children in North Wales we never took branches of this tree into the house as it was meant to cause the death of your mother – I assume because of the suffocating smell the blossoms have in a closed room. Its Welsh name is "Blodau marw mam", literally "Flowers-death-mother".'[122]

'When I married my former husband in 1973 and moved to a Tudor house in the village of Denton, Kent, he immediately removed a hawthorn tree from the garden due to a superstition handed down the generations to him. This is interesting as his background is as follows. His great-grandfather emigrated from Scotland in 1885, having been a farm labourer. He was aged 21 years. My husband was a fourth generation New Zealander who came back to the UK in the 1960s bringing this superstition with him.'[123]

All kinds of reasons and rationalisations are given for the superstition. The flowers formed the wreaths worn by human sacrifices during the Celts' spring fertility rituals (a myth for which there is no historical evidence whatever). The pollen sparks off hay fever, or aggravates spring chest disorders. The white flowers, with their red anthers and incipient red berries, suggest blood and the pallor of corpses. (Red and white flowers together are still unpopular in hospitals.)[124]

Many contributors (mostly Catholics themselves) were taught that the superstition had its roots in the era of Catholic suppression in Britain. A Benedictine nun writes: 'My mother, who was brought up in Aberdeenshire of Orcadian stock, was not especially superstitious, but held that it was unlucky to bring hawthorn into the house. May is associated with Our Lady and I wonder whether it might be an anti-Catholic connection. My mother was a non-Catholic. I also understand the superstition does not exist in Ireland.'[125]

'I was brought up in Kent in the 1920s, and there the reason given for not bringing May into the house was that it had long been seen as the Virgin Mary's plant, so if you took it into the house you might be thought to be a papist.'[126]

'Before the Reformation, when England was Catholic and known as Our Lady's Dowry, people used to make in their homes, during the month of May (the month dedicated to Mary and bearing her name), "May Altars", that is to say they would set up in a prominent place a statue of Mary, and surround it with flowers, particularly hawthorn, May, which was flowering in abundance.

Double hawthorn, painted by James Sillett, 1803.

During and after the Reformation, with the imposition of the new form of worship and Puritanism, such acts were forbidden by law so that anyone being seen taking such decoration into their homes was immediately branded "Papist" and subject to heavy fines, imprisonment and, in some cases, death. You may like to know that Catholics still make their "May Altars" without worrying about any superstitions attached to the May.'[127]

'When I was at boarding school in North Yorkshire in the 1940s, I was told that the old superstition of it being unlucky to bring May into the house dated from the custom of Catholic recusant families of placing a sprig of May blossom in the window to indicate that a priest would be saying Mass there, a practice which was forbidden in those post-Reformation times.'[128]

Yet Marina Warner, in her book on the cult of the Virgin Mary, argues that the association between the month of May and Mary began only in the eighteenth century in Italy, from whence it spread to the rest of the Catholic world.[129]

A more immediately plausible reason for the superstition is that the triethylamine responsible for the stale element in hawthorn's complicated smell is one of the first chemicals produced when living tissue starts to decay. In some areas it is still believed to be 'the smell of the Great Plague'.[130] Nurses who have worked in Africa say it is reminiscent of the smell of gangrene. Perhaps its malign associations simply echo the time when corpses were kept at home for up to a week before burial, and people were much more familiar with the smell of death. (Ted Hughes bluntly describes may blossom's smell as that 'aniseed, corpse odour'.)

Yet triethylamine's fishy scent is also the smell of sex – something rarely acknowledged in folklore archives, but implicit in much of the popular culture of the hawthorn. The Cambridge anthropologist Jack Goody suggests that this may be the reason for the different degrees of tolerance of may blossom inside and outside houses: 'The hawthorn or may was the special object of attention at May Day ceremonies that centred on the woods, the maypole and the May queen ... In contrast to Christmastide greenery and Easter willow, it is a plant kept outdoors, associated with unregulated love in the fields rather than conjugal love in the bed.'[131]

There is some circumstantial evidence for this in the *doubles entendres* of a May Day folk-song still remembered in some Hertfordshire families:

We've been rambling all this night
And the best part of the day,
And now we're returning back again
We've brought you a branch of May.

Arise, arise you pretty fair maids
And take your May Bush in,
For if it's gone before morning comes
You'll say we've never been.

I have a purse upon my arm,
It draws with a silken string,
And all it wants is a little silver
To line it well within.[132]

It may be common hawthorn's preference for open country (in contrast to woodland hawthorn, below), for heaths and rocky places for example, that also helped make it such a significant boundary tree. Historically, hawthorn seems to have been *noticed* more than any other species. In his survey of 658 Anglo-Saxon charters and boundary descriptions Oliver Rackham found that it is, by a considerable margin, the commonest tree mentioned as a feature, representing 38.7 per cent of all trees specified. (Oak is next with 13.1 per cent.)[133] It also heads the list of trees mentioned in English place names, with 18 per cent.

Hawthorn is customarily thought of as an ephemeral tree, springing up on ungrazed commons or downland to form a brief, scrubby presence before 'proper' woodland supersedes it. In practice, thorn thickets or 'spinneys' (from the Latin *spinetum*) can be remarkably resilient, and reluctant to give way to other types of woodland, as on some of the clay-with-flint commons of the Chilterns. A particularly long-lived spinney is on the Dry Tree Barrow on Goonhilly Down, Cornwall, which is named as *cruc draenoc* ('thorny barrow') in a Saxo-Cornish perambulation of AD 977.[134] Perhaps many of the boundary thorns in charters were persistent, self-perpetuating groves of thorn, rather than individual trees.

Vaughan Cornish, who suspected there was a Belgic or Romano-British thorn-cult in early Britain, traced several specific landmark thorns or thorn clumps, at, for instance, Spelthorne in Middlesex, Shimpling Thorn in Norfolk, and his own village of Salcombe Regis in Devon.[135] Here, a memorial stone by the current tree carries this inscription: 'A thorn tree has been maintained here since Saxon times when it marked the boundary between the cultivated field of the coombe and the open common of the hill. It has given the name Thorn to the adjacent house where the Manor Court was held and to the surrounding farm. Vaughan Cornish, Christopher Tomkinson, Trustees of the Thorn Estate.'[136]

Near Bracon Ash in central Norfolk, there is a boundary and meeting-place thorn which is believed to be at least 700 years old. The Hethel Old Thorn, or Witch of Hethel, close to the isolated Hethel Church, is mentioned in a thirteenth-century charter. In the mid-eighteenth century, Robert Marsham, one of Gilbert White's correspondents and a great lover of trees, found that its girth was more than 20 feet.[137]

The most celebrated old hawthorn is the Holy Thorn at Glastonbury in Somerset, which produces flowers and young foliage in midwinter, as well as blossoming again at the normal time in May. (It is now conventionally regarded as a sport, *C. monogyna* 'Biflora'.) The tree is first mentioned in an anonymous early sixteenth-century poem entitled 'Here begynneth the lyfe of Joseph of Armathia'. The poem described three thorn trees growing on Weary-All Hill, just south of Glastonbury, which

Do burge and bere greene leaues at Christmas
As freshe as other in May when ye nightingale
Wrestes out her notes musycall as pure glas.

But an explicit connection between the thorns and St Joseph of Arimathea was not made in print until the seventeenth century. Then a variety of explanations – some based on theological argument, others on oral tradition – began to surface. Local legend had it that Joseph was an uncle of the Virgin Mary and had come to Britain with 11 other disciples some time between AD 30 and 63. He had travelled to Glastonbury and had thrust his staff in the ground, where it took root and grew to become the original Christmas-flowering thorn. A variant of the story suggested the tree sprang miraculously from a fragment of Christ's Crown of Thorns, brought to Britain by Joseph; another that the second blossoming coincides with Easter,

A descendent of the original Christmas-flowering 'Glastonbury Thorn' on Windmill Hill, Somerset, with Glastonbury Tor in the background.

so that the two flowerings symbolically encompass the whole of Christ's earthly existence. (It is quite possible that the tree was brought from the Middle East at some time, as there are Mediterranean populations of *C. monogyna* that do put out blooms in early winter.)[138]

During the sixteenth and seventeenth centuries pieces of the thorn were repeatedly cut off, either as souvenirs or as cuttings to grow on. In one way this was fortunate, as the Glastonbury tree received a more drastic and eventually fatal hacking from the Puritans, who abhorred idolatry, especially of trees. But a cutting was soon established in its place, as were similar cuttings all over England. There have been notable and long-established Glastonbury scions at, for instance, Quainton and Shenley (both Buckinghamshire) and at Houghton-le-Spring, Tyne and Wear – the 'Gilpin Thorn'.[139]

After the calendar change of 1752, the trees were the subject of widespread superstitious attention, to see if they would bloom according to the old or the new calendar. Roy Vickery documents many examples of huge gatherings at the trees, stretching into the early years of this century, including a crowd of more than 2,000 at Quainton on Old Christmas Eve, 4 January 1753.[140] These days the blossoming of the trees is extremely variable,

Hawthorn leaves and fruit carved in the Chapter House at Southwell Minster.

sometimes, after a mild autumn, as early as November, sometimes not until early March if there is a severe winter. But each year, just before Christmas, sprays from one of the thorn trees which grows in St John's churchyard in Glastonbury are sent to the Queen and the Queen Mother. The Queen is said to place hers on her breakfast table on Christmas morning.[141]

In the village of Appleton, Cheshire, what is presumed to be a descendant of the Glastonbury Thorn is still dressed with red ribbons and flags at midsummer. 'Bawming the Thorn' ('bawming' is a local dialect word meaning adorning or anointing) has clear affinities with other pre-Christian tree ceremonies (cf. the Aston Flag Tree, p. 137, and Oak Apple Day rituals, p. 76). But the official tradition is that it commemorates the planting of a Glastonbury Thorn scion by Adam de Dutton on his return from the Crusades in 1125. For a supposedly 'holy' thorn it has had a rather luckless history. What was probably a second-generation successor was blown down in 1965. Its replacement quickly withered and died, and was replaced two years later. The Bawming ceremonials have been erratic, too, and were suppressed during Victorian times because they became too rowdy. Originally they took place on Old Midsummer's Day (5 July), but recently they have tended to be held on a Saturday close to the new Midsummer's Day in June.[142]

In the 1990s, a self-sown Christmas-flowering thorn (20 feet high) was discovered on Saltwells Local Nature Reserve in the West Midlands, which rather supports the theory that it is a naturally occurring sport of the native race of common hawthorn.[143]

There are other landmark thorns, both sacred and secular: 'A lone hawthorn, by the side of a rough track near Kilmeford, Argyll, is the local equivalent of a wishing well, with coins hammered into the bark.'[144]

'In 1811, at the time of the Napoleonic Wars, Selkirk Town Council agreed to billet French prisoners of war. They were housed in a crow-stepped building now used as the town library. The prisoners had a lenient incarceration, perhaps due to "the Auld Alliance". They had the freedom to walk as far as "the Prisoners' Bush", about two miles distant. The bush was a hawthorn. Sadly it became frail and was blown down about 18 months ago [1993] but it has been replaced by another hawthorn which is growing well.'[145]

Some old thorn hedges and individual trees have been pressed into domestic use: 'In our village till recently, there was a double hawthorn hedge, planted on a bank with a path between the hedges, which years ago was used for spreading linen on to dry.'[146] In the Isle of Man hawthorn bushes are sometimes trimmed to shape in order to hold milk pails while they are drying.[147] A tree at Ravenshead in Nottinghamshire is shaped in the form of a table top.[148] Hawthorn is the symbol of West Bromwich Albion Football Club: 'When the site was developed in 1900, the area was full of hawthorn bushes, and these in turn were full of thrushes. The locality was known as The Hawthorns. The stadium took its name from the same source, and also adopted the thrush (known locally as the throstle) as part of its crest.'[149]

Young hawthorn leaves are often the first wild green leaves that children eat, and they are universally known as 'bread and cheese'. This is usually explained as referring to their rudimentary culinary qualities. But some children have eaten the berries (whose flesh is a little like overripe avocado pear or, more fancifully, a whey cheese) together with the autumn leaves. These are just about edible, even in early October, and are certainly no worse than very stale bread: 'We would pick the red berries and green leaves in the autumn. These were known as "bread and cheese" – the leaf the bread, the berry the cheese.'[150] A more interesting use of the young leaves is the recipe given to Dorothy Hartley by a farm labourer's wife in Wymeswold, Leicestershire, in the 1930s. It was a 'spring dinner', made by covering a suet crust with young hawthorn leaf-buds and thin strips of bacon, and rolling and steaming as a roly-poly.[151]

Woodland hawthorn, *C. laevigata*. The botanical historian David Elliston Allen believes that hostility to having hawthorn in the house, and many of the other Maytime customs and superstitions associated with the blossoms, may have originated with another species altogether, the woodland hawthorn (unfortunately and misleadingly often called the Midland hawthorn). This is a shrub that immediately seems better qualified as a harbinger of misfortune. It blooms earlier, close to May Day, and its flowers smell nauseating, certainly much more like rotting flesh than the decadently sweet odours of common hawthorn.

Allen's theory was prompted by a passage in an unpublished letter from R. P. Murray to E. F. Linton, in the library of the Department of Botany at the British Museum, dated 21 May 1900: 'When in Switzerland we had plenty of *C. monogyna* and *C. oxyacanthoides* [now

Hawthorn is an opportunist shrub, beginning woods wherever it is protected from grazing.

C. laevigata]: the latter flowering a week or two earlier than *C. monogyna*. But I often gathered a lot of *C. oxyacanthoides* for decorative purposes: and tho' in smell quite like the other form when gathered, it used to absolutely *stink* of putrid flesh soon after: – sometimes within about half an hour. I do not remember that this *ever* occurred with *C. monogyna*.'

In the early Middle Ages, when much of our traditional plant-lore was evolving, woodland hawthorn could well have been the commoner and more familiar of the two species. Today it is a shrub restricted to ancient woodlands and very old hedge-banks on clay soils – both much more frequent habitats in pre-medieval times. 'Common' hawthorn was almost certainly a scarcer plant. '*C. laevigata*,' Allen suggests, 'would seem to have been the original May-flower.'[152]

This theory is, I suspect, partly invalidated by the calendar change of 1752 (see above). At that time common hawthorn was very much a flower of May Day, and woodland hawthorn of mid-April. Nevertheless, woodland hawthorn is an intriguing shrub, recognisable in its pure form by its shiny, three-lobed leaves, twin stigmas in the flowers and twin seeds in the haws, and much laxer habit. In old hedgerows it tends to form a complete spectrum of hybrids with common hawthorn.[153]

Cockspur-thorns, *C. persimilis* and *C. crus-galli*, are North American species, with large berries and vivid autumn foliage. They are widely planted in parks and on roadsides, and naturalised in hedges and scrub in southern and central Britain. In Cannock Chase, Staffordshire, cockspur-thorn is known as the 'cank tree'.[154]

Wild foods

Since the mid-1970s, foraging for wild plant foods has become almost as common in Britain as on the continent. Even mushroom-hunting has ceased to be regarded as an outlandish and dangerous eccentricity, and you must be out very early in British woods these days to be sure of getting your share of the first ceps and chanterelles. Fortunately the practice has lost most of its early and sometimes over-hearty associations with survivalism and (though it can reduce your food bills) its more fantastical links with self-sufficiency. Foraging is now simply fun, indulged in for the pleasure of new taste experiences, for insights into the history of our cultivated foods, and for more intimate and sensuous encounters with growing things. It can even be rewarding done as little more than wayside sampling: a wild red currant here, a sweet cicely seed there – what the 1930s fruit gourmet Edward Bunyan described delightfully as 'ambulant consumption'.[1]

Since the late 1980s wild plants have also begun to make a perceptible impact on the commercial food business. In France, which has rather more raw materials at its disposal, there is a distinct school known as *cuisine sauvage*. And in Britain it is now not uncommon to find samphire, nettles, dandelions, bitter-cress, borage, wild strawberries, bilberries and ramsons served in some form or other in both smart metropolitan restaurants and local

Seasonal snacks: (left) young stinging nettles – here growing by the old Roman Wall at Silchester, Hampshire – are used in soups in the spring; marjoram (above), the oregano of Mediterranean countries and a common summer-flowering herb of chalk country; sweet chestnuts (top right) are sweetest after the first frosts; windfall wilding apples (right), often still edible deep into winter.

pubs. Perhaps most encouraging is the fact that they are no longer regarded simply as rough peasant foods, but are being used as ingredients for modern *styles* of cooking: wild herbs and fruits flavouring oils and vinegars (and even Danish-style schnapps); spring greens – garlic mustard, nettle, sorrel – stir-fried; flowers added to salads. A contribution from a Hampshire woman of American ancestry illustrates the inventiveness of the modern approach to wild foods:

'I take so many of these foods for granted that I often forget that many people don't eat such things as dandelion pasta, nettle gnocchi, wild garlic and cheese soup or dandelion and mozzarella pie! I guard any dandelion greens that spring up in my garden with as much enthusiasm as some look after their most prized tomato plants. My latest invention is to chop a load of assorted greens I've gathered, sauté them with garlic and onion, fold in ricotta cheese, some fine bread crumbs and an egg and use this as a stuffing for pasta shells or home-made ravioli ... My grandmother's cousin, Amber, had all her workmates gathered round the bench in the mill one day to sample her dandelion and mozzarella pie. She had layered cooked dandelion greens, cheese, sliced black olives and a little tomato sauce made with garlic into flaky pastry and baked it into a pasty. It became her favourite lunch-box treat ... Have you ever had pickled greens? They are wonderful. Mustard greens, turnip greens, wild greens, the patient and persistent dandelion greens – all potted up in a spiced, slightly sweet vinegar. They make a cheese sandwich sing.'[2]

Pea family
Fabaceae (or *Leguminosae*)

False-acacia or **Locust-tree**, *Robinia pseudoacacia*. Introduced from North America in the early seventeenth century, false-acacia has been much planted as an ornamental tree in town parks and streets, and to a lesser extent in large gardens and village greens in the countryside. (At my old school in Berkhamsted, there is an ancient specimen in the quad. It is surrounded by a seat, which in my day was given rather untypical respect for an American import and reserved for the senior boys.) In urban areas in the south of England it is quite widely naturalised, chiefly by suckers, but occasionally from seed.

It is a handsome species when full-grown, with deeply furrowed bark, airy foliage, and pendulous clusters of cream, sweet-scented flowers, which are used in the perfume industry. But the comparative abundance in this country of old trees owes less to its natural good looks than to the energies of William Cobbett in promoting the virtues of its timber. Cobbett became a convert to the tree during his stay in America in 1818–19, and subsequently mounted a campaign on its behalf that came perilously close to being large-scale confidence trickery. 'The wood,' he wrote, 'is very hard, and close and heavy; it is yellow almost as box, as hard as box, but the grain not so fine. The *durability of this wood* is such, that *no man in America will pretend to say that he ever saw a bit of it in a decayed state.*'[1] The wood is certainly long-lasting when in contact with the ground, and is ideal for fence-posts, tool handles and the like. But Cobbett argued that the quick-growing wood was also superior to oak for ship- and house-building; and, being canny enough to use the American name, 'locust-tree' (then practically unknown here), he found a public eager to invest in an apparently new wonder tree. Before long he was exploiting the market's gullibility more directly, buying up trees from nurserymen (unasked for when labelled 'Robinia') and selling them on, at a considerable profit, under his seductive new brand-name. In the end he sold more than a million.[2] The timber from British-grown trees is never sufficiently straight to be remotely suitable for the building works he had extolled, but the number of trees dating from the mid-nineteenth century shows just how persuasive his campaign was.

Bean species and cultivars. Many of the beans cultivated in gardens escape to become briefly naturalised in waste places and rubbish-tips, including French bean, *Phaseolus vulgaris*, from South America; mung-bean, *Vigna radiata* (the 'bean sprout' of Chinese restaurants), from Asia; and soya-bean, *Glycine max*, which occurs especially where seed is spilled near docks and factories. Runner bean, *Phaseolus coccineus*, from tropical America, is perhaps the most frequent as a casual, and when it is seen growing as a bright-flowered scrambler, without support or pruning, it is easy to understand why the first 'runners' were introduced to Britain principally as ornamental plants.

Bladder-senna, *Colutea arborescens*. A shrub native to southern Europe, bladder-senna is one of our more puzzling naturalised species. It was certainly being grown in English gardens by the sixteenth century, but seem-

Flowers of false-acacia, which are used in the perfume industry.

Bladder-senna. The pods were once used as a cheap alternative to true senna.

ingly for purely ornamental purposes. The only utilitarian virtue the normally inventive Gerard could trace for what he called 'bastard Sene' was a vague recommendation by Theophrastus that the plant was 'good to fatten cattle, especially sheepe'.[3]

Yet by the late eighteenth or early nineteenth century it had become popular as a cheap substitute for true senna (*Cassia acutifolia*, from the Middle East), whose leaves and fruits – 'senna pods' – were widely used as a purgative. Not long after, it began to naturalise in a curious pattern, which in London largely followed railway embankments in the East End. Perhaps it was cultivated in railside gardens and plots as a home-grown laxative. Certainly its seeds are very amenable to scattering by birds. They are held in inflated, papery bladders up to three inches long, which follow the yellow flowers in summer. I have seen blue tits tearing their way into the pods (looking for insects, presumably, not the seeds) and almost entirely vanishing inside.

Goat's-rue, *Galega officinalis*, is another naturalised plant seen frequently on urban railway embankments. It is a bushy perennial with short sprays of white or purplish-lilac flowers (an old name was 'French lilac'), and was introduced to this country in the sixteenth century as a vegetable and medicinal herb, and later grown for ornament. It is especially abundant around Sheffield (see p. 324).

Wild liquorice, *Astragalus glycyphyllos*, is a conspicuous plant, stout and sprawling, with heads of yellowish-green flowers, scattered in rough grassland sites throughout Britain, mostly on chalk or limestone soils. True liquorice, *Glycyrrhiza glabra* (native to south-east Europe and western Asia), does not occur wild in Britain, and is no longer found even as a casual relic of cultivation. But it was once an important crop around Pontefract in Yorkshire, where imported roots are still used in the manufacture of the local speciality, Pontefract cakes, and it has left a spicy legacy of history and anecdote. This has recently been collected together by a group of local women in a book entitled *Talking Spanish*. Liquorice is still known as 'Spanish' in the area, because of a legend that the plant was originally introduced as 'twigs' washed ashore from an Armada galleon. It is a very tall story. The twigs were supposedly found by a Pontefract schoolmaster on holiday and used to beat boys with, who then chewed bits and discovered they were sweet!

Wild liquorice, a scarce native of chalk and limestone sites and a cousin of true liquorice.

The liquorice industry was one of the most important employers of women in the area: 'It was always lads down pit, lasses into liquorice round here,' the book's editor, Sharron Cocker, reminisced. Workers remember taking contraband liquorice 'laces' out of the factories tied round their waists, and moulding liquorice phalluses on the production lines to initiate embarrassed newcomers. Sharron Cocker's team was able to track down one woman, Emily Money, who won a fancy-dress competition dressed entirely in liquorice: 'All the lads started grabbing the strands off my skirt and eating them. It was just as well I'd black knickers on.'

The liquorice-making industry is now concentrated in two highly automated factories. The older traditions survive in Pontefract only in the form of a commemorative liquorice 'hedge' in the town and an enormous Pontefract cake, made for the town museum in May 1992. It is three feet across, lacquered and stamped with the industry's traditional 'owl and gate' symbol.[4]

Sainfoin, *Onobrychis viciifolia*, is an exquisite flower, with spikes of silky pink blooms streaked with deeper-purple veins. It was named from the French 'St Foyn' and known as 'Holy-hay', and a more erect, paler-flowered form was introduced as a fodder plant from the continent in the middle of the seventeenth century.[5] This strain is widely naturalised and has in recent years been sown on new road embankments. The show on the banks of the M25 shortly after it was opened was one of the ringway's few uplifting features. The native form is more or less prostrate, has deeper-pink flowers and is native in grassland and bare ground on the southern chalk-hills.

Kidney vetch, *Anthyllis vulneraria* (VN: Granny's pincushions, Lady's slipper). A very variable species, quite

common on chalk grassland, sand-dunes and cliffs throughout Britain, kidney vetch can monopolise areas of bare ground where the soil is right. The flowers are set in a calyx with a dense, woolly coating (which may partly explain why it was widely valued as a wound-herb). They are usually yellow, but a variety (var. *coccinea*) with red flowers appears on cliffs in Cornwall and Pembrokeshire, and another with pink flowers in Caernarvonshire.

Common bird's-foot-trefoil, *Lotus corniculatus* (VN: Eggs and bacon, Ham and eggs, Butter and eggs, Hen and chickens, Cocks and hens; Tom Thumb, Fingers and thumbs, Granny's toenails, Lady's fingers; Lady's slipper, Cuckoo's stocking, Dutchman's clogs). This is an abundant and rather sprawling species of short grassland on lawns, downland, road-verges, heaths and dunes throughout the British Isles, in flower for much of the summer. Geoffrey Grigson collected over 70 local names, of which at least a dozen seem to have survived.[6] Some refer to the shape of the individual flowers, which are reminiscent of a medieval slipper or sabot; others to their colour, which is a delicious suffusion of egg-yolk orange and yellow. The group of 'finger' names (especially the wickedly accurate 'granny's toenails') come from the long, claw-like seed-pods, which can be up to seven in number.

Grigson reasonably asks why such a diminutive plant should have attracted so much attention and such a

Left: sainfoin, on a roadside verge, Dorset.
Above right: common bird's-foot-trefoil, known variously as eggs and bacon, butter and eggs, lady's fingers and granny's toenails.
Right: kidney vetch, with its woolly-coated calyces.

plethora of names, whilst its more showy relatives (e.g. tufted vetch, below) are known by just a handful. I suspect the reason may be precisely its modesty and closeness, a taken-for-granted domesticity which, when one becomes more conscious of the flower, attracts the kind of affection that always attaches to the truly familiar. Richard Jefferies wrote of this feeling in his essay 'Wild Flowers' (1885): 'The bird's-foot lotus was the first. The boy must have seen it, must have trodden on it in the bare woodland pastures, certainly run about on it, with wet naked feet from the bathing; but the boy was not conscious of it. This was the first, when the desire came to identify and to know, fixing upon it by means of a pale and feeble picture.'[7]

Dragon's-teeth, *Tetragonolobus maritimus*, is a casual from central and southern Europe, which is also well naturalised in rough chalk grassland in a few sites in southern England. It is well named from the surprising, large, solitary flowers which seem to ride just above the top of the grass. **Horseshoe vetch**, *Hippocrepis comosa*, is a neat plant of short, dry turf on chalk or limestone, with heads of five to twelve yellow flowers, each becoming a pod which breaks up into a number of horseshoe-shaped segments.

Wood vetch, whose flowers are scented of sweet pea.

Bush vetch, *Vicia sepium*, is a common, scrambling vetch of rough grassland, road-verges and wood margins, with mottled purple-blue flowers. A variety with apricot flowers has been found in Dumbarton.[8] **Tufted vetch**, *V. cracca* (VN: Tares), is a more showy relative, climbing up to seven feet over hedges and banks, or up the tall herbage in rough pastures. It has long, one-sided clusters of bluish-violet flowers and finely divided leaflets. **Common vetch**, *V. sativa*, is a common, sprawling annual of rough ground and waysides, with single or paired pink to purple flowers, occasionally white or cream.

Wood vetch, *V. sylvatica*, is perhaps the most beautiful of all the vetches and, for me, a plant of parables. In 1981 I bought a small wood in the Chilterns, and there was one stunted, non-flowering specimen in a shady corner. Fifteen years on it lines many of the new trackways, and from June to August is covered in bunches of exquisite lilac-striped and sweet-pea-scented flowers.

Wood vetch is a scarce plant of rocky woods, scree and maritime cliffs, and occurs in a scatter of sites across Britain. It prefers sunny positions and a bank, rocks or scrub to scramble up. There are large colonies on the limestone scarp overlooking the Avon valley in Wiltshire, for instance, and on steep river-banks in Northumberland. Its resurgence in my wood in the Chilterns was almost certainly due to our creating an artificial version of one of its favourite habitats. We had cut a new pathway into a steep slope, and on its top-side there was a flinty, south-facing bank, which within two years was draped with wood vetch, sprung perhaps from long-buried seed.

The light and soil disturbance were probably important, too, as has been found by a nature reserve manager elsewhere in the Chilterns:

'In 1981, when I first got involved with Dancers End Reserve, wood vetch was growing in three different places along one of the rides through Bittams Wood. It seemed to be doing quite well, although the ride was obviously becoming more shaded each year, and nowhere was it "luxuriant" as it was described in 1964. In 1982/3 a new, much more active, management plan was drawn up and work began to re-establish coppicing, extend and create clearings, scallop woodland edges, establish rotational cutting of scrub blocks and so on. In 1983 a new patch of wood vetch appeared in an area of Ant Hill where we had cleared some thick scrub. Over the next two years it spread to other areas and flowered well, sprawling over areas of low dogwood and willow scrub.'[9]

Sea pea, *Lathyrus japonicus* ssp. *maritimus*, is a perennial which forms quite large patches on shingle, especially on the south and east coasts. The bunches of deep purple flowers eventually fade to blue, and are succeeded by pods containing quite large seeds. There is a story that during a seventeenth-century famine on the Suffolk coast villagers kept themselves alive by eating these peas.

Bitter-vetch, *L. linifolius* (VN: Dobbie horses). Bitter-vetch is a plant largely of northern and western distribution in Britain and grows on road-verges, on woodland edges and even in mountain grassland. In eastern and central England it is confined mostly to ancient woodland. Its fleshy root-tubers were formerly gathered for food. In Scotland it was occasionally used for flavouring whisky, and by the more upright – such as the Revd Angus Macfarlane, writing in 1924 – as a substitute for chewing gum: 'It has long underground roots, strung with nodulous lumps at frequent intervals. These, after they are dried, are chewed as wild liquorice. For chewing purposes I consider them superior to and far less deleterious than common chewing gum. The taste lingers in the mouth for long after the last shred is chewed. They are said to ward off hunger for a long time. The taste is both acid and sweet, and never palls.'[10]

Meadow vetchling, *L. pratensis*, is a climbing yellow-flowered perennial, widespread on grassy waysides and in waste places throughout Britain. **Tuberous pea** or **Fyfield pea**, *L. tuberosus*, is a bright-crimson clambering pea from southern Europe which was first discovered in the wild at Fyfield, north Essex, in 1859. It presumably escaped from cultivation, as it has edible tubers, like bitter-vetch, and was sold as a commercial root-vegetable in Dutch markets in the early nineteenth century.[11] By the end of the century it was being grown for purely ornamental reasons, and the journal *The Garden* (July 1886) recommends it for planting at the foot of a hedge or fence, where 'this pretty little species scrambles over and about it in a pretty way'. It has occurred as a casual in a scatter of other sites throughout

Sea pea, which forms dense mats on some shingle beaches.

Britain, but rarely persists for long. **Narrow-leaved everlasting-pea**, *L. sylvestris*, is a climbing or scrambling perennial with dull pink-purple flowers. As a native plant, it is probably confined to old woodland and maritime scrub, though introduced plants crop up in hedges, on wood margins and on roadsides.

There are a number of other conspicuous climbing peas from southern Europe which have become naturalised. **Broad-leaved everlasting-pea**, *L. latifolius*, is an aggressive scrambler with bright, almost garish magenta flowers, widely established on waste ground, especially along railway banks. **Two-flowered everlasting-pea**, *L. grandiflorus*, is a persistent escape in waste ground and hedge-banks near gardens, spreading like the previous species. **Norfolk everlasting-pea**, *L. heterophyllus*, established in damp hollows in the sand-dunes at Burnham Overy Staithe, Norfolk, probably escaped from beach-hut gardens in the early years of the century.[12] **Sweet pea**, *L. odoratus*, is a frequent casual in waste places, as is **garden pea**, *Pisum sativum*.

Grass vetchling, *Lathyrus nissolia*, is a great favourite with botanists, perhaps because of its unpredictable occurrence and the difficulty of finding it amongst the long grass that its foliage resembles. It is a medium-tall annual cropping up in old pastures, rough waysides and marshland edges. It has one or two brilliant crimson flowers on long stalks, so Geoffrey Grigson coined for it the name 'crimson shoe'.[13]

Common restharrow, *Ononis repens*, is a low-growing perennial of rough grassland on well-drained soils, especially on chalk and limestone and around the coast. The leaves are greasy to touch, and exude an oil that smells remarkably like petroleum jelly. The name 'restharrow'

comes from the long, thick rhizomes, which were tough enough to stop a horse-drawn harrow in its tracks. They were sometimes dug up and chewed like liquorice (see bitter-vetch, above). **Spiny restharrow,** *O. spinosa*, is a pert and spiny shrublet, like a woody sweet pea bred for the rockery. It has a more restricted distribution than common restharrow, but otherwise shares many of its characteristics – though its leaves are less oily. The purple-pink flowers of both species are unscented.

White melilot, *Melilotus albus*, and its yellow-flowered relatives, **tall melilot**, *M. altissima*, and **ribbed melilot**, *M. officinalis*, were originally introduced to this country from Europe as fodder plants, and are now well naturalised at the edges of arable fields, on roadsides and in waste places. They can grow up to four or five feet in a season, and when growing *en masse* can make a conspicuous show in late summer. As well as being relics of cultivation, melilot seeds sometimes arrive as impurities in seed of the more commonly sown fodder plant, **lucerne** or **alfalfa**, *Medicago sativa* ssp. *sativa*. This blue- to lilac-flowered perennial was introduced from the Mediterranean about 1650. Since then improved strains have been developed in many European countries and the United States and imported to Britain. So naturalised colonies on field margins and waysides can vary widely in leaf-size and flower colour – and also (depending on the provenance of the seed) in the weeds that accompany them.[14]

Two European immigrants: the scrambling broad-leaved everlasting-pea (left) and ribbed melilot (above). Both are quite common in waste places.

White clover, *Trifolium repens* (VN: Milky blobs, Sheepy-maa's, Bee-bread). This is an abundant perennial of grassy places throughout Britain. Almost all children learn two traditions about white clover: that the white flowers can be pulled out of the heads and sucked for a bead of honey (hence 'bee-bread' and the delightful, but now obsolete, 'honeystalks'); and that four- and, even better, five-leaved clovers are lucky, though you must ideally come across them by accident. In parts of Buckinghamshire they were pressed and used as bookmarks in prayer books.[15]

In the United States there are clover farms that specialise in growing four-leaved clovers. Apparently, a secret, genetically-engineered ingredient is added to the feed to encourage the aberration. About 10,000 leaves are harvested daily, sealed in plastic and sold as 'Good Luck' charms.[16]

There are some thirty other native, naturalised or casual clovers that occur in Britain. The following are some of the more interesting species. **Strawberry clover**, *T. fragiferum*, which has strange pink flowers that become like overripe strawberries when in full bloom, is scattered, often on heavy or brackish soils, as far north as southern Scotland. **Hop trefoil**, *T. campestre* (VN: Tom Thumbs), has small yellow flowers a little like miniature hops and, like hops, these turn brown as they age. It is common in short grassland and bare places throughout Britain. **Lesser trefoil**, *T. dubium*, is a smaller version, and probably the true Irish 'shamrock' – the *seamrog*, or 'cloverlet'. **Red clover**, *T. pratense*, is a perennial, common throughout Britain, both in its native form and as the more robust varieties sown as fodder crops, which frequently naturalise. **Sulphur clover**, *T. ochroleucon*, is another perennial, but with pale yellow flowers, confined to waysides and grassy places on clay soils in East Anglia.

Starry clover, *T. stellatum*, is a Mediterranean species, with the individual calyces star-shaped when the flowers die; it has been naturalised on shingle in West Sussex since the early 1800s, and is 'believed to have entered Shoreham-by-Sea in the ballast of ships returning from Wellington's Peninsular Campaign'.[17] **Twin-headed clover,** *T. bocconei*, with whitish flower-heads in close-packed pairs, is a very rare native, confined to short turf on the Lizard Peninsula in Cornwall. **Hare's-foot clover,** *T. arvense*, is a short annual with slender, pinkish flower-heads covered with downy hair, like a rabbit's or hare's paw. It grows on sandy soils, both inland and by the sea, and is locally frequent in Britain north to central Scotland. **Subterranean clover,** *T. subterraneum*, is a prostrate annual, rather scarce and confined to short turf and sandy soils in England and Wales, usually near the sea. The name comes from its habit of turning its heads upside down after flowering and pushing the seeds into the soil.

Hare's-foot clover, whose flower-heads are covered with downy hair.

Tree lupin, *Lupinus arboreus*, is a shrubby lupin with honey-scented yellow flowers, introduced from California in 1793. It self-seeds readily in dry, frost-free areas, and is quite widely naturalised on light soils, especially on sand and shingle around the east and south coasts. On Blakeney Point, on the north Norfolk coast, tree lupins were planted out before the Second World War as cover for migrant birds and have expanded into dense shrubberies.[18] The species has recently become popular for municipal roadside plantings, and self-sown plants are consequently beginning to appear in urban waste patches. **Garden lupin,** *L. polyphyllus*, was introduced from North America, and is naturalised in a few waste places in Scotland. The colourful lupins seen most commonly on railway embankments and roadsides are **Russell lupins,** *L. × regalis* (a cross between *L. arboreus* and *L. polyphyllus*), a group of cultivars developed by a Yorkshire allotment-holder, George Russell, between 1911 and 1937. **Nootka lupin,** *L. nootkatensis*, is a blue- or purple-flowered species from America, naturalised since at least 1862 on river shingle in northern Scotland, especially by the Rivers Tay and Dee.

Tree lupin naturalised by a roadside in east Suffolk. The species was introduced to Britain from California in 1793.

Laburnum, *Laburnum anagyroides* and *L. alpinum*. Popular garden trees from the mountains of southern central Europe, quite often self-sown on rail- and roadsides. They are the most frequent cause of plant-poisoning amongst children, who eat the seeds in mistake for peas.[19] Animals are also occasionally poisoned by eating the fallen pods, yet there are miles of planted laburnum hedges in the upland pasture countryside of western Britain. The choice of this toxic (and far from stockproof) tree as a hedging plant in what are almost exclusively stock-raising areas remains an intriguing mystery. One of the concentrations is around the squatters' and smallholders' settlements on the commons of south and west Shropshire: 'There are many examples of laburnum hedges in the old mining areas immediately to the west of the Stiperstones, in particular around Shelve, Pennerley and Stiper-

Russell lupins naturalised on railside waste ground, King's Lynn, Norfolk.

stones villages. Due to neglect over many years they are in fact rows of laburnum trees rather than hedges, and they are a wonderful sight in early summer.'[20] They also crop up in the vicinity of Brown Clee Hill, ten miles to the south-east.[21] Some way further north, there is another nucleus in Cumbria (where Coleridge wanted to plant laburnum in the woods around Grasmere):

'In various West Cumberland parishes, Arlecdon, Haile, and the township of Ennerdale and Kinniside, for instance, there are considerable stretches of laburnum in the "dykes", as the hedges and banks are called here. I farmed at Simon Keld on Kinniside Common close to the 900-feet contour, where there were numbers of laburnums in the garth around the farmhouse. I have no idea why they were there; it seemed odd to have a poisonous tree in a place where there were so many sheep. Were they grown to feed bees or for their wood? In Arlecdon there is a tree with a girth of five to six feet. The parish was enclosed in 1823 – was laburnum planted in the dyke at the time of enclosure?'[22]

That laburnums were grown for their timber is a distinct possibility in these communities of self-sufficient farmers and craftsmen. Laburnum has a unique grain pattern, with a purple-chocolate heartwood core inside pale yellow outerwood. It is hard enough to be cut thinly and has been used in ornamental furniture work, especially for inlays and veneers. But this is unlikely to have been the reason for the much more extensive hedgerow plantings in west Wales, between Carmarthen and Llandysil, for instance.[23] James Robertson believes that, given their age, these farm laburnum hedges were more likely a vernacular echo of the ornamental plantings in the big Welsh landscape gardens of the time:

'There are many miles of laburnum hedging in Cardiganshire, predominantly *L. anagyroides*, but also *L. alpinum*. These occur both as pure hedges and as old trees in hedgerows, throughout the county. They are normally planted on banks and coppiced periodically, stools measuring up to a metre in diameter. Occasionally they are layered or pleached. Estate records may provide more information about the reasons for planting laburnum, but the trees are now valued as an amenity, and more than tolerated by farmers, despite the danger that they could poison stock. Most were planted about 150 years ago, and aesthetic considerations are likely to have been as important then as now.'[24]

Broom, *Cytisus scoparius*. Broom's common name – almost universal throughout Britain – is probably a result of a long process of specialisation. *Brom* in Old English was simply a coarse shrub, many kinds of which, including gorse, heather, holly and butcher's-broom, were at some time used for sweeping. Broom itself, with its long, whippy, thornless stems, was one of the most effective and comfortable to use, and eventually won exclusive right to the title. Its twigs were also occasionally used in basket-making and in the north for thatching. Broom's Latin name, *planta genista*, also won a kind of immortality in the 'Plantagenets', the name and emblem Geoffrey of Anjou (father of Henry II) adopted for his family. It is still a defining plant of the dry countryside around Anjou.

Broom is a common shrub of heaths, open woodland, banks and pits on sandy soils throughout Britain. Its golden flowers, scented of vanilla, can be dazzling in May and June, so much so that when you walk between the

Laburnum – or 'golden rain' – well known in gardens but also widely planted in upland hedges, despite its poisonous seeds.

Large broom bush growing by the remains of a quarry worker's cottage, on sandstone in Wirral.

sprays you may involuntarily shield your eyes. They have been used, in bud or fully open, and raw or pickled, as an ingredient in salads, especially in those baroque seventeenth-century concoctions of twenty or thirty different ingredients, known as Grand Sallets, or Salmagundies.

The green tips of the flowering branches have a long tradition in herbal medicine, and were gathered during the Second World War for use as a mild diuretic. In very large doses one of broom's active constituents, sparteine, can cause excitation and hallucination, which has sometimes been suggested as the origin of the old myth that witches rode on broomsticks. But since sparteine also causes considerable gastric disturbance at the same time, this is more likely just a piece of folklorist fancy. Broom's magical reputation was of a cryptic and less material kind.

It is also a plant of spring and amorousness – and light-hearted medicine, as witness the joke, spun out in best stage mummerset to one of our West Country contributors: 'An old lady who used to help me with my children when they were young, spotted our yellow broom in full flower. "Did 'ee know tes a master cure fer piles?" I didn't and demanded information. "Well, yu' pick an armful an' yu steep un fer twenny-four hours in a metal bucket ... then yu heat un up an' stir ... then yu cool 'un, see?" "How much do you drink?" I asked. She roared with laughter. "Oh, my dear soul, yu doan drink it – yu sit on the bucket." '[25]

Spanish broom, *Spartium junceum*, is a common Mediterranean shrub widely planted on roadside banks, and naturalised in a few places in the south. In its native territory the bark is used as a source of fibre, for making paper and cloth.

Dyer's greenweed, *Genista tinctoria*, is a broom-like shrub, locally common in rough ground and grassy places throughout England and Wales. It is a native, but was widely grown as a dye-plant, and naturalised relics of cultivation have probably extended its distribution.

Dyer's greenweed, once used to produce yellow and green dyes. Newtown, Isle of Wight.

'Dyers greenweed occurs still near Kendal, the motto of which is *Pannis mea Panis* – "wool is my bread". It celebrated the town's early prosperity due to the woollen trade. It gives a yellow colour to wool, which, mordanted with alum, is then dipped into a vat of woad, or at a later date indigo, to give the famous "Kendal Green". The excavation of Viking remains at Coppergate in York produced plant remains containing dyers greenweed, dating its use at least to the Anglo-Scandinavian era, ninth–eleventh centuries. The same site yielded residues of a clubmoss, *Diphasium complanatum*, not native to Britain but found in Scandinavia. This clubmoss is one of the few plants which can accumulate aluminium from the soil. It does so in sufficient amounts (over one per cent dry weight) to be used as a mordant in place of alum, which does not occur naturally in this country, and which, before the era of chemical manufacture, would have to be imported from Italy. It would seem that the Vikings may have introduced this clubmoss specifically to use it in the York textile industry.'[26]

'As genista makes a yellow dye it had to be mixed with an extract of woad, which is blue . . . It was later superseded by the brighter dye and cloth known as Saxon Green in the 1770s.'[27]

'Genista grows in a number of rather scattered locations in my parish [Winster, Derbyshire] along the brow of the limestone ridge where the ground is leached, and in the drier parts of some of the more scruffy acidic pastures. It is not now very common in Derbyshire as a whole, but I know that up until the beginning of the nineteenth century, it was locally gathered and used as its name suggests, and sent to Manchester to make dye.'[28]

'At Holton Park near Halesworth there is a large area of this plant covering several acres. The Suffolk Wildlife Trust have an agreement with the owners of the park, which involves an annual cut and rake-off in late summer ... This area was once a flax and hemp growing district.'[29]

Petty whin, *G. anglica*, is a similar shrub, but spiny, and scarcer, being confined to heaths and moors as far north as the Highlands. Its flowers have the curious habit of changing from lemon to green if you dry them.

Gorse, *Ulex europaeus*; **western gorse**, *U. gallii*; **dwarf gorse**, *U. minor* (VN: Furze, Fuzz, Fuzzen, Vuzzen, Whin, Whinny luck). 'When gorse is in blossom, kissing's in season' is a saying known throughout Britain. Unromantic botanists are apt to explain it away by pointing out that, except in Scotland, most gorse colonies are a mixture of common gorse (in flower chiefly from January to June, though often sporadically through the year) and either western gorse (July to November) or, in the south and east of England, dwarf gorse (also July to November), so there is always likely to be one species in flower. But I suspect that there is also a 'where' implied in the 'when'. Gorse is one of the great signature plants of commonland and rough open space, places where lovers can meet, walk freely and lose themselves, if need be, in its dense thickets.

Being an abundant, fast-growing plant, it has also been pressed into a multitude of more functional roles: as fuel (especially in bakers' ovens), as cattle food, as a convenient anchor for washing, as a chimney brush, and (in flower) as a source of colour for Easter eggs.

'Gorse and heather were bound together to make besom brooms, which were then tied with the same jute string used for binding straw bales.' (Whitby)[30]

'Some local gardeners place chopped gorse or "fuzz" over germinating or emerging peas to deter mice and pigeons.' (Plymouth; also Ashridge)[31]

'Branches of gorse (and some other shrubs), known as "brobs", were used by fishermen to mark safe routes out into Morecambe Bay.'[32]

'In Wales many farmers remember gorse mills, and how important a food gorse was, especially for horses. Fields were devoted to growing gorse as a crop, and at least one smallholder in Anglesey made his living cutting gorse for other farmers, at five shillings an acre. I am told that in one corner of Anglesey, European and Western gorse were thought to be male and female of the same species. The female, Western gorse, was the much more desirable and softer source of food.'[33]

On commonland there were quite strict rules about when and how much gorse could be cut for fuel: 'In Cumnor, Oxfordshire, under the 1820 Enclosure Award, parishioners had the right to go to Cumnor Hurst to cut gorse and broom (for burning, often in bread ovens due to the fierce heat) but they were allowed only as much as they could carry on their backs.'[34]

The gorse and stone 'hedges' of central Dartmoor. The stones were gathered from the moor; the gorse either found its own way in or was deliberately planted.

'Furze was regularly gathered for firing in the last century from Harpenden Common [Hertfordshire]. To quote a 90-year-old man I interviewed: "We gathered it carefully, not haphazardly, remembering there was a tomorrow." '[35]

On Berkhamsted Common, a few miles south of Harpenden, there were regulations prohibiting the cutting of gorse for sale outside the parish and the digging-up of entire bushes. In 1725, to prevent over-exploitation, the Manor Court even specified the type and size of the cutting implements: 'Imprimis wee order that no person or persons shall cut any furze growing on the comon or heath or frith belonging to this Manor nor any of the Wast thereto adjouning with any other weapon or working tool than an one handed Bill with a stale helve or handle thereto affixed of the length of twelve inches & no longer upon pain and penalty for every person cutting any furze contrary to this order every time offending five shillings.'[36] (One year later, exceptions were made in favour of persons over sixty or under fourteen, and of the disabled, who might use 'Hows or handbills but not longbills'.)

Gorse is also an important ingredient of many turf- and hedge-banks in western Britain: 'Above the River Eden near Egremont, hedges of whin stand on the top of very considerable turf-stone banks (locally known as kests).'[37] In the New Forest: 'A family with a very small smallholding had a gorse hedge around it. Their ruse was to prune back the inside of the hedge and let the outside grow, gradually increasing their acreage.'[38] On the Isle of Man:

'The native gorse is *Ulex gallii*. It grows everywhere except on the Calf of Man. No gorse of any sort grows there, which gives rise to the catch question, "What colour is the gorse on the Calf of Man?" Large quantities of *U. europaeus* seed were imported and planted on top of the old sod hedges to make a more effective barrier to cattle, but the gorse was also planted on small patches of rough ground for use as fodder. Here and there on upland farms one can find the remains of gorse-mills which were used to bruise the new growth and make it palatable. The older and dead stems were used for fuel.

'When gorse is in blossom, kissing's in season.' Western gorse and heather in full bloom, South Stack, Anglesey.

Sticks like this are known as Bonns, and people still gather them on country walks. There was even a special tool evolved for cutting down the bushes called a Hack. No doubt they would be made by the village blacksmith. Patches of old bushes were cleared and rejuvenated by burning and this practice was also sanctioned by the May Eve custom of setting fire to the bushes to drive out witches. The gunwales of boats are still sometimes singed with a blazing branch for the same purpose. The burning of gorse on the hills is controlled now to avoid disturbance to nesting birds.'[39]

In the village of West Torrington, Devon, a traditional match-making dance involving gorse was revived in 1994. 'The Vuz Dance of Flowers' centred around an elaborate display made by the male dancer, or 'talesman'. It consisted of a tight faggot of gorse branches, supporting a hazel branch (called a 'nit-al') which was topped with another 'bush' of gorse tied with coloured ribbons. This is an account of the dance given to us by a villager, who had learned of it from his great-grandfather:

'A man would put his trade object (loaf of bread, horseshoe, nails, etc) in under cover on the top of a faggot, which he had made. There would be one dancer at least for one stand, i.e. more stands than dancers. The dance would be done with whistle pipe and drum, tune unknown. Dancers would dance singly in the fashion of "in and out the windows". When the band stopped, the girl would go to the nearest stand and pick out the talesman from the men who were standing on the side. If he liked the look of the girl who had picked [him], this talesman would go forward and stand by the girl. When all of this had been done they would do the dance together. On leaving the square, the men would take the faggots home with them, and if marriage resulted, they would use the faggot to cook their first loaf of bread.'[40]

Wherever it grows in quantity, gorse is one of the great landscape plants, especially when the blaze of yellow is mottled with purple heather. The Swedish naturalist Linnaeus reputedly fell to his knees and thanked God when he first saw furze on an English common. (It was probably on Putney Heath, in 1736, but it is a measure of gorse's popularity that half the commons in the Home Counties have claimed the honour for themselves.) It is one of the most sensual of plants – the flowers smelling of coconut and vanilla, and the seed-pods' cracking in hot sunshine hard to tell from the clacking calls of the stonechats which perch precariously on the topmost sprigs. George Meredith caught all this in the great gorse stanza in 'Juggling Jerry', where he described an old man relishing familiar scenes and scents:

Yonder came smell of the gorse, so nutty,
Gold-like and warm: it's the prime of May.
Better than mortar, brick and putty,
Is God's house on a blowing day.
Lean me more up on the mound; now I feel it;
All the old heath-smells! Ain't it strange?
There's the world laughing, as if to conceal it,
But he's by us, juggling the change.

Sea-buckthorn family
Eleagnaceae

Sea-buckthorn, *Hippophae rhamnoides*, is a spiny shrub, whose leaves and twigs are covered in silvery scales. It is a coastal species, native in sandy places along the eastern shorelines of England, but much planted elsewhere, to help stabilise dune-systems on the coast and, because of its salt tolerance, by urban roads and roundabouts.

In the autumn it bears bright orange berries, which are eagerly devoured by wintering thrushes and other migrant birds passing along the east coast. A Norfolk Flora notes: 'If it were not for bird-carriage, it would be difficult to account for the spread of this shrub along the whole of the Norfolk littoral.'[1]

The fruit has occasionally been used to make rather sharp jellies and preserves.

Sea-buckthorn berries, enjoyed by birds but underused by humans.

Purple-loosestrife family *Lythraceae*

Purple-loosestrife, *Lythrum salicaria*. 'Long purples' is an apt traditional name for this widespread plant of marshes and riversides, with its tall spikes of magenta flowers. But the 'long purples' in Shakespeare's description of Ophelia's death garland (*Hamlet* IV. vii) are, from other hints given, early-purple orchids:

> *... long purples,*
> *That liberal shepherds give a grosser name,*
> *But our cold maids do dead men's fingers call them.*

(The 'grosser name' was 'dog-stones', meaning dog's testicles, see p. 444.) But when John Everett Millais painted the drowning Ophelia in 1851, he chose the seemlier interpretation, and added a meticulously drawn clump of loosestrife beside the dog-rose and 'willow ... aslant a brook'. It was, as a classic riverside plant, a more fitting adornment for Ophelia's watery grave, and the scene was partly drawn from life. Millais and Holman Hunt had spent many days searching for the right location and found it eventually on the banks of the Ewell River in Surrey, which provided 'the exact composition of arboreal and floral richness he had dreamed of'.[1]

Purple-loosestrife with meadowsweet, Loch na Keal, Isle of Mull.

'Loose-strife' is a literal translation of the Greek name for the plant, which in classical times was believed to be so powerful 'that if placed on the yoke of inharmonious oxen, [it] will restrain their quarrelling'.[2]

Mezereon family *Thymelaceae*

Mezereon, *Daphne mezereum*. Mezereon was once known as 'paradise plant' because of the heady fragrance of its February flowers, and was widely planted close to cottage doors. It is now a national rarity, and the two facts are probably connected. Although it is conspicuous in late winter – the close-packed pink flowers appearing on leafless woody stems up to four or five feet in height – it can never have been common. It was not recorded growing in the wild in Britain until 1752, when it was found in some woods near Andover in Hampshire.[1] A few years later, Gilbert White found it on Selborne Hanger, not far from the same site, and as is evident from his journals, he transplanted some to his garden.[2]

Mezereon now occurs very scarcely in calcareous woods from Sussex and Hampshire to Yorkshire, and elsewhere as a bird-sown casual or relic of cultivation. (There is a small colony in the chalk-pit near Mildenhall, Suffolk, where the military orchids grow, see p. 448.) But it is worth saying that it is not universally admired. The scent of the flowers has been likened to 'Windolene',[3] and the bright red berries are exceedingly poisonous. A. A. Forsyth relates a sad and ironic story of how a litter of piglets was accidentally killed by them:

> 'During the last war [I] investigated the sudden death of six out of a litter of seven young pigs, about ten weeks old ... All had vomited before death and their stomachs were empty of ingesta. An evacuee child, unused to the country, had picked mezereon berries from a bush in the garden thinking that they were red currants, and after tasting one and finding that it burned his mouth and tongue had ejected it again quickly without swallowing

it. He threw the remainder of the berries into a trough, from which the pigs were feeding at the time. The surviving pig was the "rickling" or "runt" of the litter, which had apparently been kept away from the trough by the others while the berries were eaten.'[4]

Spurge-laurel, *D. laureola*, is one of the most handsome plants of woodlands on the chalk (though it can occur on quite acidic soils in the extreme south and west). It is at its best in winter, glimpsed in dark, evergreen drifts against pale beech trunks, or under a tangle of old-man's-beard. It also flowers in winter, in clusters of small greenish-yellow blooms that appear amongst the topmost of the shiny leaves, sometimes as early as mid-January. They have a faint, but teasingly musky scent, stronger in the evenings, and are a lure for any early-flying moths and bees, tempted out by unseasonable warmth. It is a good enough shrub to be tolerated in gardens by those who appreciate understated plants: 'I think spurge laurel should be called "wild daphne" – a far nicer name. It is found bird-sown in local gardens [in the Chilterns] and many people are unaware of its delicious honey scent. This is superior to the heavy, cloying scent of *Daphne mezereum* ... I love it because it takes you by surprise. We have a bush near the gate and coming in on a miserable February evening you are suddenly enveloped by this heavenly scent. It is a great lift – a first sign of spring, coming at the same time as the frogs start croaking in the pond.'[5]

One of our two native daphnes: spurge-laurel, often the earliest-blooming species in chalky woods.

Willowherb family *Onagraceae*

Great willowherb, *Epilobium hirsutum* (VN: Codlins-and-cream). This is abundant in all kinds of damp places – riversides, ditches, marshes, woodland clearings, even at the foot of damp walls. The popular name 'codlins-and-cream' was probably suggested by the petals, rosy on top (like codlins or cooking apples), with a trace of creamy whiteness beneath.

Rosebay willowherb, *Chamerion* (or *Epilobium*) *angustifolium* (VN: Bombweed, Fireweed, Ranting widow). Something has happened to rosebay over the past century. What had been a comparatively scarce woodland plant has turned into one of the most successful and colourful colonisers of waste places – car parks, railway embankments, roadsides, even cracks in chimneys. The records track the change, but do not by themselves explain it.

Up to the mid-eighteenth century most writers seemed to regard rosebay as a garden plant which occasionally escaped into the wild. The first convincingly wild records are all from rocky or riverside sites. In Northumberland in 1769, for example, rosebay could be found 'Among the rocks and bushes under the Roman wall on the west side of Shewing-sheels, and by the Crag-Lake ... It is introduced into some of our gardens under the name of French willow; but being a great runner, it makes a better figure in its more confined situation among the rocks, than under culture ... It is reputed a scarce plant.'[1] In Hertfordshire, in the 1840s, it was described as 'rare' in 'woods on a moist sandy soil, and in osier beds'.[2] At about the same time, one of the women from the Clifford family of Frampton on Severn in Gloucestershire, who collectively painted a remarkable and normally reliable Flora of their neighbourhood, produced a meticulous illustration of rosebay, but misidentified it as 'Great hairy willowherb or Codlings and Cream' (see above). From the amendments to the label it is clear that she was not entirely sure of what she had found.[3] In Hampshire, too, the Revd C. A. Johns, author of the seemingly immortal *Flowers of the Field* (1853), described 'the Rose-bay, or Flowering-willow' as 'not often met with in a wild state, but common in gardens'.[4] In Wiltshire in 1888, the Revd T. A. Preston had it confined to 'gravelly banks' (with a first county record as late as 1864) and considered it 'very doubtfully native'.[5]

But by 1867, the Worcester botanist Edwin Lees had begun to notice a change in its habits: 'Quite recently the Rosebay Willow-herb has become numerous in several parts of the Vale of Severn, and promises to spread, incited to take possession of new-made roads and embankments. I have observed it by the side of a diverted road near Shatterford, and in the cutting of the Birmingham and Gloucester Railway, near Croome Perry Wood.'[6]

During the First World War, rosebay's populations exploded, especially in the extensive areas of woodland that had been felled (and often burned) to supply timber for the war effort. In World War II, there was a second wave of expansion. Rosebay relishes areas where there have been fires, and the summer after the German bombing raids of 1940 the ruins of London's homes and shops were covered with sheets of rosebay stretching, according to some popular reports, as far as the eye could see. There is, alas, no record of how Londoners themselves felt about this purple haze rising from the rubble. Did they see it as a symbol of life triumphing over destruction, or as a weedy invasion that simply added insult to injury? But it did generate one popular name – 'bombweed' – that became current throughout the south-east.[7]

In Gloucestershire – where rosebay had been a confusing rarity just a century before – there were certainly mixed feelings about the plant. In 1948 the new county Flora edited by H. J. Riddelsdell could say: 'This species has spread with great vigour since about 1914 owing to the clearing of woods ... The seed is easily carried, of course, and the railway has been a great agent in its spread. Beautiful as the plant is in its flowering season, when it is in seed it creates desolation and ugliness over the whole of its area.'[8]

In 1953, Professor T. G. Tutin attempted to relate the new expansiveness of rosebay to the existence of two apparently different varieties: var. *macrocarpum*, the true native confined chiefly to rocky places and damp woods;

Drifts of rosebay willowherb on a Dumfries hillside.

One of the earliest portraits of rosebay, painted in Frampton c. 1840 and misidentified as 'Great Hairy Willowherb or Codlings and Cream'.

and var. *brachycarpum*, an alien strain from Scandinavia or perhaps Canada, which thrived on disturbed ground and had been the kind most commonly cultivated as 'French willow'. It was this second variety, he suggested, which had expanded to take advantage of the new open areas created by bombing and forest clearance.[9] But subsequent research has failed to find any physical or genetic differences between the undoubtedly wild populations and the supposedly naturalised colonies in the lowlands. It now looks as if the Gloucestershire Flora's suggestion that the spread was due to the railway system opening up new corridors of expansion may have been correct. Rosebay's seeds – and each plant produces about 80,000 – are certainly equipped to take advantage of such opportunities. They are fitted with plumes of featherweight hairs which, in warm, dry conditions, open like parachutes and enable the seeds to drift long distances on the breeze – or in the slipstream of trains. These days most people would take a more tolerant view of the resultant blizzards than the Revd Riddelsdell, and the difficulties they present are strictly local: 'I have seen a Gloucestershire woodland in early September as though in a summer snowstorm with the multitude of plumed seeds that appeared to fill the air.' (1961)[10] 'The local huntsman does not like letting the hounds go through stands of this when cubhunting as the downy seeds get up his hounds' noses so that they are unable to smell.' (1993)[11]

In parts of the north of England the rosebay's seeds (along with other airborne downy seeds) are known as 'sugar stealers'.[12] There has been no explanation for this name as yet, and I wonder if it might stem from their resemblance to the downy mould that often forms on the top of home-made jam.

There are numerous other willowherbs (*Epilobium*) which occur in Britain, many of them hybrids and beyond the identification skills of most amateurs. Stace identifies 14 true species, including four introductions, and nearly 40 hybrids.[13]

Hampshire-purslane, *Ludwigia palustris*, is a very rare waterplant confined to acid pools in the New Forest and Epping Forest.

Large-flowered evening-primrose, *Oenothera glazoviana*, is a tall biennial with elegant poppy-like yellow blossoms that open fully in the early evening (one expressive personal name for it is 'dusk beacon').[14] It was introduced from North America in the early seventeenth century as a decorative plant, although the roots were occasionally eaten. It has since spread to become a widely naturalised plant of waste ground and waysides, especially on sandy soils.

More recently, evening-primroses of various species (including **common evening-primrose**, *O. biennis*) have been cultivated on a small scale in Britain. Their seeds are the source of evening-primrose oil, widely used in complementary medicine as a source of gamma linoleic acid, GLA.

One family use evening-primrose flowers as an alternative to poppies in the making of 'ballroom gown' dolls.[15]

Fuchsia, *Fuchsia magellanica* (VN: Drop tree). Introduced from South America as an ornamental shrub, and widely planted as hedging (even on farms) in western Britain, fuchsia persists where planted and can become naturalised as a throw-out, but rarely self-seeds. It is one of the best-loved plants on the Isle of Man: 'Within ten years of arriving in the British Isles in 1823, fuchsia was widespread in the Isle of Man. It grows freely up to the 750-feet level and beyond. I have a friend who has a cottage on the slopes of South Barrule at about this height which had an avenue of fuchsia meeting overhead in an arch some 20 feet high. It is widely used for hedging and scarcely a cottage garden is without it. "Beneath the fuchsia tree" is a poetic metaphor for rural life in the recent past. Many Manx people would like to see it adopted as the national flower.'[16]

In North Yorkshire one child found an ingenious way of turning the flowers into dolls: 'I used to remove all the stamens except two from fuchsia flowers and use them as puppet ballet dancers.'[17] A use which is echoed in one Cambridge family's coining, 'dancing ladies'.[18]

Enchanter's-nightshade, *Circaea lutetiana*. This modest willowherb is not a plant you would obviously associate either with enchanters or nightshades. From a distance it is drab, stark, small-flowered, and haunts what Gerard described as 'obscure and darke places' – damp woods, hedge-banks, heavy soil in garden corners and the foot of old walls.[19] The sixteenth-century Flemish botanist Matthias de l'Obel reported that the botanists of Montpellier identified bittersweet as the charm that Homer's witch Circe used to turn Ulysses' crew into pigs, but that the Parisian botanists ('Lutetiani') favoured this species, which thus became *Circaea lutetiana*, the Parisian *Circaea*, and, in English, enchanter's-nightshade.[20]

No great powers or medicinal properties were ever claimed for it in this country. But, looked at more closely, it has another kind of enchantment. The tiny blossoms are mounted like butterflies on pins up the stems. They are formed from two palest pink heart-shaped petals, mounted around two deeper stamens. Almost every other part of the plant also grows in twos: the leaves in opposite pairs, the sepals, and the seeds inside the bristly egg-shaped cases.

Dogwood family *Cornaceae*

Dogwood, *Cornus sanguinea*. A common shrub of hedgerows, scrub and woodland edges, chiefly on calcareous soils. The first-year stems are dark red, and in early autumn the leaves turn a rich claret, the first colour change in the spectacular autumn display of chalk scrubland: 'As warden of a woodland nature reserve my most unusual request for material came from a wood turner. He turned wooden cherries, and prized the red, forked ends of dogwood to make the stalks for them.'[1]

Maurice Young has done research on dogwood in the Chilterns and believes that the component 'dog' is not a disparaging prefix (as in dog's mercury, for instance, see p. 256), but refers to an early use for the shrub: 'The name dogwood may derive from the term "dags", which was used in some places for butchers' skewers – three other old local names for dogwood reflect this: prickwood, skiver wood and skiver timber. H. E. Edlin supports the

A fuchsia hedge in south Devon, where the South American shrub is hardy and occasionally naturalises.

Dogwood, common in hedgerows and chalk scrub.

view that "dogwood" arose from a corruption of "dag-wood". A dag is a spike or a skewer. Dogwood shared these common names with three other calcareous shrubs – wayfaring tree, guelder rose and spindle. Spindle was certainly used in the same way for skewers.'[2]

Dogwood leaves – if pulled slowly from each end – split, leaving a number of elastic tissues joining the two pieces. Many children know this trick, and use it to make notional musical instruments (cf. plantains, p. 320).

Red-osier, *C. sericea*, is a North American species widely planted by lakes and roadsides for its red twigs, which are very showy in winter. It is widely naturalised by suckers. **Cornelian-cherry**, *C. mas*, is a European shrub, grown in gardens and on roadsides and often persisting. The small clusters of yellow flowers appear well before the leaves, and the bright red berries are edible, if fully ripe. **Dwarf cornel**, *C. suecica*, is a short perennial locally frequent on upland moors in Scotland. The small red berries were reputedly once eaten by Highlanders to stimulate their appetites.

Mistletoe family *Viscaceae*

Mistletoe, *Viscum album.* Mistletoe traditions are amongst northern Europe's last surviving remnants of plant magic. Everyone knows the custom of kissing under the mistletoe, even if they don't indulge in it. Whether a sprig is hung traditionally above the door and a berry removed with each kiss, or carried in a pocket or button-hole and flashed opportunistically like a calling card, the obligation is the same: a kiss can be claimed. Even when the real plant seems to have been forgotten – as it often is in television dramas, for instance, replaced by some notional pastiche made from bay leaves and white plastic balls – the old ritual itself is plainly bedded in the national folk-memory. And if we sometimes assume that it is all just a frivolous, albeit ancient, party game, it is worth pondering that, at least until the 1960s, the inclusion of mistletoe in church decoration was frowned on in many parishes. A more complicated and maybe darker past is also lurking in the folk-memory.

Looking at mistletoe against a low winter sun – the great tresses glistening the colour of tarnished brass, the tiers of twigs like wishbones, the whole plant's unearthly vitality in the leafless trees – it is not hard to imagine how it became one of the most revered plants of early herbalists. We know it to be a partial parasite, making some of its own food, but taking minerals from its host tree. But to early people, especially the fearful medievals, it was entirely magical – a plant without roots or obvious sources of food, that grew way above the earth and stayed green-leafed when other plants were bare. It seemed the supreme example of spontaneous generation and continuing life. It is no wonder that it was credited with extraordinary powers. In the Middle Ages, it was believed capable of breaking the death-like trances of epileptics, of dispelling tumours, divining treasure, keeping witches at bay, and protecting the crop of the trees on which it grew. And with its milk-white berries suggestively held between splayed leaves, it seemed 'signed' as a human fertility potion and aphrodisiac too. Women who wished to conceive would tie a sprig round their waists or wrists.

It may seem a long jump from these ancient beliefs to Christmas kissing. Yet in mistletoe's heartland, the border country between England and Wales (and Saxon and Celt), the plant has never entirely lost its magical role. Although mistletoe is widely scattered across southern England and Wales, it needs both a mild, humid climate and a good concentration of trees with soft bark in which its sticky seeds can be deposited by berry-eating birds. These requirements are met most successfully in a wide circle of land around the Severn estuary, where the valleys are moist, mild and sheltered from the worst of the west winds, and where there is a long tradition of fruit-growing. In the mid-nineteenth century, Dr H. G. Bull of the Woolhope Naturalists' Field Club, found that mistletoe grew on no less than 34 per cent of the apple trees in Herefordshire orchards.[1]

In this region arcane fertility rites involving the plant persisted until the early years of this century. In most Herefordshire cottages the practice was to cut the mistletoe bough on New Year's Eve and hang it up as the clock struck twelve. The old bough, which had hung there throughout the previous year, was taken down and burnt at the same time. But on many farms the more elaborate and mysterious custom of Burning the Bush persisted until the outbreak of the First World War. The Bush was a globe made of twisted hawthorn twigs and mistletoe. Early on New Year's Day it was taken to the first sown wheatfield, and burnt on a large straw fire. At Brinsop, north-west of Hereford, the globe was filled with straw and set alight, and a man ran with it over the first twelve ridges of the field. In other villages the ceremonies had their own quirks. At Birley Court near Leominster two globes were used. These were thrown on the fire together, the smaller inside the larger. Whatever routine was followed, it was regarded as an omen of bad luck if the flames went out before the end of the run, and it was thought that the soil would not then have been purified and made fertile. In almost all villages the ceremony ended with cider-drinking and general carousing.[2] While these rites went on outside, a new Bush was made indoors to replace the old one. E. M. Leather heard one farm-worker from Shobdon tell her that cider was poured onto the globe of thorn and mistletoe 'to varnish and darken the bush like'.[3] In neighbouring Worcestershire, mistletoe formed part of the Christmas greenery hung in halls and over doorways, and was dressed with apples and ribbons. It was almost certainly kissed under as part of the general medieval enthusiasm for embracing.

But it was the eighteenth-century fad for Druidism that turned these local customs into a national fashion, and revived echoes of the plant's old aphrodisiac magic. The eccentric antiquarian the Revd William Stukeley, prompted by little more than a few remarks in Pliny's *Natural History* about how the Druids revered mistletoe grown on oak, elevated the plant into the most important accoutrement of the whole religion, and argued that its priests were religious philosophers and the harbingers of Christianity in Britain. By 1728 he was creating a Druidic temple in the orchard of his Lincolnshire garden, around the centre-piece of an 'antient appletree oregrown with sacred mistletoe'.[4]

One of the practices he popularised was 'inoculating' (bud-grafting) trees with mistletoe. There was a good deal of argument at the time about how mistletoe was propagated in the wild. Some writers, like Pliny, believed that the seeds would not germinate until they had been

Lime and apple are mistletoe's favourite hosts. The lime avenue at Burton Pinsent, Somerset (left), and an apple orchard in Aymestrey, Hereford (right).

'ripened' by being passed through a bird. Sir Thomas Browne disputed that the plant grew from seed at all. He thought it 'an arboreous excrescence bred of a superfluous sap which the tree itself cannot assimilate'. But soon gardeners and botanists were successfully propagating mistletoe by 'incising' the seeds under the bark of trees such as poplars and apples.

They had little luck, though, with growing it on oak – which, thanks to the fascination with all things Druidic that Stukeley's writings were helping to fuel, was the kind the public wanted. So the market was ripe for botanical fakes, and Philip Miller, curator of the Chelsea Physic Garden, remarked that supplies reaching London were frequently passed off as 'oak-mistletoe' regardless of their origins. Such mistletoe, he wrote, was so rarely met with that 'whenever a Branch of an Oak-tree hath any of these plants growing upon it, it is cut off and preserved by the Curious in their Collections of Natural Curiosities, and of these there are few to be seen in England'. He also discovered that simply smearing the sticky seeds onto a suitable tree was sufficient to establish it; this, after all, was how the eponymous mistle-thrush spread the plant about: 'for the viscous Part of the Berry, which immediately surrounds the seed, doth sometimes fasten to the outward Part of the Bird's Beak; which to get disengag'd of, he strikes his beak against the Branches of a neighbouring Tree, and thereby leaves the Seed sticking by this viscous Matter to the Bark; which, if it light upon a smooth Part of the Tree, will fasten itself thereto, and the following Winter will put out and grow.'[5]

Border country farmers must have welcomed the growing interest in this orchard familiar, and no doubt much inoculating, incising and smearing went on to supply the growing demand. By the early nineteenth century, it was even possible to buy small trees already sporting mistletoe bushes. The mythology surrounding the plant had also become available off-the-peg. The Druidical and Celtic fertility rites, involving (probably apocryphally) golden sickles and white-robed virgins, were sanitised to a Christmas kiss, and in 1842 even that austere poet Tennyson could write: 'The game of forfeits done – the girls all kiss'd/ Beneath the sacred bush and past away…'[6]

The Victorians also resurrected the Scandinavian myth of Balder the Beautiful, in which Balder is killed by a spear of mistletoe guided by the jealous god Loki – for which his grieving mother Frigg banished it for ever to the tops of trees. (Read Enid Blyton's version of this story – and her accurate, unsentimental and uncompromising botanical and historical preface.)[7]

But the mythology was worn increasingly lightly. By

the second half of the century the Woolhope Naturalists' Field Club, a third of whose 150 members were clerics, were able, without any apparent embarrassment or Christian guilt, to act out an affectionate parody of the rites of their priestly ancestors. On their spring field outing on 24 May 1870 they assembled under a mistletoe oak near Aymestrey:

'The bunch of mistletoe in the oak was so large that it could be exceedingly well seen from the adjoining lane, notwithstanding the foliage of the tree. "There's no mistake about it," said one gentleman, as if he thought there possibly might have been, its portrait and the description in last year's volume of the Club notwithstanding! A ladder had been placed against the tree, with the same thoughtful consideration to every detail that could add to the pleasure of the visitors that prevailed throughout the reception, and it was soon mounted. There was no white yearling bull with garlanded horns to sacrifice beneath the tree for the festivities, nor was there an Archdruid to cut the mistletoe with a golden sickle – indeed the Druidical programme was rather reserved on the present occasion – but anyway the mistletoe bunch was reached and gathered amidst three rounds of applause that were given by the assembled multitude below, and small sprays of the "heaven born plant unpolluted by any touch of earth" were distributed to the ladies present and to all others who wished for it.'[8]

Today a similar scepticism exists hand in hand with a willingness to play along with the old beliefs. A member of an old Scottish family wrote to us in the same spirit, and with more than a touch of *Cold Comfort Farm*:

'The plant is the ancient badge of Clan Hay. Frazer's *Golden Bough* relates how the fate of the Perthshire Hays was influenced by mistletoe. A sprig cut by a Hay with a new dirk on All-hallowmass Eve was a sure charm against witchery and against wound or death in battle. The two most unlucky deeds that could be done in the name of Hay, was to kill a white falcon and cut down a limb from the oak of Errol. These beliefs were set down by Thomas the Rhymer:

But when the aik decays
And the mistletoe dwines on its withered breast,
The grass shall grow on Errol's hearthstone
And the corbie roup in the falcon's nest.

Well the oak crashed down and the grass grew and I'm down here in Derbyshire – with dandruff.'[9]

These days, not even churches prohibit the use of mistletoe in Christmas decorations, and maybe the disapproval was always ambivalent. Stukeley himself wrote that the plant had been carried to the High Altar of York Minster on Christmas Eve. And it is carved, with holly, in a nineteenth-century addition to the galaxy of medieval plant decoration in Southwell Minster.[10] A more modern representation is an embroidered poem, 'The Mistletoe Bough', done entirely with black and white threads, on a wall at Bramshill Police College in Hampshire.[11]

Mistletoe for sale at Christmas is now mostly imported from northern France, especially from poplars in Picardy and the cider-apple orchards of Normandy and Brittany. The home-grown trade is largely channelled through Tenbury Wells in Worcestershire. For a month before Christmas, part of the market is given over wholly to mistletoe and holly sales, and smallholders bring in their bundles to be auctioned to wholesale greengrocers from the Midlands.

Home-grown mistletoe has also cropped up at Weston-super-Mare Sunday Market, and at Stroud: 'All along the Gloucestershire part of the A38 there are the remains of ancient perry orchards, and many of them have great clumps of mistletoe growing high up in their branches. In Stroud market in Christmas 1991, local mistletoe was being sold along with that from France.'[12]

Sometimes rather unconventional gathering techniques are used: 'At the Berkshire College of Agriculture, there are numerous lime trees, which are over 200 years old. Many of these support large mistletoe plants, most of which are in the higher branches ... [this] is harvested by shooting down with a shotgun.'[13]

But in a few places mistletoe is still nurtured as a catch crop. At the Evans family's farm, for instance, at Lower Rochford near Tenbury Wells, mistletoe grows in the ancient fruit trees in their mixed orchards, above grazing sheep and free-range hens. They are happy to have it there, and harvest it for the Tenbury Wells market. But they do not cut it every year. Mistletoe is dioecious, with separate berryless male and berried female plants, and stocks of the latter need building up periodically.

Sadly, old orchards are increasingly being grubbed out, and native mistletoe is a declining plant in Britain's

A mistletoe lot for sale at the Tenbury Wells mistletoe market.

farming landscape. But there is some compensation in its increasing colonisation of soft-barked trees in parks and gardens. The border country can again boast a great variety of hosts. Mistletoe has been recorded on japonica and walnut in Ledbury and Ross-on-Wye, on cotoneaster and laburnum in Hereford, on a weeping ash in Bridgnorth, and on almond in Westbury-on-Trym, north-west of Bristol.[14]

The rose family, as can be seen here, forms an important group of hosts. (Mistletoe is very rare on the wild rose itself, though John Morton reported one from a copse near Kirby, Northamptonshire, in 1712.)[15] Cotoneaster mistletoes have been found in Gwent, at Shillingstone, Dorset, and at Upton St Leonards, near Gloucester. Mistletoe grows on rowan in Taunton, and on an amelanchier in Essex.[16] And hawthorn is one of the most frequent hosts: 22 per cent of Shropshire mistletoe, for instance, was found on hawthorn in the 1980s;[17] and over the Welsh border, at Fforest Coalpit, 'there is a hawthorn bush near our farm with a good clump of mistletoe growing on an exposed mountain side at 1,200 feet'.[18] (Hawthorn mistletoe must have been thought to pick up some of the host tree's magical potency. A woman from Somerset remembered: 'I am in my seventies; a few old-fashioned cures were used for various complaints when I was a child ... Mistletoe from a hawthorn bush for measles; this was made into a tea.')[19]

Other hosts include false-acacia, in many places, including Sible Hedingham, Essex, Warborough, Oxfordshire, and Cowes, and, unusually, a cemetery cypress in Stratford-upon-Avon, which is one of the very few records for a coniferous species.[20] Mistletoe oaks remain rare (except in Epping Forest),[21] and the commonest large tree hosts are much as they have always been: apple, hawthorn, lime, poplar, field maple, elm, sycamore and ash.

In 1991 Angus Idle surveyed all the mistletoe growing in the vicinity of High Wycombe, Buckinghamshire. This is outside the species' heartland, but the general pattern of hosts was repeated:

'I asked for information from people who knew of the location of mistletoe. This produced a number of replies, which included known plantings, but also sightings which turned out to be things like witches' brooms, rooks' nests and squirrels' dreys. Nevertheless it was found that mistletoe was locally very common, with a great deal growing in lime trees, particularly those planted up to 100 years ago in large estates such as West Wycombe Park, Wycombe Abbey, Cliveden and Bulstrode. Mistletoe also occurs on apple, especially when close to the invaded limes, so it seems probable that these have been the sources from which it has spread to other tree species in the district. I have found that mistletoe does in fact grow on a number of other species in this area, even if not commonly. I have found two lots of mistletoe growing on false acacia. In addition to growing on apple in gardens near The Rye, a large grass recreation area near the centre of Wycombe, it also grows on ornamental *Prunus* and hawthorn and poplar. I have also found it on horse chestnut and turkey oak, and I have an as yet unconfirmed report of it on *Hamamelis* "Red Glow" at Bulstrode Park.'[22]

It is finding mistletoes that are exceptions to the rule, which grow on ill-matched trees, or in incongruous (or pleasantly appropriate) places, that makes mistletoe-hunting such an agreeable pastime. The most enchanted mistletoe I have seen was in the dim heart of a hazel bush in Wales. The most aptly sited are the many bunches that grow in a lime overhanging the entrance turnstile to the Oxford Botanic Garden. More grow enticingly from trees overhanging the nearby punt-routes along the River Cherwell. But the most romantic must be in what remains of the Revd William Wilks's (of Shirley poppy fame, see p. 54) garden near Croydon. Here there are two old apple trees with their lower branches fused into a natural arch, which is densely draped with mistletoe.[23] The most dramatically beautiful festoon the long avenue of limes at Kentwell Hall in Suffolk. The trees were planted in 1678, and are covered with large fig-shaped swellings produced by centuries of mistletoe growth. On moonlit nights in winter the clumps glisten in the upper branches like balls of mist. It is easy here to understand the awe which early physicians felt about the plant.

And maybe it wasn't an entirely superstitious awe. Recent medical research has suggested that some of the chemicals in mistletoe may indeed have anti-tumour and sedative properties. The Druidical panacea certainly worked for one family:

'Winkburn Park, Notts, has always been known for mistletoe. It grows on thorn, lime, poplar ... One of the interesting things about Winkburn mistletoe is that it cured a boy in the village of epileptic fits. He could not attend school regularly, and wasn't safe to go out unattended. A gypsy woman visiting the village told his father to boil some mistletoe and give the boy a wine glass of the water to drink. His father did this, but before giving it to the boy, drank some himself as a precaution, but felt no ill effects. He then gave the boy a wine glass full, and afterwards he never had another fit. He is now sixty-six [1993].' [24]

Spindle family *Celastraceae*

Spindle, *Euonymus europaeus*. The spindle tree shares its name with the weighted stick that was used for hand-spinning raw wool before the invention of the spinning wheel. The tree's young branches were ideal for the purpose, being heavy, smooth enough to rotate between the fingers and often as straight as dowels.

But spindle does not seem to have been an especially favoured wood for this purpose in Britain, and the name appears to have been imported by the sixteenth-century botanist William Turner: 'I haue sene this tree oft tymes in England and in moste plentye betwene Ware and Barkwaye, yet for al that I coulde neuer learne an Englishe name for it: the Duche men call it in Netherlande Spilboome that is Spindel tree, because they vse to make Spindels of it in that contrey and me thynke it may be so wel named in English.' [1]

It is curious that this specialised foreign name stuck, and replaced a host of popular names that more accurately reflected its uses here. Spindle's hard, pale yellow wood made it ideal for skewers, toothpicks, pegs and knitting needles, and before Turner (and after him, in country districts) it was known, for instance, as prickwood, skewer-wood (or skiver) and pincushion shrub.

Shocking pink spindleberries. The seeds inside are bright orange, producing one of the few violent colour clashes found in our native vegetation.

Today, spindle is best known for its extraordinary berries, like miniature shocking pink pumpkins, whose four lobes contain – and reveal as they open – round, bright orange seeds. (Though a sport in a hedge at Burtons Green, Essex, has white, not pink, fruit cases.)[2] The fruits are strongly purgative, and were occasionally used for this purpose in folk medicine. They were also baked and powdered, and rubbed into hair as a remedy for head-lice.

Spindle is a slender shrub, with a southerly distribution, and is found chiefly in ancient woods and hedgerows, and in chalk scrub. It still occurs in the chalky country between Ware and Barkway in Hertfordshire, where Turner saw it four centuries ago.

Holly family *Aquifoliaceae*

Holly, *Ilex aquifolium* (VN: Hulver, Holm, Hollin). Compared to mistletoe and its cryptic links with pagan magic, holly seems an uncomplicated festive plant. There is nothing intrinsically mystical or mysterious about gathering bright red berries and shiny leaves to decorate a house in the darkest days of winter. We would probably find ourselves doing exactly the same if holly was a new plant-breeders' sensation, instead of our commonest native evergreen.

Yet scratch below the surface familiarity, and holly too has a complex and paradoxical history. Although lopping boughs for Christmas is 'allowed', for instance, there is still a widespread belief that cutting down whole holly trees will bring bad luck. We use the berries, too, in a kind of informal divination, seeing bumper crops as an ominous sign of hard weather to come rather than as a result of a good spring.

Holly occurs throughout north-west Europe – even, sparsely, in the mountain regions of the Mediterranean. But true holly-woods, of the kind that can be found in, say, Epping Forest, the Welsh Marches and in the groups of huge, unpollarded trees near Coniston in Cumbria,[1] are a British speciality, so perhaps it is not surprising that such a rich culture has grown up around the tree here. Ecologically, too, these woods are exceptional, a rare temperate-zone equivalent of the evergreen cloud forests of South America and northern China.

There is immense variety in both woods and individual trees. In the ancient wood-pasture at Staverton in Suffolk gigantic holly trees grow among thousands of contorted oak pollards (see p. 75). There is what is believed to be one of the biggest hollies in the kingdom (74 feet high and 7 feet 2 inches in girth in 1969), as well as some of the most elevated: in many of the crucks of the oak pollards, holly saplings have taken root amongst the ferns and humus, 20 feet above ground level.

'The holly bears a berry' ... *but only on female trees.*

Some hollies carry yellow berries, or variegated leaves. (I once found a prostrate, golden-leaved tree sprawling across a rock in the remote limestone wastes of the Burren in County Clare.) Occasional specimens have nothing but spineless leaves, of the kind that are normally confined to the top, unbrowsed reaches of the tree. This is called 'free' or 'slike' holly in Shropshire: 'When dealers at Christmas were buying holly, they'd rather have slike holly – there was just a bit of prickle at the end. They'd give more for slike holly than real prickle holly.'[2]

A few trees sucker weakly, forming dense multi-stemmed clumps, or root along the line of low, whippy shoots which become buried under leaf litter. Some have down-curving, 'weeping' branches, which can take root at the tip and create a kind of bower round the trunk. Birds and mammals often use these as shelters in the winter. Humans, too: 'One den I remember well was a domed holly tree, whose branches touched the ground leaving a spacious room in which to play. We painstakingly paved the floor with a mosaic of stones.'[3]

The weather in spring and summer seems to affect the flowering and fruiting of some individuals, and there are always trees which hold on to their berries into summer, or flower in winter (as in the Derbyshire Dales in 1992, and the Chilterns in 1994 and 1995).[4] One idiosyncratic flowerer, which has a girth of over eight feet, is in Broaks Wood, Suffolk: 'This venerable tree has some interesting – though largely illegible – graffiti on it (none of it recent). Another curiosity is its habit of starting to flower in January.'[5]

But it is the clusters of scarlet berries and the darkly monumental foliage that make holly such a dramatic and conspicuous tree, and one that would inevitably lend its name to places. Hollington in Derbyshire and Hollingworth in Cheshire come from the Old English *holegn* – and both still have plenty of trees.[6] Cullen in Banffshire is probably a derivation of the Gaelic word for holly, 'cuillioon'.[7] The village of Hulver in Suffolk shares the Middle English name for the tree, though there are now 'only 18 trees left in the village',[8] and Hollybush in Herefordshire, despite being in one of the holly heartlands, seems to have virtually none remaining. Holmstone (from *holm*, also

Middle English) in Kent is a unique wood of stunted hollies growing on the shingle beach at Dungeness, which was documented as early as the eighth century. There were 224 bushes at the last count in 1992: 'As the holly was used to construct sea defences (groynes) it must have been a very highly valued commodity, and this could be the reason why Lydd folk, who have always been threatened by inundation by the sea, have always and indeed still are striving to keep their Holly Forest.'[9]

But holly elements in place names can be difficult to distinguish because of their similarity to other common components such as holy, hollow, and 'holm' meaning 'island'. Hollytreeholme in Yorkshire looks like the ultimate in holly place names, but actually means 'island with a holy tree'. Hollingbourne, Kent, is probably 'the stream of Hola's people'. Christmas Common in the Chilterns, on the other hand, may well be 'holly common' as Christmas tree is an early local name for the species, which is common in the region.[10]

Another puzzling suite of names crops up elsewhere in the Chilterns, between Chipperfield and Sarratt. This is an ancient and densely hollied landscape. Hollies spread in dark copses and commons. Thick holly hedges line the narrow switchback lanes. One of these lanes, threading through this woody enclave for a mile and a half, is called Holly Hedges at one end and Olleberry at the other. This seems straightforward enough – except that there is a farm called *Hollow* Hedges half way along the lane, and records of a seventeenth-century family called Olbury living in the area. Like many persistent place names, the roots of these names are probably complex, converging and reinforcing each other, and maybe encouraging the preservation of holly as well as being prompted by it.

The most common and reliable ancient holly name is the Middle English *hollin*, which also came to stand for a group or grove of hollies that were regularly lopped for cattle feed. Holly seems an improbable and unpalatable form of browse. But feeding it to stock (sheep especially) during the winter is an ancient practice that doubtless goes back into prehistory. Its leaves have one of the highest calorific contents of any tree browsed by animals, and are rich in nutrients.[11] Martin Spray, who has done a historical survey of the use of holly as cattle food, believes that it was a widespread, if not always well-documented practice

A group of hollies lopped for cattle food (a 'hollin') in the Olchon Valley, Herefordshire.

up until the eighteenth century. It seems to have been particularly prominent on the grits and sandstone of the Pennine foothills, roughly in the triangle formed by Derby, Leeds and Manchester. Spray has traced a conspicuous concentration of surviving 'hollin' and 'holly' place names in this area, and even an abundance of the same words as family surnames (noted from telephone directories) which exactly matches the place-name distribution.[12]

Reminders of old hollins are frequent in place names in the north, as for instance in this cluster near Oldham in Lancashire:

'As you can see from my address [Hollingreave Farm, Holly Grove], hollies are reasonably common in this area. But, although the moorland peat contains roots of prehistoric trees and there were said to be many more trees in the past, it is not an area of many old trees. The hollies by the cottages have very thick trunks, sometimes divided at the base, so possibly cut back at some time. Although our

Hollies are often retained when hedgerows are cleared or lowered. They are used as guide-marks by combine drivers in many areas.

house was only built about 1800, the hamlet of Holly Grove is very old – there are church records of a Richard de Hollingrave in 1272.'[13]

'Next to my house, Hollin Hall, there stood a "Red Hall", and if you look over Ordnance Survey Maps you will often find abodes called Hollins and Red Hall close to each other, and often a tannery close by, as the holly wood was used in the tanners' fires.'[14]

'The name of my farm is "Hollinroyd", meaning holly clearing. The land was cleared *c.* 1450, but we still have holly which cows eat in winter.'[15]

But there are large surviving hollins too. One well-dispersed group is in the Olchon valley in Herefordshire, where the gnarled pollards stand in a landscape of Celtic fields and ancient stones.[16] Another is in Needwood Forest, Staffordshire – despite 150,000 trees being felled to make bobbins for the Lancashire cotton mills in 1802.[17] The most intact is probably the open woodland of almost pure holly at the southern edge of the Stiperstones in Shropshire. This is a remote area of heather ridges and splintered quartzite, where the commoners were part miners and part graziers (a way of life described in Mary Webb's novel *Gone to Earth* in 1917). The local trees, some of which are probably 400 years old, were a source of winter browse, and they have been repeatedly cut at between three and five feet above the ground. Down in one of the valleys there are the remains of the old grinding machines once used to make the spinier leaves more palatable (as with gorse, see p. 230).

Ancient coppiced holly, Yarner Wood, Devon.

Holly boughs are still cut for sheep and occasionally cattle in Dumfries, Derbyshire and Cumbria.[18] But the most extensively used hollins are in the New Forest, where an abundance of holly place names – e.g. Holmsley, Holmhill, Holly Hatch – testify to the important role holly has played in the landscape and local economy. Groups of self-sprung hollies – known locally as 'holms' or 'hats' – mixed with thickets of gorse, often provide the cover through which new oak and beech seedlings grow, in the slow process by which the Forest regenerates itself. And holly branches are still cut for the ponies and cattle when there is snow on the ground, though not as extensively as in the past.[19]

But it is a measure of just how confined plant-based traditions can be that a farmer from Eardiston, Worcester, just 20 miles from the famous hollins on the Stiperstones, could express disbelief about this particular practice: 'I have lived on a farm most of my life, apart from the last war. I have reared pigs, sheep, cattle and horses, but have never known of holly being fed to farm animals . . . The holly was known to be shade in the summer, and shelter in the winter. Many years ago the true gipsy women were known to give birth under a holly tree.'[20]

It is a custom which the stock themselves often followed: 'We have a very old holly tree. Over 100 years ago it was grafted with three different types of holly on the same trunk. We are 859 feet above sea level [in the Mendips], and have a lot of holly about around the farm. We always protect it because as a livestock farmer it is the best shelter for the animals. We find if a cow is calving out, she will always calve under a holly tree. But we have never fed holly to the livestock and have never known them eat it. We find it is the cold winds which will kill it back.'[21]

But on some farms a more superstitious attitude persists, as on the fells around Lancaster: 'In a new building or one where an animal has died a sprig of holly should be hung up (too high for the animals to reach) to remove and keep away evil spirits. The sprig should be at least two feet long and should be changed from time to time and the old one burnt.'[22]

This echoes a practice in France, in which the 'evil spirits' are replaced by more tangible parasites: 'In France, many farmers will hang branches of holly just above the height of their cattle when they are housed. The aim is to prevent ringworm.'[23]

Holly has had other practical uses, many of them touched by the tree's slight aura of magic. Like the elder, holly was believed to have power over horses, and its white, pliable wood made it the favourite timber for whips: 'The second largest use of holly in the late eighteenth and all the nineteenth centuries was as the best stocks for driving whips. When one realises that the carriage, coach, van, gig and pony and trap were used (as we use cars) in their hundreds of thousands, some 210,000 holly whips were made in the kingdom at the peak of the horse-drawn era each year.'[24]

'In the time of plough horses, holly was used for handles for horse whips by the ploughmen. The trees were established in deep burnsides, coppiced to produce many long straight stems. This would seem to be a Stirlingshire tradition, according to my forestry colleagues ... (Incidentally, roe deer crop the lower leaves in a neat cut but red deer chew the next branches up in an easily identified, ragged manner.)'[25]

There are echoes of this in the Hertfordshire mummer's practice of using holly twigs as the horns of his hobby horse;[26] and in the notes of 'Ratcatcher' (an aspiring 'rune-master') on his apprenticeship: 'I am trying to become a rune-master. Holly is one of the best woods to make runes with. The holly is a very powerful and magical tree. If you want to use the wood, you must first ask the tree if you can use some of it. (You will get a feeling of it if the tree says "yes".) Then you must provide a useful gift for offering to the tree. Then you will be given all the power you need in the wood. It is also important to cut your holly on a night when there is a full moon. This adds more power to your runes.'[27]

There are several customs in which holly seems to be regarded as a proof or deterrent against fire (despite the fact that it is extremely inflammable in leaf, and as green timber): 'I remember being told when I lived in a village on the Isle of Wight with constant thunderstorms, that holly trees were often planted on either side of a building as a form of lightning conductor.'[28]

Up to the end of the last war, young hollies were used to sweep chimneys by being hauled through them on a rope. 'In Essex I had to struggle to keep my little tree from a neighbour who wanted it for that purpose.'[29]

'When I was a young gamekeeper in south Devon, I was told that the large number of holly trees locally were grown to meet the demand for tea-pot lids and handles. I believe they were rarely cracked by heat.'[30]

In Culmhead in Devon, there is even a pub called the Holman Clavel, which means a 'holly-beamed open fireplace'.[31]

But the most persistent, *trusting* use of holly is as a boundary tree. Across Britain, in every kind of landscape, hollies are looked on as constants in the landscape. One walker has noted how often they grow close to stiles, perhaps 'assisting in locating the stile and hence indicating the line to walk'.[32] Another used the evergreen leaves as reliable shelters on a fixed route to school: 'A life-long friend of mine spent his childhood in the New Forest, for several years walking some two miles to school. Between the two locations every holly tree was known, and on days of frequent showers a dash from tree to tree was made, for it took a long time for the rain to penetrate the canopy.'[33]

In Cornwall, an inspector of mines has recorded holly trees being used to mark the overground course of a tin lode: 'Back at the beginning of the nineteenth century a mine called Wheal Pool (wheal is work in Cornish) near Helston was closed, and to mark the course of its lode, holly trees were planted along it. Most of them have long since disappeared, but what appear to be replacement trees are on the site ... As far as I know this is the only recorded case in Cornwall of holly trees being used to mark the line of a lode; the common custom in the Helston area was to use thorn or elder trees, depending on whether the mine was wet or dry.'[34]

Holly trees sometimes appear to be deliberately left standing when hedges are grubbed out, and a down-to-earth explanation for this – suggested from several arable areas – is that ploughmen use their conspicuous, dark shapes as sightlines during winter ploughing.[35] At Inverary a particularly venerable holly was saved in 1861, when the Duke of Argyll 'insisted that an awkward bend be put in the line of a public road to avoid the necessity of cutting it down'.[36] And in the absence of any firmer physical or legal evidence, even Ordnance Survey map-makers regard mature hollies as being the best pointers to the course of old boundaries. Holly, it seems, is widely regarded as capable of outliving changes in ownership and farming practice, and of echoing the contours of ancient estates. (Though the enduring protection given to one tree by a family in Gloucester was exceptional. When they sold off part of their garden, they inserted a restrictive covenant in the contract, prohibiting the felling of the male holly tree it contained, so that the female tree that grows in their own garden would continue to be pollinated and set berries.)[37]

Why should holly have this special indemnity? It is, of course, a reliable, stock-proof tree, and very visible as a marker. But it seems that many hedge and boundary hollies survive because of the stubborn persistence into the late twentieth century of the belief that cutting down holly trees brings bad luck. In Suffolk and Worcestershire, for instance, professional forestry contract workers are still reluctant to fell hollies, just in case.[38] I have seen the same reaction from scientifically trained ecologists in my own wood. And from every part of the country people have written with stories of illness, heartbreak and disaster which ensued – with a time-lag of as much as forty years – when the taboo was broken and a holly was cut down.

In farming areas there sometimes seems to be a clear religious rationale behind the superstition – either orthodox or pagan:

'There is a suspicion here [Buckinghamshire] about felling hollies. Legend still has it that witches appear instead.'[39]

'[In East Sussex] holly is left as standards, in hedgerows, to prevent the passage of witches, who are known to run along the top of hedges.'[40]

'When I was rector of Iping and Linch near Midhurst, Sussex, I noticed hollies left to grow above hedge level. One answer was given to me by a Linch parishioner who was a professional woodman as his father had been before him. He told me that his father had taught him that he must not cut hollies "because they are the King's tree". He considered the "King" to be the King of England, though it did cross my mind to wonder whether the reference was to the King of Kings, cf. in "The Holly and the Ivy": "Of all the trees that are in the wood, the holly bears the Crown" '[41] (see ivy, p. 276).

The odd thing is that this superstition should coexist with the sanctioning of the cutting of holly *branches* for Christmas. Bringing in evergreen boughs to deck out barns and houses in midwinter is a custom which goes back to pre-Christian times. Holly, especially, with its sharp spines and red berries held throughout the winter, was seen as a powerful fertility symbol and a charm against witchcraft and house goblins. (It was, ironically, also seen as a masculine plant, despite the berries being carried solely on female trees.) The custom was easily accommodated by Christianity, holly standing for the crown of thorns and the berries for Christ's blood. Yet echoes of the old religion linger, and there is still a fixed – and widely respected – routine for taking down the Christmas greenery, though the date has shifted from early February (now Candlemas) to Twelfth Night. These are some of the customs – often contradictory – in different families' holly calendars:

'A holly branch is often used instead of a Christmas tree in Cornwall.'[42]

'My father (from Dorset) would never allow holly to be put up as decoration before 25 December.'[43]

'In my childhood home [Yorkshire] every Christmas, leaves were taken from the decoration holly and dispersed around the house, one leaf to each room, usually into ornaments or vases.'[44]

'Our family custom in Shropshire involves leaving a sprig of holly to dry after all the Christmas decorations have been taken down on Twelfth Night. This holly is kept. Usually it is stuck behind the clock or over a picture rail. Then, come Shrove Tuesday, it was taken down and put onto the fire over which the first pancake was cooked.'[45]

'If a leaf falls out of a vase of holly never put it in the fire to get rid of it.' (Lancashire)[46]

'I was always taught that any holly used in decorations should be burned in the garden afterwards, for continual good luck through the year.' (Hampshire)[47]

Even commerce is not immune. At Brakspear's Brewery in Henley, a bush of holly and mistletoe is suspended from the eaves each Christmas and left there all year to ward off misfortune. No one knows when this custom began, but the bush is visible in a photograph from 1910.

The commercial trade in holly itself has long had one of its main centres in the market at Tenbury Wells in Worcestershire, where during the four weeks leading up to Christmas there is a special arena devoted to holly and mistletoe sales (see mistletoe, p. 239). Much of the holly taken to market is cut on a small scale on local farms and commons. A woman brought up on the Shropshire borders remembers taking over her grandfather's patch when he was ill after the war:

'Months before, during the summer, my grandfather would go to various farms and look over the trees, and the bargain would be made; which trees he could cut, and how much he would pay the farmer. Holly trees are very slow growing, so having cut one this year, it could be three or four years before the tree would be ready for cutting again. And cutting holly trees, let me say, is quite an art. It is no good going in and cutting great lumps out of the side. You can easily ruin a tree by hacking at it, and not treating it with care and respect ... Down below, my husband was busy packing pieces of holly still covered with frost. The sack must be filled just right, not too loose, or the branches move about and the berries fall off; and not too tightly, or they all jam together and knock the berries off. The merchant in Manchester, to whom the holly was sent, would grumble if it didn't arrive in the right condition and would reduce the price ... My grandfather was full of questions on our return. All these farms and fields and trees had been a big part of his life every Christmas time. He knew the names of all the fields and the very trees which grew there. He knew when to expect two sacks from a tree and half a sack from a much bigger tree, the tree by the stile in Cae Mawr, the two trees by the gate in Ty Gwyn, and on and on, he knew them all.'[48]

Intertwined oak and holly, Pinnick Wood, New Forest. Holly provides protection for regenerating trees in the Forest.

Alas, this modest husbanding is becoming a thing of the past. In some areas holly is illegally 'poached' on a large scale, with whole trees often being chain-sawed just for their berries. Yet in contrast to the 1960s, when holly was widely believed to be becoming scarcer, it is now unquestionably spreading. In hill-country woods and commons where grazing has ceased, the browsed clumps have started to sprout into more substantial trees. In the south, especially in the Chiltern and Hampshire beech-woods, the slowing-down of thinning has favoured this shade-tolerant tree. Bird-sown seedlings are proliferating (in my own wood they occur most commonly under a pigeon roost in a stand of beeches) and in many woods holly is now the commonest shrub in the understorey. In a few decades, given its persistence and ability to form a closed canopy, it may become the dominant tree, and we may have a new generation of pure holly-woods.

Holly has held a place in popular affection through all its swings of fortune. It has been a talisman for the woman who found a young holly on a child's grave at Walkden Moor church, near Manchester, sprung from a berry on a holly wreath.[49] And for Mrs Berry too, who named her daughter Holly, 'so we don't burn the dead holly after Christmas (nor the Rosemary – that's my name). I have no worries cutting the holly to bring it in at Xmas, but I never burn it, always compost it.'[50] And for the teacher in Leeds, one of 'hollin's' ancestral haunts, for whom the tree has been a kind of mascot throughout her life: 'One of my first teenage Christmas dresses – holly-green with green velvet trim and red buttons – really felt right for the time of year – holly on cakes – 1970, moved to Leeds, and a holly was growing in our garden – 1984, moved to Adel, Leeds, and though I shouldn't have, I brought a small growing shoot back from Adel Wood to plant in our back garden – At least six "hollins" addresses in Leeds 16 ...'[51]

Midwinter greenery

The use of evergreen plants to decorate houses at the midwinter solstice is a custom which long antedates Christianity in Europe and Asia. Evergreens, flourishing when all other plant life seems dead or dormant, were regarded as symbols of the continuity of life through the dark season.

In ancient Rome, for example, garlands were made from Mediterranean bay, box, rosemary, pines and evergreen oak. In Britain the native holly, ivy and mistletoe were (and still are) the favoured plants, and their various roles are discussed under their separate entries. But they were often used together, in wreaths hung on the door or over the porch, for instance. A favourite decoration in late medieval England was the kissing bough, which was a garland of greenery shaped roughly like a crown and adorned with fruit, coloured paper rosettes, candles and, most importantly, a bunch of mistletoe hanging from the centre.

There have been strict rules about when the midwinter greenery should be put up, when it should be taken down, and how it should be disposed of. Twelfth Night (6 January) has long been a watershed. But in some areas the greenery was kept until Candlemas Eve (see snowdrop, p. 421). In some it was ceremonially burned, in others fed as a charm to cattle. Most of these local customs have faded and been absorbed into the national pattern. But there are still places where the rituals associated with midwinter greenery kept a distinctive local flavour until recently:

'The parish church of Hest Bank, Lancaster (its full name is St Andrews, Slyne-with-Hest), kept until recently a curious and almost certainly pre-Christian custom, of the men of the church processing in company carrying small conifer trees, about three feet tall, on their shoulders. About a dozen in all, on a Sunday near Christmas.'[1]

'When I first went to live in Peel, on the Isle of Man, I was surprised to see that the holly and other Christmas greenery was still in place after Twelfth Night. I commented on this, and was told that here they are left up until Shrove Tuesday. There was a practical reason for this in addition to any regard for custom. The fierce heat which they gave when burnt in the range was just what was required for the cooking of pancakes.'[2]

The holly and mistletoe market at Tenbury Wells.

LOT
NITRAM

Box family *Buxaceae*

Box, *Buxus sempervirens*. Box is an anomaly amongst plants which have lent their names to settlements. It is a drab, malodorous and not especially useful shrub confined to the southern chalk, but in southern England it has as many places named after it as the elm. There are some 20 English place names which begin with the 'Box' prefix – 25 if you include the formations Bexhill and Bexley, Kent; Bexington ('settlement amongst box'), Dorset; Bix, Oxfordshire; and Bixley ('box woods'), Norfolk.[1] (Though half a dozen of these almost certainly have no connection with the shrub: e.g. Boxholme in Lincolnshire – beyond its possible natural range; Boxted Green and Cross in Essex – perhaps from OE *Boc-hamstede* ('homestead among beeches'); Boxworth, Cambridgeshire – 'Bucc's worth' (enclosure); Boxford, Suffolk; and Box near Minchinhampton in Gloucestershire – perhaps after Julia de la Box.[2]) Box also appears as a boundary shrub in medieval charters for Ecchinswell and East Meon, both in Hampshire.[3]

The dark outlines of box bushes on the Chequers Estate, Buckinghamshire.

Why should such a subfusc shrub have proved so compelling to the people who settled near it? One myth can be put aside at the outset. Box is not some exotic import from the Mediterranean, whose presence here can be traced to escapees from Elizabethan knot gardens or to coverts planted by Victorian landscapers. It is unquestionably native in scrub and open woodland on calcareous soils in southern England, forming dark, dense, elfin thickets amongst the paler-leaved deciduous shrubs, which are visible from long distances. This, together with the distinctive smell of the leaves (politely likened to foxes, but really more like tom cats' urine) and the remarkable hardness of the wood, is enough, I think, to account for its fascination. The best evidence of all is the surviving stands of wild box in places which have borne the shrub's name for the best part of a thousand years. Boxley, in Kent, is one of the famous sites, and local people have unearthed some fascinating details of the village's connections with the box tree:

'It does appear that the "ley" in Boxley is derived from the Anglo-Saxon word "leah" which has the fairly definite meaning of a permanent glade or clearing in woodland. This, linked to the fact that the village still supports fairly natural-looking populations of box in the surrounding countryside, proves the link between the village name and the box tree fairly conclusively. Indeed such Anglo-Saxon place names provide some evidence that the box is a native tree – more recent evidence comes from finds of box-wood charcoal in association with Neolithic camps on the South Downs.

Conditions for the box tree are almost perfect at Boxley with its position on the south-facing slopes of the calcareous North Downs. A few straggly natural-looking trees may still be found even within woodland on the steep slopes above the village ... A pair of box trees also flank the war memorial in front of the church (and there are several large ones inside the churchyard) and though they were almost certainly planted around 1919 it does show that a perception of the historical links between the village and the tree has existed for some time ... As for the future, it was agreed at a recent Parish Council meeting [1993] that as part of our centenary celebrations we will plant a box grove somewhere in the parish – well away from our precious natural population of course and perhaps in combination with other trees and shrubs characteristic of the locality such as whitebeam and wayfaring tree.'[4]

Sixty miles to the west, at the Surrey end of the North Downs, is Box Hill, with the best-known of the native boxwoods. They cover the cliff-like banks above the River Mole with a density that permits very little else to grow underneath. John Evelyn wrote about the hill in 1706: 'The *Ladies*, *Gentlemen* and other *Water-drinkers* from the neighbouring *Ebesham-Spaw* [Epsom Spa], often ... *divert* themselves in those *Antilex* natural Alleys, and shady Recesses, among the *Box-trees*; without taking any such offence at the Smell, which has of late banish'd it from our *Groves* and *Gardens*; when after all, it is infinitely to be preferr'd for the bordering of *Flower-beds*, and Flat *Embroideries*, to any sweeter less-lasting *Shrub* whatever ...'[5]

A century later, William Gilpin wrote of 'shivering precipices, and downy hillocks, everywhere interspersed with the mellow verdure of box, which is here and there tinged, as box commonly is, with red and orange'.[6]

In West Sussex, box is quite common, native or naturalised in woods and on the downs. But Boxgrove (whose name appears in the Domesday Book) is the only settlement named after it.[7] Box must once have been common in the Chilterns too, from Bix, near Henley in the south, through Boxmoor, Hertfordshire, to, perhaps, Box End in Bedfordshire. There are still scattered individual bushes in apparently wild situations throughout this range – for instance near Ivinghoe Beacon in Buckinghamshire. These may be remnants of the much larger populations reported between Tring and Dunstable by many writers, including the philosopher John Stuart Mill in 1855. A century before, Pehr Kalm had stated that these were planted by the Duke of Bridgewater, and the wood sold to London craftsmen. (The exceptional hardness of box timber made it a valuable raw material, and it was used for chessmen, rulers, rolling pins, pestles, and especially for printing blocks: the nineteenth-century engraver Thomas Bewick claimed that one of his blocks was still sound after 900,000 printings. The wood was also used for cleaning rings.)

The dark, sprawling hummocks on Shirburn Hill, east of Watlington, look natural, though the presence of exotic species such as cork oak nearby suggest they too may have been part of a planted landscape. But there is not much doubt about the provenance of the most extensive stands of box in Britain, eight miles to the north-east of Shirburn in Ellesborough Warren and the Chequers Estate. The trees are tall and ancient and occupy three steep-sided coombes. The atmosphere amongst the twisted trunks is extraordinary: humid and dark from the closed, evergreen canopy overhead and filled with that pungent but powerfully nostalgic smell. In one of the coombes, known as Happy Valley, there is an area of young regenerating box on a steep, stony slope.

The early history of the Ellesborough and Chequers colonies is not well known. But, up to the beginning of the nineteenth century, the area was apparently commonland, and local people coppiced the box for firewood. The Enclosure Award of 1805 stopped this practice, and a local tenant wrote that 'the box increases in beauty and value and forms a very picturesque appearance'.[8] Now

Box Hill, Surrey. The hardwood of box is used to make woodcut blocks.

some coppicing has been reintroduced, to try to prevent the box shading out too much of the chalk grassland. But this long rotational cutting is not entirely compatible with what has become the commonest modern use for box: the harvesting of young shoots for florists' wreaths.

The grove at Boxwell in Gloucestershire, also in a steep-sided coombe, is much smaller than those in Ellesborough, but the trees are older and the site has a much longer documented history. John Aubrey described the Boxwell trees in 1685 as 'a great wood ... which once in ... [unspecified] yeares Mr. Huntley fells, and sells to the combe-makers in London'.[9] A distinctive feature of the Boxwell boxwood is that it merges with the nearby beechwoods, forming an understorey as the tree often does on the continent. And close to the drive up to Boxwell Court the wild trees are progressively trimmed, so that by the Lodge they are in the form of a neat box hedge. (In southern France, where box is a common shrub, topiary is practised even on wild trees. A couple holidaying in the Corbières Hills in the Languedoc in 1992 were mystified to find artfully clipped box bushes out in wild *maquis*, miles from any settlement. It was some days later that they traced the sculptors: local road-menders with time on their hands after lunch.)[10]

Box, like other sombre evergreens, has long been a plant of grave decorations and funerals. A considerable thicket of box – mixed with yew and snowdrops – surrounds the memorial cross which marks the site of the execution of Piers Gaveston at Blacklow Hill in Warwickshire. The inscription on the cross reads: 'In the hollow of this Rock, was beheaded, On the 1st Day of July 1312, By Barons lawless as himself, PIERS GAVESTON, earl of Cornwall, The Minion of a Hateful King: In Life and Death, A Memorable Instance of Misrule.'[11]

Wordsworth describes a north-country funeral custom of filling a basin with sprigs of box, and placing it by the door of the house from which the coffin was taken. Each person who attended the funeral would take a piece of box and throw it into the grave after the coffin had been lowered.[12] The custom persisted in the region well into the late nineteenth century. In 1868, a *Daily Telegraph* reporter sent to cover the aftermath of a colliery disaster at Hindley Green near Wigan wrote: 'I find an old Lancashire custom observed in the case of this funeral. By the bedside of the dead man, the relatives, as they took their last look at the corpse, have formed a tray or plate, upon which lay a heap of sprigs of box. Each relative has taken one of these sprigs, and will carry it to the grave, many of them dropping it upon the coffin. Ordinarily the tray contains sprigs of rosemary or thyme: but these poor Hindley people not being able to obtain these poetical plants, have, rather than give up an old custom, contented themselves with stripping several trees of boxwood.'[13]

Spurge family *Euphorbiaceae*

Dog's mercury, *Mercurialis perennis* (VN: Boggard posy). Dog's mercury is a common woodland plant, whose spear-shaped leaves and spikes of small, greenish flowers are vaguely similar to those of some members of the *Chenopodium* family – the 'true' mercuries (see p. 96). But *Mercurialis* is highly poisonous; hence it became the 'bad', 'false' or 'dog's' mercury.

Given dog's mercury's usual habitat, its bright green, rather than mealy foliage, and a host of other differences, it is hard to see how the two kinds could be confused. But mistakes are made, even by modern foragers equipped with field guides. In 1983, the *British Medical Journal* published an account of a couple who ate a large quantity of dog's mercury in the belief that it was brooklime, an edible plant of the speedwell family (which at least bears more resemblance to it than any of the *Chenopodiaceae*). Four hours later they were admitted to hospital with vomiting, pain and gastric and kidney inflammation, and what their doctors described as a 'curious malar erythema' (reddening of the cheeks and jaw). Fortunately, after supportive treatment, both patients made a full recovery.[1]

Their doctors could find no record of malar erythema in the previous human cases (happily few) of poisoning by dog's mercury. But it was noted in what is probably the first account of the plant's toxic effects, in Shropshire in the spring of 1693:

'About Three Weeks ago, the Woman [wife of W. Matthews] went into the Fields and gathered some Herbs, and (having first Boyled them) Fryed them with Bacon for her own and her Families Supper: After they had been about Two Howrs in Bed, one of the Children (which is Dumb and about Seven Years Old) fell very Sick, and so did the other Two presently after; which obliged the Man and his Wife to Rise and take the Children to the Fire, where they Vomited and Purged, and within half an Hour fell fast asleep. They took the

Children to bed as they were asleep, and they themselves went to bed too, and fell faster asleep too than ever they had done before. The Man waked next Morning about Three Hours after his usual Time, went to his Labour at Mr. *Newports*, and so by the strength of his constitution carried it off; but he says, he thought his Chin had bin all the Day in a Fire, and was forced to keep his Hat full of Water by him all the Day long, and frequently dipt his Chin in it as he was at his Work.'[2]

In this instance one of the children died, after four days of unconsciousness – a grim warning against eating wild plants without being certain of their identity.

Dog's mercury is a problem plant in another way. It can grow in such dense, leafy colonies in woods that it shades out other, more light-demanding, species, such as oxlip, fly orchid (see p. 449) and even young ash seedlings.[3] And, though it is predominantly a plant of ancient woods and old hedgerows, it can colonise new deciduous woodland at a rate of more than three feet a year, particularly where the soil is calcareous and dry. Mercury spreads by underground rhizomes, forming large 'clonal' patches in which all the stems share some peculiarity of leaf size, shape or tint. Some clones appear to be almost evergreen in their capacity to retain leaves through the winter.

Dog's mercury often carpets the ground in old ashwoods.

Currently, dog's mercury appears to be going through a period of expansion in some areas of Britain, in the boulder-clay woodlands of East Anglia, for instance.[4] Colonies are growing in extent, and are composed of larger plants. Whether this is due to increasing shade in woods where coppicing or thinning has declined (which favours mercury's early leafing), or is a response to, say, some chemical changes in rainwater, it is hard to say.

Annual mercury, *M. annua*, is a very similar-looking plant, quite common as a weed on lighter soils. It is probably native in East Anglia and southern Britain, but has spread further afield, sometimes by unconventional means. Its arrival in west Somerset, where it was unknown before the 1880s, was almost certainly via commercial trade into the small shipping port of Watchet:

'On one occasion a cargo of sacks of Russian wheat was brought in from Danzig. It was the custom that dockers would sweep up any loose and dirty wheat which might otherwise be left in the hold of the ship and this they usually took home to feed their chickens. Bob [Williams] however decided to sow his dirty Danzig wheat in his smallholding. In due course the Russian wheat broke through the ground but it was accompanied by hundreds of pretty little weeds which had never been seen at Watchet before. Many people gathered around to see the pretty weeds and laughingly gave them the name of Bob Williams' Weeds. By the next year however they had stopped laughing for the weeds had spread into all their adjoining fields and gardens. The name of the weed was rapidly altered to "That Bloody Bob Williams' Weed". Some years later apparently it had spread over the hill to the little town of Williton but was known by the gardeners there as "That Bloody Watchet Weed". It still flourishes at Watchet ... Subsequent enquiries to the Somerset College of Agriculture have ascertained that the correct name for the weed is Annual Mercury.'[5]

Sun spurge, *Euphorbia helioscopia*, is a frequent annual of waste and cultivated ground throughout the

Cypress spurge, possibly native on chalky soils in the extreme south-east of England.

lowlands, with golden-green foliage and reaching a foot or so in height. Like other spurges, the leaves and stem exude an intensely irritant milky sap when broken, which has been used to treat warts. In the Isle of Man, where it is known as Lus y Bwoid Mooar, its therapeutic role has been more exciting, if perilous: 'I had always understood that the "big knobs" (Bwoid Mooar) referred to were those in the flower- or seed-heads, but recently an antiquarian incomer demanded to know what the plant with this name was. He had been told that as late as the 1930s the youthful Port St Mary fishermen used to rub the milky juice on their penises to "get themselves a bit excited". The remedy for over-use was to immerse the organ in sour milk.' [6] (The practice may not have been confined to the Isle of Man. Grigson quotes, without comment, 'Saturday night's pepper' as a local name for the species in Wiltshire.) [7]

Caper spurge, *E. lathyris*, is a quite large, bluish-green biennial, whose fruits have sometimes been mistaken for the true caper, *Capparis spinosa*. (It is a bad mistake: caper spurge is a drastic purgative.) It may be native in old woodland in southern England, and occasionally appears – presumably from long-dormant seeds – when an area is opened to the light by coppicing or clear-felling. But most specimens in the wild are naturalised escapes from gardens, where it is still widely planted, both as an ornamental and because it has a reputation for repelling moles.[8]

Cypress spurge, *E. cyparissias*, is a low, feathery-leaved species, also much planted in gardens. Where it is naturalised, for instance in chalk grassland on the South Downs and in field borders and roadsides on the sandy soils of the Breckland, it can spread quite extensively. (The Shropshire Flora contains the intriguing note: 'Established as an arable weed in Attingham Park: a relic of a former scientific experiment by an officer of the Nature Conservancy Council.') [9] It is possibly native in some of its southerly sites. **Twiggy spurge**, *E. × pseudovirgata*, and **leafy spurge**, *E. esula*, are both spreading perennials from Europe, quite widely naturalised on roadsides, grassy waste places, etc.

Wood spurge, *E. amygdaloides*, is an attractive perennial of broad-leaved woodland and shady banks in southern Britain, with evergreen, often red-tinged leaves, topped by bright yellowish-green flowers held in saucer-like bracts (whence the Somerset name 'Devil's cup and saucer').[10] It often flowers in abundance in newly cut coppice. The variety known as 'Mrs Robb's Bonnet' (*E. amygdaloides* ssp. *robbiae*) was discovered by the intrepid Victorian adventurer Mary Anne Robb of Liphook whilst attending a wedding in Istanbul in 1891. Having no other means of transporting the plant, she packed it in the box usually reserved for her best bonnet. The name stuck.[11]

Wood spurge, the ancestor of some garden varieties.

Buckthorn family *Rhamnaceae*

Buckthorn, *Rhamnus cathartica*. A tall, spiny shrub of hedgerows, fens and scrub, chiefly on chalk soils in England. It is best known for its shiny black berries, which are a fierce purgative. Gerard used the daunting name 'laxatiue Ram' for them, and Henry Lyte, in 1578, said that they 'do purge downeward mightily ... with great force, and violence, and excesse', and so were suitable only for 'young strong and lustie people of the Countrie, whiche do set more store of their money then their lyues'.[1] They also seemed sufficiently chastening for monastic purges: when the latrine pits of the Benedictine Abbey at St Albans were excavated in the 1920s, great numbers of buckthorn seeds were found mixed up with the fragments of cloth the monks used as lavatory paper.

Alder buckthorn, *Frangula alnus*, is a smaller, spineless shrub, preferring damp peaty sites, heathy woods, carr and riversides, in Britain most often on acidic soils. It is 'alder' buckthorn because the roundish leaves bear some resemblance to those of the true alder. The charcoal made by burning the wood is reckoned to be the finest for making gunpowder, and the shrub has been cultivated in wartime.

Grape-vine family *Vitaceae*

Grape-vine, *Vitis vinifera*. A casual on rubbish-tips and waste ground (from discarded grapes), grape-vine is now becoming naturalised in hedgerows and scrub close to the increasing number of vineyards in southern England.

Virginia-creeper, *Parthenocissus quinquefolia*. This popular creeper from North America is naturalised here and there on old walls and in scrubby hedges near habitation.

Flax family *Linaceae*

Flax, *Linum usitatissimum*. Probably originating in cultivation from unknown parents, this is the flax that has long been bred for fibre and oil. Its blue flowers – whose petals drop in the early afternoon – have begun to reappear *en masse* in the British countryside as farmers look for new crops to replace surplus grain. The variety of flax currently being grown in England is used to produce linseed

Perennial flax, a scarce native of chalk and limestone grassland.

oil for the manufacture of linoleum and some cattle foods. It is an annual and persists at the edge of fields and on nearby road-verges.

Fairy flax, *L. catharticum* (VN: Purging flax), is a demure little annual of dry grassland, with small white flowers and wiry stems. It was used as a rather milder alternative to the fierce purging buckthorn (see above). Thomas Johnson, the apothecary who edited the 1633 edition of Gerard's *Herball*, learned about its uses from a colleague in Hampshire:

'I came to know this herbe by the name of Mill-mountaine, and his vertue, by this meanes: On the second of October, 1617, going to Mr *Colsons* shop an Apothecarie of Winchester in Hampshire, I saw this herb lying on his stall, which I had seene growing long before; I desired of him to know the name of it; he told mee that it was called Mill-mountaine; and hee also told me, That beeing at Doctour *Lake* his house in Saint Crosse a mile from Winchester, seeing a man of his haue this herb in his hand, he desired the name: hee told him as before; and also the vse of it, which is this: Take a hand full of Mil-mountain, the whole plant, leaues, seeds, floures and all, bruise it and put it in a smal tunne or pipkin of a pinte filled with white Wine, and set it on the embers to infuse all night, and drinke that Wine in the morning fasting, and hee said it would giue eight or ten stooles ... *Iohn Goodyer.*'[1]

Milkwort family *Polygalaceae*

Common milkwort, *Polygala vulgaris*, and **heath milkwort**, *P. serpyllifolia*, are rather low-growing, sometimes scrambling perennials quite common on both calcareous grassland and acid heaths throughout Britain. The flowers, which come in blue, pink or white, are shaped like miniature udders, which may account for the name and the carrying of the plants in the traditional processions of Rogation Day. This was the occasion on which the bounds were beaten, crops blessed and prayers offered up for the 'preservation and multiplying of the fruits of the earth'.

Milkwort was also prescribed by medieval herbalists for nursing mothers. There was doubtless a belief in the 'signature' of the plant behind this practice, but it can also be traced to herbalists identifying the plant with the 'polygalon' or 'much milk' of classical writers.[1] They assumed this meant 'mother's milk'; but more likely it referred to milk from the cattle grazing on milkwort-rich pastures.

Heath milkwort, which also occurs with pink or white flowers.

Horse-chestnut family *Hippocastanaceae*

Horse-chestnut, *Aesculus hippocastanum* (VN for fruits: Conkers, Cheggies, Obblyonkers). For a tree which is still a greenhorn in this country (it was introduced in the late sixteenth century), the horse-chestnut has made a huge contribution to popular culture. It produces 'sticky buds' (or 'cackey monkeys')[1] for vases in February and exquisite candelabras of blossom in May. The 'spreading chestnut tree' has been a symbol of village peacefulness, as well as the theme of music-hall songs and a 1930s dance craze. ('On the word "spreading" you spread your arms, on the words "chest" you put your hand on your chest ...')[2] Chestnut is one of the commonest components of street names (56 in the London *A to Z* alone). Its autumn fruits, conkers, are the raw material for what is still the most widely played children's game with plants. And these glossy red-brown nuts have added their own tally of words and metaphors to our vocabulary: 'chestnut' (shared with *Castanea sativa*) as a colour – especially for horses and hair; 'conk', slang for bash and also for head, which, in a neat conkery circle, is also called your 'nut'. And past-their-prime, over-played conkers are, of course, 'old chestnuts'.

The horse-chestnut is native to the Balkans and was first raised in northern Europe by the botanist Charles de l'Écluse in 1576, from seeds brought from Constantinople. John Gerard describes the tree well in his *Herball* of 1597 (though he is unlikely to have seen a live specimen and was relying on Lyte's description of 1578):[3] 'The Horse Chestnut groweth ... to be a very great tree, spreading his great and large armes or branches far abroad, by which meanes it maketh a very good coole shadow. These branches are garnished with many beautifull leaues, cut or diuided into fiue, sixe or seuen sections or diuisions, like to the Cinkfoile, or rather like the leaues of *Ricinus* [castor-oil-plant], but bigger. The flowers growe at the top of the stalkes, consisting of fower small leaues like the Cherrie blossom, which turne into round, rough and prickley heads.'[4]

Thirty-six years later a note by Thomas Johnson, in his edition of Gerard, indicated that the tree was now growing in the South Lambeth garden of the plant collector John Tradescant.[5] It was already called horse-chestnut

The spreading chestnut tree – a symbol of village peacefulness and a superb shade tree. Horse-chestnuts were introduced from Turkey in the sixteenth century.

in English, and the modern interpretation of the name is that it is analogous to horse-radish, cow parsley and dog's mercury, and signifies that this chestnut (and its fruits) are an inferior version of the sweet or Spanish chestnut. But it is called horse-chestnut in its native Turkey, too, and given to horses for food and medicine. So perhaps the name is a functional one, analogous to 'motherwort'. At the very least the trees are appreciated as shelter by horses and cattle. On the common pastures of Boxmoor, Hemel Hempstead, horse-chestnuts grow in groups of four, like wooden horses themselves. (There is also a horse 'sign' in the tree, spotted by adherents of sympathetic magic: when the leaves fall, the stalk makes a horseshoe-shaped scar on the branch, which carries small nail-like marks.)

Although horse-chestnuts now naturalise freely in woods and hedgerows and on waste ground, they were slow to escape from cultivation and were originally regarded as mysteriously romantic and powerful trees. An early planting is recorded in a local legend from Herefordshire: 'On the Kidderminster side of Bewdley near the site of the old Wribbenhall church, there are three horse-chestnut trees (now in an electricity sub-station) which are reputed to be planted on the site where the black plague [*c.* 1660s] victims were buried.'[6]

The majority of early plantings were in parks and the grounds of big houses. Capability Brown arranged 4,800 in the Tottenham Park estate in Wiltshire. Oxford colleges have been planting them for at least 200 years. The spectacular mile-long Chestnut Avenue at Bushy Park, north of Hampton Court, was designed by none other than Sir Christopher Wren and planted in 1699. It was originally intended as a carriage drive for William III from Teddington to the palace. But in 1838 Queen Victoria opened Hampton Court to the public, and the Park rapidly became one of the most popular spring playgrounds in London. Not long afterwards the tradition of 'Chestnut Sunday' began, and vast numbers of people, from all classes, would gather on a Sunday in mid-May, parade up and down the Avenue and picnic under the trees. The trees were nearly two centuries old by the end of Victoria's reign and deeply impressed one participant: 'the sight is certainly remarkable and worth seeing – a wide mile of lofty walls of foliage, bespangled with countless white spires – like tapering candles – and the boughs laden and almost sweeping the ground.'[7]

The custom flagged in the 1920s and died altogether with the onset of the Second World War. But it was revived again in 1977, the year of Queen Elizabeth's Jubilee, and has been formally observed by a group of enthusiasts ever since: 'Chestnut Sunday is [now held on] the nearest Sunday to 11th May, when the trees should be at their best. At 12.30 p.m. anyone who wishes to join in meets at the Teddington gate, says "The candles are alight", and walks down the avenue to the Diana Fountain where a picnic is held.'[8]

Despite the great fashionability of horse-chestnuts in the eighteenth and nineteenth centuries, and the numbers planted on big estates, the first record of the nuts being used in the game of conkers is from the Isle of Wight in 1848. There is an oddity about the lateness of this date, given that almost identical games had been played with objects on strings for centuries. There are descriptions from the seventeenth century of cobnut-fights, in which each player had hazel-nuts strung 'like the beads of a rosary' and exchanged strikes with his opponent.[9] In the eighteenth century boys played a game called 'conquerors' (which was even called 'conkers' in some places) with snail shells, sometimes with the unfortunate snails still inside. The shells were pressed against each other until one was smashed. The survivor was 'the Conqueror' and a tally was kept. The poet John Clare kept his shells threaded on a string and called the game 'cock-fighting' (cf. ribwort plantain, p. 322).[10]

Given the obvious suitability of horse-chestnuts for the game, why did it take children a century and a half to discover them? Jeff Cloves, author of *The Official Conker Book*, believes that the reason lies in the pattern and date of the early chestnut plantings. Most trees were nurtured

'Chestnut Sunday', a 1930 poster by C. Burton for London Transport, advertising the annual celebrations in Bushy Park.

'Obbly obbly onker, my first conker.'

on private estates, during a period of massive landscape rearrangement and draconian gamekeeping, and few children got the chance to gather the nuts with impunity. But, from the beginning of the nineteenth century, horse-chestnuts began to adorn the streets of spa towns such as Bath and Cheltenham and, from 1835, the public parks that were being created in big industrial cities. By the middle of the century the horse-chestnut had become a public tree, as common in urban open spaces, suburban streets and village greens as it was in big private estates.

Since then conkers played with horse-chestnuts has flourished, amongst girls and adults as well as boys. And, despite some regional variation in the jargon, the rules and rites, as the archivists of children's games, Iona and Peter Opie, discovered, are pretty standard throughout the country.[11] Gathering the nuts is the beginning. Some are just collected from the ground, but the best are believed to be at the top of the tree and are invariably helped down by barrages of sticks and stones. Prising the shiny mahogany fruits from their prickly cases is also part of the fun and is usually done with the help of some light pressure from a shoe. Flat conkers, which often grow in pairs, are widely known as 'cheesers' or 'cheese-cutters'; under-ripe ones (in Yorkshire at least) as 'water-babies'.[12] The most promising fruits are often artificially hardened, by baking in the oven or soaking in vinegar. (Though one Putney boy rinsed his in water afterwards: 'If you did not put them in water the smell would keep on the conker and then people would not play you because they would think it was harder than theirs.')[13] These days deep-freezing and microwaving have been added to the battery of favourite hardening tricks. But more patient children simply put a few conkers into a dark cupboard until the following autumn, by which time they are shrivelled and tough, and known as 'yearsies'.

To prepare a conker for combat, a hole is made through it with a skewer, and it is threaded onto a knotted string or shoe-lace. There are elaborate rituals, shouts and rhymes to determine who has first swing, but in essence the game proceeds by the two players having alternate strikes at each other's conker, or up to three shots in a row if the first two miss. The winner is the one who finally breaks his or her opponent's nut so that no pieces remain on the string. The triumphant nut becomes a 'one-er'. If it then breaks another first-time nut, it becomes a 'two-er', and so on. But if a two-er breaks, say, a tenner, it absorbs the other's score and becomes a twelver. (In a contest staged by BBC TV in 1952, the winner emerged as a 7,351-er.)[14]

Since 1965, a World Conker Championship has been held at the village of Ashton in Northamptonshire on the second Sunday in October. It is an apt site, as Ashton is a 'model' village created by Charles de Rothschild in 1900, with a horse-chestnut avenue a mile long up to the mansion and a fully grown tree transplanted to shelter the new smithy. From small beginnings the championship has grown into a major event, attracting crowds of over 4,000 and participants from all over the world. In 1976, it was won by a Mexican, R. Ramirez – the only time the title has left the country and a remarkable achievement at an occasion which is about as eccentrically English as it is possible to get.

Given the lively activity around roadside conker trees in autumn, it was predictable that some killjoy local authority (Lowestoft) would eventually plant a commemorative avenue of a horse-chestnut variety that bears no fruits. Less misanthropic kinds include very early leafers in, for instance, Calverton, Nottinghamshire, and Ilkley, Yorkshire,[15] and the red-flowered hybrid, *A.* × *carnea*.

Conkers also have less strenuous uses in children's play. Model-making is widespread: 'For me, as a child, the conker season meant collecting the glossiest and then making a suite of furniture for my dolls' house. This involved sticking four straight pins in the underside of the conker to make legs and an arc of pins around the top. With a length of coloured wool my mother showed me how to weave in and out of these pins to make the back of the chair.'[16]

Conkers are also still used to deter moths, and the soap-like chemicals (saponins) they contain are added to enhance the 'natural' image of proprietary shampoos and shower-gels. Conkers are mildly poisonous in excess (though most children nibble the hard, bitter flesh without ill effects), but in Victorian England there were recipes for making a 'strictly agreeable and edible flour', by grinding them and then leaching out the bitterness with hot water. During the two World Wars conkers were gathered for their starch, which was converted to acetone by a process invented by Chaim Weizmann. More recently, German scientists have discovered that aescin, extracted from the nuts, is a powerfully effective remedy for sprains and bruising – precisely the ailments in horses that the Turks have used conkers to treat. The British Forestry Commission expect eventually to be helping to establish 10,000 acres of plantation horse-chestnuts to supply the needs of the pharmaceutical firms – a strange twist of fortune for a homely tree of village greens and schoolboy games.

Field maple turns the brightest yellow in autumn of any native tree.

Maple family *Aceraceae*

Field maple, *Acer campestre*. This is a handsome tree, common across much of England and Wales, but often curiously overlooked. A mature maple on a hedge-bank, with its pale, furrowed, bossy trunk and dense crown of delicate, lobed leaves, is a picture of elegance and compact strength. A contributor from Somerset regards it as the signature tree of the local limestone woods: 'The autumn tone in the woods and hedges here is painted with brilliant colours as the maples turn first golden yellow and then vivid orange brown – truly a landmark tree.'[1]

'This lady and friends who were in their 20s at the time [1930s] often went to Whippendell Woods near Watford. One of the young lads found a lump of puddingstone [a local rock, like natural concrete] and lifted it up into the crook of a field maple for a lark. They often went back to look at it, but then must have forgotten. When this lady was much older, she and her husband showed me the tree, where the lump of puddingstone could still be seen, though the tree had done its best to grow round it. When I looked for it this year [1992] it was almost hidden.'[2]

Maple wood is tough and fine-grained and is used chiefly for high-quality carved or turned work, in musical instruments or ornamental bowls, especially the medieval drinking bowls called 'mazers'. Oliver Rackham's Cambridge college is proud of its 'Swan Mazer described in an inventory of *c.* 1380 as "one maser with lid . . . with broad silver bindings on the circumference and base of the bowl ... and in the middle of the bowl there is a silver-gilt column on which sits a gilded swan ...". The wooden bowl measures about 5 x 1¾ inches, and was doubtless a rarity because maple-trees seldom have bosses with such a large solid core ... The word "mazer" was apparently first applied to the bosses of any kind of tree, including birch. The bosses were thought to be pathological; etymologists derive "measles" from the same root. Later the name was associated with the maple-tree.'[3]

Maple has been the favoured wood for harps. A maple harp has been excavated from a Saxon barrow at Taplow in Berkshire, and another maple harp-frame, wrapped in a sealskin bag, was part of the treasure unearthed from the Sutton Hoo ship burial in Suffolk.

Representations of field maple leaves, which have that fingered shape so beloved of medieval carvers, can be found in Southwell Minster and on the pew ends in the Lackham aisle of St Cyriac's Church, Lacock, Wiltshire.[4]

Norway maple, *A. platanoides*, is widely planted, chiefly for ornamental reasons, and self-seeds in scrub, woodlands and hedgerows throughout lowland Britain. **Cappadocian maple**, *A. cappadocicum*, is a south-west Asian species, planted by roadsides and in parks and spreading by suckers in Surrey and west Kent.

Sycamore, *A. pseudoplatanus*, was probably introduced to this country from central Europe some time during the fifteenth or sixteenth century and has been squabbled over ever since. Even in the late seventeenth century, when the tree was still comparatively scarce, John Evelyn was making the kind of complaints that have become familiar throughout Britain in the past few decades: 'The *Sycomor* ... is much more in reputation for its *shade* than it deserves; for the *Hony-dew* leaves, which fall early (like those of the *Ash*) turn to a *Mucilage* and noxious *insects*, and putrifie with the first moisture of the season; so as they contaminate and marr our *Walks*; and are therefore by my consent, to be banish'd from all curious *Gardens* and *Avenues*.'[5]

Since then it has spread rampantly across southern and western Britain, especially after extensive planting at the end of the eighteenth century. Sycamore 'Mucilage' is almost certainly 'the wrong sort of leaves' which has caused British Rail's trains to skid to a halt every autumn. And the 'Hony-dew' (produced by aphids) makes tacky coats on the windscreens and bonnets of parked cars. Sycamore's newest antagonists are the nature conservationists, who see this exotic tree invading sacrosanct habitats – medieval churchyards, green lanes, even the ancient English greenwood – and shading out native species. As one Home Counties contributor (something of an admirer himself) reflects: 'The sycamore is often called a weed species. It certainly has one weed quality; it is a prolific seeder when mature. The winged, aerodynamic fruits are known as "locks-and-keys". If you would like a wood in your garden, then a nearby sycamore will oblige, without your lifting a finger ... Rattling into London along the suburban rail lines here it is again, defying the diesel fumes on trackside embankments and urban wasteland. The adaptability of the tree in Britain is so complete it is almost comic. Here is a Johnny-come-lately out-nativing the natives in almost every situation.'[6] In consequence many local conservation groups spend many working days pulling seedling sycamores by the thousand from their woodland nature reserves.

Yet the tree has an alternative and less disreputable history in Britain. There is an unmistakable carving of the leaves on the shrine to St Frideswide in Oxford Cathedral (carved in 1282), alongside native species such as maple, hawthorn and oak. Perhaps the carver knew the tree from

Sycamores, sheltering an isolated Pennine farm.

elsewhere in Europe. But it is possible that a few individuals were brought here two or three centuries earlier than is usually assumed. It may have been a case of mistaken identity at first. 'Sycomore' is properly a Middle Eastern species of fig, *Ficus sycomorus*, celebrated as a shade-tree in the Bible, and *A. pseudoplatanus*, no mean shader itself, may have been popularly believed to be the same species. Gerard, who knew the difference, described 'The great Maple' as 'a stranger in England, only it groweth in the walkes and places of pleasure of noble men, where it especially is planted for the shadowe sake, and vnder the name of Sycomore tree'.[7]

And planted as individual shade-trees, on village greens and parkland and close to isolated farms in the hilly areas of Wales and the Pennines, sycamores lose their 'weedy' aura and take on something of the grandeur they show in their native mountain habitat.

There are many surviving landmark sycamores. The comparatively young tree that breaks up the long curve of Oxford High Street between All Souls and Queen's Colleges was called by the planner Thomas Sharp 'one of the most important in the world'.[8] Another lone tree, featured by Sir Peter Scott in his painting *The Sycamore at Eastpark* (1974), still stands at the edge of the wet and windswept pastures of Caerlaverock, Dumfries, where barnacle geese winter.[9] And two sycamores at Haworth in Yorkshire 'are a worldwide landmark for the thousands of visitors to the moorland above the Brontës' village ... The trees are at the location written about by Emily Brontë in *Wuthering Heights*.'[10]

Less exalted, but no less valued locally, are the Posy Tree at Mapperton in Dorset (a village named after field maples, incidentally), on which a plaque reads: 'It was past this tree that the local victims of the great plague were carried to a common grave by the surviving villagers';[11] the Poo Hill Tree, used as a landmark on the pack-horse way between Lower Denby and Clayton West near Huddersfield;[12] and the Wishing Tree on Helm Common, Kendal. The tree grows by a bridle-path, and the custom amongst children when they walk under its branches is to make a wish, pick up a stone from the path, spit on it, and then add it to the adjacent drystone wall.[13]

'A well-known local tree ... grows alone on the top of Oker Hill [near Matlock, Derbyshire]. The story goes that years ago two local brothers each planted a sycamore on top of the hill when they were young, and hoped to see the trees grow to maturity. Unfortunately only one tree

Sycamore blossom, one of the brightest flowers of April.

survived and now is a fine specimen (two people can only just touch fingers round the trunk) whilst the other withered and died. This apparently reflected the brothers' futures – one flourished in business whilst the other was unsuccessful. Some versions say the second brother died just as his tree had done.'[14]

'At Dulverton, Somerset, there is a sycamore of enormous girth and a great age for the species. Called the Belfry Tree, it stands beside the Church Tower. It was struck by lightning during a hurricane in 1845, and lost the top of its crown. But the branches were braced, and recently they have been again strengthened with steel wires, while a drainpipe has been inserted by a tree surgeon to let excess moisture escape from inside the trunk.'[15]

The most celebrated sycamore of all is the Martyrs' Tree on the Green at Tolpuddle, Dorset. In the 1830s Tolpuddle was the birthplace of what was probably the first agricultural trade union, and it was under the village sycamore that the local farm labourers used to meet and talk – an illicit practice then that ended in the transportation of six of them to Australia. There is a story that their leader, George Loveless, took a leaf from the tree with him to Australia, pressed between the pages of his Bible. By the 1960s (when it was well over two centuries old) the tree had become a rather sorry specimen, seemingly dying of heart-rot, and it was adopted by the Trades Union Congress. They had the hollow trunk filled with vermiculite and bound with iron hoops and it now looks in vigorous health again. In 1984, on the 150th anniversary of the deportation, Len Murray, then General Secretary of the TUC, planted out one of its seedlings a few yards away on the Green.

(A more modern 'moot' sycamore is in Sheriff Hutton in Yorkshire, where 'a speculative builder was anxious to buy the Glebe Field and develop it as a building site. The villagers were determined that this should not happen, so called a meeting on the mound under the tree – and bought the field.'[16] It is now established as a parish conservation area in perpetuity.)

Hostility towards the tree because of its alien origins and aggressive habits meant that its positive contributions to indigenous wildlife were rather slow to be acknowledged. But it is now known, for instance, that, despite the initial sliminess of its fallen leaves, they decay very quickly and boost earthworm populations. Flower species characteristic of ancient woodland such as woodruff and wood anemone also thrive perfectly well on soils under sycamore. It is not particularly rich in insect species, but it makes up for this by having the highest insect productivity by weight (chiefly of aphids) of any widespread tree – 35.8 grams per square metre, as compared with 27.76 for oak and 11.15 for ash.[17] This is especially important in urban areas, where sycamore may be the only significant source of food for airborne insect-feeders such as house martins. But, because they are so frequently destroyed before they mature, very little is known about what would happen to a pure sycamore wood if it were given free rein. The seedlings are very vulnerable to fungal attack when growing under a sycamore canopy, and it seems likely that at a certain point a pure sycamore wood would begin to break down and admit other species.

Sycamore, in William Inchbold's painting 'In Early Spring', 1854–5.

The Martyrs' Tree at Tolpuddle, Dorset, the sycamore under which the local farmworkers held their union meetings in the 1830s, a practice for which they were later deported.

At present sycamore is better managed by coppicing than by attempted eradication. The regrowth is very fast, and it can be cut on a cycle as short as eight years. The wood is clean and pale, and liked by turners and makers of kitchen furniture: 'Sycamore is the best wood for kitchen tables. It has no odour to taint food processed on it, and a fine grain so that it can be easily cleaned. Rolling pins and baking boards also used to be made from it, and the best bread boards and wooden spoons still are.'[18] In Wales sycamore is carved into clogs, and is still the favourite wood for 'love-spoons'. These betrothal gifts, with their symbolically linked rings, were traditionally whittled out of a single piece of wood. (Today they have become part of a thriving if small-scale souvenir industry.)[19]

In the West Country sycamore leaves were used as bases on which to bake small cakes, either for Easter ('Revel Buns') or at harvest-time. The thick veins on the leaves make a distinct pattern on the underside of the buns.

'I do not know why they must be baked on sycamore leaves, but as a child I should not have thought them right if they had not had the imprint of the leaf under them.'[20]

'Years ago, my friend's grandmother [in Cornwall] used to send the children out to collect large sycamore leaves on which she baked the harvest cakes which were taken out to the fields for the harvesters.'[21]

Children have, as usual, been ingenious in their uses of all parts of the plant: 'In spring, when the sap was rising, we would take short lengths of sycamore twigs, about the thickness of one's little finger, and make whistles out of them. They would only last for a day before they dried out and became useless, so we used to make them more or less continuously until the bark became firmly attached to the underlying tissues.'[22]

The winged seeds – known as 'helicopters' in England, and 'backies' in Scotland, from their resemblance to small bats[23] – are used in flying competitions and model-making.

'Three generations of my family have made miniature water-wheels by impaling four single seeds of sycamore on a long blackthorn spike and resting them across two Y-shaped twigs over a stream.'[24]

'Edinburgh children play a game called "noses" in which you stick one of the wishbone-shaped seeds of sycamore on the end of your nose. The winner is the one who manages to keep it on longest.'[25] A similar game in England is called 'sticky noses'. The seeds are split and plastered down each side of the nose with the stalk pointing outwards, Pinocchio-style.

The 'Corstorphine Plane' is a sycamore sport with initially bright yellow leaves. The original tree, of unknown origin and planted out *c.* 1600, is preserved by the Corstorphine Trust in Edinburgh.[26]

Sumach family *Anacardiaceae*

Stag's-horn sumach, *Rhus hirta,* is a colourful shrub from North America, much planted in parks and on urban road-verges. It spreads freely by suckers and can build up quite large thickets, especially along railway lines. Its autumn colours are spectacular. The drooping leaves turn a fiery orange-crimson and the fruits, like unopened cones, a deeper red. ('Stag's-horn' refers to the texture of the shoots in spring, which are similar to a stag's horn 'in velvet'.)

The vivid autumn foliage of stag's-horn sumach, a North American shrub naturalised on many railway embankments.

Tree-of-heaven family *Simaroubaceae*

Tree-of-heaven, *Ailanthus altissima,* is a handsome tree, with ash-like leaves and panicles of strong-smelling creamish flowers, introduced to Europe from China in the mid-eighteenth century. The first British specimens were grown in about 1751 from seed sent to Philip Miller, of the Chelsea Physic Garden, by Père Nicholas d'Incarville. (Though until the tree fruited Miller obstinately main-

Tree-of-heaven. London specimens are a favourite source of nectar for urban honey-bees.

tained it was one of the Chinese lacquer trees.)[1] In parts of southern Europe it naturalises so freely that it has become a common part of open woodland and *maquis*. In Britain it is a much more restrained coloniser, but in south-east England increasing numbers of self-sown seedlings are appearing – and surviving – on waste ground and railway embankments. In London the oldest known wilding (on a river wall at Kew) dates from 1936. An opportunist sapling was even seen rising from a lidless bin left uncleared during one of the dustmen's strikes of the 1970s.[2]

It is especially numerous around Kensington, where planted trees are frequent in the nearby Royal Parks. And, though the flowers' nectar is regarded as acrid-smelling by many humans, it is relished by bees, and believed to be responsible for the muscat-flavoured honey that is occasionally found in west London beehives.

Wood-sorrel family *Oxalidaceae*

Wood-sorrel, *Oxalis acetosella* (VN: Alleluia, Cuckoo's bread and cheese, Granny's sour grass). Wood-sorrel, whose delicate, veined, white flowers usually appear between Easter and Whitsun, has long been called Alleluia, a plant which joins in the celebration of the Resurrection and Ascension.[1] But it was left to the Victorians to tease religious meanings out of its minutest habits. Charlotte Clifford, one of the family responsible for *The Frampton Flora*,[2] copied into her diary for 1860 a passage on the species from *The Garland of the Year*. This was a popular botanical annual at the time, and the passage is an extravagant but typical example of the contemporary custom of using the lives of plants (well observed, it must be granted) as moral parables:

'A more beautiful floral emblem of praise could not be selected than this exquisitely sensitive little plant. Coming forth at the first summons of spring it continues to adorn the woods with its bright triple leaves, until the fading foliage of autumn consigns it to a living grave. Even then, the flower-searcher may discover here and there a delicately-folded leaf looking out from the desolation and death by which it is surrounded. For the alleluya, fragile though it be, can brave the roughest gales, and weather the wildest storms, bowing its meek head beneath the clouds, and looking up with joy, to greet the sunshine. Sweet and precious are the lessons which this little woodland plant may teach us – lessons of humble faith, and constant loving praise. Teaching us that, as the shrinking wood sorrel finds protection in its triple leaves, so our souls, strengthened by the three-fold gifts of the Holy Ghost, should bow in meek submission to the trials of their mortal existence. Ever praising, never repining, bearing all sorrow; thankful for all joys!'

The trefoil, shamrock-shaped leaves are wood-sorrel's most distinctive feature. They lie in layered clumps in shady woods and hedge-banks, often growing directly on leaf-mould or cushioned by moss on fallen logs. When they first open they can look an almost luminous viridian, and they are folded back (their religious associations seemingly inexhaustible) like an episcopal hat. Then they open flat, three hearts with their points joined at the stem.

Gerard Manley Hopkins thought they were like 'green lettering', but their symmetrical clusters have more of the look of fretwork.

As might be guessed from its other surviving names, wood-sorrel has been used as a green vegetable, though it is slightly toxic if eaten in large quantities. The sour, lemon-sharp leaves are reminiscent of sorrel, *Rumex acetosa* (see p. 110). John Evelyn recommended them as a salad.[3] One forager from Lancashire adds them to cream-cheese sandwiches.[4]

Colour variants are sometimes found. A deep pink form has been seen near Abbey St Bathans in Berwickshire, and a purple in Torver, Cumbria.[5]

Increasing numbers of wood-sorrels from different corners of the globe are naturalising from gardens, chiefly in dry and stony places. **Procumbent yellow-sorrel**, *O. corniculata*, a species of uncertain origin, is widely established in gardens, on walls and in waste places, as are – though less prolifically – **least yellow-sorrel**, *O. exilis* (from New Zealand and Tasmania), and **upright yellow-sorrel**, *O. stricta* (from North America). The largest of the yellow-flowered wood-sorrels, **Bermuda-buttercup**, *O. pes-caprae*, comes not from Bermuda but from South Africa, and is a widespread weed of bulb-fields in the Isles of Scilly.

Crane's-bill family *Geraniaceae*

The crane's-bills, or wild geraniums, are an exceptionally attractive group of plants, albeit with rather similar habits and leaf and flower forms. They make an important contribution to natural landscapes, and many exotic species are grown in gardens, whence they have become naturalised in the wild. But beyond these not inconsiderable roles, they have not entered much into folklore or social life, despite the distinctive seed-cases shaped like a bird's bill.

Meadow crane's-bill, *Geranium pratense* (VN: Jingling Johnny), is the most widespread and striking of the larger-flowered crane's-bills. In the south of England it is commonest on, but by no means confined to, limestone

Wood-sorrel, blooming around Eastertime and popularly known in much of Europe as 'alleluia'.

Meadow crane's-bill, common on waysides in limestone areas.

grassland, and in June its bright-blue flowers, tinged with violet veins towards the centre, colour mile upon mile of roadsides in the Cotswolds and Wiltshire. Further north it is less choosy about soils, and can be found in damp meadows as well as hill pastures. In autumn the deeply cut leaves turn a rich red-brown.

Double-flowered forms and colour variants (including white) are available as garden plants from some nurseries. A pale-flowered form occurs 'rather plentifully' in the wild along the banks of the Whiteadder Water in Berwickshire, and a variety with flesh-coloured flowers more rarely near Helensburgh in Dunbarton.[1] The hybrid with *G. himalayense* is the famous 'Johnson's Blue'.

French crane's-bill, *G. endressii*, is a pink-flowered species from the Pyrenees, frequently and persistently naturalised throughout Britain, but rather less than might be expected from its aggressive spread in gardens. Self-seeded colonies in my own garden contain a wide range of

Three geranium species now frequently grown in gardens. Top: French crane's-bill. Centre: pencilled crane's-bill. Bottom: bloody crane's-bill.

colours, from crimson to very pale pink, and occasional plants with thin, sharply pointed petals. **Druce's crane's-bill**, *G.* × *oxonianum*, a natural cross between *G. endressii* and **pencilled crane's-bill**, *G. versicolor* (a cultivated and naturalised Mediterranean species), appears spontaneously at times. It is naturalised more rarely than French crane's-bill. **Wood crane's-bill**, *G. sylvaticum*, native in meadows and on river-banks and waysides mainly north from the Yorkshire Dales, is a beautiful, upright plant, usually with bluish-violet flowers. But colour variants occur, from pure white to pinkish-purple. It is sometimes known as 'thunder-flower' in north-west England.[2]

Bloody crane's-bill, *G. sanguineum*, is one of the great signature plants of the northern limestone, especially the Yorkshire and Derbyshire Dales, though it does grow on limestone cliffs elsewhere (e.g. the Avon Gorge, Great Orme and Gower in Wales) and on calcareous sand-dunes. There seems to be no other English name (and evidently none better: it has a pedigree of four centuries) than 'bloody cranesbill', but it is misleading, originally referring to the colour of the stalk-joints. The flowers themselves are a brilliant magenta, as big as a fifty-pence piece and visible a long way off, especially when the plant is scrambling amongst the creamy-white blooms of burnet rose (often in flower in the same places at the same time in June, see p. 193) and both are set against grey limestone. As the flowers fade, they go through a subtle colour change that was vividly described by John Gerard in 1597: 'the flowers are like those of the wilde mallowe, and of the same bignesse, of a perfect bright purple [corrected to 'red' in the 1633 edition] colour, which if they be suffered to growe and stande vntill the next day, will be a murrey [mulberry] colour; and if they stand vnto the third daie, they will turne into a deepe purple tending to blewnesse; their changing is such, that you shall finde at one time vpon one branch, flowers like in forme, but of diuers colours.'[3] A pale-pink-flowered variety, var. *striatum* (still known to gardeners as var. *lancastriense*, the name given to it by Philip Miller in the 1730s), grows very locally on sand-dunes in north-west England.

Cut-leaved crane's-bill, *G. dissectum*, is a common native species of stony places and cultivated ground, with small pink flowers. **Purple crane's-bill,** *G. × magnificum*, is a hybrid of garden origin, something like a compact, more deeply coloured meadow crane's-bill, which is quite widely naturalised close to habitation. **Hedgerow crane's-bill,** *G. pyrenaicum*, first recorded in 1762,[4] has pinkish-purple flowers and rounded leaves and is quite common in hedgerows and grassy places, chiefly in southern and eastern England. **Dove's-foot crane's-bill,** *G. molle*, another round-leaved species, with smaller pinkish-purple or off-white flowers, is common in hedgebanks, arable fields and waste ground. **Rock crane's-bill,**

Herb-Robert, mouse-scented and ubiquitous.

The 'mourning widow', naturalised from central Europe, often close to churchyards.

G. macrorrhizum, is a species from southern Europe, with sticky, pungent, almost peppermint-scented foliage. Several varieties are grown in gardens, and very occasionally naturalise close to habitation. **Shining crane's-bill**, *G. lucidum*, is a pert annual of walls, rocks and stony banks, chiefly on calcareous ground. It is the leaves that are shiny.

Herb-Robert, *G. robertianum*, is a very common plant of woods, waysides, walls and cultivated ground, known as much for the acrid, mousy stench of its leaves as for its perky pink flowers, which can be found in almost any month of the year. Its name probably derived originally from an ancient association (perhaps because of its smell) with the house goblin, Robin Goodfellow. White-flowered forms are not uncommon – e.g. in the Yorkshire and Derbyshire Dales, Upper Teesdale and sites in Dyfed and Clwyd.[5] Graham Rice points out that an increasing number of variants of white herb-Robert are becoming available as garden plants: a pure white form known as 'Celtic White', with fresh green leaves that lack any of the red pigment that usually suffuses the stems and leaves; 'Album', trailing, strongly pigmented in the leaves, and with large white or pale pink flowers; and 'Cygnus', with white flowers and only slight pigmentation in the leaves and stalks.[6]

Little-Robin, *G. purpureum*, a smaller version of herb-Robert, as its name suggests, is very scarce in south-west Britain on shingle, rocky places and cliffs near the sea.

Dusky crane's-bill, *G. phaeum*, named 'mourning widow' by gardeners because of its dramatic dark-purple flowers, has the nearest thing to black blooms in the wild-flower world. It is a species from central Europe, widely grown in gardens and quite often naturalised in shady places, especially in or close to churchyards – perhaps because the funereal associations of its colour have made it a grave ornament. Two London sites where it is well established are the churchyards of St James's, Piccadilly, and St Stephen's in South Kensington.[7] It occurs occasionally in a pinkish-mauve form (var. *lividum*) and more rarely still in white.

English Munich crane's-bill, *G.* × *monacense* var. *anglicum*, is a hybrid between *G. phaeum* var. *lividum* and *G. reflexum* which may have originated in England, and which was first discovered in the wild by Patrick Roper on a roadside in Hurst Green, East Sussex, in 1975. It has vanished from that site but has been grown on in his garden, and he believes it should be re-established in its original location.[8]

Stork's-bills, *Erodium* species, are similar to the smaller crane's-bills, but with longer seed-cases (from which they derive their name), and are generally found in dry, bare places on sand. There are four native species.

Balsam family *Balsaminaceae*

Touch-me-not balsam, *Impatiens noli-tangere*, is the only native member of this family, which acquired its name from the habit of the ripe seeds of bursting violently when touched and hurling their seeds away. (Hence Erasmus Darwin's excruciating lines: 'With fierce distracted eye Impatiens stands,/ Swells her pale cheeks and brandishes her hands,/ With rage and hate the astonished grove alarms/ And hurls her infants from her frantic arms.') Touch-me-not is a scarce and local plant of damp woods in the Lake District, the Shropshire–Montgomery border and the Dolgellau area, and was first discovered by George Bowles on a botanising trip in 1632 on the banks of the River Camlad in Shropshire (it is still there). Thomas Johnson called it 'codded or impatient Arsmart', believing it to be related to the persicarias.[1]

Orange balsam, *I. capensis.* This attractive waterside plant from North America has the distinction of being first noticed in the wild in Britain by the philosopher John Stuart Mill. In the summer of 1822 he found it growing on the banks of the River Tillingbourne near Albury in Surrey – naturalised, it is believed, from the gardens of Albury Park upstream.[2] Over the next 20 years it found its way to Hertfordshire and Middlesex, especially along the Grand Junction Canal, and by the end of the nineteenth century it had reached Glamorgan in the west and Sussex in the south.[3] Today it is found as far afield as Yorkshire and Devon.

Orange balsam, first discovered in the wild in Britain by the philosopher John Stuart Mill in 1822.

Orange balsam, painted 21 years later. The sadly obsolete name 'quick in hand' refers to the explosive seed-pods.

Orange balsam was able to achieve this rapid colonisation of the waterway system because of the family's highly efficient seed dispersal mechanism. Its seeds are light and corky enough to float on water, and, once launched, they float off like tiny coracles until they lodge in a muddy bank. Where there is no way through by water, birds' feet and anglers' boots no doubt help them on their way. Orange balsam is known as 'jewel-weed' in its native America, and it must count as one of the most welcome additions to our flora. It is hung with large numbers of nasturtium-like blooms – orange lanterns flecked inside with reddish-brown.

Small balsam, *I. parviflora*, is a yellow-flowered balsam introduced from Russia about the middle of the nineteenth century, and now naturalised in damp, shady places throughout Britain. It can build up quite large colonies, even sometimes in long-established woodland.

Indian balsam, *I. glandulifera* (VN: Policeman's helmet, Stinky pops, Jumping Jacks, Bee-bums, Poor-man's orchid). Indian, or Himalayan, balsam is the most recently arrived of the family but already the most widespread and conspicuous. It was introduced to gardens from the Himalayas in 1839, and by the end of the nineteenth century had become widely naturalised, especially along rivers in the West Country. A. O. Hume found it in the Looe Valley in Cornwall, when it was still something of a novelty in the wild, and wrote the classic description:

'Growing in the soft warm south-west, with the base of the stem in the clear running stream, it is a magnificent plant, 5 to 7 feet or more in height, stalwart, with a stem from 1 to 2 inches in diameter just above the surface of the water, erect, symmetrical in shape, with numerous aggregations of blossom, the central mass as big as a man's head ... masses of bloom varying on different plants through a dozen lovely shades of colour from the very palest pink imaginable to the deepest claret, and with a profusion of large, elegant, dark green, lanceolate leaves, some of them fully 15 inches in length. Stunted specimens of this Balsam are common in Cornwall in orchard and cottage gardens; but in the Upper Looe the plant has become thoroughly naturalised, and I have never seen it quite as fine even in its native habitats.'[4]

Indian or Himalayan balsam. It can hurl its seeds up to 12 yards.

Aggressive, colonial and frequently growing up to 10 feet tall in a year, it now dominates areas of river-bank throughout Britain, especially close to towns, sometimes to the point of swamping other summer-flowering species. The reason for its rampaging spread is the same as for other balsams: its seeds are fired off explosively and carried along by water. But, for all its territorial ambitions, many people still have a soft spot for Indian balsam, and it has been given a deliberate helping hand in some places, for instance in Somerset: 'My favourite plant. I am not quite sure if it is indigenous to this area, as I was given mine from somebody's garden. I have thrown the seeds in various hedges and ditches while walking the dog, so it soon will be! ... I was told that it was known locally as bee-bums and having noticed that bumblebees are attracted to it, and that their bums are all that is seen of them while they are on it owing to the shape of the flower, I thought it was an extremely good name.'[5]

'A great friend of mine married a Nobel prizewinner who was also a botanist. He took seeds from our Himalayan Balsams and would scatter them around – some on the banks of the Sheal where it flows through Sheffield, and some along the river banks in Derbyshire where he took his family for country walks. He *may* have been the one to introduce them to the Derwent Valley, *c.* 1946.'[6]

'When my sister dropped a rattle in the garden it split open and what looked like black seeds fell out. They were obviously old, but we planted them and they grew into large balsam plants. Ever since, the garden has been full of them, and this has been going on for almost 33 years.'[7]

It is, like many expansive foreign plants, viewed with mixed feelings, as in Worcestershire: 'Himalayan balsam has become a conservationists' nightmare in Worcestershire – the local Wildlife Trust regularly organises "balsam bashing" parties at our marshland reserves. However, in the Black Country, around Dudley and Stourbridge, it is a welcome addition to many industrial streamsides, where its gaudy pink flowers clash with the seepage of iron oxide from the factories. In the late 1960s my grandmother used to call it "poor man's orchid" [see lady's-slipper orchid, p. 439] – a name also used by people in the Halesowen and Stourbridge area.'[8]

Even the scent is debatable. In the Calder Valley in Yorkshire, for instance, where the Countryside Service also runs 'balsam bashes', one walker enjoys the 'heavy scent' of the mass of plants in the woods.[9] But in Cheshire, 'the River Mersey and its banks are infested with "Mersey Weed". This seems to be completely impervious to occasional industrial pollution, and has a pervasive evening scent reminiscent of Jeyes fluid.'[10]

Despite its comparatively recent arrival, children have rapidly learnt to exploit the potential of 'popping' the seed-pods, and at least one of the modern vernacular names ('stinky pops', Hertfordshire) was coined by schoolchildren. The secret of spectacular explosions is to choose a pod that is on the point of bursting spontaneously under its own tension. In the Lake District some children achieved what must be a record 12-yard throw with the seeds from one pod.[11] But more sophisticated, and sensuous, games have been invented, too: 'If a bunch of these seed-pods are held in the hand before they explode, and the hand is clenched, squeezing the seeds, there is a horrible "squirming" effect felt in the hand, like a lot of wriggling insects.'[12]

As for adults, if they feel above such things, they can always munch the pods: 'We always eat the seed-pods and ripe and unripe seed which have a pleasant nutty taste (but first catch your pod!). We first learned of this plant from an old friend who had undertaken an exploration of the Himalayas for the RGS. They had nothing but tinned food, except that he and his porters picked and nibbled balsam seeds. His wife, who would not eat them, became more and more ill and on reaching civilisation was found to be in an advanced state of scurvy.'[13]

Ivy family *Araliaceae*

Ivy, *Hedera helix*. Attitudes towards ivy have been ambivalent since classical times. In the days of sympathetic magic, ivy's ability to smother grape-vines persuaded early herbalists that its berries could overcome the

Ivy, a gothic, softening cover for a dead tree (Gresham, Norfolk) ...

malign effects of alcohol. Goblets were sometimes made out of ivy-wood to neutralise the toxic effects of bad (or poisoned) wine. In medieval times ivy and drink became such bosom companions that ivy-covered poles, known as 'ale-stakes' or 'bushes', were used to advertise taverns. The bigger the ale-stake, the more ambitious the inn-keeper. (Hence the expression 'Good wine needs no bush'.) In 1375 an Act of Parliament was passed to restrict their height to seven feet.[1] But, though it was a toper's mascot, ivy was also seen as a weak, 'feminine' plant, contrasted in mythology and poetry with the red-blooded, prickly holly.

During the fashion for the Picturesque in the eighteenth century, ivy became a token of melancholy, especially when draped over ruins. But at the same time it was included in the festive garlands brought into houses at Christmas time. Today, it divides geographical regions and professions. In the gale-dashed West Country and the territories of commercial foresters it is regarded as a curse and a killer of trees. In less turbulent eastern regions and amongst more easy-going gardeners and naturalists it is looked on benignly, as an ornament to buildings and woods and a boon to birds and bees.

Myths about the plant persist, and the facts of its life-style are worth setting down at the outset. Ivy is our only evergreen liana, a clinger and hanger-on. But it is not even a partial parasite and manufactures all its own nourishment in the same manner as other plants. Although it is most often seen as a climber, it is perfectly happy to creep about the ground and in its maturity to stand on its own roots. It uses trees and walls simply as scaffolding, clamping itself on by means of a mat of adhesive suckers. Only when these encounter soil or deep crevices does it put out true, feeding roots. But when ivy reaches the top of a tree (and it has been known to climb more than 100 feet) it begins to bush out and can then begin to cause trouble by the sheer weight of its foliage or by shading out the tree's own leaves. This is when it also begins to bring out our own loyalties and prejudices:

'I was raised on a moorland farm on the Bodmin Moors, and I was taught that there were two types of ivy. The type that competed with the tree for food below ground, but only used the tree for support. And the type that not only competed for food below ground but also sunk its aerial roots/clingers into the tree and sucked the sap from the tree, thus weakening it.'[2]

'It is a somewhat sweeping statement that ivy does no harm to trees. Here in the West Country [Cornwall], one important factor is the wind, and in a gale a tree that is heavily laden with ivy is like a fully-rigged ship, unable to lower its sails. Over it goes!'[3]

'When clearing ivy from fallen trees I have found that the weight of ivy exceeds the weight of its host.' (Dorset)[4]

'As a Tree Surgeon I dislike ivy and she doesn't like me. It will take 30 years to smother a tree and I have seen trees killed off by too much ivy. Its roots will strangle its host and it will in time almost ring-bark it, causing the branch to swell up on either side. The dead leaves can't drop from the host and the ivy will put its roots into the rotting debris. Any cavities are covered over ... It grows nearly eighteen inches a year. Its dark growth stops the light coming through even in winter. As you can see I like taking ivy off trees to give them freedom.'[5]

'I spent several weeks climbing into the yew trees in our churchyard and cutting down the ivy, which had such a grip that in some cases I had to use a crowbar to remove it . . . It had left deep weals where it was wrapped round the branches.' (Dorset)[6]

... a garden gate (Coverack, Cornwall) ...

One parish priest from central Wales – perhaps seeing similarities between ivy and the serpent – would go even further: 'Throughout England and Wales, hundreds of thousands of trees can be seen in various stages of total destruction by ivy. How futile it seems for people to be asked to plant a tree when the present trees are being killed. I challenge honest people to open their eyes as they travel through every part of England and Wales to see this terrible destruction of precious trees ... It would be splendid and so merciful to trees if there could be a campaign to kill the ivy around our trees to save them.'[7]

But there are more sympathetic views, even amongst tree surgeons: 'Although ivy is a superb groundcover, creeper and wildlife habitat, it can and does cause problems for trees ... Our solution is: where ivy is heavy within the crown, we remove it. On the trunks, we trim it back and leave it. Only in certain situations where there is a possible risk to the public do we remove it completely from the tree. In parkland and where the public do not often go, it is quite alright to leave ivy on trees as a wildlife habitat.' (Derbyshire)[8]

'As a bee-farmer, I have found that ivy provides the last main source of nectar and pollen for bees to top up their winter stores in mid-September, when the heather has finished flowering.'[9]

The late-flowering of ivy from September to November (when it can roar with bees as loudly as a lime tree in July) and the dark berries that last through until spring have made it attractive to gardeners as well as to insects and birds. In the eighteenth century it was regarded as the classic Gothic plant, and the landscaper Thomas Whately's description of Tintern Abbey graphically captures its contribution to the genius of the place: 'The shapes even of the windows are little altered; but some of them are quite obscured, others partially shaded, by tufts of ivy, and those which are most clear, are edged with its slender tindrils, and lighter foliage, wreathing about the sides and the divisions; it winds round the pillars; it clings to the walls; and in one of the isles, clusters at the top in bunches so thick and so large, as to darken the space below ... No circumstance so forcibly marks the desolation of a spot once inhabited, as the prevalence of nature over it.'[10]

'The Victorians loved ivies and encouraged them to cover their summer houses and garden bowers. They trained and trailed them all through their homes, growing them outside and then encouraging them to grow through an open window ... Queen Victoria as a young wife wore a wreath of real Osborne Ivy, interwined with diamonds, in her hair.'[11]

'House-ivy' is now viewed as ambivalently as that on trees. One taken into a house in Manchester as a tiny cutting has 'grown right across the fireplace, round the pictures. We are wondering if it will ever stop.'[12]

'I live in a semi-detached house which has been "taken over" by ivy. It has grown up the walls over many, many

... and a headstone (Burnham Overy, Norfolk).

John Ruskin's 'Study of Ivy', c. 1872.

years and is now covering the roof. I cannot get rid of any of it, as it is firmly attached to the pebble-dashing. If I saw off the trunks at the base it will be brown and look unsightly, and I cannot remove the individual branches as they pull off the pebble-dash in great lumps.' [13]

'We have a rather superb ivy hedge; it must be about eight to nine feet tall; it was obviously planted to cover a fence. It is now holding the fence UP! It is wide; a fox has been seen to climb up through it and sunbathe on the top.' [14]

'[We have] an "ivy tree" – about nine feet in diameter and about ten times the height of a man. It is surrounding a hazel tree – hidden inside it, but still healthy, because its branches stick out all around it. Everyone round here calls it the "ivy tree".' [15] (The ability of ivy to survive as a free-standing shrub has long been known by gardeners, as is the fact that it can be deliberately propagated in this form. Cuttings taken from an ordinary, ivy-leaved climbing or scrambling shoot produce a climbing or scrambling plant. But cuttings from the flowering branches produce a so-called 'tree-ivy', which makes a small, upright, self-supporting bush with simple oval leaves.) [16]

Ivy's heartwood is a glossy cream and dries to the tint and surface texture of ivory. Branches are often used to give a driftwood effect in flower arrangements. In several craftshops in Wales, sculpture is made from thick sections of the trunks. [17] Being naturally forked, it is often used to make rough pitchforks, especially on the continent.

'Some time ago I came across a large trunk of ivy cut through at ground level and hanging from an oak tree. A friend told me that ivy was once used for pastry rolling pins, so as he used a lathe I asked him to spin three and keep one himself. They are still in use and pastry does not stick to them as much as to other wood.' [18]

Ivy berries and leaves have also had decorative and domestic uses: 'As a child during the war when cultivated holly was difficult to obtain, and the wild species practically non-existent in the Fen country, it was common practice to collect ivy for Christmas decoration and paint the berries red. However, I have discovered this painting of ivy berries was a much earlier practice in this area. John Clare, the Northamptonshire poet and a great recorder of the flora and fauna of his native heath and fen, makes the

Ivy, painted with flowers and berries in Frampton, Gloucestershire, c. 1840.

following observation: "hasten to the woods to get ivy branches with their chocolat berries which our parents used to color with whiting & the bluebag sticking the branches behind the pictures on the walls".'[19]

In Chudleigh, Devon, ivy from woods and hedges was called 'coloured ivy' because of the different leaf-shades, which were sought after for home decoration.[20]

'When my mother was born in 1885 at Inverkip, Ayrshire, her father was a policeman. I was told that my grandmother used to boil up ivy leaves to use as a colour restorer when his uniform looked shabby.'[21]

'In the early years of my married life (in the 1940s) I used to press my husband's suits with a damp cloth wrung out in a solution made from boiling water and ivy leaves. It certainly took the shine off suits, which had to last for ever – clothes coupons and finances making that necessary then. I think Gilbert Harding in a Brains Trust Programme on the wireless passed on that tip.'[22]

Ivy completely capping the wall of a derelict church at Ufton Nervet, Berkshire.

'As a child I attended a school in Wales where we had to wear navy blue serge gym slips as a school uniform – saddle, yoke, and three box-pleats back and front. I never remember these being washed, but during the school holidays we gathered the darkest green ivy leaves and our mothers poured onto them boiling rainwater, left it to steep overnight and sponged down our tunics with the liquid. Garments were cleaned of all grease and looked like new.'[23]

'I am 70 years of age. When I was about 10 or 12 years of age I had verrucas on one of my feet. Each morning I was sent to a neighbour's back yard to get two ivy leaves. These I put inside my white school sock. I wore this all day at school. This continued for about two weeks, and then there were two small very clean holes in the sole of my foot where the verrucas had been, no pain at all.'[24]

'Ivy was used to charm warts away in Essex up to the 1950s. A hole was pricked for each wart in an ivy leaf, which was then impaled on a thorn in a hedge or bush.'[25]

But the most frequent uses of ivy which still seem touched by ancient beliefs about its magical powers are in cattle-farming. Ivy has always been browsed by domestic animals, and it is sometimes still used as emergency winter food, but it is seen as having a protective role as well:

'In the Highlands and Islands ivy was plaited into wreaths with rowan and honeysuckle as a good-luck charm. It was especially good at keeping evil away from milk, butter and cows, and would be hung up on the lintel of the byre or hidden beneath the churn. In the Hebrides, people went to great lengths to collect the necessary plants, one man in the Uists swimming out regularly to an island in the Loch.'[26]

'My husband (a livestock farmer) tells me that sick animals suffering from poisoning, e.g. through eating yew or ragwort, will eat ivy when they refuse all else. We wonder, therefore, whether ivy has a medicinal or purgative property. In his experience livestock when healthy will only eat ivy when no other forage is available.' (Shropshire)[27]

'In the sick pen of the lambing shed I place a branch of ivy leaves to entice a ewe to find her appetite after a difficult birth or illness.' (Devon)[28]

'I have found that my cattle do exceptionally well when they have access to ivy. Also about four years ago I had an outbreak of New Forest Eye, and the cattle were moved into a wooded area with a lot of ivy and the eye trouble cleared up very quickly.' (Isle of Wight)[29]

'[In the 1930s] on Xmas morning, as soon as milking

Ivy flowers are one of the few plentiful sources of nectar for late-flying butterflies and bees in September and October.

was over, we had to go out collecting ivy, and every animal, young stock as well, had to have a piece before 12 o'clock. The belief was that kept the Devil away for 12 months.' (Shropshire)[30]

The inclusion in Christmas greenery and rituals is the high point of ivy's ceremonial year. But it has been used in children's games and divinations:

'An Ivy leaf left on New Year's Eve in a dish in water and left untouched till 6th Jan. had some significance. If it was fresh or green the year would be happy. Black spots meant illness – according to where they were found – near the point, pain in feet and legs; near the middle, pain in stomach, etc. If it withered, the one who played it would die.'[31]

'When I taught at a school in Ruthin, Clwyd, on May 1st I found many children wearing an ivy leaf: "First of May is pinching day." The ivy leaf was to prevent being pinched by another child.'[32]

At Denbigh (also Clwyd) 'you could hide your ivy leaf and, if someone pinched you, you could reveal your leaf (usually hidden down a sock) and pinch the pincher ten times!'[33]

'When I was young in Rosneath village, schoolgirls used leaves of Irish ivy [see below] as a forecaster of their marriage prospects, singing the verse collected by the Opies from all over England and Scotland, and beginning "Ivy, ivy, I love you".'[34] (The full rhyme, as quoted by the Opies, is 'Ivy, ivy, I love you,/ In my bosom I put you./ The first young man who speaks to me/ My future husband he shall be.')[35]

But the best known of all ivy rhymes, the carol 'The Holly and the Ivy', remains something of a conundrum:

The holly and the ivy,
When they are both full grown,
Of all the trees that are in the wood,
The holly bears the crown.

A 'false spruce' – the framework of the dead tree totally cloaked by evergreen ivy.

'Coloured ivy' – often used in flower arrangements.

On the surface it is almost a nonsense verse. Holly is one of the smaller woodland trees; as a model for the crown of thorns it is beaten easily by blackthorn and hawthorn. Ivy is not a tree and makes no other appearance in the full carol, which is devoted to the Christian symbolism of holly. The editors of *The New Oxford Book of Carols* believe that the best-known stanza was the chorus from an older carol, tacked on as a verse by a Birmingham broadside publisher around 1710.[36]

A white witch from Yorkshire suggests, on the other hand, that the carol is a satire on the battle of the sexes; also that ivy is banned as a decoration in churches.[37] Although this was undoubtedly true in a few parishes because of ivy's traditional associations with drink, there are fine medieval carvings of it on supporters in both Westminster Abbey and Wells Cathedral.[38]

But there are secular medieval poems which do suggest that the red-berried, festive holly was seen as a man's plant, and the entwining, black-berried ivy as a woman's. One contains the stanzas:

Ivy bereth beris
As blak as any sloe.
There commeth the woode colver [pigeon],
And fedeth her of tho;
She lifteth up her taill
And she cakkès or she go;
She wold not for an hundred pound
Serve Holly so.

Holly with his mery men
They can daunce in hall;
Ivy and her jentell women
Can not daunce at all,
But like a meine of bullokès
In a water fall,
Or on a hot somers day
Whan they be mad all.[39]

How would the medievals have coped with the knowledge that holly's berries are borne on the female tree?

Atlantic or **Irish ivy**, *H. helix* ssp. *hibernica*. This is the commoner plant in many places in the extreme west and south-west of Britain. Its large and uniform leaves, rapid growth and lack of inclination to climb walls and trees have made it popular as ground cover with gardeners and landscapers. In the wild it can cover cliffs, and even grow on shingle banks.[40]

Carrot family
Apiaceae (or *Umbelliferae*)

Marsh pennywort, *Hydrocotyle vulgaris*. This creeping perennial of damp meadows, marshes and watersides throughout Britain was formerly believed to cause liver-rot in sheep (cf. bog asphodel, p. 400). The leaves are roughly circular in outline, with the stems emerging from the centres.

Sanicle, *Sanicula europaea*, grows in deciduous woods, especially of ash or beech on calcareous soils. Its tight, round clusters of white or pinkish flowers don't obviously resemble the flat, open flower-heads – 'umbels' – usually associated with this family until you look at them closely. The name 'sanicle' derives originally from the Latin *sanus*, meaning 'whole' or 'sound', and it was once popular as a wound-herb.

Sea-holly, *Eryngium maritimum*, is an unmistakable

Marsh pennywort, New Forest.

plant of sand-dunes and shingle. It rarely grows more than a foot or so tall, but its holly-like leaves are a pale, frosty blue, as if they had been coated with a thin layer of ice. In fact they are covered with a waxy cuticle, a device to help the plant retain water in salt winds and seaside sunshine. The flowers are bright powder blue, contrasting wonderfully with the leaves, and it is no surprise that the species – unimproved – is popular in gardens on dry soils. It may be that digging-up from the wild has been partly responsible for sea-holly's gradual decline around our coasts (it has virtually vanished from Scotland and north-east England), though it was always a local plant.

Cow parsley, *Anthriscus sylvestris* (VN: Queen Anne's lace, Lady's lace, Fairy lace, Spanish lace, Kex, Kecksie, Queque, Mother die, Mummy die, Step-mother, Grandpa's pepper, Hedge parsley, Badman's oatmeal, Blackman's tobacco, Rabbit meat). Cow parsley is arguably the most important spring landscape flower in Britain. For nearly all of May, almost every country road is edged with its froth of white blooms. Regions where bluebells are embattled in private woods and buttercups sprayed out of the fields are still ornamented by mile upon mile of this indomitable, dusty smocking. It is odd that its rather dismissive English name – which simply means (in reference to the leaves) an inferior version of real parsley – has stood out against all comers. Queen Anne's lace sometimes makes a bid to replace it but has never become widely used, despite no end of elaborate stories to explain its origin as a name:

'Queen Anne's Lace is so called I understand because when Queen Anne travelled the countryside in May the people said that the roadsides had been decorated for her.'[1]

'The story is that Queen Anne, who suffered from asthma, used to come out to the countryside around Kensington, then open meadow and farmland, to get fresher air. As she and her ladies walked along the country lanes in spring sunshine, they carried their lace pillows and made lace. The flowering cow parsley, with its beautiful, lacy flowers, resembled the court ladies' lace patterns, and so the country folk began to call it Queen Anne's Lace, a name which persists today.'[2]

'Queen Anne's Lace is generally understood to refer to its lace-like appearance, but also to her (Queen Anne's) tragic child losses.'[3]

All the explanations have a rather contrived feel about them, and it is more likely that the name is an import from North America, where cow parsley is widely naturalised. A Warwickshire woman's experience lends wry circumstantial evidence to this possibility: 'At the Warwickshire Trust gift shop where I work as a volunteer, we sell silver

The frosted leaves and powder-blue flowers of sea-holly, now a popular border plant.

pendants and brooches containing pressed wild flowers. One morning two Alaskan men came into the shop, and selected pendants containing tiny white flowers on a black background – very striking. The problem came when they wanted the name of the flower. It was cow parsley. They were doubtful if this would convey the right message to Alaskan women. We ventured the alternative name of Queen Anne's Lace (hoping we were right). Big smiles. The flowers immediately looked *much* prettier and we had two satisfied customers.'[4]

At any rate Queen Anne's lace hasn't caught on, and perhaps the unpretentious 'cow parsley' is the best name for this truly vernacular blossom. It certainly has not diminished its rising reputation as a decorative flower. The sprays work very well in a vase, looking spacious and balanced, and keeping both their shape and blossom for more than a week. They are becoming popular in church decoration, too;[5] and I have seen them used in the flower arrangements for a May wedding at Selborne, Hampshire.

The old dialect term 'kex', which has several derivatives (see above), is still in use, but it is also applied to other umbellifers, including hogweed and hemlock (see below), and especially to the dried stalks of these plants. Its origins are unknown.

'Mother die' remains the most widely current – or at least most remembered – alternative name, the implication being that if you pick the plant your mother will die. Perhaps this is guilt by association, resulting from the slight similarity between the scents of cow parsley and may, to which a similar superstition is attached. But several potential victims have suggested that for them it is simply a useful warning tag, which they have passed on to their children to discourage the picking of *any* umbellifers – a family full of deceptively similar edible and toxic species. Just how confusing the family can be is illustrated by this story from Yorkshire, of cow parsley and pignut (see below) being taken for the same species: 'When I was a child we used to dig for hours (with penknives) underneath "Mother Die" plants, otherwise known as Queen Anne's Lace. If we were *very* lucky, we found something resembling an oddly shaped nut – a "cat nut" – which was similar in taste to hazel nut or celery. Incidentally I have

Cow parsley: humbly named, but no other plant gives so much character to country roads in May.

Shepherd's-needle, a scarce arable weed named from its long seed-cases.

dug under Mother Die this year, but never found the delicious cat nut of childhood.'[6]

Properly identified, in fact, young cow parsley leaves (it is a relative of the herb, garden chervil, *A. cerefolium*, which occurs occasionally as an escape) are a fresh and mildly aromatic addition to salads and omelettes.

The sequence of wayside 'parsleys', fondly memorised by many wild-flower lovers, is that cow parsley is followed by the slightly scarcer, less floriferous and wirier **rough chervil**, *Chaerophyllum temulum* (May to July), and the **upright hedge-parsley**, *Torilis japonica* (July to September).[7]

Shepherd's-needle, *Scandix pecten-veneris*, is a rare annual of arable fields, distinguished by its long, splayed seed-cases, which John Gerard likened to 'packneedles, orderlie set one by another like the great teeth of a combe'.[8] It was regarded as common throughout the corn-growing areas of Britain up until the 1950s. Since then it has declined dramatically, like many other cornfield weed species. In this case one cause may have been the practice of stubble-burning after harvest, and since the beginning of the 1990s, when stubble-burning was banned, there have been signs that shepherd's-needle is making a comeback. On a farm in Tuddenham St Martin in Suffolk, shepherd's-needle began to return to barley fields when the farmer ceased burning in 1984. The plant has proved impervious to herbicides and the needles impossible to separate from the grain, and it has continued to spread.[9]

Sweet cicely, *Myrrhis odorata* (VN: The Myrrh, Liquorice plant). Although it is probably an ancient introduction to Britain, sweet cicely looks thoroughly at home by rivers and lanes in the north – especially where its extravagantly feathery leaf-sprays and pure white flowers are standing against a drystone wall. All parts of the plant taste and smell pleasantly of aniseed, though the adjective 'sweet' may also apply to a sugary undertone in the taste. John Gerard enjoyed the roots, boiled like parsnips, and the leaves were sometimes added to stewed fruit dishes to reduce the amount of sugar needed. But today, it is the long seed-cases which are most frequently used. They have something of the appearance of small gherkins whilst they are still green in June and have become a popular

The feathery leaves and gherkin-like seed-cases of sweet cicely are pleasantly aniseed-flavoured.

wayside nibble and unusual salad ingredient.[10] A wood-turner in Middlesbrough regularly uses the oily seed-cases to polish his finished pieces (which echoes the old Westmorland practice of using the leaves to polish oak panels).[11]

Like other species of uncertain origin, myths have gathered around sweet cicely's provenance. One contributor from the Welsh borders believes that its distribution in some places may reflect ancient historical boundaries: 'On the roadside verges around Selattyn, sweet cicely grows in such profusion that it has completely taken over from the other common umbellifers as the dominant hedgerow plant; yet south and east of Oswestry – and over Offa's Dyke – it becomes a rare plant. I cannot think that weather, soil or climate can change so greatly in three miles as to cause this difference, and wonder if this is a cultural boundary rather than a physical one, perhaps between Celtic use of the plant against the plague, not held by, or passed on to, the invading Saxons beyond the Dyke.'[12]

In the Isle of Man sweet cicely is believed to bloom on Old Christmas Eve, and there has been a small revival of a custom of searching for blooms on this day:

> 'The Myrrh, as it is known here, was used as a proof that the new calendar was not to be trusted. It was looked on to flower on Old Christmas Eve, 5th January [cf. the Glastonbury thorn, p. 214], and was often recorded as being brought into church on that day, in flower. It was as often said that it was not truly in flower by the embarrassed minister, and it was only a very new leaf that was unfurling. However I have seen, on a number of occasions, that after a mild winter it is indeed possible for a small bud to develop this early ... The Myrrh is often found near the doors of old houses. There are many clumps of it at Cregneash, the Manx Museum. Within the last ten years a local newspaper showed a photograph of it in flower at Old Christmas.'[13]

Coriander, *Coriandrum sativum*, is a herb and spice annual from the eastern Mediterranean, increasingly naturalised in waste places, especially close to ethnic communities from the Middle East and Asia, which make extensive use of both leaves and seeds in their cooking. Occasionally coriander can be found sprouting from pavement cracks in city centres, presumably from spilt seed.

Alexanders, *Smyrnium olustratum*. The glossy green leaves of alexanders are often the first new foliage of the year to appear on hedge-banks close to the sea. It was originally a Mediterranean plant, the 'parsley of Alexandria', and was probably first introduced to this country by the Romans as an all-purpose spring vegetable and tonic. Almost every part of the plant was used, from the root to the young flower-buds, which were pickled like miniature cauliflowers. It was cultivated in monastic herb gardens in the medieval period, which perhaps explains the colonies often found on the sites of old religious foundations, for instance on Steepholm in the Bristol Channel, where there was an Augustinian community in the twelfth century, and at an unusual inland site in Bedfordshire: 'At Elstow the plants occur not far from the Abbey Church and the remains of the former Benedictine Abbey, founded around 1078, which suggests that Alexanders was first grown in this area in the Abbey gardens as a vegetable for the nuns, prior to the Act of Dissolution in 1539.'[14] It also sometimes occurs near castles.

It continued to be cultivated in cottage gardens until the early eighteenth century, when it was supplanted by celery. By then it was becoming well naturalised, with a clear preference for sites close to the sea, where large colonies of bushy plants up to four feet high can develop. (Though it does occur sporadically inland. I have seen it, for instance, near villages on the eastern edge of Dartmoor.)

Alexanders has a pungent, angelica-like savour that is probably too pronounced for it to make a comeback as a vegetable. But it is worth trying the thicker stems, where they have been blanched by the plant's own leaf-sheaths, cooked like celery.

Pignut, *Conopodium majus* (VN: Ground nut, Cat nut, Earth nut, Earth chestnut, Yennett, Jog-journals). Digging for the dark-brown tubers of pignut used to be a common habit amongst country children. The nuts are usually between six and eight inches under the earth, and, eaten raw, their white flesh has something of the crisp taste of young hazel-nuts: 'Between the wars my uncle, an ex-sailor, would take me walking through the fields of the West Riding. He would regularly stop everything to plunge his knife into the ground and dig up a "pig nut"

Alexanders at Church Knowle, Dorset. Probably a Roman introduction originally, as a pot-herb, it is now thoroughly naturalised, especially (as here) near the coast.

Pignut, once harvested for its edible nut-like tubers which have the flavour of parsnips.

about the size of a walnut or chestnut, peel it and offer it to me to eat. He deplored the poor supply, saying that 40 years previously you could dig up enough to feed four people in a half an hour. These would be cooked in a "Dutch Oven" with rabbit joints. Pignuts cooked have the texture of and a milder taste than parsnips.'[15]

If they are less common now, it is a result not of over-zealous foragers, but of loss of habitat. Pignut, a modest-sized umbellifer with fine, thready foliage, is a plant of long-established grassland and open woodland, except on chalk, and has declined just as these places have. It is, sadly, probably fortunate that digging for the tubers is now illegal except where the landowner has given permission.

Burnet-saxifrage, *Pimpinella saxifraga*, is a summer-flowering, parsley-like umbellifer of rough grassland and waysides, chiefly on calcareous soils. It is a rare example of a plant named after two other, unrelated, families: 'burnet' from the shape of the leaves, and 'saxifrage' from its traditional herbal use against kidney and bladder stones. **Great burnet-saxifrage**, *P. major*, is a scarcer, taller species, largely confined to southern England and the Midlands. Oliver Rackham coined the term 'circumboscal' to describe this, and a few other species, which tend to grow in the vicinity of woods (on old lanesides, for instance) but rarely inside them.[16]

Ground-elder, *Aegopodium podagraria* (VN: Goutweed, Bishop's weed, Dog elder; Goat's foot, Devil's guts, Seven-toed Jack, Housemaid's knee; White ash). Ground-elder is best known as one of the most ineradicable of garden weeds, and many of the popular names refer to the daunting persistence of the roots: 'Devil's guts is another name for ground elder. My father comes from Dartford [a quarrying area]. After digging 30 feet deep they could still find the roots. It even comes through paved paths.'[17]

'Housemaid's knee' is, I imagine, a fairly modern gardener's tag, referring to the likely effects of trying to weed away a plant which can occupy three square yards in a season.

Ground-elder was almost certainly introduced to Britain from continental Europe as a pot-herb and a medicine against gout. Its herbal use – and popularity as a vegetable – eventually declined, but the plant itself did not. Even Gerard adopted an untypically desperate tone about it, saying that, 'where it hath once taken roote, it will hardly be gotten out againe, spoiling and getting euery yeere more ground, to the annoying of better herbes'.[18] Eating the leaves, boiled like spinach, is one way of reaching a *modus vivendi* with the plant. They make a stringy but tangy dish. The other way is to learn to accept it in the garden. In shady corners, the shoals of white umbels over neat elder-like foliage (or ash-like: Henry Lyte called it 'Aishe Weede', and the name 'white ash' is still current in Gwent)[19] are undeniably attractive, and, it hardly needs saying, incomparable as ground cover.

Rock samphire, *Crithmum maritimum*, is a yellow-flowered perennial, growing chiefly on cliffs and rocks by the sea, and occasionally on shingle beaches. The leaves smell sulphurous when crushed (some people find the scent similar to furniture polish), but it has been used as a vegetable, chiefly in pickles. In the nineteenth century rock samphire from Dover and the Isle of Wight was sent in casks of brine to London, where wholesalers would pay up to four shillings a bushel for it. Shakespeare knew the plant from the south coast, and in *King Lear*, in a scene near Dover, has Edgar say to Gloucester, 'half way down/ Hangs one that gathers samphire, dreadful trade!'

Hemlock water-dropwort, *Oenanthe crocata*, is a conspicuously large and locally common species of marshes, pondsides and ditches, with large flower-heads composed of almost spherical umbels. It is one of our most poisonous species and has caused several deaths in recent years, particularly amongst tourists from the continent, who mistake it for wild celery, *Apium graveolens*. A dramatic seventeenth-century poisoning was reported by John Ray, quoting a letter from an Irish physician, Dr Francis Vaughan:

'Eight Young Lads went one Afternoon a fishing to a Brook in this Country [Clonmell, Tipperary], and there meeting with a great Parcel of *Oenanthe Aquatica succo viroso*, (in Irish *Tahow*) they mistook the Roots of it for *Sium aquaticum* [fool's water-cress] Roots, and did eat a great deal of them. About four or Five Hours after going home, the Eldest of them ... on a sudden fell down backwards, and lay kicking and sprawling on the Ground, his Countenance soon turned very Ghastly, and he foamed at the Mouth. Soon after Four more were seized the same way, and they all died before Morning, not one of them having spoken a Word from the Moment in which the venenate Particles surprised the *Genus Nervosum*. Of the other Three One ran stark Mad, but came to his right Reason again the next Morning. Another had his Hair and Nails faln off ...'[20]

The biter bit: ground-elder ambushing an abandoned seed drill. Church Farm, Rockland St Peter.

The scent of the plant is deceptively pleasant, and when Georg Ehret, the eighteenth-century botanical illustrator, was drawing it, 'the smell, or effluvia only, rendered him so giddy that he was several times obliged to quit the room, and walk out in the fresh air to recover himself; but recollecting at last what might probably be the cause of his repeated illness, he opened the door and windows of the room, and the free air then enabled him to finish his work without any more return of his giddiness.'[21]

There were a number of cases of cattle being poisoned by water-dropwort in the West Country in the drought of 1995. They were being driven to grazing in the ditches because of the shortage of grass in the fields.

Six other species of water-dropwort occur in Britain, and all are poisonous.

Fool's parsley, *Aethusa cynapium*, is a common annual of cultivated and waste ground, with parsley-like leaves, but distinguishable by the green bracts that hang like threads below the white flowers. It is poisonous, though its nauseating smell when crushed deters most people from mistaking it for true parsley.

Fennel, *Foeniculum vulgare*. For all its aggressive growth in some places, fennel never looks convincingly native on British waysides. Its hair-like plumes of bright green leaves seem foppish and exotic by the side of rustic native relatives such as hogweed. In fact, it was probably introduced by the Romans as a medicinal and culinary herb. It is now widely naturalised across Britain, especially near the sea, where it grows on roadsides, sand-dunes and sea-walls. Inland it is a plant more of waste ground and tips, originating often from discarded seeds.

The leaves smell strongly of aniseed, and are widely used in cooking, especially with oily fish. A sauce of chopped fennel and gooseberry is a classic accompaniment to mackerel. The seeds can also be used as a cooking

Rock samphire below the cliffs at Kimmeridge, Dorset. The leaves (despite their sulphurous smell) were once gathered for pickling.

herb, or as an after-dinner *digestif* (they are often served at the end of a meal in Indian restaurants). This practice has a history going back at least to the Middle Ages:

> *In Fennel-seed, this vertue you shall finde,*
> *Foorth of your lower parts to drive the winde.*[22]

In Gwent, children make pens out of the hard stalks.[23]

Dill, *Anethum graveolens*, is an aniseed-scented herb from Asia, which occurs rather frequently as a waste-ground casual.

Pepper-saxifrage, *Silaum silaus*, is a yellowish-flowered species, rather local and largely confined to old meadows, pastures and commons on clay soils, where it is often regarded as an indicator of ancient grassland. **Spignel** or **Baldmoney**, *Meum athamanticum*, is a pleasantly aromatic species of mountain grassland in the north. The roots (also aromatic) were sometimes eaten in Scotland. The origin of the English names is unknown.[24]

Hemlock, *Conium maculatum* (VN: Mother die, Kexies, Woomlicks). Hemlock is notorious as the poison which was given to Socrates at his execution. It would have been a slow death: paralysis, respiratory failure and stupor, but not necessarily loss of consciousness until the very end.

Despite its extreme toxicity (due chiefly to an alkaloid called coniine), cases of hemlock poisoning are very rare. One sniff of the crushed foliage soon explains why: it is repellently sour and mousy, and even livestock normally avoid it. The few human poisoning cases have occurred early in the year, when the young leaves have less scent and can be mistaken for parsley. Later in the year, when hemlock's purple-spotted stems often exceed six feet in height, it can hardly be confused with any other species. It is a biennial and can build up large, unbroken colonies on damp waste ground, along river-banks and in roadside ditches. The pale, ferny drifts are a distinctive landscape feature, though no pleasure to walk through. Anne Pratt tells a story about an eccentric keeper of Kent's South Foreland lighthouse in the 1850s who, defying its odour and reputation, made a bower out of hemlock. It must have had a strange, underwater atmosphere, with its hanging tassels of swaying green-weed:

Fennel, probably introduced by the Romans as a herb for digestive problems, now widely naturalised, especially near the coast.

'A large bed of Hemlock grows there, and the man occupied in the charge of the Marine Telegraph at that station has availed himself of its abundance to deck with its stems and branches his little cave in the cliffs. This has a sloping entrance, and all about it he has planted the hemlock, which attains there a great luxuriance, and is in summer six feet high, affording by its numerous branches a shelter alike from sun and shower. The owner of the cave, an intelligent man, has an eye for grace and beauty, and prizes the elegant foliage, taking care to preserve its verdure by cutting off the fruits as they appear; while the robustness given by an out-of-door life, by airs and sounds from the sea, have rendered his nervous system too strong to be injured by the odour. To him the faint smell gives no disgust, though he tells how a friend, an old coast-guardsman, who occasionally visits him, cautiously declines to subject himself to its influence, and seats himself on some crag at a distance, where he may see its branches wave in safety.'[25]

When they are dry the stalks lose most of their poisonous qualities and have been used by children as peashooters. In Northumberland, under the inscrutable name of 'woomlicks', 'they were also used until quite recently by fishermen as moulds for their lead weights'.[26]

False thorow-wax, *Bupleurum subovatum*, is a Mediterranean species which quite commonly springs up in parks and gardens from birdseed mixtures, especially in warm summers. It is an attractive plant, with yellow flower-clusters held in five-pointed yellow-green cups. The leaves are greyish and perforated by the stem. A very similar species, **thorow-wax**, *B. rotundifolium* (the English name was coined by William Turner in 1548, 'because the stalke waxeth thorowe the leaues'),[27] was once a common cornfield weed, but is now extinct as a wild plant in Britain. Its only recent appearance has been in unexpectedly formal circumstances. In December 1990, Brian Wurzell spotted the plant as a component of wreaths on a newly interred grave in St Pancras and Islington Cemetery, north London. It had been interleaved between a conventional arrangement of chrysanthemums, irises and carnations, and the 'large tufts of a yellow green colour ... glowed with astonishing vividness for an overcast day'.[28]

Hemlock, a highly poisonous plant, distinguished by its purple-spotted stems and rank mousy smell.

Sickle-leaved hare's-ear, *B. falcatum*, though common in parts of northern France, is in Britain an exceptionally rare and local species, known only from one locality in Essex, a cluster of damp roadside verges, ditches and hedge-banks at Norton Heath. It was first found here in 1831, was eradicated during hedgerow clearance in 1962, and reappeared in 1979.

Wild celery, *Apium graveolens*, is a native in brackish places, usually near the sea. This is the ancestor of our cultivated celeries, which were developed in the seventeenth century, and is edible itself, though too tangy for most tastes. It has a pungent smell which can often be detected 50 yards away. **Fool's water-cress**, *A. nodiflorum*, is not such a foolish plant in the kitchen as fool's parsley. It grows with and resembles (in leaf at least) true water-cress, but is bland to eat rather than toxic.

Garden parsley, *Petroselinum crispum*, is the familiar garden herb, introduced from the eastern Mediterranean, and well naturalised on sandy banks and waste places in much of England and Wales. **Stone parsley**, *Sison amomum*, a delicately wispy umbellifer about three feet high, is native but rather local in hedge-banks and grassland mainly in south-east England, and best known for Professor Tom Tutin's inspired description of the nauseous scent of the leaves when crushed, as 'somewhat resembling that of nutmeg mixed with petrol'.[29]

Garden angelica, the plant whose candied stems are used for cake decorations.

Caraway, *Carum carvi*, is a culinary plant from Europe, grown for its seeds and naturalised in a scatter of waste places: 'Caraway once sprouted in a meadow near Oxford in a wild-looking situation, but it had a curious origin. At the time it was the practice of publicans to have on their counters a tray of caraway seeds, for the benefit of people who wished to disguise the odour of their drinker's breath. Many were dropped on the floor, the sweepings from which were apparently conveyed to this field.'[30]

Milk-parsley, *Peucedanum palustre*, is a native of fens and marshes, very rare outside the fens of East Anglia. It is the food-plant of the swallowtail butterfly. The similar **Cambridge milk-parsley**, *Selinum carvifolia*, is a rare native of fens and damp meadows, confined now to three sites in Cambridgeshire, including Chippenham Fen, where it has been known since 1882. **Lovage**, *Levisticum offinale*, is a tall culinary plant from Iran, occasionally persisting as a relic of cultivation in dry places. **Scots lovage**, *Ligusticum scoticum*, is a native of rocky places by the sea in Scotland. **Wild angelica**, *Angelica sylvestris*, is common by streams and ditches and in damp woods throughout Britain. It is a tall and handsome species with purple-tinged, 'claret-dipped' flower-heads, smelling unmistakably like **garden angelica**, *A. archangelica*, the plant whose candied stems are used in cake-decoration. This native of continental Europe is occasionally naturalised on damp waste ground. But the ability of its seeds to float has also enabled it to build up quite large populations by some urban rivers – notably in east London. Here it has spread along the banks of the River Lea from Hackney Marshes to the Regent's Canal in the East End and by the Thames itself as far as Dartford.[31]

Wild parsnip, *Pastinaca sativa*, is the wild ancestor of cultivated parsnips and was probably taken into gardens to be 'manured' in the early Middle Ages. It is common on chalky grassland and roadsides in southern and eastern England, often growing in large colonies, so that its flowers seem to form clouds of yellow at a distance. The roots

Wild parsnip, the ancestor of cultivated varieties and common on chalk grassland and roadsides.

are hard and wiry but just about edible; and even the leaves have the scent of parsnip.

Hogweed, *Heracleum sphondylium* (VN: Cow parsnip, Bilders, Caddy, Eltrot, Limperscrimps, Cow-weed, Kirk, Chirk, Keks). A solid, rather hairy native, abundant in hedge-banks, waysides, rough grasslands and waste places, hogweed is the commonest umbellifer in flower in late summer, but it can usually be found flowering in all months of the year. In June, areas of rough grassland full of the large white flower discs can seem almost luminous under the full moon. And on mown waysides, it will often bloom again in winter, the flowers forming inside the new leaf-sheaths very close to the ground. (These young hogweed 'spears' are very succulent eating if cooked like broccoli.) The stems are thick and hollow, and have long been a basic raw material for children's games:

'We made water guns from hogweed (locally called "caddy"), using the large, hollow seeding stems and cutting them with the node at one end of the stem. A tiny hole was made through the node section and, for a plunger and washer, a stick was used with wool tightly wound around the bottom inch, until the necessary radius was achieved to just fit the hollow. The stick had to be short enough for it not to push through the node. They were very effective.'[32]

'We used to have "swordfights" with the dead stems of hogweed. I remember crisp winter mornings, my brother and I bashing the dry, hard hogweeds one against the other until one of the stems gave in.'[33]

Giant hogweed, *H. mantegazzianum* (VN: Giant cow parsnip, Cartwheel plant, The Hog). Giant hogweed was a long while coming into its nefarious reputation. It had been growing in Britain for more than 150 years, an awesome but apparently well-mannered curiosity of Victorian shrubberies and ornamental lakesides, with no more than a hint of troll-like mischief in its huge, looming umbels. It seemed exactly what it was – a giant cow parsnip. Then, in 1970, it broke cover. That summer large numbers of children began arriving in hospital casualty departments with unusual, circular blisters on their lips, hands and eyes, and it was not long before they were traced to a common cause – the sap from giant hogweed stems which had been used as blowpipes and 'telescopes'.[34] Overnight a plant from Asia that even botanists seem scarcely to have noticed in the wild (it is mentioned in very few Floras prior to 1940) was transformed into an aggressive and invasive public menace. The popular press immediately dubbed it 'The Triffid' after John Wyndham's science-fiction monster, and published pin-up-sized photographs of its speckled 15-foot stems and cartwheel-sized flower-heads. There were endless suggestions about how to eradicate the aliens: scythes, poison, flame-throwers, even excavators. One neighbour of mine dug out a plant to a depth of three feet, and then filled the root-hole with turpentine. None of these assaults had much effect. Except when it is young, giant hogweed is not very susceptible to weed-killers. Each plant produces up to 5,000 seeds, and, growing by rivers as it often does, is able to disperse them by water over long distances. So the expansion which had been quietly under way for nearly a century continued – though now under much more vigilant eyes.

It had seemed such an exciting and promising introduction when it was first brought to this country from the Caucasus mountains in the early nineteenth century, perhaps the temperate zone's answer to some of the titanic plants that were being discovered in the tropics. The great arbiter of gardening taste, John Loudon, certainly thought so, and praised '*Heracleum asperum* ... the Siberian Cow Parsnep' in *The Gardener's Magazine*:

'The magnificent umbelliferous plant, when grown in good soil, will attain the height of upwards of 12 ft. Even in our crowded garden in Bayswater, it last year (1835) was 12 ft when it came into flower ... Its seeds are now (July 29) ripe; and we intend to distribute them to our friends: not because the plant is useful, for we do not know any use to which it can be applied; but because it is extremely interesting from the rapidity of its growth, and the great size which it attains in five months ... We do not know a more suitable plant for the retired corner of a churchyard, or for a glade in a wood; and we have, accordingly, given one friend, who is making a tour in the north of England and Ireland, and another, who is gone to Norway, seeds for depositing in proper places.'[35]

Whether this was one of the origins of rapidly expanding Norwegian populations is a moot point. But giant hogweed – known in the north of Norway as 'Tromso palm' – is regarded with much more affection than it is in Britain, and even appears as a feature on local tourist postcards.[36]

By 1849 its seeds were being offered commercially by Hardy and Sons of Maldon, Essex, as '*Heracleum giganteum*, One of the most magnificent Plants in the World'.[37] And in 1870 William Robinson recommended them as 'very suitable for rough places on the banks of rivers or artificial water, islands, or any place where bold foliage may be desired', but added, prophetically, that 'when established they often sow themselves, so that seedling plants in abundance may be picked up around them; but it is important not to allow them to become giant weeds'.[38] Some large garden populations (e.g. a long-established colony in a 100-year-old garden in Hampton-in-Arden)[39] may date back to plantings made at this time.

By the early 1900s what was now called, with a more appropriately sinister buzz, *H. mantegazzianum* (after its Italian discoverer) was beginning to make its first sporadic break-outs. The collection in Buckingham Palace Gardens (which includes other closely related species and hybrids) edged into the Royal Parks and thence into the west London canal system.

In Colinsburgh, Fife, a large local concentration spread from an ornamental pond on an Edwardian housing estate which had used giant hogweed as a decorative plant: 'Since its introduction ... the hogweed has escaped into the outfall stream, and this meanders about two miles to the burn, which is the main watercourse. A further two miles and the burn meets the sea. The banks of stream and burn are both covered in hogweed, almost to the exclusion of anything else.'[40] Further west in Scotland, its spread along the River Ayr has been traced to two identifiable country houses with large riverside gardens created *c.* 1939–45.[41]

What is remarkable is that it took so long for the plant's dermatitic effects to be noticed. It is hard to believe that gardeners had remained immune to the sap for more than a century. Yet a mother from Lancashire describes the confusion following her children's first clashes with the plant at the start of the 1970s: 'We had a holiday at a camp site where there was an awful lot of giant hogweed on the roadside and no one seemed to have done anything about it. We had four small children and they and some school friends were all going to the doctors who said they had been fire-burned and all various things. They were covered with burn-weals at the time. No one knew the reason. Till a doctor in Preston said the giant hogweed seeds had carried down the stream where they picnicked. At the time we had them growing close by us in a private garden. They are now gone, the lady got rid of them when *she* had children.'[42]

The irritant chemicals in the sap and bristles of giant hogweed are known as furocoumarins. They make the skin hypersensitive to bright sunlight and liable to blister and redden, often for long periods. The plant's aggressive spread – especially along river systems, where it can sometimes choke out native vegetation – has added to its bad reputation, and under the Wildlife and Countryside Act of 1981 it is now an offence to 'plant or otherwise cause [it] to grow in the wild'. Agencies such as the National Rivers Authority and local countryside management services have mounted campaigns against it. Needless to say it is generating its own myths: 'We have that awful giant hogweed here in Overstrand ... It is so tall it is quite disturbing. I am told it is poison to children so should not something be done about it? I'm wondering what effect its pollen has on us.'[43]

In Richmond, Yorkshire, the story has grown that the plant originally arrived as a stowaway on Russian ships, stealing up the Tees from Yarm, when it was still a port.[44]

But even those who have suffered from the plant are not unanimous in condemning it: 'We have a giant hogweed growing in our front garden and I discovered its dangers to my peril last year when I cut it down after it had flowered. We were aware of the danger of the plant, as the previous occupant had warned us, but I was quite unprepared for the burns I received on my arms and legs. It was while I was sitting in the casualty department of the hospital that I started to wish I had left it well alone. It is so unusual and spectacular, despite its hazards. We do not want to get rid of it; we just know now to treat it with a bit of love and respect.'[45]

Many contributors feel that the onus of responsibility is on us, not to 'tamper' with the plant; and that caution and respect are fairer responses than extermination[46] – an approach which has been taken to heart by an Australian musical instrument maker: 'My husband has recently purchased a Digeridoo made from a giant hogweed stem. The stem had been varnished and the mouthpiece heavily waxed (presumably to protect the lips from the irritants). P.S. It works.'[47]

'The giant hogweed has a long history here [Chester] ... Our Drive was built on the site of the famous Dicksons Nurseries Company who existed here for about a century, closing down in the 1925–30 period. The boundary of our garden, and the west boundary of the Nursery, was Newton Brook, and the Dicksons used this for growing all sorts of exotic waterside plants – the giant hogweed being one of them. Since the demise of the Nursery, survivors have existed roughly, growing along the brook route and even spread to the Bache area. Even more interesting is a local botanist's discovery that some had colonised the nearby railway track as far as Birkenhead! Locally, most have succumbed to over-zealous Council cutting-back – ostensibly to avoid youngsters being afflicted by the corrosive sap ... For all this, I and two family generations have survived to value these remarkably handsome plants. We now keep a couple of specimens each year in the garden. I should hate to see them vanish in this over-protective age.'[48]

'I have got two growing in different parts of my garden. Last year the main one had one flower head. This year it had three, probably because of my liking for this plant. I feed it on Growmore.'[49]

'To me the giant hogweed ... is one of our finest natural sculptures outside the world of trees and comparable to the smaller thistles and teasels as a "thing of beauty" ... The simple answer is to keep your distance, or wear gloves.'[50]

'It yomps determinedly along the banks of the River Kent at Jevens Bridge. Occasionally it escapes onto the roadside verge and causes a frisson of anxiety in the local newspapers which describe the dangers of handling it ... When gas pipes were being laid across the A6 nearby, a warning notice, "Danger, Heavy Plant Crossing", was displayed at the foot of one of these escapees.'[51]

For aficionados, there are hogweed landmarks all across Britain: in a fen above sweeps of marsh-orchid on the coast at Aldeburgh, Suffolk; by the side of the Art-Deco Hoover factory west of London; around the marina at Ilfracombe, Devon, and the Launceston Recycling Centre in Cornwall; by the sides of the Toll Bridge on the outskirts of Nottingham; and a large riverside sweep along the River Usk near Abergavenny.[52]

But Scotland has the most majestic – or insidious – colonies, depending on how you view them. In Glasgow, there are huge drifts on the banks of the Clyde, especially downstream of Kelvinbridge. Jim Dickson, Senior Lecturer in Botany at the university, describes the colonies growing out of the river-bank and waste ground of the Cunningar Loop as 'one of the most remarkable natural history sights of the Glasgow area'.[53]

Glasgow citizens seem to have reached a better *modus vivendi* with the plant, too, than their southern counterparts: 'People in the Glasgow area have brought it in indoors for flower arrangement, the older terraced houses having the necessary high ceilings. People are aware of the plant and incidents of dermatitis are rare.'[54]

The irony is that more cases of photodermatitis are caused by two much commoner and certainly less vilified umbellifers – common hogweed and wild parsnip (see above).[55] Since the introduction of the strimmer, anyone clearing rough grass which includes either of these species risks spraying their skins with a fine mulch of furocoumarins. An ecology lecturer remembers what was widely known as 'dreaded plod rot' erupting routinely amongst her students when they were doing fieldwork amongst common hogweed on Skokholm Island in the late 1960s.[56]

Wild carrot, *Daucus carota* (VN: Bird's-nest). A common umbellifer of high summer, especially on dry grassy

Giant hogweed framing Ringwood church, Hampshire.

Wild carrot above Lulworth Cove, Dorset.

places on chalk soils, wild carrot is a variable plant, but normally grows up to three feet tall. The umbels are claret-coloured or pale pink before they open; then white and rounded, with a festoon of bracts beneath; finally, as they turn to seed, they contract and become concave like birds'-nests. The leaves smell of carrots, as do the roots, but these are thin and wiry and bear little resemblance to the thick, orange tap-roots of the cultivated vegetable. It is believed these were developed from a distinct subspecies, ssp. *sativus*, probably native to the Mediterranean, and brought to Britain in the fifteenth century.

Gentian family *Gentianaceae*

Common centaury, *Centaurium erythraea*. This is a short, upright biennial of dry, grassy places – sand-dunes, heaths, woodland rides, quarries and the like. The pink flowers are borne in small clusters and usually close up early in the afternoon. Centaury had a reputation for controlling fevers in early herbal medicine. **Yellow-wort**, *Blackstonia perfoliata*, is an unmistakable annual of calcareous grassland and dunes in England with eight-petalled yellow flowers, again shutting in the afternoon. The leaves are a waxy, glaucous grey, and the upper ones are fused around the stem like shallow cups. Its Latin name commemorates the London apothecary and botanist John Blackstone (1712–53).

Fringed gentian, *Gentianella ciliata*. This continental species, with dramatic blue flowers up to two inches across and fringed with fine silvery hairs, is known from just one chalk hill near Wendover in the Chilterns. It was found by a Miss Williams in 1875, but her report was dismissed imperiously by George Claridge Druce: 'There must be some gross carelessness in such a record, as *ciliata* is not likely to occur in England.'[1] It was refound (and verified) at the same site more than a century later, and Miss Williams's reputation was salvaged.

Autumn gentian or **Felwort**, *G. amarella*, is a late-flowering (often into October), purple-bloomed biennial of calcareous pastures and dunes. It sometimes grows in large troops, especially on slightly disturbed or thin-turfed chalk. **Chiltern gentian**, *G. germanica*, is a taller, larger-flowered species and one of the specialities of the

Top: autumn gentian, the commonest species.
Above: the diminutive early gentian, often no more than an inch or two tall. Both species grow in chalk grassland.

Chilterns, where it often grows deep amongst chalk scrub, though its range does stretch into the downs of north Hampshire. Hybrids often occur where *G. amarella* and *G. germanica* grow together. **Early gentian**, *G. anglica*, is a jewel of a plant, often no more than an inch or two tall in short chalk grassland, but with the same delicate lipstick-like buds and stiff, sharp-petalled flat flowers as larger species of *Gentianella*.

Marsh gentian, *Gentiana pneumonanthe*, is a rare native of wet heathland with flowers of a clear sky-blue with five green lines outside. A white variety has occurred in the New Forest.[2] **Spring gentian**, *G. verna*, has beautiful, intense blue flowers; it is confined in England to Upper Teesdale. **Alpine gentian**, *G. nivalis*, is the snow gentian; in Britain it occurs only on mountains in Perth and Angus above 2,000 feet, and opens its brilliant deep-blue flowers only in very bright light.

Periwinkle family *Apocynaceae*

Lesser periwinkle, *Vinca minor*, and **greater periwinkle**, *V. major* (VN: Blue betsy). Both species are popular garden plants, widely naturalised on hedge-banks and in woods,

Greater periwinkle in Droop churchyard, near Hazelbury Bryan, Dorset.

especially close to houses. The purplish-blue-flowered greater periwinkle, from the Mediterranean, is unquestionably an escape. The lesser, however, looks so thoroughly at home in some woods in the south, spreading its sky-blue flowers across the floor by suckers, that some botanists believe it may be native. It occasionally occurs with white flowers in the wild.

Periwinkles were traditional plants for making celebratory and funereal wreaths, whose form echoes the plant's natural habit of twining amongst other low-growing plants – a practice which helped give them their English name (cf. periwig).

Nightshade family *Solanaceae*

All native and most cultivated members of the nightshade family are poisonous to some degree, even when part of the plant – e.g. the fruits in the case of tomatoes and the tubers in the case of potatoes – is edible. The toxins are alkaloids, chiefly hyoscyamine, hyoscine and atropine. These have similar properties, drying the mucous membranes of the mouth and throat, dilating the pupil of the eye, dampening spasms of the gut, and speeding the heartbeat to the extent that it can often be heard several feet away. In larger quantities they produce hallucinations, coma and sometimes death, but in controlled doses they are still important drugs in conventional medicine, especially as gastro-intestinal sedatives.

Duke of Argyll's teaplant, *Lycium barbarum*, is a scrambling shrub from Asia, grown as hedging and often naturalised in hedges and scrub close to habitation. Its name supposedly records a botanical muddle involving Archibald Campbell, the third Duke of Argyll and a famous plant-collector in the first half of the eighteenth century. (Horace Walpole called him the 'Treemonger'.) He was sent a true tea plant, *Camellia sinensis*, and a *Lycium* with their labels mixed and, unwittingly or as a joke, continued to grow them under their wrong names. The story surfaced in 1838, more than two generations after he died, but the plant has continued to keep its ironical title.[1]

It is an intriguing shrub, with small purple flowers followed by oval scarlet berries, which are relished by birds. But these (as well as the leaves, despite their inviting 'tea' name) are probably mildly poisonous to humans.

Deadly nightshade, *Atropa belladonna* (VN: Dwale). Deadly nightshade is a handsome plant whose appearance belies its toxicity. It is a bushy perennial, growing up from ground level each year, its multiple branches carrying pale green, ribbed leaves. The flowers are purplish-brown bells, inside which the berries form – green at first, then a shiny black. They look as succulent and seductive as cherries, yet as few as three have been fatal to children. All parts of the plant contain quite high quantities of hyoscyamine (but very little atropine, contrary to popular belief and the suggestion in the scientific name), which acts in ways described above. Anne Pratt reported, somewhat improbably, that 'paralysis of the hand is said, on good authority, to have arisen from carrying it for some length of time'.[2]

Extracts of the plant are still used in stomach sedatives, and in tinctures for dilating the pupil of the eye. This is an ancient practice, and believed to be the origin of the specific Latin name 'belladonna'. Italian women used water distilled from the 'beautiful lady herb' as a cosmetic, to enlarge their pupils.

A useful plant, then, but one that landowners such as the National Trust are chary of tolerating, at least close to footpaths and car parks. But the disturbed soil in such places is precisely where deadly nightshade flourishes. On chalky and limestone soils (the only sites where it is native) it relishes rabbit warrens, old quarries and new forest tracks. After the great storm of 1987, large numbers of deadly nightshade plants sprang up – presumably from long-buried seed – in the root-holes of windthrown beeches in both the Chilterns and the South Downs.

Its frequent and somewhat sinister liking for grave-

The tempting, cherry-like, but poisonous berries of deadly nightshade.

The sinister flowers of henbane, the source of the poison used by Dr Crippen.

yards (it grows directly out of tombs at St Cross, Holywell, in Oxford) is more curious.[3] Perhaps in such places it is a relic of ancient herbal cultivation, which was sometimes carried out at the edges of churchyards. A physic garden was almost certainly the source of the colony which grew at Guy's Hospital, London, until it was destroyed by building works in 1978.[4] And there have always been so many plants among the ruins of Furness Abbey in Lancashire, that the area is known as the Vale of the Deadly Nightshade.[5] This is a name which goes back at least to Wordsworth's day. He knew it as 'Bekang's Ghyll – or the dell of the Nightshade – in which stand St Mary's Abbey in Low Furness'.[6] In Felixstowe there are a number of sites of a variety with very pale lilac flowers, which may be a relic of an introduction by the Romans, who had a garrison in the old town. (The Suffolk botanist Francis Simpson used to remove the berries from these plants, to avoid their being 'found by some over-zealous person and destroyed'.)[7]

Henbane, *Hyoscyamus niger*, is a sinister, malodorous species that even those with no knowledge of plants might suspect of being poisonous. The leaves are grey-green and densely covered with sticky hairs, the flowers liverish-looking, their pale yellow petals netted with purple veins.

Henbane contains considerable quantities of hyoscyamine and hyoscine, and was the poison with which Dr Crippen chose to murder his wife in 1910. Much earlier, it was prescribed as a specific remedy for toothache, because of a strong resemblance between the seed-heads and a row of molars. It certainly would have dulled the pain, but not the sufferers' imaginations, and Gerard describes how the hallucinatory effects of henbane were exploited by quack herbalists to bolster its reputation: 'The seede is vsed of mountibancke toothdrawers which runne about the countrey, for to cause woormes come foorth of mens teeth by burning it in a chafing dish with coles, the partie holding his mouth ouer the fume thereof: but some craftie companions to gaine money conuey small lute strings into the water, perswading the patient that those small creeping beasts came out of his mouth or other parts, which he intended to ease.'[8]

Henbane is native in sandy places by the sea, and on disturbed areas (rabbit warrens, for instance) on the chalk. Elsewhere it is a casual (as with the specimen that grew close to London's Festival Hall in 1966), or a relic of cultivation for medicinal uses. Long-dormant seed was almost certainly the origin of the many hundreds of plants which appeared in the summer of 1993 on set-aside land at Shrewton on Salisbury Plain.[9]

Cape-gooseberry or **Chinese-lantern**, *Physalis peruviana*, is an ornamental and culinary plant from South

Henbane is a casual that occasionally crops up on disturbed chalk grassland – as here, at Lulworth, Dorset.

Apple-of-Peru, a poisonous casual of waste ground and tips.

America, appearing as a quite common casual on tips. The orange berry inside the papery, lantern-shaped fruit-case is edible. (But beware of confusion with the superficially similar **apple-of-Peru**, *Nicandra physalodes*, also a frequent casual of tips and cultivated ground, whose *brown* berries, in more open papery cases, are poisonous.) Other culinary and ornamental members of the nightshade family, grown in gardens but frequently escaping to waste ground, include **tomato**, *Lycopersicon esculentum*; **potato**, *Solanum tuberosum*; **sweet tobacco**, *Nicotiana alata*; and **petunia**, *Petunia × hybrida*.

Bittersweet or **Woody nightshade**, *Solanum dulcamara*, is popularly known as 'deadly nightshade' in many parts of the country. This is not only a misidentification but a misnomer: it is one of the less poisonous members of the family. And, though it is common in shady corners of gardens and has rather tempting scarlet berries, like miniature plum tomatoes, cases of poisoning even amongst children are very rare. The intense bitterness that gives the species the first part of its name (the sweetness is an after-taste) causes most curious nibblers to spit the berries out immediately.

In ornamental terms it is rather an attractive plant, a perennial scrambler that has about it an echo of rain-forest vines. The flowers are purple, with the petals reflexed behind a yellow cone, and the stems will wind over hedges, fences and bushy plants in woods. The leaves, if you crush them, have a disagreeable smell of burnt rubber.

Several colonies with white flowers have been found in Dunbarton.[10]

Black nightshade, *S. nigrum*, is a frequent annual of cultivated and waste ground and farmyards. Near Chesham, Buckinghamshire, it has gained a foothold in commercial greenhouses, 'growing into bushes four or five feet wide under the glass. It was in company with cherry tomatoes, another member of the Solanaceae.'[11] Black nightshade is like a potato plant, with white reflexed flowers and green berries which ripen to a shiny black. Two South American annuals, **green nightshade**, *S. physalifolium*, and **leafy-fruited nightshade**, *S. sarachoides*, are cropping up increasingly as casuals on tips and waste ground.

Thorn-apple, *Datura stramonium* (VN: Jimsonweed). By far the most dramatic of the family to appear in Britain, thorn-apple is a hefty, thick-stemmed annual from warmer regions, with large jagged leaves. And, like other *Datura* species, it has graceful, swan-necked white or purple flowers. The problem lies in the fruits, which are sufficiently like conkers to have attracted the interest of some children (though one specimen in a South Humber-

Woody nightshade, one of the less poisonous members of the family and an attractive scrambler in late summer.

side garden was mistaken for a teasel).[12] But their formidable spines and nauseating smell must deter all but the most adventurous from exploring them more intimately, and there have been very few cases of human poisoning.

The atmosphere of suspicion stirred up by the giant hogweed scare in the 1970s soon spread, however, to unfamiliar nightshades, especially on public land:

'Hard on the heels of the giant hogweed reported in August Rail News, comes another rare and dangerous plant – the thorn apple. It was spotted at New Milton [Hampshire] station by a retired chemist and local botanist. He told his next-door neighbour, one of Waterloo's assistants. Apparently the drug stramonium, which came from the plant, was once used for the alleviation of asthma. And it is believed wizards in medieval times used it. New Milton's Stationmaster was told that the plant was in his station car park and arranged for it to be destroyed. After hacking down the three-foot plant they dug up its roots and burned the lot.'[13]

British Rail's staff were probably erring on the side of caution, but they had some of the facts about the plant correct. It was used by herbal 'wizards' (though not medieval ones: it didn't arrive in this country until the late sixteenth century) – and perhaps 'witches', too. At the end of the seventeenth century John Pechey maintained that 'Wenches give half a dram of it to their Lovers, in beer or wine. Some are so skill'd in dosing of it, that they can make men mad for as many hours as they please.'[14]

And 'stramonium' extracted from the flowers and leaves (again, a mixture of nightshade family alkaloids such as atropine and hyoscine) had an honourable place in the treatment of asthma up to the end of the Second World War. Although some was grown commercially in this country, the bulk of our supplies had been imported from eastern Europe and were cut off in 1939. Wild specimens became valuable then, and the County Herb Committees were asked to gather leaves and flowers to augment the increased production from farms.

These days many thorn-apples originate from impurities in bags of South American fertiliser. But some could be relics of the days when the plant was more widely grown in this country. The 500 or so seeds which are scattered when the spiny capsule breaks open can stay dormant for exceptionally long periods. At the end of the nineteenth century, for example, occasional plants were

Thorn-apple's conker-like seed-cases are preceded by white swan-necked flowers.

seen at Woolwich Arsenal in London. When the eastern end was demolished before the building of Thamesmead in 1969, hundreds of thorn-apples reappeared in the churned-up soil. And when a local botanist found a plant on Wimbledon Common in 1935, he dismissed it as a casual, but it had been seen near the windmill on the same site a century before.[15] Hot summers (like 1975 and 1976) always produce a rash of records, often from suburban gardens.

A rich hunting ground for all manner of *Datura* species and other *Solanaceae* is proving to be the campus of Nottingham University. Here the Department of Pharmaceutical Sciences has for some years been investigating the medicinal properties of alkaloids from the nightshade family, which has involved the cultivation of live plant material for analysis and breeding. Over the past few years a remarkable number of these species have found their way into waste ground and cultivated beds on the campus, including six exotic *Solanum* species, five *Datura* species (and some spontaneous hybrids) and the aggressive South American sprawler **cock's-eggs**, *Salpichroa origanifolia*, with its pineapple-scented fruit.[16]

Bindweed family
Convolvulaceae

VN: No great distinction is made between the different species in popular nomenclature. Barbine, Bellbine, Bethwine, Cornbine; Withybine, Withywind, Waywind; Lady, Lady-jump-out-of-bed, Granny-jumps-out-of-

Large bindweed ornamenting the fence by a wayside pulpit, High Wycombe, Buckinghamshire.

bed; Robin-run-the-hedge; Snake's meat; Devil's guts; Sussex lily, Wild lily.

Hedge bindweed, *Calystegia sepium*. The feelings of gardeners about bindweed's strangling abilities are well expressed in the surviving vernacular names. All the species have the capacity to root from the smallest fragments and to spread at a prodigious rate: 'In the garden of our house in Cornwall stands a *Cupressus macrocarpa* which I estimate to be about 55 feet high. A huge liana-like growth of *Convolvulus* clothes the tree, nearly to its topmost twig – perhaps a carpet 52 feet high would describe it best.'[1]

Yet hedge bindweed, with its large white trumpet-flowers and mats of arrow-shaped leaves, is a handsome plant and, in urban areas especially, does good service in cloaking wire fences and derelict brickwork. It is a native plant that can be found across Britain in all kinds of habitats – scrambling about hedges, ditches, wood margins, reedbeds and the tall vegetation of river-banks. The flowers, both in bud and in full bloom, have been ingeniously exploited by children:

'My mum remembers using bindweed buds. When she was little she used to pick them for pretend lipstick.'[2]

'Children sharply pinch the base of the calyx, causing the whole corolla to pop out and float to the ground, looking like an old-fashioned nightgown, and say "Grandmother, grandmother, pop out of bed." '[3]

Hedge bindweed occurs in several subspecies and forms, some pink, with or without white stripes. **Large bindweed**, *C. silvatica*, a naturalised species from southern Europe, is like a large-flowered *C. sepium* and seems to be spreading, especially around London. **Hairy bindweed**, *C. pulchra*, has more delicate flowers of deep pink with thin white stripes. Its origin is uncertain, but it may be a garden hybrid. **Field bindweed**, *Convolvulus arvensis*, is a common creeping perennial of waste and cultivated places, with flowers either white, pink, or striped, and would be a fine rockery plant if it were less aggressive.

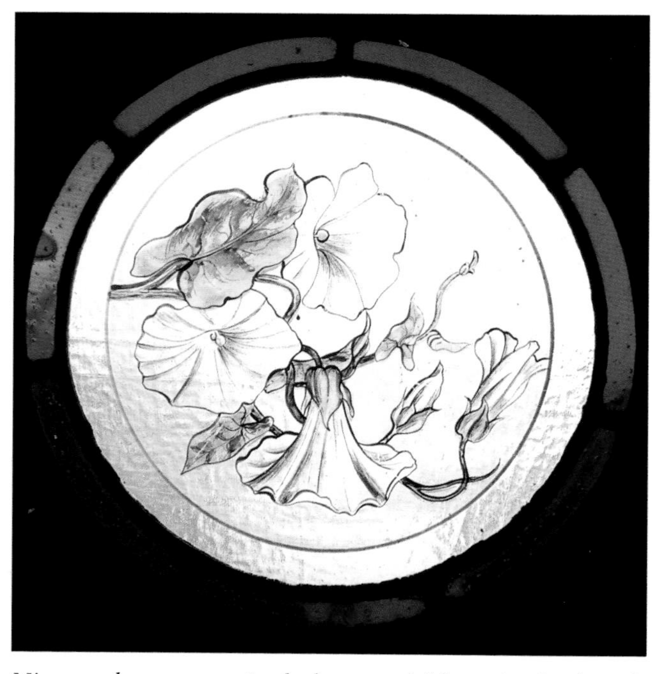

Nineteenth-century stained-glass roundel featuring bindweed.

Bogbean family *Menyanthaceae*

Bogbean, *Menyanthes trifoliata*, is locally frequent in shallow ponds, bogs and fens. The name (from the roughly broad-bean-like shape of the leaves) hardly does justice to the intricacy of the flowers. Borne on stems up to a foot or so in height, the flowers are white stars, flushed pink on the outside of the petals, which are fringed inside with long white hairs.

Fringed water-lily, *Nymphoides peltata*, is a small, yellow-flowered water-lily-like plant of slow-moving rivers, ponds and lakes, whose petals, like bogbean's, are edged with fine hairs. It is probably native in some areas – e.g. the non-tidal reaches of the Thames and the East Anglian Fens. But it is also widely planted as an ornamental in ponds (especially village green ponds) and in the absence of competing vegetation can rapidly colonise the whole surface.

Bogbean, an attractive species of ponds and bogs. The leaves were once used to treat rheumatism.

Jacob's-ladder family *Polemoniaceae*

Jacob's-ladder, *Polemonium caeruleum*. Truly native sites for this beautiful species, with its spires of bright blue (or, rarely, white) flowers, are confined to grassland, scree and rocky ledges in the Derbyshire and Yorkshire Dales and one place in Northumberland. Its most famous and long-lasting site is at Malham Cove in Yorkshire, where John Ray first saw it in 1671.[1]

Specimens appearing elsewhere are almost invariably garden escapes. Jacob's-ladder is a popular border plant, though the varieties in cultivation are usually continental in origin, with larger, earlier-flowering blooms.

The name 'Jacob's-ladder' (recalling the story in Genesis, Chapter 28) probably derives from the pinnate leaves, which have rows of rung-like leaflets.

Part of the spectacular colony of Jacob's-ladder at Malham Cove, which mounts the scree like a 'ladder to heaven'.

Borage family *Boraginaceae*

Purple gromwell, *Lithospermum purpureocaeruleum*. This is a glittering recluse of scrub and wood margins on the Mendip limestone and in south Devon and a few parts of Wales. The flowers are reddish-purple to begin with, but soon turn a bright blue and shine through the brambles with the intensity of gentians.

Viper's-bugloss, *Echium vulgare*, is a viperish plant in all its parts. The sprays of flowers that spiral up the stem are half-coiled; the long red stamens protrude from the mouths of the blue and purple flowers like tongues; the fruits resemble adders' heads. Even the 'speckled' stem (it is hairy in fact) suggested snakes' skins to early herbalists.

Viper's-bugloss prefers dry, sandy soils and can put on spectacular shows of colour. Carl Linnaeus adored fields of *Echium*, 'surpassing in splendour anything that can be imagined'.[1] Part of the Marske sand-dunes in Yorkshire is called 'the Blue Mountain', because of the abundance of bugloss.[2] It will grow on bare or disturbed patches of chalk grassland (on Salisbury Plain, on old tank-tracks especially)[3] and is prolific in the East Anglian Breckland, lining open forest rides in company with weld (see p. 157). It also seems able to colonise and tolerate polluted soils, and I have seen sweeps of it growing from tar-contaminated concrete in an abandoned gasworks at Beckton in east London. It has even found its way to Australia and America: 'I have read that in parts of the USA it is known as Blue devil. And an Australian friend saw a picture of viper's bugloss. "Crikey, we've got that. It's Paterson's Curse." '[4]

Lungwort, *Pulmonaria officinalis* (VN: Jerusalem cowslip, Jacob's coat, Soldiers and sailors, Mary-spilt-the-milk). This is a familiar early-flowering garden plant from continental Europe, widely naturalised in woods and hedge-banks, occasionally as white or purple cultivars. The name 'lungwort' derives from the days of sympathetic magic, when the blotched oval leaves were seen as 'signing' the plant for ulcerated and diseased lungs. 'Mary-spilt-the-milk' also refers to the white-splashed leaves. Most of the other double or triple names, including

Viper's-bugloss, unpredictable but sometimes abundant on disturbed sandy and chalky soils.

'Soldiers and sailors', 'Jacob's coat' and the now obsolete (at least for this species, see below) 'Abraham, Isaac and Jacob', derive from the variety of colours to be seen in the flowers, which change gradually from pink to blue, so that the plant is always 'in the motley'. Two rare and highly local native lungworts are **unspotted lungwort**, *P. obscura*, with darker green leaves, which clings on in a couple of woods in Suffolk, and **narrow-leaved lungwort**, *P. longifolia*, a speciality of the New Forest and the Isle of Wight.

Comfreys, *Symphytum* species, are a small but difficult group, variable in themselves and prone to hybridisation. The three commonest species, below, tend to be used interchangeably in herbal medicine, etc. **Common comfrey**, *S. officinale*, is a bushy perennial with bristly leaves and spear-shaped, reticulated leaves. The flowers are pale cream or purplish and hang in bell-like clusters. It is a native, locally frequent by streams and rivers, in fens and ditches and on damp roadsides and waste ground. **Russian comfrey**, *S.* × *uplandicum*, is found in similar habitats and is now the commoner plant. It is a cross between common comfrey and **Rough comfrey**, *S. asperum*, and was probably first introduced to Britain as a fodder plant. The flowers are blue to violet or purplish when open. It back-crosses with *S. officinale*, forming a range of intermediate types. **White comfrey**, *S. orientale*, was introduced from west Russia and Turkey and is naturalised in hedgerows, churchyards and waste places, chiefly in eastern and southern England.

(VN for all species: Knitbone, Nip-bone, Ass-ear.) As can be guessed from its surviving common names, comfrey is still used as a healing poultice, for sprains, bruises and abrasions, and with more apparent success than almost any other herbal medicine. In this respect its users are carrying on a tradition which goes back to classical times. Comfrey (probably a corruption of the Latin *conferva*, a healing waterplant mentioned by Pliny, whose name is related to the verb *confervere*, to grow together) contains a substance called allantoin, which promotes healing in connective tissue. The medieval herbalists knew the plant as 'bone-set', and the root was lifted in spring, grated and used much as plaster of Paris is today. The

Narrow-leaved lungwort in Swanpond Copse, Isle of Wight. A local rarity confined to the Isle of Wight and the New Forest.

whole plant was regarded as a master-healer and was used for everything from drawing splinters to easing backache.

Today, the uses are just as various and range from emergency backwoods first-aid to relieving the sting of a hard catch in a village cricket match.[5]

'After a chain-saw accident I cycled four miles to Riseley to get some comfrey roots. That ride was uncomfortable and opened up the gash even more. I dug some comfrey root and scraped the white flesh into a thick, jelly-like paste. I plastered this into the cut and topped it off with a piece of kitchen towel pushed into the jelly. After a while the jelly set into a stiff but yielding gel which held the edges of the wound together. I did nothing more with the cut, just left the comfrey in place until it and the sides of the cut had dried, at which time it more or less fell out. Within a couple of weeks only a surprisingly small scar could be seen.'[6]

Comfreys are common waste-ground and wayside species, much used in herbal medicine for healing wounds.

'I work as a GP, and one of my patients had a coronary artery bypass recently. This was a very high-tech operation at the famous unit at Harefield Hospital. The lower end of the long leg wound failed to heal despite numerous standard dressings of different medical products. After an interval of a few days I visited the house to find a beaming patient. The leg wound was healed. He was delighted to tell me he had made an infusion of comfrey from his own garden plants (he is a town-dweller) and applied soft dressings soaked in the infusion.'[7]

'Conway Valley Nurseries had large quantities of Russian comfrey growing in an unused area in 1987. The owner told me that, prior to the First World War, the previous owner had cultivated this plant and that the leaves were transported to Manchester markets, where they were sold to cotton-mill workers for lining clogs to ease tired and aching feet.'[8]

'In the Yorkshire coalfields, comfrey poultices were used to treat what was called "beet knee" – purple, painful and unbearable to walk on, the reward of crawling in the low seams of the mine. Mother's remedy was the fresh leaves, boiled, drained slightly and then placed as hot as he [Father] could stand on the offending knee, and bandaged tightly on top to hold it in place. This indeed was a miracle cure, the swelling being reduced overnight and Father back at work without too much time off. I can still remember, 35 years on, helping Mum to pick those comfrey leaves and the exact location of the plant.'[9]

'My mother is in her eighties and when she came to visit us after breaking her wrist last year [1992] announced while unpacking, "I've bought some knitbone with me, to put on my wrist, that Elizabeth picked, up the lane." From these she made an infusion which she used to bathe her wrist.'[10]

Similar stories of use on ageing and arthritic joints come from many places, including an old people's home in Staffordshire.[11] Its use even extends to household pets: 'I used it in a poultice for a dog after a road accident – his fur grew back, so I used it on a bald guinea pig, whose fur also grew back.'[12]

The old practice of taking regular comfrey infusions or concentrated tablets internally for ailments such as gastric ulcers and colitis is now discouraged, as the plant contains alkaloids which can cause liver damage in large quantities. But the occasional young leaf eaten as a salad, or fried in batter as the German dessert Schwarzwurz, is unlikely to cause any problems. Comfrey was eaten quite widely in the Second World War.[13]

But probably the most widespread use of comfrey at present is as 'green manure'. Gardeners either grow it on the spot and dig it in or make a liquid feed from it: 'People are using it to make a practical tomato fertiliser by soaking a carrier bag full of the foliage in a big bucket of water until it goes brown and smells like animal manure.'[14]

Green alkanet, *Pentaglottis sempervirens* (VN: Bird's-eye). A pert, bristly, medium-sized perennial from south-west Europe, now widely naturalised in hedge-banks and woodland edges, especially close to settlements, it has bright blue, white-eyed flowers that begin to bloom in March and carry on well into the summer. It is at its best in May, when the flowers are often caught up in the lacy nets of fading cow parsley.

The name alkanet derives from the Arabic for henna, and it is possible that this 'little henna' was first introduced to this country for the much cheaper red dye which can be extracted from its roots.

Bugloss, *Anchusa arvensis*, is an untidy, bristly species. It is a native, quite common as an annual in arable and waste ground on light soils. **Borage**, *Borago officinalis*, is an annual from southern Europe, popular in herb gardens for its leaves – which, despite their hairiness, have a fresh taste reminiscent of cucumber – and its star-like blue flowers. Both are added to summer fruit cups (an

Above: the brilliant blue flowers of borage are used to decorate summer fruit drinks. Left: evergreen alkanet, self-seeded in the High Street, Whitchurch-on-Thames, Oxfordshire.

echo, perhaps, of its old herbal use as a reviver for 'the hypochondriac and ... hard student').[15] In more gracious times, Richard Jefferies reported, borage leaves used 'to float in the claret cup ladled out to thirsty travellers at the London railway stations'.[16] A modern conceit is to freeze the blue flowers in ice-cubes. **Abraham-Isaac-Jacob**, *Trachystemon orientalis*, is a similar though smaller plant, whose bluish-violet petals arch back, leaving a prominent cone of stamens. Originally from the Caucasus and Turkey, it is naturalised in a few damp woods and shady hedge-banks.

Oysterplant, *Mertensia maritima*, is a rather scarce native of shingle or sand-banks in Scotland and north-west England. The smooth, grey, fleshy leaves have been eaten and have a taste too much like oysters for one Scottish browser's constitution: 'A shingle beach near Sunnyside carries the only local patch. I was once poisoned by an oyster and can't abide them. The taste of the leaf of this plant made me retch, so true is it to its name.'[17]

Common fiddleneck, *Amsinckia micrantha*, is something like a bristly, orange-flowered forget-me-not. Recently arrived in this country as an impurity in crop seed from North America, it is commonest – and increasing – as a weed of carrot fields on the sandy soils of East Anglia. **Scarce fiddleneck**, *A. lycopsoides*. An isolated, naturalised colony has persisted on rough ground in the Farne Islands since 1922. It is also increasing on light soils in eastern England.

'Le ne m'oubliez pas ou Vergissmeinnicht': Pierre Redouté's (1759–1840) painting of forget-me-not gives its name in French and German. It was not commonly known as such in English till the beginning of the nineteenth century.

Water forget-me-not, probably the first of the family to be taken into gardens.

Water forget-me-not, *Myosotis scorpioides*. Forget-me-not is not such an ancient, cottagey name as it appears. There is no forget-me-not in Shakespeare, and no instance of the name quoted in *The Oxford English Dictionary* between about 1532 (in a translation of the French '*une fleur de ne m'oubliez mye*') and Coleridge's poem 'The Keepsake', which it dates at 1817. Gerard (1597) describes the plant, but calls it a scorpion-grass.

It was almost certainly Coleridge who popularised the name in English. He knew German folklore and would have been familiar with the tale of the knight who picked *Myosotis scorpioides* for his lady as they strolled by a river.

The knight fell in, but, before he was drowned, he threw the flowers to his love, crying '*vergisz mein nicht*'. In 'The Keepsake', which was first published in a newspaper in 1802, Coleridge writes:

> *... Nor can I find, amid my lonely walk*
> *By rivulet, or spring, or wet roadside*
> *That blue and bright-eyed flowerlet of the brook,*
> *Hope's gentle gem, the sweet Forget-me-not!* [18]

If there is any doubt from the descriptions of plant and habitat that the plant in question is our water forget-me-not, it is dispelled by Coleridge's note to the poem: '[This is] one of the names (and meriting to be the only one) of the Myosotis Scorpioides Palustris: a flower from six to twelve inches high, with blue blossom and bright yellow eye. It has the same name over the whole Empire of Germany ... and, we believe, in Denmark and Sweden.'

Water forget-me-not may have been the source of early garden forget-me-nots. But most perennial garden varieties are now forms of **wood forget-me-not**, *M. sylvatica*. This is a rather local plant in Britain, quite common in woods and on rock-ledges in the north, but confined to damp ancient woods in the south. It is a species which is transformed when seen truly wild in an upland wood, the flowers often in steep tiers and gleaming azure-blue amongst the ferns. Escaped garden varieties, cropping up in other habitats such as hedge-banks, are often larger-flowered, and pink and white forms are quite frequent. Garden annuals are usually the large-flowered form of **field forget-me-not**, *M. arvensis* var. *sylvestris*. *M. arvensis* itself is a common annual of cultivated and waste ground, flowering from early April. This was probably the species John Clare was thinking of in his poem 'On May Morning':

> *The little blue Forget-me-not*
> *Comes too on friendship's gentle plea*
> *Spring's messenger in every spot,*
> *Smiling on all – 'Remember me!'* [19]

(Appropriately for John Clare country, a Northamptonshire 'Elders' Club' is called 'Forget-me-not, and each member has a small enamel brooch of the flower'.) [20]

Even sooner in bloom is **early forget-me-not**, *M. ramosissima*. This is a much smaller, almost dwarf species, with softly-haired greyish leaves, which is locally common on open areas of chalky or sandy soils in lowland Britain. **Changing forget-me-not**, *M. discolor*, has a similar taste in habitats – though it also occurs in damp places and is more widespread. The miniature flowers are yellow or cream at first, changing to pink, blue or violet.

Hound's-tongue, *Cynoglossum officinale*, is a biennial of open patches on sandy and chalky soils, chiefly in southern and eastern England. The flowers have a colour whose tone is more like that of a dyed fabric – worn purple velvet, perhaps – than a bloom. The seventeenth-century herbalist John Pechey described them as 'a sordid red'.[21] The leaves are long, greyish and covered with soft hairs, and sufficiently reminiscent of a dog's tongue to give the plant its name and suggest some of its early medicinal uses. They were given in cases of dog-bite and sometimes worn in the shoe as a charm to deter dogs. They also have a very unusual smell, which herbalists, doubtless eager to boost the plant's canine associations, likened to dogs' urine. It was a long way off the mark, and it needed that twentieth-century invention, the roasted peanut, to provide a real analogy.

Hound's-tongue's long and slightly furry leaves were once given as a remedy for dog bites.

Vervain family *Verbenaceae*

Vervain, *Verbena officinalis* (VN: The Herb). A rather local perennial of bare ground and rough grassland, chiefly on chalky soils in southern England, vervain was once a venerated plant, valued not just as a panacea (it was trumpeted as a cure for the plague in the Middle Ages) but as a magical charm, which could both protect against witches and demons and conjure up devilry of its own. It was traditionally associated with the gods of war, and gun-flints were sometimes boiled with rue and vervain to make them more effective.[1] As so often, the church exorcised its magic by appropriating the plant and suggesting that it grew under the cross at Calvary – though there were still incantations for picking it, albeit couched in Christian language and symbolism:

Vervain, an unprespossessing plant, but one of the Anglo-Saxons' most valued herbs.

Hallowed be thou, Vervein, as thou growest in the ground,
For in the mount of calvary there thou wast first found.
Thou healedst our Saviour Jesus Christ, and stanchedst his bleeding wound;
In the name of the Father, the Son, and the Holy Ghost, I take thee from the ground.[2]

The Isle of Man is the last British redoubt of belief in vervain's potency, though the plant is probably not native there:

'The Manx name, Yn Lus or Yn Ard Lus, gives an insight into its importance; the translation is The Herb or The Chief Herb. It has medical uses, but mere possession of it conferred all manner of protection. A person going on a journey would carry a piece, and many a Manxman would have a piece permanently sewn into his clothing. I have seen a number of plants growing in gardens, but so far I have not been successful in obtaining a plant for myself. The procedure for getting a piece is rather complicated. It cannot be asked for directly. Broad hints will be dropped and perhaps the possessor will take the hint and a plant will discreetly change hands, usually wrapped in paper. No word should be exchanged. It must always change hands from man to woman or vice-versa. It can be stolen, but I have not stooped to that yet.'[3]

The odd thing is that the plant behind this great edifice of lore is rather scrawny and nondescript. But its small pale lilac flowers, slowly opening up the spike until there is just a single one at the tip, suggested the nickname 'sparklers' to one family.[4]

Dead-nettle family *Lamiaceae*

Betony, *Stachys officinalis*. Like an elegant, late-flowering red dead-nettle, betony is widespread but rather local in old grassland, wood edges and heaths in England and Wales. It was one of the great 'all-heals' of medieval herbalists.

Lamb's-ear, *S. byzantina*, is a popular garden plant from south-west Asia, with thick, densely woolly silver leaves, which is quite widely naturalised on waste ground and roadsides near habitation.

Downy woundwort, *S. germanica*, is a rare species in Britain, but one with remarkable territorial loyalty. For

Betony growing amongst wild carrot on cliffs at Bedruthan Steps, Cornwall.

three and a half centuries, since it was first discovered by a London apothecary 'in the field joyning Witney Parke, a mile from the Towne',[1] its main colonies have all been rooted in the limestone country of west Oxfordshire. In that time its robust spires – dramatically tall for a dead-nettle – have been searched for by generations of distinguished naturalists. John Blackstone found it growing plentifully 'by the lane leading from Wychwood Forest to Charlbury' in 1737 (just north of the original Witney site). On 20 October 1767, Gilbert White, on a visit to his friend John Mulso, then rector of Witney, 'rode out on purpose to look after the base horehound, the *Stachys Fuchsii* of Ray, which, that gent: says, grows near Witney park: I found but one plant under the wall: but farther on near the turnpike that leads to Burford, in a hedge opposite to Minster Lovel [west of Witney], it grows most plentifully. It was still blowing, & abounded with seed; a good parcel of which I brought away with me to sow in the dry banks round the village of Selborne.'[2] In 1794, Humphrey Sibthorp, Professor of Botany at Oxford, listed it for waysides around Witney, Woodstock and Stonesfield. A century later George Claridge Druce saw it in some of the same parishes and on road-verges near Sturdy's Castle pub, which had long been a favoured site.[3] Today (1996), the main populations remain in west Oxfordshire, close to Witney and Wychwood – although the plant has appeared, and then in all cases disappeared, in at least 10 other counties in the nineteenth and early twentieth centuries. And if it rarely reappears in its exact historical sites, downy woundwort continues to demonstrate its ancient preference for lanes and roadsides. Of the nine sporadic colonies which have been recorded since 1970, six have been by roadsides, including the traditional site near Sturdy's Castle, a lane near Worsham, and the old Roman Road (Akeman Street) near Minster Lovell.

Downy woundwort has a remarkable loyalty to the area around Witney, Oxfordshire.

Why has downy woundwort persisted in this area, yet disappeared from other places where conditions seem to be just as suitable? Why does it have this curious loyalty to waysides – yet even here vanishes often for decades at a time? Peter Marren, who has studied the plant's history in detail, believes that the answer lies partly in downy woundwort's ecological requirements and partly in the nature of the landscape around Wychwood Forest.

In central and southern Europe, downy woundwort is a common biennial or perennial of dry calcareous banks and pastures. It is on the northern edge of its range in England and prospers only where there is light and warmth – and, paradoxically, some disturbance of the ground to enable its seeds to become established. They are very heavy, rarely travelling far from the parent plant (one observer likened their descent to the ground to 'coals falling from a scuttle'), and can't easily get a roothold amongst more aggressive species. The lanesides favoured by the plant have typically been ancient trackways, drove-roads and green lanes, characterised by tall hedges, broad verges and long continuity of use. Until quite recently they would have been kept in a condition that exactly suited downy woundwort, with some periodic churning of the soil, grazing by animals and the tolerance of longer vegetation close to the hedge. But many have fallen into disuse and are no longer cut or kept open, and downy woundwort has been smothered by rough grasses and broadening hedges. That it so often reappears when the hedges and verges are cut back shows that the seeds are capable of long periods of dormancy.

The area around Wychwood Forest has, from a very early period, been 'scored by an intricate network of narrow lanes, "like veins on a leaf", many of which survive in their original, unmetalled condition', providing an unusually high density of the marginal habitats which the woundwort requires. It is a signature plant of the geography and human history of this patch of Oxfordshire, a wayside familiar, and it is to be hoped that current efforts to conserve it will succeed.[4]

Limestone woundwort, *S. alpina*, is similar to the preceding species, but with less grey felting on the leaves. It is a very rare native, now restricted to two sites, one in Denbighshire and one in west Gloucestershire. **Hedge woundwort**, *S. sylvatica*, is a common perennial of woods, hedgerows and rough ground, with a strong, astringent smell when crushed. **Marsh woundwort**, *S. palustris*, is a more elegant species, with pinkish-purple flowers and an upright habit, quite common by the side of rivers, ponds and canals throughout Britain. This was regarded as being the most effective 'wound-herb' of the group. **Black horehound**, *Ballota nigra* (VN: Stinking Roger), is a common, bushy perennial of waste ground and waysides, with foetid foliage. **Motherwort**, *Leonurus cardiaca*, was introduced from continental Europe in the Middle Ages as a medicinal herb to ease childbirth. It survives as a naturalised plant in scattered sites across England and Wales and especially in the Isle of Man, presumably (though not always obviously) as an escape from cultivation.

Yellow archangel, *Lamiastrum galeobdolon* (VN: Weasel-snout). This plant of ancient woods and old hedge-banks on well-drained soils throughout England and Wales has golden-yellow flowers that bloom in May, just as the bluebells (with which it very often grows) are fading. 'In Suffolk yellow archangel is a good indicator of old ditches and banks – indeed its presence enabled parts of the boundaries of three medieval deer parks that covered much of the parish of Hundon to be identified.'[5]

'Archangel' is shared, as a vernacular name, with other 'dead' nettles, and may refer – albeit grandiosely – to their virtue of being non-stinging. 'Aluminium archangel' is the improbable but memorable nickname coined in Glasgow for ssp. *argentatum*, the silver-leaved (or more accurately silver-blotched) variety that was introduced to gardens in 1960 and is now escaping widely.[6]

White dead-nettle, *Lamium album*, is a common perennial of waysides and rough ground in most of Britain. Most country children know that a small drop of nectar can be sucked from the base of each flower, but a less benign game is occasionally played with this plant: 'Boys used to pick the flowers off dead nettles and chase the girls pretending they were real stinging nettles.'[7]

Red dead-nettle, *L. purpureum*, and **henbit dead-nettle**, *L. aplexicaule*, are both common annuals of cultivated and waste ground whose flowers can also be milked for a smidgen of sweetness. **Bastard balm**, *Melittis melissophyllum*, is an unusual plant, a dead-nettle with huge (up to one and a half inches long) flowers of blotched purple and white. It is more or less confined to south-west Britain, but is locally common on a few hedge-banks in Devon and Cornwall, where it has been much transplanted into gardens. It is 'bastard' balm to distinguish it from the true 'bee' or 'lemon' balm, *Melissa officinalis* (see below). **White horehound**, *Marrubium vulgare*, is a white-flowered, woolly-leaved species which can still be found as an ingredient in herbal cough medicines – a use that goes back at least 2,000 years. It is possibly native in dry grassy places near the sea and a few sandy sites inland. Elsewhere it is an escape from cultivation.

Skullcap, *Scutellaria galericulata*, is a delicate species of fens and the banks of ponds, canals and slow rivers, locally common throughout much of Britain. The plant's English and Latin names both derive from the shape of the blue flowers, which reminded early botanists of the leather helmet or *galerum* worn by Roman soldiers.

Wood sage, *Teucrium scorodonia*, is a sage in almost every respect except scent, of which it has very little. It has sage-green, crinkly leaves and upright sprays of straw-coloured flowers that suit the tawny heaths and dry woodland rides which are its favourite habitats in most of

Yellow archangel, an indicator of old woods and banks, in a green lane at Ipsden, Oxfordshire.

Left: wood sage, more usually found on acid soils, here growing in a limestone pavement. Gait Barrows, Lancashire.
Below: bugle.

Britain. **Wall germander**, *T. chamaedrys*, is a short evergreen shrub from Europe, grown in gardens and naturalised on some old walls and banks, but seemingly native in one place in the extreme south-east of England, on chalk downland near Cuckmere in Sussex.[8]

Bugle, *Ajuga reptans*, is a finely structured and tinted plant of woodland clearings and damp grassland, quite common throughout Britain and often growing in large troops. Its blue flowers (rarely pink) stand out against dark leaves whose colour is hard to pin down. They have a sheen of purple-brown on a dark-green base, rather like metal which has been tempered in a fire. A good fourteenth-century name was 'wodebroun'. The modern name bugle has nothing to do with the brass instrument, but probably derives from the dark, lustrous and long glass beads which were once sewn onto clothes as ornaments and which echo aspects of both the flowers and leaves.

Ground-pine, *A. chamaepitys*, is a very rare annual of arable fields and bare patches on the chalk in south-east England. It resembles a yellow-flowered pine shoot, even to the extent of having a faint resinous scent.

Cat-mint, *Nepeta cataria*, is a scarce species found chiefly on calcareous soils in south and east England. It has white flowers, spotted with purple, and does not much resemble the familiar blue-flowered **garden cat-mint**, *N. × faassenii* (which is occasionally naturalised), except in its distinctive smell, which, as Gerard describes,

Ground-ivy. The aromatic leaves make a good herb tea and were once used as a bittering agent in beer.

is relished by cats: 'The latter herbarists do call it *Herba Cattaria* ... bicause the cats are very much delighted herewith; for the smell thereof is so pleasant vnto them, that they rub themselues vpon it, and wallow or tumble in it, and also feede on the branches and leaues very greedily.'[9]

Ground-ivy, *Glechoma hederacea*, is a common, blue-flowered dead-nettle of woods, hedgerows and damp rough ground throughout most of Britain. A pleasant, now obsolete, name was 'blue runner', because of the plant's habit of forming large clumps by overground runners. It has a strong, rough aroma, and, before hops became widely used in brewing, it was one of the chief bittering agents in the making of beer – hence the name 'alehoof'. It is still quite widely used in tonic herb teas.

Self-heal, *Prunella vulgaris*, is abundant on short grassland and in woodland clearings, and very familiar on lawns which have not had weed-killer applied, where its bluish-violet flowers often bloom amongst the grass. It seems a quite different plant in unshorn grass or in a woodland ride, where it can often grow to a foot in height. It was a popular wound-herb in country areas until quite recently: 'In the Kentish weald during the war, a family of charcoal burners used the leaves of self-heal for cuts and bruises. The leaves were smeared with lard which acted as a binding and soothing agent.'[10]

Balm, *Melissa officinalis* (VN: Lemon-balm, Bee-balm), is a lemon-scented herb from southern Europe, frequently self-sown or naturalised in waysides and waste places. The leaves are used in teas, and the white flowers loved by honey-bees, and most feral colonies originated from herb-garden plantings. **Wild basil**, *Clinopodium vulgare*, has an odd name for a species that neither looks nor smells like true basil, *Ocimum basilicum*. But it is a cheery and eye-catching plant of dry hedge-banks, chalk grassland and heaths in high summer. It carries its pink-

Wild basil – a more accurate name is 'cushion calamint'. Unlike its culinary namesake, wild basil is almost scentless.

ish-purple flowers in whorls up the stem, with a downy, dome-shaped cluster on the top, which gave it the more descriptive but now obsolete name of 'cushion calamint'.

Gipsywort, *Lycopus europaeus*, is quite a common plant of wet places (mainly in England and Wales), with distinctive jagged-edged leaves. It is most often found in fens or by riversides, but occasionally in damp woodland clearings, and it has long been used to give a fast black dye. The name originates from an ancient belief that gipsies used the juice of the plant to darken their skins. It is quite possibly an apocryphal story, but it shows how the hierarchies of prejudice can change. The most unpleasant version was given by Caleb Threlkeld in 1727: 'Some call this the *Gipsy-herb*, because those stroling Cheats called *Gipsies* do dye themselves of a blackish Hue with the Juice of this Plant, the better to pass for *Africans* by their tanned Looks, and swarthy Hides, to bubble the credulous and ignorant by the Practice of Magick and Fortune-telling; they being indeed a nasty Sink of all Nations, living by Rapine, Filching, Pilfering and Imposture.'[11]

Wild marjoram, *Origanum vulgare*. This is the same species as the 'oregano' that is such a characteristic herb of Mediterranean cooking. It does not develop quite such an earthy fragrance in our cooler climate, but it is still an excellent herb in the kitchen and common enough in rough calcareous grassland throughout Britain for small-scale gathering of sprigs and leaves to be acceptable. It is a perennial, growing up to two feet tall, with purple, pink or sometimes white flowers in bunches at the top of the stems, and attracting many species of butterfly in high summer.

Thyme, *Thymus vulgaris*. This is the thyme normally used in cooking, a Mediterranean species which sometimes persists as a garden throw-out on old walls and dry banks. There are three other species native to Britain: **Breckland thyme**, *T. serpyllum*, confined to sandy heaths in west Suffolk and Norfolk; **large thyme**, *T. pulegioides*, a sprawling, larger-leaved species, locally frequent in short chalky or sandy grassland in southern and central England; and **wild thyme**, *T. polytrichus*. This last is by far the most widespread and abundant species. It is confined to the chalk in south-east England, but elsewhere will grow on acid, short-turfed pastures, ant-hills in meadows, cliffs, walls and rocky places. Wild thyme often grows in large mats, especially on the chalk, and provides one of the memorable scents of walks over the southern downlands in warm sunshine. Both our common wild species are fragrant enough to be serviceable in cooking.

All species of thyme contain the characteristic volatile oil, thymol, which is a reasonably powerful antiseptic. Perhaps it was some inkling of this quality which led to the herb being a key ingredient in Judges' posies and the Sovereign's Maundy Thursday posy, which were both devices originally intended to afford the carriers some

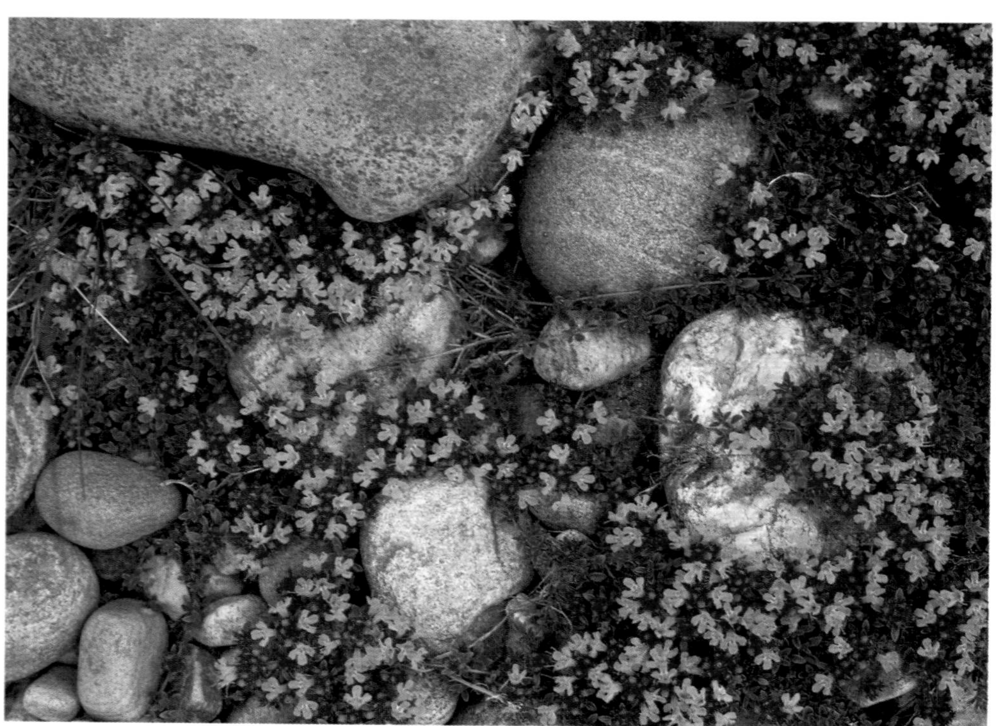

Wild thyme on river-bank shingle.

protection from the infectious diseases of the poor.

'In the Western Isles wild thyme, which grows abundantly, was put under the pillow or drunk as an infusion to prevent nightmares or otherwise give a restful sleep. Thyme tea was popular throughout the Highlands as an everyday beverage. In an area where lavender would not grow well, women used flowering sprigs of thyme to scent their clothes, handkerchiefs and household linen.'[12]

Other fragrant Mediterranean herbs sometimes become naturalised on old walls and dry ground, for instance **winter savory**, *Satureja montana*, at Beaulieu Abbey, Hampshire; **rosemary**, *Rosmarinus officinalis*, especially in graveyards, where it is planted as a symbol of remembrance; and **lavender**, usually *Lavandula × intermedia*, a hybrid of *L. angustifolia*. Self-sown lavender seedlings are increasingly common both in and outside gardens in southern England – perhaps an indication of the trend towards milder winters.

Mints. There are some 14 or 15 mint species and hybrids found growing wild or naturalised in Britain. Some are escaped garden cultivars, such as **apple mint**, *Mentha × villosa*; others are natives which have been taken into cultivation. Most grow in damp places: woodland clearings, pondsides, humus-rich waste ground, and all have the characteristic mint-scent – though there are many variations on this. In 1798, William Sole published a survey of British mints (he included 25 separate kinds), and his vivid attempts to pin down their different scents is the most entertaining feature of the book.[13] **Corn mint**, *M. arvensis*, 'has a strong fullsome mixed smell of mellow apples and gingerbread' – which is certainly more complimentary than Grigson's 'wet, mouldy gorgonzola'. (Sole's spotting of the hint of gingerbread was perceptive, and corn mint is almost certainly the ancestor of the yellow-striped garden variety known as 'ginger mint'.) His 'strong-scented mint' – which may be our **sharp-toothed mint**, *M. × villosonervata* – 'has a very strong volatile mixed smell of volatile salt and amber, camphor, and mint' and is 'an honourable relict of our venerable Gothick ruins'. The scent of **water mint**, *M. aquatica*, 'is exactly that of a ropy chimney in a wet summer, where wood fires have been kept in winter-time'. (Water mint, incidentally, has a bronze-leaved variety which seems to be confined to pond-edges and damp flushes on limestone and which possesses a distinct Eau-de-Cologne scent, though true Eau-de-Cologne mint is usually referred to var. *citrata* of peppermint.)

Peppermint, *M. × piperita*, is the most commercially exploited species, for its essential oil which is used both in medicine and in cooking. A Surrey woman remembers gathering it (and many other medicinal herbs) during the Second World War:

'At the beginning of the war, through writing to a newspaper, I was put in touch with Messrs Brome and Schimmer, druggists in London, who supplied hospitals. Mr Brome came down to see me to make sure I knew the plants, because in the First War some enthusiastic gatherers had mixed poisonous and non-poisonous plants ... In our spare time we gathered and dried the plants and sent them off by train in sacks. We didn't get great quantities. The dried weight is so little compared with the gathered plant, but the druggists were very pleased with the quality of what we sent and the money they gave us we gave to the Red Cross. Among the plants we gathered were foxglove leaves (for the heart), nettles, coltsfoot leaves (for asthma), wood betony and agrimony (tonics) and peppermint for flavouring. Our room at home had strings stretched across between the picture rails hung with bunches of drying herbs. When we had peppermint drying my mother and I went to sleep in our chairs.'[14]

Pennyroyal, *M. pulegium*, is a smaller, creeping species, with a very pungent odour. In folk medicine it was regarded as something of a panacea and widely grown in cottage gardens. It certainly has sedative and antispasmodic properties and, in large doses, had an underground reputation in the countryside as an abortifacient. Pennyroyal's habitat – the muddy edges of cattle wallows and village ponds (cf. starfruit, p. 382) – has declined dramatically over the past few decades, and it is now an endangered species, down to little more than a dozen sites, mainly in the south.

Wild clary, *Salvia verbenaca*. Not unlike cultivated sage, though less aromatic, wild clary grows in dry grassland, dunes and roadsides in southern and eastern Britain, and seems to have had a long affinity for churchyards, particularly in Suffolk and Sussex. The Revd Dr Frederick Arnold (author of *The Flora of Sussex*, 1887) suggested that this was the result of a medieval practice of sowing 'wild English sage' on graves, in the belief that it conferred immortality, and cited as his authority a twelfth-century aphorism: 'Why should he who grows sage in his garden die?' Arnold's interpretation seems rather perverse and it

Pennyroyal, once a mainstay of folk medicine, but now reduced to a handful of pond-edges and damp commons.

is more likely that the saying refers to the health-giving properties of clary for the *living*. (The Latin name *Salvia* is related to *salus*, 'health'.) In early herbal medicine it was a 'clear-eye', which became vulgarised to 'clary'. The seeds were soaked in water until they became mucilaginous, and the resulting jelly (rather like frog-spawn) was put in the eye to soothe and cleanse it.

Several other species of *Salvia* from southern Europe are naturalised in waste ground, including **sage** itself, *S. officinalis*, **clary**, *S. sclarea*, and **whorled clary**, *S. verticillata*.

Plantain family *Plantaginaceae*

Greater plantain, *Plantago major* (VN: Rat's tails, Angels' harps, Banjos). 'Rat's tail' is a perfect description of the flowering spike of this very common perennial of paths, pavement cracks, waysides, lawns, short grassland and field edges. The rosette of leaves, lying flush with the ground, also gave it a name when it migrated to North America with the early settlers. Amongst the Indians it was known as 'English-Mans Foot', not so much because its leaves are flat and broad, but because they seemed to dog the settlers' tracks, 'as though produced by their treading'.[1]

Plantain's leaves are tough, elastic and resilient, and exceptionally tolerant of trampling. This quality, interpreted according to the principles of sympathetic magic, suggested that it would be a healing herb for bruising and crushing wounds. As 'waybread' (and there could not be a more basic or reverent description than that) it was included amongst the Anglo-Saxons' nine sacred herbs:

And you, Waybread, mother of worts,
Open from eastward, powerful within,
Over you chariots rolled, over you queens rode,
Over you brides cried, over you bulls belled;
All these you withstood, and these you confounded,
So withstand now the venom that flies through the air,
And the loathed thing which through the land roves.[2]

The healing powers of plantain aren't entirely fanciful. The leaves contain tannins and astringent chemicals, which can make them useful styptics if crushed and applied to small cuts, and an alternative to dock leaves in the relief of nettle stings.

The elasticity of the leaves has also made them natural subjects for children's games:

'The stalks or leaves, if broken gently, retained a few strong fibres, which are slightly elastic, allowing one to

Albrecht Dürer's watercolour 'Large Tuft of Herbs' (1503), which clearly shows greater plantain leaves, as well as yarrow and dandelion.

"milk a cow" by pulling the leaf gently out, then relaxing it. Or seeing who could get the longest fibre before the leaf finally parted.'[3]

'I remember we used to pull off the leaves of ratstail plantain, and from the number of ribs or threads which pulled out and hung down, and by the length of them, that was an indication of how many, and how lengthy, had been the lies we had told that day.'[4]

'They were known as "Angels' Harps", because when you pull the leaves apart you get the fibres showing between.'[5]

This name has many secular – and more contemporary – variations, from 'banjos' to 'Beatles' guitars', some of which are probably confined to individual schools or even gangs.[6]

Ribwort plantain, *P. lanceolata* (VN: Fighting cocks, Short bobs, Soldiers and sailors, Black Jacks, Hard-heads, Carl doddies; Fire-weed, Fire-leaf). Ribwort plantain (also abundant in grassy places) has, in contrast with Englishman's foot, lance-shaped leaves and a short, stubby flower-head on top of a long wiry stem. This is still used in the game of 'soldiers'. The stem is wound once round itself, like a noose, just below the head. Then, by tightening the noose and pulling it sharply forward, the plantain's head is yanked off and hurled forward like a catapulted stone.

The fragrant, dusty flower-heads of hoary plantain, a typical plant of country churchyards.

A variant is to use the plantain on its long stem as a substitute for a conker and attempt to knock a rival's flower-head off. In Kent this game is known as 'dongers'[7] and in Scotland (along with the plant itself) as 'Carl doddies': 'Carl and Doddie are diminutives of Charles and George, and the name of the game is an obvious memory of the '45 Jacobite Rebellion, with Bonnie Prince Charlie and King George III trying to knock each other's heads off.'[8]

'We had a game for two with ribwort plantains, when they were in bloom. Each person pulled a stalk with a strong "head", held it out and recited:

Ma faither and your faither
Were sitting supping brose.
Ma faither said to your faither
Ah'll hit off your nose.

Then one struck a sharp blow on the rival's plantain head hoping to knock it off and so be the winner.'[9]

'As in conkers, some plantain heads are much tougher than others, and a champion soldier is greatly prized. Some veterans were like grizzled warriors and would knock out hundreds.'[10]

In hayricks, the brittleness and dryness of ribwort plantain leaves is still used by some farmers as a clue to the likelihood of the stack catching fire. It is, presumably, a rough measure of the amount of moisture in the hay itself, though one farmer, at least, believes the plantain leaves themselves could set the rick afire: 'A farmer in south Shropshire told me that ribwort plantain is called fire-weed or fire-leaf because, if not thoroughly dried, it can cause spontaneous combustion of hay.'[11]

Hoary plantain, *P. media*, is a handsome species, with white to pinkish fragrant flower-heads, midway in length between those of greater and ribwort plantain. It is locally common on grassland on chalky or neutral soils, mainly in England, and seems to have a special liking for churchyards.

Butterfly-bush family
Buddlejaceae

Buddleia or **Butterfly-bush**, *Buddleja davidii*. Since its introduction to this country from China in the 1890s buddleia has spread across almost the whole of Britain, except the far north, and could be said to have been the saving of many butterfly populations in urban areas. Its long, honey-scented purple flower-spikes, in bloom from July to October, are the favourite source of nectar for almost all butterflies and moths that haunt gardens and waste places. In my own garden in August I have regularly seen more than 50 individuals of up to ten species together on a single bush.

Buddleja davidii was first found in the mountains near the Tibetan–Chinese border in 1869 by the French missionary Père David, a discovery which is commemorated in the plant's Latin name. The first specimens were sent to Europe some 20 years later by another French missionary and plant-collector, Jean André Soulie. The early imports were apparently weak and semi-prostrate specimens, and poor in colour, and were rapidly superseded by a more vigorous and attractive variant raised in Paris in 1893 by the famous nursery firm of Vilmorin.[1]

Railside buddleia. The granite chips of the permanent way are a fair imitation of its native habitat in China.

Once established, buddleia rapidly began to colonise waste ground. Its seeds are light and winged, and can be blown some distance on the wind. And, like many species from exotic stony habitats, it seems to find the ballast along the edge of railway lines a particularly congenial habitat. The railway system provided a network of corridors along which the seeds could be dispersed over long distances, sometimes, no doubt, being drawn along by the slipstream of trains (cf. Oxford ragwort, p. 376). From its strongholds on railway embankments it spread to waste ground, bomb sites, allotments, walls, sometimes even being blown upwards to take root in chimneys. An admirer from Cheshire describes a typical invasion:

> 'In the early 1980s there were several large demolition jobs going on in Chester. In one area a large piece of derelict land was left isolated at the junction of five roads. Within a few years the whole island was covered with purple buddleia to the extent of it becoming a recognised "show". Not satisfied, this enterprising plant soon crossed the roads and is now sprouting from bridge brickwork, tiny corners of wasteland, pavements and factory buildings on the north side of the city. I see a distinct plea in all this to be recognised as British, having shown such fertility and enthusiasm for our least favoured growing sites.'[2]

In many southern cities it can form dense shrubberies that, mixed with birch and scrub willow, amount to a unique form of urban woodland. 'Its abundance in towns like Bristol is amazing; it forms thickets everywhere, colonising ledges on buildings as well as covering waste ground. A visit to the city makes it easy to visualise the description of an early visitor to China who reported that buddleia thickets on shingle beside the Satani River provided "famous harbourage for tigers".'[3]

Various colour variants, including purple and white, sometimes crop up amongst naturalised colonies, but are not usually as rich in nectar.

Urban commons

In the 1980s the phrase 'urban commons' became a widespread but informal description for the undeveloped 'white' land in and around towns and cities (which I had nicknamed 'the unofficial countryside' in a book in 1973).[1] The term covers a huge range of habitats – railway embankments, factory buffer-land, demolition sites, rubbish-tips, cemeteries, old docks, canals, even 'vertical land', such as walls and bridgework.

Dr Oliver Gilbert is the acknowledged expert on the vegetation of the urban common, and in 1993 published a fascinating acount of how this varies from city to city, depending on climate, geography and economic and social history.[2] Below, reproduced with his permission, are précis of some of the highly local and distinctive plant communities of our big cities.

Birmingham has an abundance of Canadian goldenrod, probably a consequence of its being very much an allotment plant. (The city has a long history of 'guinea gardens' and allotments.) The local Urban Wildlife Group have adopted greater bindweed as a motif for their publications, as it is conspicuous in the city, scrambling up chain-link fences and over walls.

Bristol is dominated by buddleia (see p. 323), and a slogan on one wall reads 'Buddleia rules OK'. At a site by the Royal Hotel it is beginning to be overtopped by sycamore, probably its natural successor in towns. There is a good deal of traveller's-joy in the city centre, windblown seeds from the local limestone cliffs readily establishing themselves on stony rubble, and red valerian on the older stone walls.

Fort William is one of the few urban areas of any size in north-west Scotland. The infill and rubble are largely acidic and the climate wet and cool. This has a dramatic effect on the vegetation of the urban commons, which includes heather, gorse, foxglove, Nootka lupin, rhododendron, larch and Sitka spruce.

Glasgow also has a wet climate, with moorland close by. 'The vast site formerly occupied by the Dalmarnock Generating Station is being colonised by an assemblage of woody plants that include birch, Scots pine, grey alder, common alder, broom, goat willow and cotoneasters. Nothing similiar is known in other UK cities.' There is also an abundance of elm seedlings, Dutch elm disease having spared many local trees, and a new orchid, Young's helleborine (see p. 440).

Leeds is one of the few cities where poplars regenerate from seed. In waste ground near the city centre there is a hybrid swarm of crosses between native black-poplars and balsam-poplars (see p. 139).

Liverpool has relatively few mature urban commons, owing to extensive landscaping work by environmental agencies. But there are enormous colonies of evening primrose on sand-dunes just outside the city.

Manchester is distinctive for its abundance of Japanese knotweed, major thickets of which occur at almost every site, often with the even taller giant knotweed.

Above: Rosebay willowherb on industrial patio. Derelict gasworks at Beckton, east London.
Right: the River Don corridor through the heart of the old steelworks area in Sheffield, fringed by fig trees, willow scrub and lush exotic flowers such as Indian balsam and Michaelmas-daisies.

Norwich has the most continental climate of all the cities surveyed by Oliver Gilbert. As a result of summer warmth, several colourful garden annuals such as larkspur, blue lobelia and white alyssum are becoming established on pavements and on the bare edges of waste ground.

Sheffield has large numbers of colourful garden escapes, including hillsides covered with spectacular displays of pink-, purple- and white-flowered goat's-rue. 'A Sheffield miner told me that he remembered his father recounting how in the early part of this century horticultural traders used to work the poorer parts of the city suburbs selling garden plants which only just merited that description. They were aggressive species like tansy, Michaelmas-daisy, feverfew and goat's-rue, all of which have naturalised widely in the city. He recalled his father purchasing Japanese knotweed and how friends were invited round to marvel at the spotted stem and attactive foliage and how later the plant was divided up for exchange [see p. 108].'

There are also the celebrated fig trees along the River Don (see p. 66), whose banks carry large populations of soapwort, wormwood and Indian balsam.

Southampton is a south-coast city that has characteristic communities of warmth-loving species. In several areas of railway land there is a remarkable, species-rich scrub developing amongst buddleia thickets. Up to 16 species can be involved, including firethorn, seven different cotoneasters, Norway maple and the bramble 'Himalayan Giant'.

Swansea is also dominated by Japanese knotweed, which is known locally as 'rhubarb' and 'cemetery weed'. It is often joined on waste ground by buddleia (butterfly-bush), and the combination of these two species with hemp-agrimony and pale toadflax is very characteristic. Swansea also, unusually, has large populations of the fern polypody on many walls.

Teesside has a long history as a port, and a number of its specialities are believed to have been introduced in ships' ballast. Perennial wall-rocket is unusually abundant; it was established on ballast hills by the middle of the nineteenth century and is now ubiquitous. Chalk-loving species such as carline thistle, centaury and yellow-wort, formerly associated with calcium-rich spoil-tips near steelworks, are spreading.

Ash family *Oleaceae*

Ash, *Fraxinus excelsior* (VN: Esh, Hampshire weed, Widow-maker). In Scandinavian mythology, the ash was Yggdrasil, the tree of life, 'the greatest and best of all trees. Its branches spread all over the world.' In Britain, up until the end of the eighteenth century, it was regarded as a healing tree, and Gilbert White knew Hampshire villagers who, as children, had been through an ash ritual as a treatment for rupture or weak limbs. It was an extraordinary ceremony, a relic of pre-Christian sympathetic magic. A young ash was split and held open by wedges, while the afflicted child was passed, stark naked, through the gap. The split was then 'plastered with loam, and carefully swathed up. If the parts coalesced and soldered together ... the party was cured; but, where the cleft continued to gape, the operation, it was supposed, would prove ineffectual.'[1]

These days, its mystique has disappeared. It is regarded as a second-class timber tree, neither as gracious as the beech nor as sturdily useful as the oak. The speed with which its seedlings colonise open ground on damp and calcareous soils has given it a bad name with foresters, who call it a 'weed tree' and often treat it accordingly. In a sensible world we would be grateful for its hardiness and productivity, and see that it is, albeit in a new way, still a healing plant. It is young self-sown ash trees that have largely repaired the gashes in southern England's woods caused by the great storms of October 1987 and January 1990. It is ashes that fill gaps in hedges left by dead elms and turn abandoned arable fields on the clay into woods, sometimes growing six feet in a single season.

Its resilience and rapid growth made ash an invaluable tree in the economy of small farms and cottages. It is still the commonest tree in coppice woodland across much of lowland Britain and, up till the last war, the young poles, cut on a 10-year rotation, were probably the most versatile raw material in the countryside, used for everything from firewood to fork handles. It was a sustainable resource, too. The stools from which the poles were cut could go on producing straight, stout poles indefinitely. In Bradfield Woods, Suffolk, there is an ash stool eighteen and a half feet across and showing no signs of declining vigour. Oliver Rackham estimates that it is more than 1,000 years old.[2]

Ash is still the timber of choice wherever combined strength and elasticity are needed. No other wood can be bent so safely once seasoned, or is better at withstanding sudden shocks, and it is used, for instance, in many sporting goods – oars, billiard cues, hockey sticks. There is also a continuing small-scale trade in making walking-sticks from ash wood, which has a satisfying springiness when leaned against the ground. (The smoothness of the bark in the hand is a bonus.) The wood for the commercial manufacture of 'ashplants', as they are called, is sometimes cut from two- or three-year-old coppice growth, and the handles are formed by heating the sticks in damp sand and bending them in a curved vice.

In some areas of southern England sticks are still specially grown to have naturally curved handles. Nursery-raised ash seedlings are transplanted when they are one or two years old. But, instead of being set upright, they are planted at an angle in the ground with their end buds nipped off, so that the seedling has to use a side bud if it is to continue growing upwards. This new shoot – destined to become the shaft of the stick – rises almost at right angles to the original stem, which eventually becomes the handle.

Ash's adaptability is obvious in some of its other surviving uses:

'In Northumberland, crab and lobster pots, known locally as "creeves", are still made using traditional materials. Early this century, the bases of the pots were constructed from driftwood, which was then more plentiful than it is now. Today, all the wood for pot bottoms is bought in. The frames – the arched "bows" and straight "rails" – were traditionally made from ash and hazel sticks.'[3]

'It was well known that the best source for a catapult was a young ash where the terminal bud had failed and a natural fork had developed. We made lead slugs for the catapults somewhat in the shape of humbugs by pouring melted-down lead into a mould cut from a potato.'[4]

There was also a widespread game (known as 'mud yacks' in some places) in which a ball of clay was fixed to the end of a long, whippy ash pole. The pole was either stuck in the ground and bent back like a bow, or held in the hand and flicked. Either way, the intention was to hurl the mud-ball as high and far as possible, and perhaps get it

What is reputedly the oldest ash tree in Europe, at Clapton, Somerset.

A Beadnell fisherman making a crab pot using ash sticks cut from a local plantation.

to stick on a distant window. (Getting it *over* the house was the aim in my childhood.)[5]

'In the spring when the ash shoots were young and had a purple tinge, we cut them six or eight inches long, bound them with rushes, and boiled them in burn water till soft. We sucked them like asparagus tips.'[6]

There was probably an echo of ash's old magical power, as well as pure practicality, in many of these uses. In Wales, for example, ash was always used for making 'adder-sticks', which were carried by the lengthsmen when the verges were cut by hand.[7] Foresters in Wiltshire use ash for making the handles for fire-beaters, despite the fact that it burns well even when green.[8] (And Forest Enterprise workers in north-east Essex still occasionally refer to the tree as 'the widow maker', because of its lethal habit of splitting as it is felled.)[9]

The opening of ash leaves is widely used for predicting summer weather. The conventional formula is given in a distinctively Scottish version from Roxburgh:

> *Ash before oak, the lady wears a cloak.*
> *Oak before ash, the lady wears a sash.*[10]

This is completely reversed in a rhyme from Surrey:

> *If the oak comes out before the ash,*
> *'Twill be a year of mix and splash.*
> *If the ash comes out before the oak,*
> *'Twill be a year of fire and smoke* [i.e. drought].[11]

Something of the same uncertainty is reflected in an unusual children's ritual from the Weald of Kent. It was acted out on Ash Wednesday, the first day of Lent (which of course has nothing whatever to do with ash, the tree):

'On Ash Wednesday children arrived at school carrying a twig of ash (it must have at least one black bud) to avoid having their feet stamped on by the other children. But after noon this was reversed. Then any child still carrying a twig of ash had their feet stamped on by those children who had rid themselves of their ash twigs.'[12]

It is in fact almost unprecedented for ash leaves to emerge before the oak's, and, when they are finally fully open, they are something of an anticlimax after the gothic stages that have preceded them: the sooty, angular buds; the flowers like tufts of purple coral; and finally the green fish-bone fronds of the unfurled leaves. Ashes can become grand, spacious trees, letting through more sunlight than heavier-leaved species; and ash-woods, pale-trunked and feather-foliaged, have a special, invigorating luminosity about them, even in the heart of summer.

The ash tree is ubiquitous as well as useful, and it is the commonest tree as a place name element after the thorn.[13] But standard (uncoppiced) trees rarely live as long or

The coral-like male flowers of ash, which emerge before the leaves. The more feathery female flowers sometimes occur on the same tree.

'A Study of Ash and Other Trees' by John Constable.

develop such grainy character as oaks or beeches, and ashes have not often been landmark or boundary trees. They have tended instead to become more personal talismans, as perhaps befits a species that has been a congenial domestic workhorse as well as a refuge for ancient spirits:

'We had an ash tree where my sister, friends and I regularly played. We called it Anty's tree (but we don't know why!). It had numerous holes ideal for posting letters and hiding poems we had written. And of course these also provided good hiding places for special pebbles. With quite a large hollow at the bottom that was twisted and gnarled, it was like a living doll's house ... Sometimes I would take a handful of grass and try to form a bird's nest.'[14]

'The weald and champion plain of West Sussex contain a considerable number of venerable ash stands and in former times most farming communities contained an ash wood. Some communities were named as such: Ashurst, Ashlington, East and West Ashling ... In many old Sussex communities the ash was regarded as a tree of magic and mystery. As a child, I was taught never to pass an ash without wishing it "good-day", and never must we harm an ash tree ... I still hold the ash as a tree to be respected, and find myself furtively dipping and bidding whenever I pass one.'[15]

Lilac, *Syringa vulgaris*, is a familiar garden shrub, which can persist and spread by suckers on the sites of old cottages, or where it has been thrown out. There are quite large stands in Cheddar Gorge, for instance.

Wild privet, *Ligustrum vulgare*, is a semi-evergreen shrub, quite common throughout lowland Britain, especially on calcareous soils. The panicles of ivory-white flowers have a sweetish smell, overlaid with the fishy odour familiar in hawthorn blossom. The matt-black berries are poisonous (fatally so to some children), yet wild privet was in use in hedges long before the larger-leaved and more compact Japanese species (see below) became the standard for suburban gardens. It has even, occasionally, been used in farm hedges. During the enclosure of Frampton on Severn in 1815, some of the new hedges were an unusual mixture of hawthorn and privet, with privet inserted as roughly every fifth bush.[16] On Romney Marsh in Kent, where it occurs in hedgerows along the boundary dykes, it is rather aptly known as 'private'.[17]

Garden privet, *L. ovalifolium*, is a Japanese species with rounder, and often yellower, leaves than the native species. It has been widely planted for hedging and crops up quite frequently on waste ground where garden rubbish is thrown away. Bird-sown bushes very occasionally spring up in plantation woodlands.

Figwort family
Scrophulariaceae

Great mullein, *Verbascum thapsus* (VN: Aaron's Rod, Hagtapers, Adam's flannel, Our Lady's candle). This is an impressive plant in both flower and leaf. The yellow flowers are packed on a spike that can rise to four or five feet in a season, and the large leaves are covered with soft greyish wool. Many of the obsolete names (and the surviving 'Adam's flannel') are a catalogue of soft objects: Donkey's ears, Duffle, Hare's-beard, Rag-paper. In a more modern – and practical – vein, mullein has been nicknamed 'the Andrex plant', and its leaves used accordingly.[1] John

Parkinson suggested that the plant not only resembled an enormous taper, but that the stalks, dipped in suet, had once been used as firebrands.[2]

Great mullein is a biennial, usually spending its first year as a large rosette of furry leaves, and is quite frequent in waste places, roadsides and bare ground on sandy or chalky soils.

Orange mullein, *V. phlomoides*, is a similar species, but with larger, orange-yellow flowers. It is a native of continental Europe, occasionally naturalised from gardens. **Dark mullein**, *V. nigrum*, is an altogether smaller and less robust species, also yellow-flowered, but without felt on the leaves. It is fairly common in rough grassland and hedge-banks, chiefly on calcareous soils in England. I have seen a form with white flowers growing amongst the normal yellow-flowered colonies at Maidscross Hill, Suffolk. **Hoary mullein**, *V. pulverulentum*, is a tall, much-branched species, almost as woolly as great mullein, but confined to roadsides and rough fields on sandy soils in East Anglia. Various hybrids occur between all these species.

Great mullein, known locally as Aaron's rod and Our Lady's candle.

Common figwort, *Scrophularia nodosa*, is a common perennial of woodland glades and rides and of hedge-banks. The 'fig' in figwort is an old word for piles, which both the globular red flower-buds and the root-protuberances were thought to resemble. Figwort was recommended for piles and also for the tubercular swellings of scrofula, 'the King's Evil'. **Water figwort**, *S. auriculata*, is a similar species of river-banks, pondsides and damp meadows, but with distinct wings along the angles of the square stems. An old children's game in the West Country was to rub two of these stalks together, like grasshoppers' legs. The vibrating wings make a sound like a squeaky violin, and local names for the plant were fiddlesticks and crowdy-kit (a Somerset name for a fiddle).[3] **Yellow figwort**, *V. vernalis*, is a perennial from continental Europe, increasingly naturalised in disturbed ground in shady places, including conifer plantations.

Monkeyflower, *Mimulus guttatus*, was first discovered and brought to English gardens from the damp and foggy islands off the Alaskan coast in 1812. But it crops up in damp places right down the west coast of America, as far south as the mountains of New Mexico, and it has been just as catholic in its choice of British habitats since it became naturalised in the 1820s. It now occurs by the banks of burns, streams, lowland lakes, rivers and canals throughout Britain. The golden-yellow flowers (more like nasturtiums than monkeys) brighten up many otherwise dark waterways. **Blood-drop-emlets**, *M. luteus*, is a scarcer relative, originally from Chile, occasionally naturalised in damp places in Scotland and the north of England. Its bloom is similar to monkeyflower's, but the yellow is speckled with red spots and blotches and its 'mouth' is open. The hybrid between these two species, **hybrid monkeyflower**, *M.* × *robertsii*, is probably commoner than both its parents in the uplands. **Musk**, *M. moschatus*, is a smaller-flowered species, also from western North America, naturalised in a scatter of locations across Britain. It was originally cultivated for its musky scent, but naturalised plants in Britain are entirely scentless.

Snapdragon, *Antirrhinum majus* (VN: Chooky pigs), is a familiar garden flower from south-western Europe, widely established on cliffs, walls, railway sidings and stony banks. In the wild species, the flowers are reddish-purple, and naturalised escapees usually revert to this colour.

Ivy-leaved toadflax, *Cymbalaria muralis* (VN: Mother of thousands, Travelling sailor). A delicate but aggressive creeper that trails over walls, banks and pavements, ivy-leaved toadflax was introduced from southern Europe early in the seventeenth century. William Baxter, writing in the 1830s, described 'this very pretty plant' as 'a native of Italy ... said to have been originally introduced into England by means of its seeds having been brought in some marble sculptures from that country to Oxford, where it has long established itself on the walls of the Colleges, gardens, &c. in such abundance, as to have obtained the name of "Oxford-weed".'[4] It is virtually unknown in natural habitats in this country.

Ivy-leaved toadflax has a mechanism which makes it easy for the plant to colonise walls vertically upwards. When it is in bloom, the flower-stalks bend towards the light; once the flowers are over, the seed-heads bend the other way, so that the seeds are more likely to be shed into cracks in the supporting stones. Its small, neat, purple and yellow snapdragons made it very popular as an ornamental plant between the seventeenth and nineteenth centuries, when there were many new walled gardens which it could exploit. It is no longer deliberately planted, but it is now found throughout Britain.

Ivy-leaved toadflax, introduced to Britain in the seventeenth century as a wall plant.

Monkeyflower, from the Aleutian Islands off Alaska, brightening a corner of the River Ribble, North Yorkshire.

'My daughter grew up believing that a plant growing on walls was called "I believe in toadflax". It is now a family name.'[5]

Sharp-leaved fluellen, *Kickxia elatine*, and **round-leaved fluellen**, *K. spuria*, are locally common annuals of cultivated land on light calcareous soils in the south. They are creeping species, with tiny dark-violet and yellow snapdragon flowers.

Common toadflax, *Linaria vulgaris* (VN: Butter and eggs, Bunny mouths), is a common perennial of waysides, open grassland and waste places throughout most of Britain. The flowers – orange and yellow snapdragons – are often tightly packed and can persist (or even produce a second flowering after verge-cutting) well into November. **Purple toadflax**, *L. purpurea*, is a tall (up to three feet) perennial from Italy, widely naturalised in waste places and on walls and railway embankments. Occasional pink-flowered plants occur, and this is the commoner form in gardens. **Pale toadflax**, *L. repens*, is a smaller, paler-flowered plant, commonest in the south-west.

Foxglove, *Digitalis purpurea* (VN: Fairy gloves, Fairy bells, Floppy dock, Tod-tails). (For illustration, see frontispiece.) The tapered, tubular flowers of foxglove obviously suggest the gloves of some small creature. But why, given their rosy-pink colouring, should it be a fox? (The name is not, as has sometimes been suggested, a contraction of folk's – i.e. fairies' – gloves: its root in Old English is clearly *foxes glofa*.) Perhaps it is because it grows in foxy places: amongst the bracken at the edges of heaths, on steep banks above rabbit-fields, by tracks up rough hill-pastures, in glades in acid woods. It defines, too, a particular moment of the year – the end of spring and beginning of high summer, when the landscape first begins to have a spent, tawny look.

Children, never pedantic about such things, have turned the fox's gloves into their own dolls, finger puppets and fake claws:

'The flowers were put on the finger tips, the point where they had been attached to the stalk forming a kind of hook so that the hands could then be used as pretend claws. To make the fingertip claws stay on, their size had to be carefully matched to the finger concerned.'[6]

'My mother, who was born in Derwen in Clwyd, used to "pop" foxgloves. To do this you remove a not fully opened trumpet and hold each end tightly with your thumbs to make an air-sac; then make it pop by pressing your thumbs together quickly.'[7]

The foxglove was also once widely used in folk-medicine, despite its high toxicity. Infusions of the leaves were given for sore throats and catarrh, and compresses for ulcers, swellings and bruises. But it was most frequently employed as a diuretic against dropsy (accumulation of fluid in the tissues), for which it could be dramatically but unpredictably effective, sometimes, alas, proving fatal. Ironically, it was the investigation of this powerful herb in the eighteenth century by the botanist and physician William Withering that was the turning point in the development of modern pharmacology and in its splitting away from traditional herbal medicine. Withering studied many cases of dropsy and its treatment by foxglove leaves, and he recorded his findings in his classic book, *An Account of the Foxglove* (1785).

He realised that its principal action was on the heart, whose beat it slowed and strengthened, which in turn stimulated the kidneys to clear the body and lungs of excess fluid. From this he found that the leaf could be an invaluable help in the treatment of heart failure. But the dosage was critical: a fraction too high and it could stop the heart altogether. Withering's insistence on the use of small and accurately measured quantities of dried foxglove leaf ('digitalis') led to a new discipline in the prescription of powerful plant drugs and eventually to the isolation and purification of the foxglove's active principles digitoxin and digoxin, still widely used in orthodox medicine as heart stimulants.

The drugs are now mostly prepared from imported leaves of European *Digitalis* species (usually *D. lanata*). But during the privation and blockades of the Second World War native foxglove leaves were gathered in large quantities by the County Herb Committees. These were mainly organised by the Women's Institutes, and a member of Bocking WI remembers that they used a condemned house for drying leaves, which were stretched out on netting in an upstairs room. In Montgomery a loft above a bakery was pressed into service and in Shrewsbury a clothes-drying room.[8]

In another part of Shropshire, wartime gathering revealed new sources for high-yield foxgloves: 'Foxglove was gathered in big quantities until about 1949 for the extraction of the heart drug digitalis. South Shropshire is divided diagonally by the high limestone ridge, Wenlock Edge. In many parts of this formation the surface limestone has been leached away, allowing the growth of the acid-loving foxglove. A drug manufacturer said that foxglove leaves from one site on top of the ridge were far richer in digitalis than any others known to him, British or foreign. His firm was very disappointed to learn of the small area involved.'[9]

But the old and sometimes reckless folk use of the leaves has never entirely disappeared. A GP in Oundle recalls that, when he was a houseman, he was called to examine a man treated by a herb-woman living in a wood in Repton, who had been given digitalis tea for breathlessness and was eating the 'tea-leaves' in sandwiches.[10]

Foxglove, especially the quite frequent white-flowered form, has been taken into cultivation, but has not changed very much in the process. I have seen occasional freaks in wild populations, where the top tubes have fused into a single speckled pink sunflower.

Fairy foxglove, *Erinus alpinus* (VN: Roman Wall plant). This is a tufted, purplish-pink alpine, introduced at an unknown time from the mountains of south-west Europe and now naturalised on a scatter of old walls, ruined castles, bridges and stony places (especially on calcareous rock) throughout Britain. As with many species of uncertain history and evocative settings, myths have gathered about its origins. Because of a long presence on Hadrian's Wall, it is widely known on both sides of the border as 'the Roman Wall plant', and its arrival is attributed to the Romans, despite evidence that it was probably sown in the wall in the nineteenth century.[11] One family have meticulously worked out the route the seeds might have followed, stuck to the boots of the legionaries:

'The seeds of *Erinus* could have been accidentally picked up by the men of the Twentieth Legion on their arduous march through the high passes of the Pyrenees and transported to Hadrian's Wall, where there would be plenty of suitable habitats and a climate imposed by northern latitudes not altogether different from the much higher altitudes of the Pyrenees.'[12]

Fairy foxglove is currently expanding its British range, and the most plausible explanation for its welcome arrival in new sites is simply that it has escaped from a nearby cottage garden or rockery. But what has all the appearance of an urban myth about its migrations has recently begun to be passed on in botanical circles. The story concerns two botanists who met at a fairy foxglove site, fell in love, married and ever since have carried *Erinus* seeds to sow as a memento at any site they visit together – hence the species' seemingly uncanny preference for nature reserves and romantic ruins!

Germander speedwell, once worn by travellers as a good-luck charm.

Germander speedwell, *Veronica chamaedrys* (VN: Bird's eye, Cat's eye, Eye of the child Jesus, Farewell, Goodbye). Speedwells are roadside plants which speed you on your journey. Just why they acquired this reputation is unknown, but in Ireland they were sometimes sewn onto the clothes of travellers for good luck. In the eighteenth century they acquired another odd reputation, for curing gout, and 'Sir' John Hill reported that 'the dried leaves picked from the stalks, were sold in our markets, and the people made tea of them. The opinion was so prevalent, that the plant was in a manner destroyed for many miles about London, but, like all other things that want for truth for their foundation, it came to nothing.'[13]

Germander speedwell has bright blue flowers with white 'bird's eyes' and can often form large clumps in hedge-banks and open woodlands. The nineteenth-century flower painter Caroline May found a pink-flowered variety in Breamore, Hampshire, and an off-white one in South Petherwyn, Cornwall.[14] I have seen a single plant with pale grey flowers on a laneside bank at Hawkley in Hampshire.

Thyme-leaved speedwell, *V. serpyllifolia*, is a creeping perennial of waste and cultivated places, short grassland and woodland rides. The flowers are white or pale blue. **Heath speedwell**, *V. officinalis*, is a more compact species – though often semi-prostrate. The lilac flowers are carried in a spike. It is quite common on dry banks, heaths and downland. **Wood speedwell**, *V. montana*, is a creeper, softly hairy all over, with pale grey-blue to lilac flowers. It is a characteristic species of ancient woodland and shady hedge-banks, chiefly in England and Wales.

Brooklime, *V. beccabunga*, is common in ditches, ponds and streamsides, with spikes of bright blue flowers. The fleshy leaves are edible, but bitter, and the same precautions should be taken with them as with water-cress (see p. 148). **Blue water-speedwell**, *V. anagallis-aquatica*, is a medium-sized perennial of shallow streams, ponds and marshes, usually with several branches of pale blue flowers. Specimens with pink flowers (and narrower leaves) are a different species, **Pink water-speedwell**, *V. catenata*, which usually prefers muddy places with no running water. The two species hybridise.

Three low-growing annual species – **breckland speedwell**, *V. praecox*, **spring speedwell**, *V. verna*, and **fingered speedwell**, *V. triphyllos* – are specialities of waste places and field-edges on the sandy soils of the East Anglian Breckland. **Wall speedwell**, *V. arvensis*, is a much commoner and widespread annual of walls, pavements and cultivated and waste ground with tiny brilliant blue flowers in spikes. **Green field-speedwell**, *V. agrestis*, **grey field-speedwell**, *V. polita*, **common field-speedwell**, *V. persica*, and **ivy-leaved speedwell**, *V. hederifolia*, are similar, common, sprawling annual weeds of cultivated and waste ground, all with solitary flowers. **Slender speedwell**, *V. filiformis*, was introduced from the Caucasus in the 1830s as an alpine ornamental. It began to escape in 1927, moving from the rockery to the lawn – which ensured its redefinition as a 'weed' amongst tidy-minded gardeners. In fact it has chosen the best possible setting for its silvery-blue and pale lilac flowers. They can form large, glistening pools in the grass (if they are allowed to) at the same season as the buttercups and fallen cherry blossom.

Spiked speedwell, *V. spicata*, is a rare and attractive tufted perennial of calcareous rocks and short grassland, notable for having two quite distinct populations in Britain. They are conventionally regarded as different subspecies. A few grassland sites on the chalky sands of the East Anglian Breckland hold ssp. *spicata*, whose dense spikes of deep violet-blue flowers rarely rise more than 12 inches. Ssp. *hybrida*, which occurs on limestone rocks in a scatter of sites from the Avon Gorge, through Craig Breidden in Montgomeryshire, to Westmorland, is a more robust species, reaching up to two feet in height. This species occasionally hybridises with **garden speedwell**, *V. longifolia*, from continental Europe, which is sometimes found as a naturalised garden escape in its own right.

Cornish moneywort, *Sibthorpia europaea*, is a delicate, creeping plant confined to brooksides and damp places, mainly in south-west England. The minute cream flowers, flushed with pink, are no bigger than a match-head.

Common cow-wheat, *Melampyrum pratense*, is a curiosity amongst plants of shady old woodland in being an annual. It grows generally in rather loose, sprawling patches, with pairs of lemon-yellow flowers held between the leaves – a subtle but welcome glimpse of colour which persists right through the summer. Two much rarer cow-wheats are amongst the most flamboyant of all British natives. **Crested cow-wheat**, *M. cristatum*, has crimped yellow and purple flowers stacked as elaborately as a Christmas table decoration. It is a rare annual of edges and clearings in ancient woods in East Anglia and adjacent counties, sometimes appearing after coppicing. It is also occasionally found on chalky hedge-banks. **Field cow-wheat**, *M. arvense*, is its even more extravagant equivalent in arable land, with a flower-spike 'like a purple, rose and yellow pagoda'.[15] It was formerly quite widespread as a weed of arable fields on dry and chalky soils, but now survives precariously in north Essex, Bedfordshire and the Isle of Wight.[16] On the fields above the Undercliff between Ventnor and St Lawrence, it was once so abundant that it was called 'poverty weed' because of the way its seeds reduced the market value of corn. Its first record on the Isle was in 1823:

> 'A few years later Dr Bromfield ... carefully investigated its history. Local tradition asserted that the plant was imported with wheat-seed from "foreign parts" – some said Spain, some Jersey, others, with more probability, from Norfolk. He learnt that it was the custom at harvest time to pull up the weed with the greatest care, and carry it off the fields in bags, and to burn it, picking up the very seeds from the ground wherever they could be perceived lying. The bread, he was told, made from the wheat on the farms above the Undercliff was not so dark coloured and "hot" as it used to be, and that the "droll" plant was less plentiful than formerly.'[17]

Eyebrights, *Euphrasia* species, are a large and difficult group of 20 very similar species and some 60 hybrids. They are low-growing annual plants, found in all sorts of short grassland – downs, meadows, heaths, mountains and cliff-tops. Their name and old medicinal use are a classic example of the Doctrine of Signatures. The flowers, like tiny violets in shape, are mottled with purple and yel-

Weed as landscape: slender speedwell colouring grassland at Milton Abbey, Dorset.

low blotches and stripes, not unlike the colours of a bruised eye, and compresses and tinctures made from them were prescribed for all manner of eye disorders.

Yellow-rattle, *Rhinanthus minor* (VN: Hay-rattle, Rattlebaskets, Pots and pans, Tiddibottles). This is a semi-parasitic annual of grassy meadows and pastures, found throughout Britain, chiefly on well-drained calcareous soils and sometimes growing in large colonies. The yellow flowers are followed by brown, inflated calyces in the shape of little purses or seashells, inside which the seeds rattle when they are ripe. **Lousewort**, *Pedicularis sylvatica*, is another semi-parasite, which occupies the same niche as yellow-rattle on acid grassland, especially damp heaths and moors. It has pinkish-purple flowers and is often almost stemless, the flower-spike appearing to spring straight from the ground. **Marsh lousewort** or **Red-rattle**, *P. palustris*, is similar but mainly northern and western.

Broomrape family
Orobanchaceae

Toothwort, *Lathraea squamaria*. One of the most unearthly looking of our native species, toothwort is a ghostly parasite, entirely lacking in chlorophyll and growing on the roots of trees, especially hazel, elm, alder and willow. The wax-white stems, fringed with fleshy scales, push through the dead leaves in late March, and in April and May sprout tiers of flowers which resemble dirty, mauve-stained molars. The seed-capsules are shiny ivory and even more toothlike.

Toothwort – rather surprisingly, given its sinister and suggestive appearance – never became an important herbal medicine. Probably its relative scarcity kept it out of reach. It occurs throughout the British Isles, chiefly on moist and chalky soils, but is very local and apt to shift as its host trees die or the habitat changes, as at Greenstead Green, Essex: 'I first saw it growing in the early 70s, on the roots of an elm by the Bourne Brook. When the tree died, so did the toothwort. Then I found it growing on the river bank under blackthorn. This was washed away in the floods of 1987. But more clumps have appeared (or were overlooked), growing on the roots of hybrid poplar and blackthorn along the same stretch of brook.'[1]

It tends to be loyal to its favoured regions, though, and

The waxy flowers of the parasitic toothwort.

clumps still grow, for instance, on hedge-banks not far from 'Church Litten Copse' (long vanished) in Selborne, where Gilbert White noted it in May 1772.[2]

Purple toothwort, *L. clandestina*, is a continental species, forming tufts of livid purple fangs, which sometimes grow straight from subterranean roots. It is occasionally naturalised from gardens in Britain, especially in damp sites. A remarkable colony near Crookham in Berkshire has crossed a stream and is growing several feet above ground level in the crook of a riverside willow. It is visible, like some exotic spring fungus, a hundred yards away.

Broomrapes, *Orobanche* spp. The broomrapes are curiosities – all parasitic perennials, like toothwort, and lacking their own chlorophyll. Most are brownish-yellow in colour, with something of the look of withered orchids. **Common broomrape**, *O. minor*, occurs mainly on the roots of plants of the pea family and is the most widespread. Other species that are specific to particular hosts, including yarrow, hedge bedstraw, greater knapweed, thyme, thistles and ivy, are much scarcer and more local. Occasionally broomrapes crop up in gardens on cultivated plants, which usually prove to be relatives of their native hosts.

Bellflower family
Campanulaceae

Canterbury-bells, *Campanula medium*, is a popular garden biennial from southern Europe, naturalised in waste ground and grassy places, chiefly in southern and central

Giant bellflower in its heartland in Wharfedale, Yorkshire.

England. The largest colonies are probably on railway embankments in south-east London, where the wind-blown seeds easily become established on bare patches of chalk. The flowers are deep and lipped, like small tea-cups, and may have been named after the similarly-shaped horse-bells of pilgrims to St Thomas à Becket's shrine.[1]

Rampion bellflower, *C. rapunculus*, is a European species, once cultivated for its edible tubers, known as 'rampions', and naturalised in a few damp waste patches in south-east England. A representation of it is carved on a tombstone in Friston Churchyard, Sussex.[2] **Cornish bellflower**, *C. alliariifolia*, is a white-flowered garden campanula from Turkey and the Caucasus, naturalised chiefly on railway banks in south-west England. **Clustered bellflower**, *C. glomerata*, is a locally common native of chalk and limestone grassland, and occasionally of sea-cliffs. Its bunches of trumpet-shaped violet-blue flowers on shortish stems persist until quite late in the summer – which makes its flowering period overlap with that of the autumn gentian (see p. 298). It should perhaps be called 'fool's gentian' from the number of occasions on which the two species are confused. **Adria bellflower**, *C. portenschlagiana*, and **trailing bellflower**, *C. poscharskyana*, are very similar rockery plants from the former Yugoslavia, often grown around porches and doorways. They are both aggressive scramblers and are widely naturalised on walls and banks, especially in the south-west. **Giant bellflower**, *C. latifolia*, sometimes reaches up to four or five feet in height and is one of the glories of northern England in high summer. It grows in calcareous woods and on river-banks and sheltered waysides, especially in the Yorkshire and Derbyshire Dales, spreading into Wales but petering out in the south of England. Its bellflowers range from purplish-blue to white, and point upwards. **Nettle-leaved bellflower**, *C. trachelium* (VN: Bats-in-the-belfry), is a smaller plant in all its parts and occupies the niche of *C. latifolia* in southern Britain. **Creeping bellflower**, *C. rapunculoides*, is a not dissimilar garden escape from Europe, but with bells held downwards. It is an aggressive coloniser which can form large patches.

Harebell, *C. rotundifolia* (VN in Scotland: Bluebell), is one of the most delicately beautiful of all Britain's common wild flowers, and certainly one of the most catholic in its choice of habitats. Harebell grows throughout Britain (being scarce only in parts of Devon and Cornwall), on almost any kind of dry, open and relatively undisturbed ground, from mountain-tops to sand-dunes.

Harebells, tolerant of all kinds of dry, sunny soils.

It is as happy on chalk grassland as on acid heaths, and in the shelter of tall bracken as on exposed cliff-tops. I have even seen colonies thriving in water-meadows by resorting to the tops of ant-hills. (Damp is one condition harebell cannot tolerate.) The paper-thin, almost translucent, sky-blue bells shaking slightly on their thin stalks are, fittingly, one of the last flowers of the year, blooming on into the first autumn gales.

Ivy-leaved bellflower, *Wahlenbergia hederacea*. This creeping perennial with pale blue blooms is a speciality of Dartmoor, where its slender stems support themselves on sedges and rushes in the bogs.

Venus's-looking-glass, *Legousia hybrida*, is a delicate annual of arable and disturbed ground on the chalk in southern and eastern England, now sadly declining in most places. Even where it is still quite common, as on Salisbury Plain, it can be hard to find. It is rarely more than six inches high, with narrow, crimped leaves that seem to weave between the corn stalks, and it closes its flowers in dull weather. But open, they are exquisite miniatures: purple stars carried on the end of long seed-capsules, and caught like cog-wheels between the points of the calyx. And inside the seed-capsule lies the source of the plant's name: a cluster of shining oval fruits, like brilliantly polished brass hand-mirrors.

Round-headed rampion or **Pride of Sussex**, *Phyteuma orbiculare*, is a speciality of the North and South Downs and the chalk hills of north Wiltshire. Geoffrey Grigson has given an incomparable description: 'Climb Silbury in Wiltshire on a hot August afternoon, climb to the top of this ziggurat of prehistory, and at your feet you may see an unusual insect of sharp blue or violet. Look nearer, and it is more like a violet sea-anemone – air-anemone – closing upon an incautious bee or fly. But it is vegetable, after all, a globe of incurving, tentacle-like corollas.'[3]

Ever since John Goodyer discovered it on the Hampshire downlands four centuries ago it has been a favourite amongst flower-lovers: 'My fondest childhood memory is linked to this plant which I picked in bunches on Brighton racecourse in the early 1930s. I grew up in a street of Victorian villas in London but my parents managed to afford an annual week's holiday in Brighton, where the races were a must for my father. Roaming away from the crowds, I always looked for this special purple flower with no idea of what it was ... I still feel a thrill when I find it. Of course, it is no longer on Brighton racecourse, but it is still frequent in parts of Sussex and is the emblem of their Naturalists' Trust.'[4]

Spiked rampion, *P. spicatum*, is a rare and surprising rampion with yellowish flower-heads – though in this species they are less like 'air-anemones' than bottle-brushes. It is confined to old woods and waysides in East Sussex, between Hadlow Down and Heathfield in the north and Arlington and Abbots Wood in the south.[5] **Oxford rampion**, *P. scheuchzeri*, is a native of the southern Alps, a little like a *P. orbiculare* with unfurled tentacles, which has hung precariously on to walls and pavements in Parks Road, Oxford, since 1951 (though 'scraped off' in 1988).[6]

Sheep's-bit, *Jasione montana*, is superficially similar to devil's-bit scabious (see p. 353), but with more globular

Sheep's-bit on the Lizard Peninsula, Cornwall.

and lighter blue flowers. Sheep's-bit is also more of a westerly plant, growing on heaths, acid grassland and sea-cliffs, often in quite large numbers.

Garden lobelia, *Lobelia erinus*, is a familiar garden annual from South Africa, with bright blue or sometimes pinkish flowers, widely grown in window-boxes and hanging baskets, from where it frequently seeds itself in pavement cracks and walls. **Californian lobelia**, *Downingia elegans*, is a North American annual, occasionally introduced as an impurity in imported grass-seed. Large quantities sprang up after the seeding of the banks around man-made lakes in Milton Keynes, Buckinghamshire, but persisted for only one year.[7]

Bedstraw family *Rubiaceae*

Field madder, *Sherardia arvensis*. A low, bristly annual of disturbed and cultivated ground on chalky soils, field madder has clusters of small flowers ranging from pale pink to mauve. The blooms of **squinancywort**, *Asperula cynanchica*, are similar, but vanilla-scented. Squinancywort is a more delicate plant, confined to calcareous grassland, where it has a special liking for ant-hills. The name is a variant of 'quinsy-wort', a herb for sore throats.

Woodruff, *Galium odoratum*. 'Woodruff' was 'woodrove' in the sixteenth century, and it may have been a plant that 'roved', or spread, through woods. But the name is more likely to be a simple, literal description. There are no more perfectly formed woodland plants in early May, with their tiny, chalk-white flowers on upright stems decked out with bright green six-leaved ruffs.

Whilst they are growing the plants are almost odourless. But, picked and dried, they quickly develop the fresh smell of new-mown hay, which later takes on hints of almond. The leaves will retain this scent for many months, and woodruff – traditionally called 'sweet woodruff' – was once held in high regard as an all-purpose domestic freshener. Dried bunches were hung in wardrobes to deter moths and laid amongst stored linen. The leaf-whorls were used as bookmarks, and 'during Georgian times they were placed in the cases of pocket-watches, so that their fragrance could be inhaled whenever telling the time'.[1] They also found their way into snuffs, pot-pourris, pillows, and 'sweet waters' for improving the complexion.

The scent is taken up quickly by liquids. I was taught to steep dried woodruff in apple juice by an Austrian living in the Chilterns (where the plant is abundant in the beechwoods). And it is in Austria and Germany, especially along the banks of the Rhine, that it forms a key ingredient of 'Maibowl', a punch which is drunk on 1 May, and which is made by steeping a bunch of dried woodruff in Moselle or Rhine wine flavoured with sliced orange and sugar.

Woodruff occurs throughout the British Isles, but in East Anglia it is scarce and virtually confined to ancient woodland. In central and southern England, it has a preference for old woodland and hedge-banks on chalky soils. In the west and the north it is more tolerant and can be found in shady places of all kinds, including hollow lanes and stream-banks.

It no longer grows in London, but was once hung in churches in the City (as elsewhere) on St Barnabas Day, 11 June. A small turning off Tower Hill, now Cooper's Row, was once called Woodruff Lane, a name commemorated in a tablet on the wall.[2]

Woodruff growing around – and on – a dead elm stump.

Lady's bedstraw – honey-scented when fresh, like new-mown hay when dried.

Lady's bedstraw, *G. verum* (VN: Lady's tresses). Like woodruff, lady's bedstraw dries to give the scent of new-mown hay, and its name probably derives from the old custom of including it in straw mattresses, and especially in the beds of women about to give birth. In full, frothy yellow flower it smells strongly of honey. It is a coagulant and was once employed not just as a styptic, but as a vegetable substitute for rennet (made from the stomach linings of unweaned calves) in the making of cheese. But the technique has been lost, and even research in the dairy department of Reading Agricultural College has failed to produce more than a thin layer of junket when the plant is combined with warm milk.[3]

Lady's bedstraw is a common perennial of chalk downs, dry waysides, meadows, heaths and sandy places. It is short and soft, but the spikes are so dense with tiny yellow flowers that it can colour large patches of grassland from June to September.

Other species of bedstraw also make significant contributions to summer wild-flower displays in different habitats. **Northern bedstraw**, *G. boreale*, is a rather stiff and bristly species of rocky places, stream-sides and shingle in northern Britain. Its leaf-whorls are four-leaved cruciforms. **Fen bedstraw**, *G. uliginosum*, and **common marsh-bedstraw**, *G. palustre*, wind thinly amongst the taller vegetation in damp meadows, ditches and fens. **Hedge bedstraw**, *G. mollugo*, is a much bulkier scrambler of hedge-banks and waysides, chiefly on calcareous soils. **Heath bedstraw**, *G. saxatile*, is one of the commonest small white-flowered plants of heaths and short grassland on acid soils. **Slender bedstraw**, *G. pumilum*, is its rare equivalent on calcareous grassland in the south, and **limestone bedstraw**, *G. sterneri*, on similar soils north of a line between the Severn and the Humber. **Crosswort**, *Cruciata laevipes*, is so named because of the four-leaved whorls up the stem, the upper ones surrounded by clusters of tiny yellow, honey-scented flowers. It is quite common on waysides and in rough grassland, chiefly on calcareous soils.

Cleavers, *Galium aparine* (VN: Bobby Buttons, Clyders, Clydon, Clivvers, Goosegrass, Gosling weed, Goose bumps, Gollenweed, Herriff, Hairiff, Sweethearts, Kisses, Sticky Willy, Sticky grass, Sticky weed, Stickleback, Stick-a-back, Sticky bobs, Sticky buds, Sticky William, Claggy Meggies, Robin-run-the-hedge). Cleavers is the abundant scrambling annual (it can grow 10 feet in a season) which children still wield and stick to each other's coats and hair. The whole plant is covered with hooked bristles that make it difficult to detach from any slightly rough surface. Hence the great majority of its

Cleavers or goosegrass – which children still stick to coats and hair and long-suffering pets.

surviving local names, from cleavers itself to kisses. ('It is known as sweetheart because of its clinging habits.')[4] In Scotland there is a game called 'Bleedy Tongues', in which anyone foolish enough to stick out his tongue has it cut by the rough leaves.[5]

Cleavers is still used as food for geese and chickens (and occasionally for humans: the bristles soften completely on cooking). The round fruits were also made into tops for lace pins in the Chilterns,[6] as were the seeds of the now rare arable weed, **corn cleavers**, *G. tricornutum*: 'The fruits were used by Bedfordshire lace-makers, who, to protect their fingers, covered the pin-heads on their lace-making cushions with them.'[7]

Finally, for a plant whose uses seem endless, there is an evocative account of the making of goosegrass beer in Staffordshire:

'In Stapenhill, during the summer an old lady could be seen early every morning, setting off with a large wicker basket, a man's flat cap on her head, an apron tied round her waist and a shawl over her shoulders. Her destination was the fields to collect herriff (cleavers) and nettles for the day's brewing of the favourite drink of the neighbourhood. She had two large zinc baths in her kitchen and a big copper, under which a fire would be lit to heat the water. In these containers she would make a brew, boiling the nettles and herriff together. No one knew what other ingredients she used. When it was ready and cooled, she would pour it into bottles. The brew resembled ginger beer to look at and tasted wonderful. It was non-alcoholic, but was fizzy and would go up our noses and make us laugh and splutter when we drank it. Everyone called it Granny Holden's Pop. She charged two old pennies for a bottle, but only one and a half pence if an empty bottle was taken back.'[8]

Wild madder, *Rubia peregrina*, is a dark-leaved, strongly prickly, almost rasping perennial that scrambles around rocky scrub and cliff paths along the coasts of south-west England and Wales. It is related to the Asian dye plant, madder, *R. tinctorum*, and its roots have been used to give a pink tone by English dyers.

Honeysuckle family
Caprifoliaceae

Elder, *Sambucus nigra* (VN: Boortree, Boontree, Borewood, Battery, Dog tree, Ellern, Fairy tree). It is hard to understand how this mangy, short-lived, opportunist and foul-smelling shrub was once regarded as one of the most magically powerful of plants. If you burned it, you would see the Devil. But, grown by the house, it also had the power to keep the Devil at bay. It could charm away warts and vermin. Until early this century drovers used malodorous elder switches to protect cattle from flies and disease, and hearse-drivers favoured elder-wood for the handles of their horse-whips – a telling and condensed symbol of centuries of jostling between superstition and practicality.

And this was by no means the only practical use. A

A natural elder arch. The shrub was once deliberately planted to make ornamental hedging.

mid-eighteenth-century farming encyclopaedia recommends elder as hedging in the kind of terms usually reserved for exotic spice trees: 'The Elder is the quickest of any in its shooting; and it will bear planting so large, and takes Root so easily, that it may be called an immediate Fence. To this let us add, that the Flowers and Berries bear a Price at Market; and that the Wood of the old Stumps is valuable, and of sure Sale to the Turners: and we shall find that there is great Reason for naming the Elder among the Hedge Shrubs, for that it equals any of them in Value.'[1] (And these 'elder fences' were not just workaday hedges, but were often planted in elaborate, criss-crossed double rows, so that the blossoms formed diamond patterns in June.)

Two centuries later a popular book on 'traditional' country crafts had just one sentence to spare on elder's place in the hedgerow economy: 'The unwanted wood such as elder and bramble is cut out.' The shrub is now widely regarded as little more than a jumped-up weed, a ragamuffin haunter of dung-heaps and drains.

Yet elder's image has always swung between the poles of veneration and distaste. Even its anatomy is ambivalent. It is too small to be a tree, yet too large and airy for a bush. Its root and heartwood are as hard as ebony, yet the young branches are weak, hollow and filled with insubstantial pith. The umbels of white flowers smell of honey, the leaves of mice nests. Having (like the nettle) a taste for rich, fertilised soils, it haunts graveyards and rubbish-tips. Yet it has always been regarded as one of the most bountiful sources of home medicine – 'a kind of *Catholicon* against all Infirmities whatever', as John Evelyn put it.[2] Even the name blurs magic and practicality. It is compounded from the Scandinavian tree spirit Hylde-Moer and Anglo-Saxon eldrun, derived from *aeld*, meaning fire, perhaps because the branches hollowed of their pith were used like bellows for blowing on fire, but must on no account be added to the flames.

Perhaps it was the contradictory qualities of elder that made it such a compelling and respected plant. It certainly needed a myth to account for the paradoxes, and in the Middle Ages it was declared to be the tree on which Judas hanged himself and, for the sake of mystical symmetry, also the tree of the Cross – which was why it became barely capable of supporting its own fully-grown branches. ('Bour-tree, bour-tree, crookit rung,/ Never straight, and never strong,/ Ever bush, and never tree/ Since our Lord was nailed t'ye' is a stanza from a vernacular Scots verse. 'Bour' means pipe.)[3] Most of these myths have now disappeared. But some of elder's practical uses survive (one contributor said she would choose it as her 'desert island plant'),[4] and occasionally echo older practices mentioned by contributors:

'My husband – 50 years a farmer – recalls using the pungent elder leaves on the head bands of his working horses to keep the flies away. We had a large, low elder bush by the barn in the yard, and the cows used to rub under it in the summertime on their way into the cow-shed for milking. One old cow would stay under there, given half a chance, when the flies were particularly bad.' (Kent)[5]

'As a youth my late father worked on the land, often handling horses. It was common practice to tie bunches of elder leaves to the harness to ward off flies.' (Essex)[6]

In Somerset and elsewhere, it was planted near dairies to keep flies away[7] – and by outside privies (though this may be a piece of retrospective explanation, since elder grows spontaneously in such places).

On the Isle of Man, 'tramman' 'was grown around houses to keep away evil spirits. If a girl washed her face in a lotion of tramman flowers and water it would make her beautiful.'[8]

Elder-flower water has retained a reputation as a skin-cleanser (one contributor recommends boiling the blossoms with cider vinegar),[9] and a refined commercial version is still sold as Eau de Sareau. It is also medically recognised as an eye-lotion.

But the principal use for the flowers and berries these days is as food. I find the flowers eaten straight off the bush as refreshing as ice-cream soda, but they are not to everyone's taste. In the war they were packed in barrels with salt to be used as a flavouring.[10] More popular are elder-flower fritters (known in Market Harborough as 'frizzets'),[11] made simply by dipping the freshly-opened umbels in batter and frying for a few minutes. The sweet muscat-flavoured flowers contrast wonderfully with the crisp batter.

Elder-flower cordial is made commercially at Woodchester near Stroud, using wild blossom harvested by part-time piece workers and infused with sugar-water.[12] So is elder-flower 'champagne', in Surrey, though the firm which produces this was ordered by the courts to cease using the name in 1994, after a case brought by the French Champagne association. It was felt that the use of the term for a wild-flower beverage would 'cheapen and debase'

Elder flowers can be eaten straight off the bush or fried in fritters.

the reputation of true champagne. But the stuff itself is still regarded as one of the best 'country wines': 'Many farmers' wives still make elder-flower champagne – a quick and easily made summer drink, very popular at hay-making.'[13]

Back at the house, Lady Statham favours 'Lazy girl's pudding': 'Peel, core and slice thin two or three different varieties of apples and some very firm, practically unripe pears. Pour over it the following mixture – elder flowers steeped in a mixture of honey and hot water – and let your guests guess the ingredients.'[14]

'I like to eat, in effect, the flowers and berries at the same time: something which would have seemed impossible before the invention of the deep freeze. To do this I make a sorbet using the flowers and serve some with the last of the sorbet which was made from the berries during the previous autumn.'[15]

The berries by themselves also make a rich, dark wine, a 'rob' for sore throats, and are an ingredient of 'hedgerow jam'. They are also used in the making of a relish from the nineteenth-century gentlemen's club circuit, known as Pontack Sauce. Pontack's was a famous restaurant in London's Lombard Street, and its recipe for this tangy brew of elder-berries, claret, spices and onions was taken back to many country seats and adapted to the owner's idiosyncrasies. It was supposed to be kept for seven years before use.

The catalogue goes on. The pith inside the young stems is one of the lightest natural solids (with a specific gravity of 0.09, as against cork's 0.24). It is used commercially for gripping small biological specimens whilst they are cut for sections for microscopy. The stems from which it is scraped are of a surprisingly hard, pale, shiny wood which is very satisfactory to carve. A half-inch diameter elder twig can be whittled into a paper-knife in 20 minutes, by taking advantage of its already hollowed-out interior.

'On the Northumberland coast the "thowelds" or pins for the oars were cut by the fishermen themselves from the "boontree" bushes which grew among the hedges. Apparently, this was a very greasy wood, which helped the oars to move freely. Identification of the boontree is difficult – opinion [and the name] suggests the elder.'[16]

The hollowed-out stems have also been made into a variety of peashooters and guns by generations of children:

'An elder gun was made from a hollowed-out stem, filled with a flexible twig (or a piece of clock spring) as a propellant.'[17]

' "Gun-skutes" were made. This was a sort of pop-gun made from hollow elder stems and a close-fitting stick.' (Isle of Man)[18]

'We produced guns for play from hollow elder bush stems, removing the pith except for a couple of inches at one end. The spring was a "steel" from corsets, which were in everyday use then by the village women. The steel was inserted in a loop. When firing at each other the hard seeds of goosegrass (locally named "clyders") were used.'[19]

'We used elderberries as ammunition in the pea-shooters to make a satisfactory bloody splodge. A handful of ripe elderberries squeezed on the wrist also produced a very believable stream of blood to alarm your Mum.'[20]

Finally, a note on 'touch-burners': 'In Sheffield my father introduced me to these home-made hand-warmers. A small, rectangular box of clay, say 2 by 4 by 1 inches internally, with a lid, sun-baked only, was half filled with "touch-wood" – in his case, the dry "tinder" of rotten elder wood. Onto this, a live coal or glowing piece of wood was placed; blown on until smouldering; then lidded and held in the hands. Sometimes, these hand-warmers worked for several hours.'[21]

Elder grows prolifically in hedgerows, woods (especially secondary woods), chalk downs, waste ground on enriched soils and abandoned cultivated land across Britain. It is, as one writer remarked, 'a typical product of contemporary life. It is light in construction, cheaply and rapidly produced, short-lived and either quickly repaired when damaged, or scrapped and replaced.'[22] It also has thousands of years of magical belief and peasant ingenuity behind it.

Cultivars of both the common elder and the **American elder**, *S. canadensis*, with variegated or finely cut leaves, are occasionally naturalised, chiefly in the south. In the north of England and Scotland, the **red-berried elder**, *S. racemosa*, is also widely planted and naturalised.

Ancient moss-covered elders in the Chequers boxwood. Elder trunk-wood is almost as hard as box.

Dwarf elder, or danewort, probably an early herbal introduction from southern Europe.

Dwarf elder, *S. ebulus* (VN: Danewort), is a non-woody perennial, often forming quite large colonies up to five feet tall, but dying away in winter. Its leaflets are long and lance-like, and have an unusual gravy-like smell when crushed. The flowers, which appear from July to late August, change from pink in bud to white with purplish anthers, developing eventually into black berries from September, by which time the foliage and stems of the plant have turned wine-red.

Perhaps these suffusions of purple and red helped bolster the legend that dwarf elder sprang from the blood of Danes slaughtered in battle (like pasqueflower, see p. 44). It is a curious legend, which probably originated from over-enthusiastic seventeenth- and eighteenth-century antiquarians misinterpreting the herbal name 'danewort', acquired because of dwarf elder's great effectiveness in producing the 'danes' or diarrhoea.[23] Although dwarf elder does grow at some historic sites, few have any connection whatsoever with Danes or battles. One Rutland contributor found that 'when I first moved into this area it was common knowledge locally that dwarf elder grew only where Danes had been buried after a skirmish with the Romano-British inhabitants'. But it proved to be literary folklore. The only one of the several colonies in the area at a remotely antiquarian site is on Oakham Castle Mound, where, with no great historical reverence, 'it has moved its position by 25 yards since I first saw it 30 years ago'.[24]

More typical sites for what may well have been an early herbal introduction from southern Europe are the unkempt hedge-banks round a chicken farm at Winterbourne Stoke, Wiltshire, and the enormous colonies, probably the largest in Britain, on railway embankments and waste ground at Stratford Marsh in east London. It occurs in a scatter of similarly marginal habitats across Britain.

But dwarf elder's capacity for generating spurious myths seems undiminished. In 1985, a Sussex writer pieced together, from a few suggestive quotes from old herbals and some scientific sleight-of-hand, an apparently serious account of how dwarf elder was used in the mass-production of Sussex dwarves for export to the courts of sixteenth-century Spain and Russia.[25]

Guelder-rose, *Viburnum opulus* (VN: Dogberry, Water elder), is the wild ancestor of the 'snowball tree'. The globes of sterile flowers in this garden shrub are developed from the outer ring of large china-white blooms which encircle the cluster of duller white, fertile flowers in the wild plant. The flowers are followed by heavy clusters of waxy, sticky-juiced berries. Guelder-rose grows up to about 12 feet by riversides and in fens, damp scrub, old hedgerows and woods throughout Britain. The lobed leaves turn scarlet in the autumn.

Wayfaring-tree, *V. lantana*, is a quite common shrub in southern England and Wales, chiefly in hedges, scrub and open woods in chalk country. The creamy-white flower umbels, out in May and June, are followed by berries which are red, then black. The name is curious: the shrub is no more of a wayside speciality than a dozen other commoner species (cf. traveller's-joy, p. 44). But at least its lily-scented flowers make it one of the pleasures of trackways on the Downs.

Snowberry, *Symphoricarpos albus* (VN for fruit: Lardy balls). This North American shrub was introduced to Britain in 1817. It is extensively planted in shrubberies and plantations, especially as cover for game, and widely

Above: guelder-rose, wild ancestor of the snowball tree.
Right: wayfaring-tree with unripe fruit, South Downs, Sussex.

naturalised by suckering. The flowers are pink and inconspicuous, and the shrub is best known for its marble-sized white berries:

'They are known as "Lardy balls" in Wiltshire. My grandmother told my mother that this is because lard used to be packed in pigs' bladders which looked perfectly white and spherical, like the autumn berries on this shrub.'[26]

'Local children pop snowberries beneath their feet on footpaths. They make a surprisingly loud noise.'[27]

Snowberries are readily eaten by blackbirds in hard weather. They are edible by humans, too, though they are hardly worth the effort and can cause digestive upsets in sensitive individuals and especially in children.

Honeysuckle, *Lonicera periclymenum* (VN: Woodbine). Honeysuckle flowers have one of the sweetest and best-loved scents of all British wild flowers. This is strongest at night (to attract pollinating moths, which can detect it a quarter of a mile away) but is also there, in a muted and more subtle form, in the daytime. Because it is a robust, entwining climber with soft and sometimes overwintering greyish leaves, it was a popular plant for arbours in gardens – and sought out in natural bowers too. William Bullein, in 1562, was quite carried away by its sensuous possibilities: 'Oh how swete and pleasaunte is Woodbinde, in Woodes and Arbours, after a tender soft rain: and how frendly doe this herbe if I maie so name it, imbrace the bodies, armes and branches of trees, with his long windyng stalkes, and tender leaues, openyng or spreding forthe his swete Lillis, like ladies fingers, emong the thornes or bushes.'[28]

A detail of the cotton wall hangings in the Honeysuckle Bedroom at Wightwick Manor, designed by William Morris and installed in 1893.

Shakespeare's *A Midsummer Night's Dream* features it twice: Oberon's bank, 'quite over-canopied with lush woodbine', and later, where Titania says to Bottom:

Sleep thou, and I will wind thee in my arms ...
So doth the woodbine, the sweet honeysuckle,
Gently entwist; the female ivy so
Enrings the barky fingers of the elm.

This passage has caused some puzzlement, since it can be read as suggesting that the honeysuckle was twining round itself, or that Shakespeare used 'woodbine' to refer to two different plants in the same play and here meant either traveller's-joy or, more improbably, bindweed. But the confusion perhaps arose from an unpunctuated version, lacking commas around the phrase 'the sweet honeysuckle'. Shakespeare knew his plants and may have given both names to make it quite clear he was talking about honeysuckle and *not* about traveller's-joy (which had been so christened by Gerard only a few years before).[29]

Children still pick the scroll-like flowers to suck nectar from the base. Foresters are less enthusiastic, since honeysuckle is a robust climber and can distort the young trees it uses for support. On hazel and ash, these twisted branches and trunks have been cut for use as 'barley-sugar' walking-sticks, once popular with Scots music-hall performers.

Honeysuckle grows in woods, scrub and hedgerows throughout Britain. The smaller and more delicately flowered **fly honeysuckle,** *L. xylosteum*, is confined as a native to the downs near Amberley in West Sussex, but is widely naturalised elsewhere from garden specimens, as are **Himalayan honeysuckle,** *Leycesteria formosa*, **Japanese honeysuckle,** *Lonicera japonica*, **garden honeysuckle,** *L.* × *italica*, and **perfoliate honeysuckle,** *L. caprifolium*, from southern Europe. In Cherry Hinton Chalk Pits, Cambridgeshire, this last species has been present since at least 1763.

Moschatel, whose flowers are arranged like the faces of a cube or a townhall clock. They smell faintly of musk.

Moschatel family *Adoxaceae*

Moschatel, *Adoxa moschatellina* (VN: Townhall clock, Good Friday plant). An inconspicuous but delightful plant of woods and shady banks, moschatel is one of the first spring flowers to come into bloom, nearly always by the beginning of April (hence 'Good Friday plant'). The small flowers are pale yellowish-green in colour, but are arranged in a remarkable fashion, at right angles to one another, like the faces of a town clock – except that there is a fifth on top, pointing towards the sky. At the end of the war, when I was a small child, I was told this was 'for the Spitfire pilots to read'.

Townhall clock often grows in quite large colonies, especially where the soil is damp or slightly disturbed, as along the edges of woodland rides. On warm damp days (or simply sniffed close-to) the massed flowers give off a faint but memorable scent of musk.

Valerian family *Valerianaceae*

Common cornsalad or **Lamb's lettuce**, *Valerianella locusta*, is a frequent annual of bare places in grassland, arable fields, rocky places, banks and walls. The minute lilac flowers are held in loose clusters. The oval leaves are edible, with a succulent texture but little flavour – and large-leaved varieties (which have even less) have been developed and are now popular salad plants.

Common valerian, *Valeriana officinalis*, is one of the most strikingly scented of our native species. The clusters of pinkish-white flowers, out from late June to August, have a high vanilla-like perfume, which can become overpowering where the plant is growing in large quantities. The dried roots, by contrast, have a stale, rancid smell. (Valeric acid occurs both in the plant and in human perspiration.) Cats are fascinated by the smell and react in the same intoxicated way as they do to cat-mint (see p. 316).

Valerian is a variable species (sometimes divided into two subspecies) and occurs in a wide variety of forms and habitats, from short plants on chalk downland, to five-foot-tall specimens on river-banks and waysides, and in woodland clearings and damp meadows.

The roots have quite strong sedative properties, and an extract from them is found in many proprietary herbal

The flowers of valerian, by contrast, have a strong, vanilla-like perfume.

tranquillisers. It is reputed to have been one of the drugs Hitler was fond of. Derbyshire was a traditional centre for harvesting valerian, and a contributor has sent us his grandfather's diary account, from the mid-1860s, of how the wild plants were taken into small-scale cultivation:

'He and his companion would set out early on a spring morning walking from Clay Cross towards Chesterfield. Each of the pair has an empty bag rolled up and carried under one arm, and he also has a small fork of wrought iron. They proceed together for several miles for tho' the seedling valerian plants of which they are in quest are to be found in the woods on each side of the road they are not in sufficient abundance to justify a break in their journey till near Chatsworth woods. The seedlings are now developing a couple of rough leaves and where these appear in abundance the little iron forks are applied in lifting them from the leaf mould of which the soil here mainly consists. They collect sufficient seedlings for the plot of land already set apart and prepared for the replanting. This was done in regular rows and at the right distance apart to allow for growth. After which little attention or care was required beyond keeping the ground free from weeds. However vigorous the growth may be above ground, this is all rejected and allowed to waste, the root being now the only portion of the plant of any value ... A little field, barely three-quarters of an acre in extent, in one season grew a crop of valerian which realised seventy-five pounds. The bulk of this produce is now exported to the USA.'[1]

Red valerian, or 'kiss-me-quick', Padstow, Cornwall. It is especially common on walls in the south-west.

Marsh valerian, *V. dioica*, is a smaller version of common valerian with male and female flowers on separate plants, found in marshes and fens, chiefly in England.

Red valerian, *Centranthus ruber* (VN: Kiss-me-quick, Drunkards, Sweet Betsy). This Mediterranean species was first introduced to British gardens before 1600. Towards the end of the eighteenth century it was recorded as naturalised on rocks and old walls in Devon and Cornwall.[2] It is now common on walls, banks and cliffs in England and Wales, especially in the south-west. Its abundance in cities such as Bristol, Plymouth and Torquay helps to add to their slight Mediterranean atmosphere. The flowers can range in colour from deep red through to white and occasionally occur in cream. Their habit of nodding tipsily in the wind – plus the plant's fondness for maritime areas – probably explains the 'drunken sailor' motif that has always run through the local names.

Teasel family *Dipsacaceae*

Wild teasel, *Dipsacus fullonum* (VN: Brushes and combs, Venus's basin). Teasel is named from the use of its spiny heads to 'tease' out the separate fibres of wool before spinning (a process known as carding) or to raise the pile or 'nap' of finished cloth: 'At Otterburn Mill [Northumberland], a local tweed mill, there are still many old looms and tools remaining, and some frames containing teasels for raising the nap on the cloth. These were split with all the prickles going the same way, and placed in frames which were fastened onto rotating drums.'[1]

In most mills, teasels were replaced by steel brushes in the nineteenth century. But they have proved themselves unsurpassable in the finishing of cloth that needs an

Wild teasel – a statuesque plant related to fuller's teasel, formerly used for carding wool.

exceptionally fine and evenly raised pile – as in some hats and, especially, the baize covering used for billiard tables. The reason for teasel's superiority lies in the small hooked spikes which cover the conical flower-heads. They have a 'give', as if they were mounted in rubber, and if they meet a snag or irregularity in the cloth they bend and skate gently over it, unlike steel brushes, which are apt to tear through it indiscriminately.

The teasel used most commonly in the cloth industry has been **fuller's teasel** (*D. sativus*, formerly *D. fullonum* ssp. *sativus*), a species or subspecies of uncertain origins, possibly in southern Europe, in which the stiff, spiny bracts curve down at their tips. Fuller's teasel (fullers were the craftsmen who cleaned and finished cloth) is still cultivated on the Somerset Levels and was once grown much more widely, including in damp areas of Gloucestershire. It lives on in the form of occasional naturalised specimens in ditches and waste ground, and in field names, pub signs (e.g. the Clothiers' Arms at Stroud, and the one-time inn of the same name at Nailsworth) and church kneelers (as at Witney, Oxfordshire, in the heart of the Cotswold wool country).[2]

But the wild teasel, whose spines are longer and weaker and grow straight upwards, was doubtless also used for carding wool at one time (though its spines are too weak for it to be of any use in nap-raising). It is still found adequate for small-scale carding by the Oakfield Ladies' Circle in Liverpool, and probably other hand-spinners.[3] Being a tall and statuesque plant, common on waysides and rough ground and by streams throughout much of the British Isles, it is also widely picked for flower arrangements. Children use the heads to make toy hedgehogs, and occasionally makeshift hairbrushes.[4]

Richard Jefferies caught the architectural elegance and detail of the plant wonderfully in an essay written in the 1870s:

'The large leaves of this plant grow in pairs, one on each side of the stem, and while the plant is young are connected in a curious manner by a green membrane, or continuation of the lower part of the leaf round the stem, so as to form a cup. The stalk rises in the centre of the cup, and of these vessels there are three or four above each other in storeys. When it rains, the drops, instead of falling off as from other leaves, run down these and are collected in the cups, which thus form so many natural rain-gauges. If it is a large plant, the cup nearest the ground – the biggest – will hold as much as two or three wine glasses. This water remains there for a considerable time, for several days after a shower, and it is fatal to numbers of insects which climb up the stalk or alight on the leaves and fall in. While the grass and the earth of the bank are quite dry, therefore, the teazle often has a supply of water; and when it dries up, the drowned insects remain at the bottom like the dregs of a draught the plant has drained. Round the prickly dome-shaped head, as the summer advances, two circles of violet-hued flowers push out from cells defended by the spines, so that, seen protruding above the hedge, it resembles a tiara – a green circle at the bottom of the dome, and two circles of gems above.'[5]

Water or 'dew', held or gathered by plants (cf. sundew, p. 125, and lady's-mantle, p. 189), has always been thought to have rejuvenating powers. In the eighteenth century, teasel-water was believed to remove freckles.[6]

Two hundred and fifty years later it is still being used as a soothing cosmetic: 'Every year I get hay fever and my eyes itch. I relieve this by bathing them in teasel water collected from their deep leaves.'[7]

Cut-leaved teasel, *D. laciniatus*, is a magnificent species from central and southern Europe, pink- to white-flowered and growing up to 12 feet tall, which first appeared beside the track at Charlbury railway station in 1989. Five years later, the Oxfordshire naturalist W. D. Campbell (d. 1994) explained how they got there – and, incidentally, what an influence botanists themselves sometimes have on plant distribution. Campbell had found the teasels in a refuse dump in a worked-out gravel pit at Stanton Harcourt in August 1978. Knowing that the site was soon to be cleared and levelled, he took seeds and grew them on successfully in his own garden. Soon they were forming hybrids with native wild teasels. Four years later he was brought a specimen of a strange weed to identify from a garden a quarter of a mile away. It turned out to be his alien teasel. Ten years later it had reached Charlbury railway station, just under a mile away, probably by bird-ferried seed.[8] In 1992, Brian Wurzell found 14 specimens in scrub by the Parkland Walk in Harringay, north London.[9] They are unlikely to have come from the same source, but networking amongst botanists is so intimate you can never be sure.

Field scabious, *Knautia arvensis* (VN: Pincushion flower, Lady's pincushion, Blue bonnets). This is a common summer perennial of waysides, meadows and downland, usually on calcareous soils. It is very variable in both size and coloration, the flowers occurring in shades from pale lilac to purple. 'Scabious' is derived from *scabiosa herba*, the herb for scabies. **Small scabious**, *Scabiosa columbaria*, is a similar but smaller- and usually paler-flowered species, confined to calcareous grassland in England and Wales. **Devil's-bit scabious**, *Succisa pratensis* (VN: Bobby bright buttons). With its roundish violet-blue flower-heads and protruding reddish anthers, devil's-bit is easily mistaken for sheep's-bit (see p. 339), but prefers damp places – meadows, stream-banks, rocky grassland – chiefly on calcareous or slightly acidic soils. It is 'devil's-bit' from the short, bitten-off look of the rootstock.

Field scabious, a familiar plant of grassy waysides. Pale lilac and sometimes pinkish flowers occasionally occur. It was once given for the scabies because of its rough stalks.

Daisy family
Asteraceae (or *Compositae*)

Carline thistle, *Carlina vulgaris*. A distinctive plant of short, dry grassland, usually on calcareous soils, carline thistle is like a small thistle in habit, but the flowers are surrounded by a fringe of long, shiny yellow bracts, which open in warm, dry weather and close in the cool and wet. It is an 'everlasting' flower, drying out and persisting through the winter, but, unlike its giant cousin – the 'cardabelle' or 'chardon soleil', *C. acanthifolia*, of southern Europe – there is no tradition of nailing the dry flowers to doors for good luck (see p. 7).

Burdock, *Arctium minus* and related species (VN: Bachelor's buttons, Button sourees, Beggar's buttons, Love leaves, Sticklebacks, Sticky bobs, Sticky Jack, Sticky Willy, Cleavers, Velcro plant). The inventor of Velcro fastening reputedly got the idea from the seed-heads of burdock, the familiar 'burs' whose hooked bristles clamp them to any rough surface they come in contact with, including each other. Modern children have returned the compliment by nicknaming burdock 'the Velcro plant'.[1] Many older local names survive, and testify to the plant's adhesive reputation. (Some are shared, incidentally, with that other clinging plant, cleavers, see p. 341.)

The games played with burdock are usually quite basic. Children throw them at each other and stick them surreptitiously on the backs of shirts and jumpers or onto the fur of tolerant pets. (Giving the seeds a chance to

Carline thistle heads dry out and persist over the winter on chalk grassland.

hitchhike on passing mammals is the chief function of the hooks.) But in Lancashire a more elaborate game was devised: 'They were of course individually used in fights, but we also used to collect them and construct large balls, also thrown at each other or rolled around like sticky bowls. We would also coat our jumpers with them, much to the annoyance of our Mums of course, because getting them off again could pull a jumper to shreds.'[2]

This game echoes one of the most extraordinary and elaborate plant-based rituals in Britain – the Burry Man parade held each summer in the Royal Burgh of Queensferry, Edinburgh. No one knows how old the ceremony is, but since 1687 it has been associated with the annual Ferry Fair, now held on the second Friday in August. On that day a man dressed from head to ankle in burs perambulates about the town, visiting houses and receiving gifts and greetings. The following notes are condensed from an account prepared by the Edinburgh City Museum:

Burdock's large leaves were, like butterbur's, popular foreground fillers with seventeenth- and eighteenth-century landscape painters.

'For some days before the festival, large numbers of burs are gathered locally (though they are increasingly hard to find in the developing edges of cities). They are spread out on tables and allowed to dry. This also makes it easier to extract pieces of stalk, leaves and grass and encourages the departure of various hedgerow insects which the Burry Man has no desire to carry with him in his costume. A couple of days later, the process of making the "patches" begins. These are rectangular panels, each composed of 500 or so burs, and of course require no adhesive to stick to each other, or to the Burry Man's clothing.

The Burry Man's day begins at around 6 a.m. in the Town Hall, where the making of his costume takes place. Onto undergarments made of white flannel, the dresser fits the patches from the ankle upwards, and sensitive areas, such as the back of the knees, the crotch and the armpits, are filled in by the careful placement of individual burs. The process, which takes two hours, continues until the Burry Man's entire body is encased in bur-patches. The overall effect is of a suit of chain-mail. The only exceptions are his hands and his shoes, and the crown of his head, which is covered by a hat garlanded with flowers. Apertures in his face-mask are made for his eyes and mouth. The costume is completed by small sprays of flowers being pushed between the burs at the shoulders, hips and outside of the knees, and by a folded Scottish standard being wound round his waist.

At 9 a.m. the Burry Man emerges into Queensferry High Street, carrying two staves bedecked with flowers. He walks slowly and awkwardly with his arms outstretched sideways, carrying the two staves, and two attendants, one on each side, help him to keep his balance by also holding on to the staves. Led by a boy ringing a bell, the Burry Man and his supporters begin their nine-hour perambulation of South Queensferry.

The first stop is traditionally outside the Provost's house, where the Burry Man receives a drink of whisky through a straw.

Occasional offerings like this must keep him going throughout the day. At about 6 p.m., the Burry Man returns to the Town Hall, exhausted by his efforts and usually somewhat inebriated by his intake of neat whisky. Although it occurs only once a year, the task of being Burry Man is extremely demanding, requiring stamina, a strong bladder, an indifference to the discomfort caused by the more penetrative burs, and a

Edinburgh's Burry Man – entirely clothed in burs – sets out for his day-long procession around the city.

conviction that this ancient custom should not die out.'[3]

The origins of this custom are not known for certain. It has obvious affinities with the May-time fertility rites, especially the parades of the Green Man (see p. 175). But it is possible that the Burry Man also played the part of the ritual scapegoat, a figure who wore an exaggerated 'hair-shirt' for the rest of the community and carried away on his clinging back any evils afflicting them. At Fraserburgh, until the middle of the nineteenth century, an almost identical Burry Man ritual was enacted to 'raise the herring'. And at Buckie on the Moray Firth, when the fishing season was bad, a man wearing a flannel shirt stuck all over with burs was paraded through the village in a hand-barrow, to bring better luck to the fishing.[4] (There are suggestions of sympathetic magic here, with the burs representing both fish-scales and fish-hooks.)

Burdock has had many more immediately practical uses. Its large leaves have been used as butter-wrappings, like butterbur (see p. 377), and as alfresco lavatory paper.[5] Its young shoots, peeled of their outer skin, can be eaten raw and have a taste reminiscent of young new potatoes. The roots can be roasted or stir-fried (as related species are in Japan) but are better known in this country as an ingredient in various 'near-beers'. The burdock for that one-time favourite fizzy drink 'Dandelion and Burdock' is these days imported in the form of hot-water extracts from eastern Europe.[6] But similar brews are still occasionally home-made. In Tredegar, for example, one mother used a concoction of burdock, cleavers, dandelion, blackberry tops and nettles.[7]

Although burdock is a rather gawky plant, it has the virtue of being sturdy, upright and large-leaved, and it has been a favourite foreground filling for landscape painters, George Stubbs for example. Claude Lorrain, one of the fathers of pastoral painting, insinuated its leaves into many shady corners.

Thistles. The thistles are widely regarded as weeds, but they play a bigger role as landscape plants than perhaps we realise. Most have also been used as human food at some time, the young shoots being stripped of spines and eaten as a salad, and the hearts or 'kernels' of the flowers used like miniature globe artichokes.[8] (VN for thistles in general: Dashels, Milky Dashels, Dicels, Donkey's breakfast; for seeds: Fairies, Sugar stealers.)

Spear thistle, *Cirsium vulgare*, a common tall and upright thistle of cultivated and grazing land, is the most likely candidate for the true 'Scotch thistle'. (Peter Marren discusses the identity of this contentious national emblem on p. 455.) It certainly figures in some Scottish ceremonies, in Dumfries and Galloway, for example, especially in its occasional white-flowered forms. (This, incidentally, is the only form to occur on the Brough of Birsay in Orkney, a site associated with ancient Christian churches and St Magnus the Martyr.)[9] 'This thistle is always carried at the ceremony of "Riding the Marches", held in the town of Langholm in July. The 1992 specimen was six feet tall. The thistle is reputed to be a warning for lairds and others not to meddle with the privileges of the people.'[10]

Spear thistle is also the emblem of the English village of Newton Regis, from its old name of 'Newton in the Thistles': 'It has been used as a symbol carved on the village noticeboard and in the name of the bowls and cricket club. A thistle is depicted in a modern stained glass window in the church. Actually the thistle in the village's former name may have been teasel, used for carding flax,

Spear thistle, the most likely candidate for the role of the true 'Scotch thistle'.

Woolly thistle, a species of chalk soils in central England.

but the thistle depicted now is always the wild or Scottish thistle rather than the teasel.'[11]

Creeping thistle, *C. arvense*, is the commonest species, its creeping rootstock often enabling it to develop large colonies in rough grassland, waysides and cultivated and waste ground. **Woolly thistle**, *C. eriophorum*, is a scarce but conspicuous biennial of chalk scrub and grassland, chiefly in central England. It has woolly stems and leaves and large flower-heads, with the pink blooms resting on an almost spherical mass of cotton and spines. **Melancholy thistle**, *C. heterophyllum*, has single, initially drooping flower-heads (occasionally white) on spineless stems, and leaves densely white-felted beneath. It is a very characteristic plant of high summer in upland pasture country, where it haunts damp grassland, open woods and streamsides. Its southern equivalent is the scarcer **meadow thistle**, *C. dissectum*, largely confined these days to old hay meadows and fens. **Tuberous thistle**,

C. tuberosum, is similar but even rarer, though it does grow in large numbers in its few remaining sites on old chalk grassland in Wiltshire. One local farmer, prone to malapropisms, calls it 'tubercular thistle'. **Dwarf thistle**, *C. acaule*, grows as a flat rosette with a single flower-head in the centre and is widely known as 'picnic thistle' because of its fondness for favourite beauty spots on calcareous grassland and for giving no warning of its lurking spininess even when the flowers are out. It forms hybrids with tuberous thistle. **Marsh thistle**, *C. palustre*, and **welted thistle**, *Carduus crispus*, are rather alike and favour damp situations in meadows, marshes, ditches and woodland rides and clearings. Both have spiny 'wings' up the stem, but welted thistle's stop just below the flower-heads, leaving a distinctive 'bald patch'. **Slender thistle**, *Carduus tenuiflorus*, is a medium-sized thistle, with

Melancholy thistle, looking anything but melancholy by a drystone wall in Upper Wharfedale, Yorkshire.

Musk thistle, in the field (left) and on an old Derby porcelain dish, painted from life (as a 'lady thistle') by 'Quaker' Pegg, c. 1800.

continuous, spiny wings up the stem. It is locally frequent on waysides and in sandy and waste places close to the sea. **Plymouth thistle**, *C. pycnocephalus*, differs from slender thistle in having more dense white cotton on stems and leaves. It is a southern European species naturalised on Plymouth cliffs since 1868. It now 'occurs on Plymouth Hoe and is protected under a local by-law passed *c.* 1936'.[12] **Musk thistle**, *C. nutans*, is common in rough grassland and on waysides, chiefly on calcareous soils. It is a familiar species, with a single, large, nodding flower-head on top of each stalk.

Cotton thistle, *Onopordum acanthium*, is an unmistakable species, growing up to six feet tall and covered with silver-grey cottony hairs. It is what English people call the Scotch thistle – though it is probably an introduction from the continent, is rarely found north of the border and is most common (and possibly native) on the sandy soils of East Anglia. John Gerard (1597) says that the down which covers the leaves and stems was gathered for stuffing pillows and cushions. **Milk thistle**, *Silybum marianum*, is a Mediterranean species introduced as a medicinal herb before the sixteenth century and naturalised in a few places, mostly close to the sea or historic buildings. The leaves are etched with a network of white veins and spots and, by the Doctrine of Signatures, were given to increase the milk output of nursing mothers. The young leaves were also eaten as a green vegetable, once the spines had been removed.

Saw-wort, *Serratula tinctoria*, is a wiry, somewhat thistle-like perennial of old grassland and open woodland on well-drained soils. It is rather western in its distribution and only just reaches Scotland. The English name comes from the bristle-tipped teeth of the leaves, described by Gerard as 'somewhat snipt about the edges like a sawe', and the Latin *tinctoria* from its former use as a dye-plant.

Common knapweed, *Centaurea nigra* (VN: Hardheads, Paintbrush), is a very common tough perennial of all kinds of grasslands, from lawns to cliff-tops. John Clare described a love-divination game that was played by village girls, using the pinkish-purple flower-heads:

> *They pull the little blossom threads*
> *From out the knapweeds button heads*
> *And put the husk wi many a smile*
> *In their white bosoms for awhile*
> *Who if they guess aright the swain*
> *That loves sweet fancys trys to gain*
> *Tis said that ere its lain an hour*
> *Twill blossom wi a second flower*
> *And from her white breasts hankerchief*
> *Bloom as they had ne'er lost a leaf.*[13]

Greater knapweed, *C. scabiosa*, is a larger species with longer outer florets and more deeply divided leaves. It is more restricted in its distribution, being found on waysides, downs and cliffs mainly in England and chiefly on calcareous soils. A single plant with white flowers was seen around 1980 on a road-verge in Pershore, Worcestershire.[14] Greater knapweed flowers are a great favourite of chalk-country butterflies.

Cornflower, *C. cyanus* (VN: Bluebottle, Blavers). Fields full of cornflowers (one of the plants striking enough to have given its name to a colour) were last seen in Britain between the wars. Sir Edward Salisbury recalled 'a cornfield near Oxford even in 1926, the year when the Prince of Wales was President of the British Association meeting there, which looked as though it had been sown for the occasion, since it was red, white, and blue with Poppies, Mayweed, and *Centaurea cyanus*'.[15] But it had been regarded as a pestilential weed since the seventeenth century, and, when even someone as sympathetic to weeds as the poet John Clare could write of its large sheets 'troubling the cornfields with their destroying beauty', it was plain that it would eventually get its comeuppance.

By the end of the 1970s, it had become a nationally scarce species. But, in the late 1980s and early 1990s, it began to reappear on unsprayed set-aside land and in the disturbed ground created by the Department of Transport's extensive road-building programme, which was bringing long-buried seed to the surface (e.g. by the A41 bypass at Berkhamsted). And in one place at least, it was possible to see a field of corn something like the one Edward Salisbury had witnessed in 1926. At College Lake Nature Reserve at Pitstone, Buckinghamshire, created in an abandoned chalk quarry, the warden found the cache of topsoil that had been removed before quarrying had begun in the 1930s. It was soil from the era before herbicides and, when he spread it over a field at the edge of the reserve, dormant seeds of cornflower, corncockle and pheasant's-eye exploded in a riot of colour. The field is now maintained as an arable weed reserve.

Juice from the flowers, mixed with alum, is used by water-colourists.[16]

Common knapweed – a form with rayed flowers. A love-divination game was played by pulling out the rays.

Chicory, quite common on roadsides and rough grass on the chalk.

Perennial cornflower, *C. montana*, is a southern and central European species, much planted in gardens and widely naturalised along roadsides and in waste places. Curiously, these are often a long distance from habitation, despite the fact that the plant rarely produces fertile seed in this country.

Chicory, *Cichorium intybus*, is a tall perennial with vivid, sky-blue flowers held close to the stems, quite common on roadsides and in rough grassland on calcareous soils in central and southern Britain. As 'succory', it was under cultivation by at least the early sixteenth century, and gardeners knew many tricks for producing blanched spears of the kind familiar in commerce today, from forcing them in dark cellars or under flower-pots to tying the emergent leaves together with string. The unblanched leaves of wild chicory (which these days is sometimes naturalised from cultivated varieties) are bitter, but tolerable in a mixed salad; and the dried and ground roots have long been used as an adulterant of (and sometimes a substitute for) coffee. A form with white flowers was found in 1989 in Holt, Wiltshire.[17]

Dandelion, *Taraxacum* species (VN: Jack-piss-the-bed, Pissy beds, Pittley beds, Tiddle-beds, Wet-the-bed; Dog's posy, Old man's clock, Peasant's clock, Swine's snout; seeds: Fairies, Parachutes, Sugar eaters).

Dandelions mean carpets of golden-yellow flowers, jagged green leaves (the *dent de lion* – 'lion's tooth') and clouds of featherweight seeds blowing in the wind. Albrecht Dürer saw the appeal even of the closed and withering flower-heads and included them in his extraordinary close-up portrait (painted in 1503) of a square foot of meadowland, *Large Tuft of Herbs* (illustrated on p. 321). Shakespeare understood the popular familiarity with the whole plant, as well as local Midlands names for it,[18] and included them in the elegy in *Cymbeline*: 'Golden lads and girls all must/ As chimney-sweepers come to dust.' Keats imagined 'The soft rustle of a maiden's gown/ Fanning away the dandelion's down'.[19]

Children still blow dandelion down from the round dandelion clocks – the 'chimney-sweepers' – to 'tell the time'. The number of blows needed to remove all the seeds gives the hour. If you can catch one on the wing,

'Golden lads and girls all must/As chimney-sweepers come to dust.' (Shakespeare)

you can make a wish.[20] Other children's games with dandelions robustly ignore the myth passed down in some families that even touching the plant will make you wet your bed:

'Take a long stalk of dandelion, remove the head, and split one end downwards approximately half an inch. Place the split inside your mouth and blow gently. A raspberry sound should be made.'[21]

'We made bracelets of dandelion stems, tucking the narrow end into the wide end.'[22]

'We used the white latex from the stem to draw on pavements.'[23]

But dandelion is still widely known as wet-the-bed or pissy bed by children across Britain, and it has long been used as a herbal diuretic and laxative. Its reputation has been confirmed scientifically in a number of studies (e.g. at the North London Polytechnic in 1978), and as a bonus it contains high levels of potassium, an element which is removed from the body when urine production is stepped up. A Cambridgeshire man whose family have been seed-merchants since the 1890s recalls the high repute in which dandelion's medicinal powers have been held in this country:

'In the 1920s, when times were very bad for this district, it was quite usual for families to go out and dig up dandelion roots in the fields, the wild ones, and sell them to the chemist to buy bread. But also, particularly in the 1880s, it was customary for the gentry to grow dandelions in their unheated greenhouses for winter salads. As you know, they are very beneficial for flushing out the kidneys and helped to prevent the rich port-drinking inhabitants from getting gout. Dandelion salads were very popular in those days; also sandwiches of thin brown bread and butter filled with dandelion leaves were served by the ladies for afternoon tea.

My father was very friendly with the well-known Snell family who ran the famous Royal Nurseries in Broad Street, Ely [now defunct]. Gradually we took over the postal distribution of dandelion seed but at first it was locally grown ... One point worth making is that in the 1880s it was difficult to grow lettuce throughout the winter months. Today it is easy with the modern varieties. But of course dandelion has a very distinctive flavour and it seems to be growing in popularity. We sell more seed each year. This may be a reaction against the modern greenhouse-produced, rather tasteless lettuce.'[24]

Andy Goldsworthy's spring stamp design, made entirely from dandelion heads.

Certainly dandelions are now much more familiar to British cooks. The traditional French dish *pissenlit au lard*, fried bacon scraps and croutons served on a dandelion salad, is found increasingly on menus, and I have noticed dandelion leaves in ordinary pub salads. New treatments and recipes – for dandelion pasta, pickled dandelions, dandelion and mozzarella pie (see Wild foods, p. 217) and stir-fried dandelion – are being developed all the time: 'I couldn't come to terms with any of the dandelion family; the bitterness was just too unbearable to my taste. I tried, before passing final sentence, lightly frying the leaves of dandelion and ratstail plantain in a little fat for a minute or two, and the result was exquisite. It turned the tough leaves into wafer-thin crisps, the bitterness into a gentle edge, and brought out the flavour which had always been masked by the bitterness.'[25]

The flowers can be added to dandelion salads and provide a welcome dash of colour and a soft, honey-flavoured foil to the leaves' tartness. But they are most usually gathered for making dandelion wine, the classic account of which is in Laurie Lee's *Cider with Rosie*. The roots provide a coffee substitute when roasted and ground, and they were much used during the Second World War when real coffee was unobtainable. Even the seeds are eaten – at least by cage-birds: 'A bird fancier of my acquaintance encourages dandelions in one corner of his allotment. He singes the down off the clocks with his cigarette lighter and feeds the seeds to his finches.'[26]

More than 200 microspecies of dandelion have been recognised in Britain, and all can be used in similar ways.

There are also large numbers of yellow-flowered

hawkbits, hawkweeds and hawk's-beards, etc, some of whose leaves are quite palatable but which have never been differentiated much (or used) at a popular level.

Fox-and-cubs, *Pilosella aurantiaca* (VN: Grim the collier), is a frequently naturalised garden escape from continental Europe, whose nestling clusters of tawny-orange flowers and buds easily explain its name. 'Grim the collier', though, awaits a satisfactory explanation: even in the early seventeenth century it was thought 'both idle and foolish'.

Nipplewort, *Lapsana communis*, is a common and rather lax annual of rough ground in gardens, open woods and hedge-banks. The name derives from the shape of the flower-buds. **Bristly oxtongue**, *Picris echioides*, is a rough, bristly waste-ground annual or biennial with pronounced pimples on the leaves. It is known in Buckinghamshire as 'Milton Keynes weed' because it is so abundant and expansive in the city.[27] **Goat's-beard**,

Fox-and-cubs, named from its tawny-orange clusters of flowers and buds.

A goat's-beard 'clock'.

Tragopogon pratensis (VN: Jack- or Johnny-go-to-bed-at-noon), is an annual or short-lived perennial with dandelion-like flowers, most notable for its spherical seed 'clock' which is as elaborate as an astrolabe. The flowers close up at midday inside the long-pointed bracts. It is common on road-verges and in rough grassland. **Salsify**, *T. porrifolius*, is a purple-flowered Mediterranean native, cultivated as a root vegetable and occasionally naturalised on waste ground and waysides.

Smooth sow-thistle, *Sonchus oleraceus* (VN: Milky disels, Dindle, Silver and gold). This is a common annual of waste, cultivated and disturbed ground throughout Britain. The seeds, which are attached to short plumes ('flights') of silvery-green hairs, change colour as they ripen, and this habit has been noticed by Wiltshire children in what must be the most botanically sophisticated of all plant games: 'We played a game called "silver and gold". Each child would choose a seed-head and at the given sign slit it open with their finger nails. The one who had chosen a head where the seeds were yellow (gold) and the other part silver was the winner. Too young and the flights would be green and the seeds a greeny yellow; too old and the seeds would be brown. We always called the plant "silver and gold".'[28]

Perennial sow-thistle, *S. arvensis*, and **prickly sow-thistle**, *S. asper*, are found in almost identical habitats. The young leaves of all these three sow-thistles can be used as salad vegetables (they are fleshier and less bitter than dandelions) once their spines have been trimmed off. Pliny had Theseus dining on a dish of sow-thistles before going to finish off the Minotaur. **Marsh sow-thistle**, *S. palustris*, is a giant, statuesque perennial often growing to seven feet in height, with long halberd-shaped leaves. It is a speciality of the Norfolk Broads, but is also occasionally found in marshes and by rivers and dykes in Suffolk and Kent.

Common blue-sow-thistle, *Cicerbita macrophylla*, was introduced in 1915 from the Urals. It is a tall, lilac-flowered perennial now quite widely naturalised on roadsides and waste places, often forming large patches.

Prickly lettuce, *Lactuca serriola*, is a tall annual with panicles of yellow flowers (closing up early in the day) and spiny grey-green leaves, which clasp the stem at their base. The upper leaves are held erect in a roughly north–south plane, hence the occasional name – clique botanical rather than vernacular – of 'compass plant'. It is probably native, but has spread considerably since the war and is now common on disturbed and open wasteland, especially in urban areas in south-east England. An eastern Mediterranean form was probably the ancestor of the garden lettuce, *L. sativa*. **Great lettuce**, *L. virosa*, is similar but scarcer. The bitter white latex that seeps from the stem is mildly poisonous. Concentrated into 'lactucarium', it was prescribed as a sedative in the nineteenth century. **Wall lettuce**, *Mycelis muralis*, is quite common on shady banks and walls and in open woods, chiefly on calcareous soils or rocks, over most of England and Wales. The dandelion-like leaves are deeply lobed and often purplish.

Marsh cudweed, *Gnaphalium uliginosum*, is a grey, woolly-leaved annual of damp or compacted soils on arable and new grassland, especially cart-tracks and footpaths.

Mountain everlasting, *Antennaria dioica* (VN: Cat's foot). A short mat-forming perennial of alpine habit, it is unusual for this family in having male and female flowers on separate plants. The male flowers are like clusters of small daisies with pink centres; the female are pinkish throughout and brush-like. Both sexes have downy grey foliage and grow in short grassland chiefly on calcium-rich rocks in upland Britain. In the southern lowlands one of its few remaining sites is at the old quarry nature reserve of Barnack Hills and Holes, near Stamford, where it has been known for nearly a century and a half.

Elecampane, *Inula helenium*, is an Asian species originally introduced to Britain for its medicinal properties – especially for respiratory disorders. The aromatic root, 'being candied, or dried, and powder'd, mix'd with Hony or Sugar, is very good in a Difficulty of Breathing, an *Asthma*, and an old Cough'.[29] The chemical inulin, found in this species and the related sunflowers, has recently been accepted by mainstream medicine as a useful treatment for asthma. Elecampane's broad spear-shaped leaves and large yellow daisy-flowers can sometimes be seen naturalised on waysides, especially close to old cottage gardens, where it may be a relic of herbal – or even veterinary – cultivation: 'More recently it has reputedly been used to

Mountain everlasting, a plant of alpine habit that still grows in a few lowland sites.

treat lame horses. I know of only two local sites near Oswestry, Shropshire. One is opposite a packhorse inn at a spot where a track across the Berwyn Mountains from the Dee Valley falls down into the shelter of the Ceiriog Valley. The other site is an unlikely spot in a ditch by a house by the old racecourse at Oswestry. Horse races were held there from the early eighteenth century till 1849.'[30]

Ploughman's-spikenard, *I. conyzae*. True spikenard, or 'nard', was an expensive, spicy perfume made from the roots of a Himalayan plant, *Nardostachys jatamansi*. Ploughman's-spikenard was the rustic English equivalent, a yellow-flowered perennial of woodland edges and scrub on calcareous soils, whose roots have a strong aromatic smell. They were sometimes dried and hung up in cottages as room-fresheners.

Above: elecampane, an Asian species whose roots are used in herbal medicine and which is naturalised in a few places. Right: the native goldenrod by the River Tees.

Common fleabane, *Pulicaria dysenterica*, is a frequent perennial of ditches, damp hedge-banks and meadows, with wrinkled, downy-green leaves and neat marigold flowers. The leaves have a curious scent, with hints of chrysanthemum and carbolic soap, and it is easy to understand why they were believed to repel fleas. (Bunches were hung in rooms or dried and burned as a fumigant.) Its insect-deterrent powers may not be entirely fanciful. It is a comparatively close relative of the species which supplies the insecticide 'pyrethrum'.

Small fleabane, *P. vulgaris*, is a rare, button-flowered annual of winter-flooded sandy sites. It is now virtually confined to pond-edges, cart-ruts and the damp, pony-grazed 'lawns' of the New Forest, but was previously much more widespread. It is one of a cluster of species (including pennyroyal and starfruit, see pp. 319 and 382) which seem to have depended on a combination of muddy soil and disturbance (by grazing animals or horse-drawn traffic) to help their seeds germinate. Tony Hare has studied the decline of this 'plant of seasonal hollows':

'It is easy to imagine that it takes little to change a habitat from suitable for Lesser [Small] Fleabane to unsuitable; a ceasing of disturbance and grazing would soon see it and its fellow species of disturbed ground replaced by sward-forming grasses or taller herbs.

This is exactly what seems to have happened to most Lesser Fleabane sites. They have become overgrown or been drained and tidied up. Muddy tracks have given way to metalled roads as the horse has given way to the car. Saddest of all is the fate of the village green: once the heart of the village, busy with horses and wagons, well churned up in winter and full of plants such as Lesser Fleabane and Pennyroyal Mint in summer, they have been sanitised and made home to dull swards of ryegrass, flowerbeds full of garish plants and folk celebrations stripped of meaning.'[31]

Goldenrod, *Solidago virgaurea*, is a bright, late-summer-flowering perennial of hedge-banks, open woods and rocky places, which is common in the west and north, but absent over much of central England. In the sixteenth and seventeenth centuries it was in great demand as a wound herb but, according to John Gerard, fell from favour when it was discovered to be rather common: 'It ... hath in times past been had in greater estimation and regarde than in these daies: for within my remembrance, I haue knowne the drie herbe which came from beyond the seas,

solde in Bucklers burie in London for halfe a crowne an ounce. But since it was founde in Hampsteed wood, euen as it were at our townes end, no man will giue halfe a crowne for an hundred weight of it: which plainly setteth foorth our inconstancie and sudden mutabilitie, esteeming no longer of any thing (how pretious soeuer it be) than whilest it is strange and rare.'[32]

Canadian goldenrod, *S. canadensis*. Our native goldenrod was upstaged in another way when its tall and showy relative from the New World was introduced to this country in 1648. Canadian goldenrod was a popular border plant, until its tremendous capacity for self-seeding and spreading was realised in the Victorian garden, when – along with the Michaelmas-daisy (see below) – it was consigned to the rubbish-heap or the wild garden. This century it and **early goldenrod**, *S. gigantea*, have become two of the dominant flowers of urban wasteland in late summer and early autumn.

Michaelmas-daisies, *Aster* species. Michaelmas-daisies of all kinds are widely naturalised in waste places. They have a particular liking for railway embankments, and the drifts along some lines, in pastel shades that range from palest blue to mauve, mark the year's last great show of wild-flower colour, often continuing well into November.

They were first introduced to this country from North America at the start of the eighteenth century. But it was the taste for wilder gardens in the mid-Victorian period that prompted their much wider spread. William Robinson, as usual, was perceptive and unwittingly prophetic about the role they might play in the landscape. In 1870 he wrote of 'Starwort, or Aster': 'they form a very good example of a class of plants for which the true place is the copse, and by wood-walks, where they will grow as freely as any native weeds, and in many cases prove charming in autumn ... Associated with the Golden Rods (*Solidago*) – also common plants of the American woods – the best of the Asters or Michaelmas Daisies will form a very interesting aspect of vegetation. It is that which one sees in American woods in late summer and autumn when the Golden Rods and Asters are seen in bloom together. It is one of the numerous aspects of the vegetation of other countries which the "wild garden" will make possible in gardens.'[33]

And it is precisely with naturalised goldenrods that Michaelmas-daisies provide such an adornment to urban wasteland – though here they are echoing their other North American native habitat, the plains and prairies, rather than the woods.

Mexican fleabane naturalised on a wall in Portloe, Cornwall.

Sea aster, *A. tripolium*, is our native Michaelmas-daisy, a smaller and less floriferous species common in saltmarshes throughout Britain. The flowers usually carry thin bluish florets around a yellow disk. But a form without florets, var. *discoideus*, is increasing and is now the commonest form in some East Anglian saltmarshes.

The light windblown seeds occasionally cause isolated specimens to spring up at damp inland sites, but they rarely persist. Recently (1993), quite large numbers were seen growing along the edges of the A63 west of Kingston-upon-Hull. The plants were at their greatest density at the foot of an inclined slip-road and around signs and crash barriers, suggesting that accumulation of salt used for de-icing may have produced localised facsimile saltmarshes which favoured this maritime species.[34]

Mexican fleabane, *Erigeron karvinskianus*, is a long-stemmed daisy, naturalised on walls and banks in scattered sites in England. It was introduced to this country in the nineteenth century. There are well-established colonies on the walls of Merton College, Oxford, and natural rock-faces at Powis Castle near Welshpool. **Canadian fleabane**, *Conyza canadensis*, is a rather nondescript annual with small whitish flowers, whose seed was reputedly brought to this country in the seventeenth century, in the stuffing of a bird imported from North America. Now it is widely naturalised in pavements, waste ground and walls, chiefly in south-east England. **Guernsey** or **Sumatran fleabane**, *C. sumatrensis*. The original English name – 'Sumatran fleabane' – may not have been the most helpful for this originally Peruvian species, but it does catch its tropical luxuriance – it can grow up to seven feet in a single summer – rather better than the now accepted 'Guernsey fleabane'. It was first discovered in London's Dockland by the indefatigable alien plant hunter Brian Wurzell in 1984. Since then it has spread throughout the capital and outwards into the Home Counties, even into the exclusive estates of Ascot in Berkshire, where its favourite habitat appears to be cracks in patio paving.[35]

Daisy, *Bellis perennis*. There is a saying that spring has not arrived until you can cover three, or nine, or a dozen daisy flowers with your foot. If there is some disagreement about the requisite number, it is because there is scarcely a day in the year (except during freezing weather) when there is not a daisy in flower somewhere. In his Journal for 1824, John Clare notes that he 'gatherd a hand ful of daiseys in full bloom' on Christmas Day.[36]

In the short turf of paddocks and lawns they can grow in constellations so dense that there is no space between the flowers, and it is this sheer availability and abundance that is partly responsible for their popularity in children's games. These go some way beyond the familiar daisy chain, made by threading the wiry stalks through each other, via slits made with a fingernail. In Wales there is a custom of making daisy 'caterpillars': 'We used to make caterpillars by taking a daisy with a long stem and then threading more daisy heads onto that. This was done by pushing the head of the first daisy (the long one) through the yellow part of the daisy heads.'[37]

A variant on threading games was the manufacture of inverted – 'Irish' or 'Australian' – daisies:

'An Irish daisy is one where the head has been turned upside down and re-threaded on its stalk, making it look as if it has grown with the head the wrong way round.'[38]

'This is how to make an Australian daisy. Pick a daisy. Pick the stem as low as you can, right next to the flower. There will be a sort of covered hole where the stem was. Stick the stem back through the hole till it comes through the yellow disc florets.'[39]

In Clwyd children play a game with daisy heads reminiscent of the flicking or 'flirting' games played with cigarette cards (or milk-bottle tops when 'fag-cards' were scarce).[40] This is an account by a 12-year-old girl:

'We used to play a game called Flacks in primary school. We would use a lot of daisy heads to make a circle, filled. Then we would throw dandelion heads into the circle and see how many flowers you could hit in the circle. The flowers which you did you won. The one with the most daisies won!'[41]

'We made daisy chains and also daisy plaques, using mud on old plates and sticking the heads in patterns. Although we knew and saw well-dressing in the Peak District I don't remember using any flower except daisies.'[42]

Seven daisies underfoot. Enough for spring?

Recognising daisies as attractive flowers despite their humble status is probably an adult response, and William Hazlitt in his lecture 'On Thomson and Cowper' suggests: 'The daisy that first strikes the child's eye in trying to leap over his own shadow, is the same flower that with timid upward glance implores the grown man not to tread on it.'[43]

'The down-to-earth nature of the daisy has penetrated the language, where "daisy roots" is now slang for "boots" and "kicking up the daisies" a term used to describe those who have given up earthly gardening once and for all.'[44]

But the name is quite unequivocal in its affection. It is, as Chaucer wrote in the best tribute to the daisy, the day's-eye, which opens with the dawn and reflects the sunrise in the pinkish flush on the underside of its petals:

Now have I thanne eek this condicioun,
That, of al the floures in the mede,
Thanne love I most thise floures white and rede,
Swiche as men callen daysyes in our toun.
To hem have I so gret affeccioun,
As I seyde erst, whanne comen is the May,
That in my bed ther daweth me no day
Than I nam up and walkyng in the mede
To seen this flour ayein the sonne sprede …
And lenynge on myn elbowe and my syde,
The longe day I shoop me for t'abide
For nothing elles, and I shal nat lye,
But for to loke upon the dayesie,
That wel by reson men hit calle may
The 'dayesye,' or elles the 'ye of day,'
The emperice and flour of floures alle.
I pray to God that faire mote she falle,
And alle that loven floures, for hire sake![45]

Feverfew, *Tanacetum parthenium* (VN: Bachelor's buttons, Featherfew). As its name suggests, feverfew was a medicinal herb given for colds and fevers. In fact it was almost the classical and medieval world's aspirin, recommended for headaches, rheumatism, and general aches and pains. It reached Britain from its home in the Balkans some time during the early Middle Ages and is now widely naturalised by walls and old buildings and on waste ground and waysides close to cottage gardens. It has also become one of the great success stories in herbal medicine and had its reputation vindicated by work at the City of London Migraine Clinic. In 1978, after a newspaper story about a woman who had successfully rid herself of persistent migraine attacks by chewing feverfew leaves daily, many patients at the clinic began self-medicating with the leaves. Dr Stewart Johnson decided to undertake a long-term survey of some 270 feverfew-takers (partly to ensure that they were not doing themselves any harm). The results were remarkable. After a leaf a day for three months, 70 per cent reported a significant decrease in the

Left: feverfew, an ancient remedy recently vindicated by orthodox medicine. This is a golden-leaved variety. Right: sea wormwood by the Wash.

frequency or severity of their attacks. A third appeared to have abolished their attacks altogether. These were far better results than had been obtained with any other form of preventative treatment, and they were confirmed by a more rigorous double-blind trial on a smaller sample of patients, using measured quantities of dried leaf made up into capsules.[46] The active chemicals have now been isolated and work by stopping blood vessels in the brain going into spasm, which is believed to be the immediate 'cause' of migraine attacks. Many sufferers now make their own feverfew pills or sandwiches[47] or simply nibble the leaves: 'I am a migraine sufferer who takes feverfew and believes it really helps to reduce the frequency of attacks. I walk to work with my dog over the Malvern Hills each day, deliberately taking a route past a stone wall through the Whyche cutting (so called because it was the old salt road from Droitwich through to Wales) where feverfew grows abundantly. I browse on a fresh leaf or two every morning on my way down to the office.'[48]

Even without its medicinal properties, feverfew would be a welcome wayside plant – often becoming naturalised in its golden-leaved or double-flowered forms (the latter one of many plants known as 'bachelor's buttons') – with its foliage smelling bracingly of camphor.

Tansy, *T. vulgare*. Tansy's leaves are pungent and bitter, and at one time they were eaten at Eastertide, to kill off the 'phlegm and worms' which the Lenten fish diet gave rise to. They were mixed with eggs, milk and flour, presumably to make them more palatable, and from the fifteenth to the nineteenth centuries, a 'tansye' was a generic term for any omelette or pancake-like dish flavoured with bitter herbs. The concentrated oil is quite toxic and has been used in the treatment of worms and as an abortifacient.

But its ferny foliage and golden button-flowers make it an attractive plant of rough grassland, river-banks and waysides, enough for it to become a landmark plant in one village: 'Tansy Green Public House is a comparatively modern pub built on a private housing complex. Before the pub was opened the local people were asked to suggest a name for it. The winner chose the name 'Tansy Green' because the field on which the pub was built was well-known for the number of tansy plants that grew there.'[49]

Sea wormwood, *Seriphidium maritimum*. Sprays of this silver-leaved perennial adorn the drier reaches of salt-marshes around Britain (especially in the east) blending

with the pastel shades of sea-lavender and sea aster in high summer. The leaves have a strong, bitter but refreshing fragrance.

Wormwood, *Artemisia absinthium*, is a handsome, silver-leaved perennial of waste ground, waysides and railway embankments, especially in the Midlands. Possibly native, or an ancient medicinal introduction from southern Europe, it is a bitter and pungent herb and, as its name suggests, was once used as a powerful worm-dispeller and deterrent. The sixteenth-century farmer Thomas Tusser even recommended it for deterring fleas:

while woormwoode hath seede, get a bundle or twayne,
to saue against March, to make flea to refraine.
where chamber is swept, & ye wormwood is strowne
no flea for his life, dare abyde to be knowne.[50]

Wormwood also provides the bitter principle in that potent liqueur, absinthe – and perhaps accounts for its notorious reputation: in excess wormwood oil is corrosive and hallucinogenic, making objects appear as if they have changed colour. The plant's Russian name is 'Chernobyl'.[51]

Mugwort, *A. vulgaris* (VN: Gipsy's tobacco, Muggar). A very common perennial of waysides and waste places, growing to four feet tall, mugwort's leaves are deep-cut, greyish below and glossy green above, smell only faintly of wormwood, and often have a rather dusty and bedraggled look when growing close to roads. (It is called 'Council weed' in the Bickerstaffe/Melling area of Lancashire, as 'it always appears after the Council have been out'.)[52] Once *Mater Herbarum* (the Mother of Herbs), it was in widespread use as a charm and medicinal plant (stuffed in the shoes to prevent travel weariness, for instance). But today it is chiefly used as a children's smoking leaf.

'Village boys said they knew how to make cigarettes, but this was just rolling mugwort ("muggar") flowerets inside newspaper strips.'[53]

'We would gather mugwort, which we called "cosi", to dry and smoke in big acorn cups, making a hole in the side and putting a stout straw through for the stem. Many a time we went home feeling groggy but never telling why. Our dear old Mum would have had a fit.'[54]

Mugwort is still an important symbolic plant in Manx folklore, where it is known as Bollan Bane or Bollan Feaill:

'Sprigs are worn at the July 4th annual open-air parliamentary assembly on Tynwald Hill. (Tynwald, the Manx Parliament, has survived continuously from the period of the Norse kingdom of Man.) The wearing of mugwort was revived about 1924, having gone into abeyance in the latter nineteenth century. It is now a conspicuous feature of this National Day, although some born-again Christians objected to it in the mid-1980s on the grounds that it was a pagan idea.'[55]

'Until quite recently no law had force until it had been read from the Tynwald Hill. The turves of the hill are said to have been taken from each Parish on the Island. It is the most revered spot in Man to Manxmen all over the world. Bollan Bane (the Manx name means "White Wort", referring to the white underside to the leaves) is worn as a charm by almost everybody present.'[56]

Chinese mugwort, *A. verlotiorum*, is an Asian mugwort with darker leaves first noticed in this country in 1908, and now frequently naturalised in the London area, especially near the Thames. A hybrid between this species

Yarrow, painted in its three colour varieties by Caroline May at Breamore, Hampshire, in July 1834.

and *A. vulgaris* has been discovered on Tottenham Marshes by Brian Wurzell.

Yarrow, *Achillea millefolium* (VN: Yarroway, Staunchweed, Poor man's pepper). Yarrow was regarded as a powerful herb as far back as Anglo-Saxon times and was used in divination rituals and as a charm against bad luck and illness,[57] as well as for staunching wounds. (It was called carpenter's grass in places.) But, with the paradox often found in magic-based herbalism, it was also believed to cause nosebleeds or at least sneezing if a leaf was put up the nose.

It is a resilient perennial which is common in all kinds of grassland, including lawns. When mown regularly, its feathery leaves can build up a fine if wiry sward, and some obsolete local names are translations (or maybe the origins of) its Latin specific name, *millefolium*, 'thousand-leaf'. The flat flower-heads are usually white, but quite often pale (and occasionally dark) pink. **Sneezewort**, *A. ptarmica*. True sneezewort is a widespread plant of damp grassy places scattered throughout Britain, with small white daisy flowers with brownish-green disks.

Chamomile, *Chamaemelum nobile*. Chamomile seats, mossily soft and smelling of apples (or bubblegum to some modern young noses),[58] were a favourite feature of Elizabethan herb gardens,[59] and are currently enjoying a revival. There were even chamomile lawns, and it was on one such fragrant pitch that Sir Francis Drake is reputed to have played his famous game of bowls. But the feathery foliage needed regular clipping and de-flowering if the plant was to develop into a tight sward. It was not until the discovery of a non-flowering and less scrambling variety ('Treneague') in Cornwall earlier this century that chamomile lawns became a practical possibility in ordinary gardens.

But the chamomile lawn is not a human invention. The plant's natural habitat is tightly grazed cliff-top grasslands, sandy commons and damp woodland clearings, and it adapts to grazing by adopting a more or less prostrate form. Previously it occurred over much of England, though it was always commoner in the extreme south. But now, as a consequence of a general decline in grazing, especially of village greens and commons, it has become a

Corn chamomile, one of the classic 'rayed' composites.

scarce plant, confined to the south-west peninsula and the Hampshire and Sussex Wealds. Its natural strongholds are some of the commons and greens on the edge of Dartmoor and damp, pony-grazed glades in the New Forest, which are actually known locally as 'lawns'. But Heather Winship, who has studied the history and fortunes of chamomile, has found that, in Sussex particularly, it is flourishing in a comparatively new habitat – the cricket pitch. Very often these have been created on village greens which hold residual colonies of chamomile, and where the regular summer mowing and rolling mimic the kind of grazing pressure under which chamomile swards flourish. As a result the cricket grounds at Heyshott and Westbourne now support some of the healthiest populations of chamomile in West Sussex. It has also been found recently at the ground at Hartley Wintney, Hampshire, one of the oldest pitches in England, established on the village green in 1776.[60]

Several species of annual chamomiles and mayweeds, all with feathery leaves and white flowers, appear as weeds in cultivated ground and on waysides. (Place names such as Maghull, Lancashire, and Mayfield, Sussex, may indicate sites where *maegde*, OE for mayweed, grew.)[61] **Corn chamomile**, *Anthemis arvensis*, has fragrant flowers and slightly scented leaves and occurs rarely on calcareous soils, mainly in England. **Stinking chamomile**, *A. cotula*, has foetid foliage and prefers heavier clay soils. **Scentless mayweed**, *Tripleurospermum inodorum*, is the commonest and most widespread of the group, often occurring in large masses in arable and set-aside fields. **Scented mayweed**, *Matricaria recutita*, is superficially similar to *T. inodorum*, but quite strongly and pleasantly aromatic when fresh and less widespread.

Pineappleweed, *Matricaria discoidea* (VN: Apple virgin). A widespread annual of bare places, paths and even city pavements, whose foliage and rayless yellow flowers smell unmistakably of pineapple when crushed. It is a native of north-east Asia and came to this country via North America in 1871. Its subsequent spread across the country owes a lot to the rapid growth of motor transport – and especially to the pneumatic tyre. In the early years of this century, before the introduction of tarmac, most roads were morasses of mud for much of the winter and after periods of rain. The ribbed treads of motor tyres would pick up mud, perhaps containing pineappleweed seeds, and carry it long distances before it dried and fell off. Then, together with its cargo of seeds, it would be sluiced to the edge of the road by rainwash, where the seeds had a chance of taking root.[62]

In 1968, an experiment was carried out in the Midlands to quantify this happenstance hitchhiking by plant seeds. A car with scrupulously washed tyres was driven along 65 miles of road after a period of rain. On the way it was turned into field gateways, lay-bys and passing-places. At the end of the journey, the tyres were hosed down, and the sediment collected and incubated in sterilised compost. In a few months 13 different species of flowering plant had sprung from it – including 220 seedlings of pineappleweed.[63]

Leptinella, *Cotula squalida*, is a member of the Buttonweed genus, which, along with its close relative, *C. dioica*, is becoming increasingly naturalised in lawns: 'Some years ago a small creeping plant appeared on the edge of my lawn and rapidly started to spread. It was identified as *Cotula squalida* ... It is a New Zealand plant and used for lawns out there. In New Zealand itself I was surprised to see it in the public gardens in Christchurch, and two bowling greens laid out with it.'[64]

Oxeye daisy, *Leucanthemum vulgare* (VN: Dog daisy, Horse daisy, Moon daisy, Moonpenny, Marguerite). Moon daisy or moonpenny was always a better name for this bright, brisk flower that can seem to glow in the fields on midsummer evenings. It is one of the first meadow flowers to colonise unsprayed grassland and is consequently making a welcome return to waysides, set-aside land and even lawns, after having been driven out of most agricultural grassland. It occurs throughout Britain.

Left: sneezewort in a fen by Loch Insh in the Spey valley.
Above: pineappleweed – true to its name when crushed.

Oxeye or moon daisies, on a Dorset cliff-top.

Shasta daisy, *L.* × *superbum*, is a much larger version of garden origin, widely naturalised on roadsides and waste ground.

Corn marigold or **Gold**, *Chrysanthemum segetum* (VN: Dunwich Buddle, Johnny Buddle). It is ironic that for centuries corn marigold was commonly named after the most precious of metals at the same time as being regarded as one of the most noxious of cornfield weeds. In the twelfth century, Henry II issued an ordinance against 'Guilde Weed', which was probably the earliest enactment requiring the destruction of a weed. In *A Boke of Husbandry*, 1523, John Fitzherbert included 'Gouldes' in his blacklist of plants that 'doe moche harme'. A few parishes in arable England – Golder, Oxfordshire, Goldhanger, Essex, and Goltho, Lincolnshire – may even have been named after it.[65] (It is also tempting to wonder if the still-used East Anglian names of 'Boodle' and 'Buddle' have any connection with the slang word for money or burgled jewellery.)

In the late nineteenth century, when bright colours came into vogue, corn marigolds became briefly fashionable for table decoration. Matthew Arnold wrote to his sister in 1883: 'I thought of you in passing through a cleared cornfield full of marigolds. I send you one of them. Nelly gathered a handful, and they are very effective in a vase in the drawing-room.'[66]

Gold was common on light soils throughout Britain until after the war. A Welsh farmer remembers: 'teams of women in the fields towards Hereford with special long-handled tools with a sort of fork on the end, walking along the rows of corn and hooking out "the yeller daisies". They are a more disliked weed because they don't dry out easily and tend to rot the straw in the bales.'[67]

Between the 1950s and 80s corn marigold retreated in the face of modern weedkillers. But, more recently, set-aside policies and reduced spraying on field edges have resulted in some localised resurgence, often from long-dormant seed, as in Corby Hill, Cumbria: 'In 1983 there was an explosion of corn marigold growing in short barley in a field near my home. How long the seeds have remained dormant, God only knows, as I have not seen this before during my lifetime – 59 years.'[68]

It has been most successful in the sandy soils of East Anglia, by Sizewell B nuclear power station for instance: 'The rough verges on the side of the road were bright with corn marigolds, right up to the power-station fence. How's that for ancient and modern?'[69]

Silver ragwort, *Senecio cineraria*, is a shrubby perennial from the Mediterranean, with finely hairy and deeply toothed silver leaves, which is quite well naturalised on cliffs and rough ground near the coast, chiefly in south-west England.

Fen ragwort, *S. paludosus*. This striking fenland plant – which can grow up to six feet tall, with large, saw-toothed leaves and flat-topped yellow daisy-flowers – once grew in several East Anglian fens. It was first recorded by John Ray in 1660[70] and still occurred in the region until the middle of the nineteenth century, when the extensive drainage works under way in the Fens drove it into its last-ditch redoubts and finally into extinction. The last record backed up by a specimen is from Wicken Fen, Cambridgeshire, in 1857. Then, in July 1972, David Dupree found a small clump in a freshly excavated fen ditch inside the plant's ancestral territory in Cambridgeshire.[71] It is possible that fen ragwort recolonised the area by seed blown in from Holland, where it is still comparatively widespread. But the most likely explanation is that the plant sprang from seed lying dormant in an undisturbed layer of the peat in which it had once grown. When the new ditch was dug, it brought the seeds to light, so to speak, and in the process created a damp habitat sufficiently like a fen to ensure the plant's survival after germination.

Common ragwort, *S. jacobaea* (VN: St James' wort, Staggerwort, Stammerwort, Yellow tops, Stinking Willie, Mare's fart). Ragwort is regarded as the great enemy by those who keep horses, and summer weekends spent laboriously hand-pulling and removing the plants are a regular chore to check the plant's spread. Neither horses nor other grazing animals will normally eat the growing plant, unless it is so dense that it is difficult to graze without ingesting some, but they will when it has died and dried out. Green or dry, it causes insidious and irreversible cirrhosis of the liver.[72] A Ministry of Agriculture adviser has seen its effects at close quarters: 'The plant is responsible for half the cases of stock poisoning in Britain. It is susceptible to spraying and cutting, but when wilting or distorted by the spray, cattle will eat it with fatal results. But sheep can consume small quantities of the mature plant with impunity and apparent relish. Coming to this area [Montgomeryshire] of few cattle and thousands of sheep I was impressed by the total absence of ragwort except in those places inaccessible to grazing sheep. By contrast, in North Staffs, a dairy farming area with few sheep, ragwort posed the principal weed problem, many fields being heavily infested.'[73]

The still-used local names of mare's fart in adjacent North Shropshire and Cheshire and of stinking Willie in Scotland suggest how loathed the malodorous foliage is

Corn marigold, an arable weed which is recovering in some areas. The Black Isle, Scotland.

by cattle farmers.[74] Ironically, paddocks and pastures full of the golden flowers, with swallows dipping amongst the ragged-edged leaves, are one of the most beautiful sights of high summer and have become more common with the spread of set-aside schemes on arable land. It is no real surprise that, as the farming writer Robin Page reported, at least one farmer found a more positive way of dealing with ragwort in the hot summer of 1993: 'I have discovered a farmer in the West Country who has solved his ragwort problems. He has been picking it and selling it in his farm-shop as "summer gold". The townies love it. Can you imagine visiting a flat in Birmingham and seeing a tasteful vase of ragwort on the sideboard – bought here for a very modest price. He surely deserves a diversification award.'[75]

It is not such a preposterous idea. John Clare, very familiar with the tribulations weeds caused to farmers, could also see the beauty of 'summer gold':

> *Ragwort thou humble flower with tattered leaves*
> *I love to see thee come & litter gold ...*
> *Thy waste of shining blossoms richly shields*
> *The sun tanned sward in splendid hues that burn*
> *So bright & glaring that the very light*
> *Of the rich sunshine doth to paleness turn*
> *& seems but very shadows in thy sight*
>
> (*c.* 1831)[76]

In the Isle of Man, ragwort is called 'Cushag' and is the national flower, though its ambivalent reputation in the countryside is reflected in the fact that it is also used as a satirical emblem: 'Among the many Manx exiles living in America and elsewhere there was an expression used to curb over-enthusiastic and nostalgic praise of the "li'l island": "We know – Ta airh er ny Cushaghyn er shen" (There's gold on the Cushags there).'[77]

Oxford ragwort, *S. squalidus*. Although it is now the most abundant ragwort in most British cities, this south European species has a fair claim to its Oxford title. A plant (reputedly gathered from the volcanic rocks of Mount Etna) was certainly growing in the University Botanic Garden in the eighteenth century, and was noticed there by Sir Joseph Banks in the 1770s. Linnaeus is believed to have described and named the species from specimens sent to him from Oxford.[78] But by the turn of the century its downy seeds had wafted out of the Garden and begun to colonise the city's old walls. By the 1830s it had arrived at Oxford Railway Station, and from there it set off down the Great Western Railway. It found the granite chips and clinker of the permanent way a congenial substitute for its natural dry habitats in the southern European mountains, and by the end of the nineteenth century it was well established in many southern English counties. The slipstream of trains seemed to help the seeds on their way. George Claridge Druce described a journey he took with some of them, which floated into his carriage at Oxford and out again at Tilehurst, in Berkshire.[79] Now it is distributed over almost the whole of England and Wales, even down to the tip of Cornwall: 'Oxford ragwort has long since reached the end of the [old] GWR line, growing all along the track on the last couple of miles to Penzance.'[80]

Yet it has remained very much an urban plant, sticking close to railways, factory walls, motorway verges, building sites and car parks, and rarely invading village walls, for instance, or grasslands where common ragwort grows; for example, in Herefordshire: 'The main railway line

Oxford ragwort, putting on a show in its own city.

runs from Abergavenny to Hereford and any ragwort growing on waste ground within a quarter of a mile on each side of the track is usually Oxford ragwort. Further afield it is common ragwort.'[81]

Perhaps this taste in habitats lay behind the specific scientific name *squalidus*, for the plant itself is anything but squalid. It is smaller and more compact than common ragwort and brightens all the wastelands it graces, especially in the company (as it so often is) of rosebay willowherb.

London ragwort, *S.* × *subnebrodensis*, is a hybrid between Oxford ragwort and sticky groundsel (below) which crops up occasionally on wasteland in the south-east. J. E. Lousley first discovered it in London in 1944, when he named it *S.* × *londinensis*.[82] **Sticky groundsel**, *S. viscosus*, is a similar plant to common groundsel, but unpleasant-smelling and covered with very sticky hairs. It may be native on sandy and gravelly soils. (John Ray first recorded it in Britain in 1660 as plentiful in the Isle of Ely.)[83] But in this century it has behaved very like an alien, spreading rapidly across the country in a similar pattern – and into similar wasteland habitats – to Oxford ragwort.

Common groundsel, *S. vulgaris* (VN: Charlie), is an abundant annual of rough and cultivated ground throughout Britain. The name groundsel derives from an Old English root meaning 'ground-swallower'. But, for once, the Latin name of the genus is more graphic: *Senecio* derives from *senex*, an old man. As a sixteenth-century herbalist wrote: 'The flower of this herbe, hath white heere, & when the winde bloweth it away, then it appereth like a balde headded man therfore it is called Senecio.'[84] Pull the white seed-heads from a groundsel and you will see exactly what he meant – they leave a round, slightly shiny scalp, dotted with follicles. Groundsel is still widely used as food for cage-birds.

Groundsel has some highly local varieties and hybrids, mostly formed by crosses with Oxford ragwort, the most notable being **Welsh groundsel**, *S. cambrensis*, now regarded genetically as a new species. It was first recognised in 1948 in Flintshire and then in 1982 in Midlothian. It is one of Britain's few endemic species.

Leopard's-bane, *Doronicum pardalianches*, is one of several related continental European species with large yellow flowers, familiar in gardens and well naturalised in woods and on shady waysides.

Colt's-foot, *Tussilago farfara* (VN: Son-before-father, Foal's foot, Disherlagie, Dishylaggie, Tushylucky, Tushies, Baccy plant, Coughwort, Cleats). This is one of the earliest spring flowers, its cheerful, yellow blooms and scaly stems often appearing in February, a month or two before the leaves – hence 'son-before-father'. The plant's whole story is told by its common names. 'Colt's-foot' itself describes the hoof-like shape of the leaves, which are mealy above when they first appear and covered with white felt beneath. The Scottish 'tushylucky' and its variants are corruptions of the Latin *tussilago* – a name used by Pliny, related to *tussis*, a cough – which records the use of the leaves as a cough medicine (perhaps because of a slight similarity to lungs in their shape). Ironically, it has also been used as a substitute for tobacco. The dry felt on the leaves certainly smoulders well and has been used as tinder.

'It is used as herbal tobacco, and known as "Baccy plant" in Somerset.'[85]

'I have vivid memories of an itinerant farmworker who used to appear cross-country to gather the leaves of the coltsfoot which was prolific in just one triangle of our 20-acre orchard, presumably for selling as herbal tobacco.'[86]

'I remember eating coltsfoot rock in wartime because it came from the chemist and counted as a cough sweet and was therefore not "on the ration".'[87]

Colt's-foot grows in all kinds of waste, rough and cultivated places, especially where there is poor drainage.

Butterbur, *Petasites hybridus* (VN: Wild rhubarb, Butcher's rhubarb). The huge rhubarb-like leaves of butterbur really were used for wrapping butter in the days before refrigeration. Handle a leaf and you will understand why. It is not only large and pliable enough to fold without breaking, and thick enough to cushion butter from bruising and soak up any seepage, but actually feels cool to the touch because of the soft grey down on the underside. It can still make a serviceable wrapping for picnic left-overs or a cache of wild berries.

But butterbur leaves are most often used today as umbrellas or sunshades – as their scientific name suggests: *Petasites* derives from the Greek *petasos*, meaning a broad-brimmed felt hat. Gerard agreed that the leaves were 'of such a widenesse, as that of it selfe it is bigge and large inough to keepe a mans head from raine, and from the heate of the sunne'.[88]

By the River Dove in Derbyshire I have seen quite young children spontaneously picking butterbur leaves to protect themselves from a summer downpour, as have contributors in many parts of the country.[89] One botanist's grandchildren regularly dress up entirely in butterbur leaves when playing in their hop-vine wigwams.[90]

Like colt's-foot, butterbur's flowers appear before its leaves, often as early as February. When they first push through the soil the spikes look sufficiently like flushed button mushrooms to have earned the now obsolete country name of 'early mushrooms'. When fully emerged, the tassled blooms give the flower-spike the look of a dwarf pink conifer. Butterbur is a species of stream-banks and damp waysides almost throughout Britain, and is evocative of shady places by water in high summer: 'There is a fine patch of [butterbur] along the River Lea near Hertford. I identified it as a child (as so much else) from Brooke Bond tea-cards. I find it often grows in a haunting situation.'[91]

White butterbur, *P. albus*, is a smaller, white-flowered species from continental Europe, quite widely naturalised on roadside verges, woodland edges and river-banks, mainly in Scotland and northern England.

'It was brought to Staffs by Canon Hawksford before 1890. He was the headmaster of a boys' school situated at the head of the valley of Cotton Dell. The [white] butterbur has spread down the stream which flows through Cotton Dell and Star Wood, and is even found along the River Churnet in places.'[92]

Winter heliotrope, *P. fragrans*, was introduced from Mediterranean North Africa as a garden plant in 1806. It has spikes of pale mauve flowers which appear around Christmas (a few weeks before the leaves) and have a welcome vanilla-like fragrance. But the plant is extremely invasive, spreading rampantly by means of underground stems, and its many naturalised colonies on roadsides tend to exclude all other species.

But the leaves are not entirely worthless. In Truro, they have been used to wrap violets, ready for market.[93]

Pot marigold, *Calendula officinalis*, is a familiar garden annual, possibly native in southern Europe and frequently found as a self-seeded casual or throw-out on waste ground and waysides. **Ragweed**, *Ambrosia artemisiifolia*, is an annual from North America, where it is one of the commonest causes of hay fever. It is becom-

Butterbur by a Peak District river. The pale flower-spikes, which appear before the leaves, were once known as 'early mushrooms'.

Claude Lorrain's 'Landscape with Narcissus and Echo', with butterbur and wild narcissus (daffodil) in the foreground.

ing more frequent in Britain. **Sunflower**, *Helianthus annuus*, is a tall and familiar annual, frequently appearing on rubbish-tips and in unexpected places in gardens, usually from birdseed.

Gallant-soldier, *Galinsoga parviflora*, was brought to Kew Gardens from Peru in 1793, bearing a name that commemorated the Spanish botanist Don Mariano Martinez de Galinsoga. The plant itself was rather less imperious, being a thin, lax and greenish-flowered daisy with weedy habits. In the early 1860s it escaped from Kew and became widely established in gutters, gardens and waste places around Richmond. For a while it became known as Kew weed. But, once its airborne seeds began arriving in less salubrious areas, it was ripe for a more down-to-earth name, and *Galinsoga* was corrupted to 'Gallant soldier', a name which has doubtless stuck partly because it is so ironically inappropriate. (In Malawi, where it is also naturalised, it is known as 'Mwamuna aligone' – 'My husband is sleeping'!)[94]

Hemp-agrimony, *Eupatorium cannabinum*, is a tall and bushy perennial of riversides, damp woods and waysides, chiefly in England and Wales. The plant is no relative of either hemp or agrimony (though its leaves have a resemblance to those of *Cannabis*, see p. 62). The flowers – pink froth over darker sepals – have something of the look of a whipped strawberry mousse.

Galinsoga quadriradiata, a relative of the true gallant-soldier.

Plant medicine

The use of wild plants in herbal and folk medicine is on the increase in Britain again – though it is, as ever, a highly subjective business. Personal faith, idiosyncratic sensitivities and mode of use all profoundly influence the results. Anecdotal accounts of successful remedies show that a large number of British plants have been used almost interchangeably to treat an equally large number of common disorders: almost anything will work with *someone* when it comes to dispelling bruises, warts and wind.

But there have been some discernible traditions, which have shaped the names, distribution and fortunes of many species. Most remedies must first have been discovered by trial and error, and species with quick, dramatic effects – purgatives such as buckthorn and narcotic sedatives such as henbane, for example – would have made their mark early. In more recent times, trial and error as a scientific discipline has brought a steady trickle of plant-based drugs into mainstream medicine – for example, aspirin from willow bark, colchicine (for gout) from meadow saffron, digoxin from foxgloves, and menthol from mints.

Yet from at least the early historic period many other plants with no obvious immediate effects on human physiology were also believed to have healing properties. It is hard for us to imagine the awe with which plants were held in a pre-scientific age. They could appear by apparently spontaneous generation. Barely distinguishable species could feed you, poison you or drive you mad. It is no wonder that all manner of theories were developed to explain and predict their effects. Astrology, scriptural interpretation and numerology were all pressed into service. But the most popular system was sympathetic magic. This was based on a search for analogy, association and pattern within nature, in the belief that like (or sometimes unlike) would cure like. Exterior similarities in shape, colour or texture were regarded as clues to inner resemblances. Processes of apparent cause and effect in the larger, visible world might generate similar processes deep inside living things. So parasitic plants such as mistletoe might 'overcome' human 'parasites' such as cancer. Ivy berries would cure drunkenness, because ivy strangles vines. Some classical authors even believed that notoriously windy food-plants such as lentil could protect (by repulsion) a garden from gale damage.

It is easy to mock these beliefs as primitive and superstitious, but at least they were based on observation and an ecological outlook – of a kind. Sadly, in the expansive, market-driven climate of the seventeenth and eighteenth centuries, they began to be vulgarised by a new breed of commercial herbalists into the notorious Doctrine of Signatures. This decreed that all plants had been 'signed' by the Creator with some physical clue as to their medicinal qualities. Yellow flowers were marked out for jaundice. The blotchy, oval leaves of lungworts were ordained for diseased lungs. Plants which rooted in stone (such as parsley-piert) would break through kidney stones just as effectively. Much of this, I suspect, was little more than an extravagant brand of sales talk aimed at the gullible, analogous to the brandishing of scientific terminology in modern advertising. (Though sometimes it seems to have been a form of *aide-mémoire*, or a way of rationalising the properties of a plant whose effectiveness had already been proved by trial and error.) But the Doctrine was responsible for probably the bulk of the *materia medica* in the written herbal tradition, and its influence is still obvious on the lengthening shelves of herbal remedies in chemists' shops.

Yet there is a third strand of indigenous plant medicine, often overlooked in the written herbal tradition, in which both the above strands were rooted. 'Folk medicine' relies heavily on native plants, but is essentially an oral tradition, derived from hard-won experience mixed with family and local customs and a dash of superstition. Although the medicinal history of many plants is included in *Flora Britannica*, the evidence given here of current use is chiefly confined to this last category. Given the intensely local nature of the folk medicine tradition, it is surprising how much agreement there is about effective plants. There were some half a dozen species – including comfrey as a poultice for bruises, greater celandine as a wart-remover, dandelion as a laxative and diuretic, and feverfew for migraine – which were recommended from personal experience from all over Britain. And, it should be added, by mainstream practitioners, too.

Lesser celandine or 'pilewort'. The knobbly root tubers resemble piles and, according to the Doctrine of Signatures, were given as a treatment for them. Herbal plants which sprang up in the sanctified ground of the churchyard were regarded as being especially powerful.

Flowering-rush family
Butomaceae

Flowering-rush, *Butomus umbellatus*, is a tall and handsome plant of ditches, fens, reedbeds, the edges of slow-moving rivers, even moats. The head is an umbel – a bouquet almost – of pale pink flowers striped with darker pink veins, reminiscent of a huge *Allium* or lily. Grigson suggested 'Pride of the Thames' would be a prettier and more appropriate name;[1] but flowering-rush is now scarce in our most famous river, as it is in many other waterways, thanks to the dredging and canalising of their edges. The two main strongholds are probably the Somerset Levels and the Norfolk Broads.

Water-plantain family
Alismataceae

Arrowhead, *Sagittaria sagittifolia*, is a handsome aquatic plant, with arrow-shaped leaves and white blooms, each of whose three petals has a purple wash at the base. It is found in canals, ponds, ditches and slow rivers, mainly in England, but is nowhere common.

Water-plantain, *Alisma plantago-aquatica*, is found in similar habitats to arrowhead, but is much commoner and more widespread. Its flowers are usually pale lilac and its leaves plantain-shaped. Its chief claim to fame is that John Ruskin believed that the particular curve of its leaf-ribs (along with the southern edge of the Matterhorn, the sinuous lip of a nautilus shell and the side of a bay leaf) represented a model of 'divine proportion', one of those shapes on which 'God has stamped those characters of beauty which He has made it man's nature to love'. It helped him construct his theory of Gothic architecture. (Unfortunately for his credibility, he also made a critical bloomer over the plant, misidentifying the arrowheads in Charles Collins's painting *Convent Thoughts* as his heavenly *Alisma*, which he believed he had never seen 'so thoroughly or so well drawn'.)[1]

Starfruit, *Damasonium alisma*, is one of that group of damp-loving annuals (cf. chamomile, small fleabane, pennyroyal) which have all suffered from the decline of the working village green and its ponds. It once grew in many English counties from Sussex north to Shropshire, but by 1990 was reduced to three ponds, two in Buckinghamshire and one in Surrey. A local Flora of 1931 remarked: 'The records for this species in Surrey are extremely numerous but the great majority of them are more than 60 years old.'[2]

In recent years it has begun to make a recovery, thanks to a better understanding of its needs and the conservation body Plantlife's sterling work at a number of its old sites. The turning point was the chance discovery of a single starfruit flower in a Surrey pond in 1989, soon after it had been cleared of silt, leaf litter and overhanging trees by the local commons preservation society. Starfruit, it seemed, was one of those species that needs open, well-lit, shallow water to grow in and regularly churned-up mud for its

Above: water-plantain.
Right: the striking foliage and isolated flowers of water-soldier, now a nationally scarce plant, found most plentifully in the Norfolk Broads.

seeds to germinate. Accordingly, Plantlife, working closely with English Nature and the various parish councils, began pond restoration work at four former locations in the Buckinghamshire Chilterns, with encouraging success. At the village pond at Coleshill, flowering starfruits reappeared in 1992. In the same year, two out of four rejuvenated ponds on Downley and Naphill Commons were graced by the plant's return. In June, there were some 360 plants – more than had been seen in the whole of Britain for decades. The following year work began on Latchmore and New Ponds, on Gerrards Cross Common.[3]

The plant was well enough known once to have shared the vernacular name 'thrumwort' with other water-plaintains (*OED*). A 'thrum' is an old word for the bundle of uncut threads that protrude from the edge of a woven fabric and, in plants, usually refers to a cluster of thread-like stamens in the centre of the flower. These are present in starfruit, but are nothing like as obvious as the stiff, star-like seed-cases which have given it its official English name.

Frogbit family
Hydrocharitaceae

Water-soldier, *Stratiotes aloides*. This strange and rather sinister aquatic is now a nationally scarce plant, confined as a native chiefly to a few areas of eastern England (the Norfolk Broads especially) and the north Midlands. It grows beneath the water for much of the year, surfacing to flower between June and September. The leaves are stiff, spear-shaped and saw-edged, like an aloe's (hence *aloides*), but, as they begin to surface, they most resemble the top foliage of pineapples. Arising from this rosette there is sometimes one solitary three-petalled white flower, held just above the surface.

Canadian waterweed, *Elodea canadensis*, is the best known of a growing catalogue of exotic aquatic plants which have become naturalised in British waterways by way of aquaria and garden ponds. Canadian waterweed was first detected in the wild in mainland Britain in 1842, in the lakes of Duns Castle, Berwickshire, and over the next 25 years its dense, tangled underwater leaves spread through the ditch, canal and river system with startling speed. William Baxter described its breakout from the Oxford Botanic Garden in 1853:

> 'On taking a walk on Sunday Aug. 28, 1853, across the meadow between St Clement's and Iffley, I was surprised to find *Anacharis Alsinastrum* [as it was then named] in a ditch in Long Meadow. It formed an uninterrupted dense mass from one end of the ditch to the other. Three or four years ago the plant was not known in a wild or even cultivated state [in Oxfordshire] but about 1850 a small specimen was introduced to the lower aquarium of the Botanic Garden. It is also abundant in the ditches on each side of the Canal, Jericho to the railway, Port Meadow, also in the Isis close to the Railway Station. The Plant in the Canal and Isis must have travelled out of Warwickshire.'[1]

It seemed to reach its maximum extent in the 1870s and then began to decline, probably because of an absence of male plants.

Lords-and-ladies family
Araceae

Sweet-flag, *Acorus calamus*. The long, iris-like leaves of this water plant are not exactly sweet: crushed, they have a citrous odour, most like that of tangerines, but with a slightly cloying undertone of vanilla. It was a popular scented plant in the sixteenth century and was introduced from its native Asia to Europe in 1567. It was chiefly used for strewing on floors as a kind of disposable carpeting. By 1610 it was being cultivated in the Norfolk Fens. Fifty years later it was spreading rapidly of its own accord in the county, and Sir Thomas Browne wrote that 'this elegant plante groweth very plentifully and beareth its Jules yearly by the bankes of the Norwich river, chiefly about Claxton and Surlingham, and also between Norwich and Hellsden Bridge, so that I have known Heigham Church in the suburbs of Norwich strewed all over with it. It hath been transplanted and set on the sides of the Marish ponds in severall places of the county.'[1]

The practice of strewing rushes in the aisles and chapels of churches (even Norwich Cathedral was carpeted with them in summer) was widespread until the boarding of floors became the norm. Before then, church floors were often simply beaten earth or, at best, stone flags, and a congregation at prayer were resigned to cold, damp knees. Sweet-flag was something of a cosmetic bonus amongst the motley bundles of reeds, rushes and hay that were used in the annual strewing, which was the focus of elaborate 'rush-bearing' ceremonies in many parishes (see p. 390).

The Norfolk Broads are the only place where sweet-flag could be called common, but it persists in small colonies in shallow water at the edges of canals, rivers and lakes over most of England. The 'Jules' referred to by Thomas Browne are the yellowish-green flower-cones (spadixes) which emerge from the stems. But flowering is infrequent in Britain, and fruiting unknown.

American skunk-cabbage, *Lysichiton americanus*, is an ornamental arum from North America, grown in gardens, where it can be persistent. It is spreading as a naturalised plant in swampy places. **Altar-lily**, *Zantedeschia aethiopica*, is a white trumpet-flowered 'lily' from South Africa, commonly used in funeral bouquets. It is naturalised in damp hedge-banks and ditches in the south-west.

Lords-and-ladies or cuckoo-pint, a very common species.

Lords-and-ladies, *Arum maculatum* (VN: Cuckoo-pint, Cuckoo flower, Jack in the pulpit, Parson in the pulpit, Devils and angels, Red-hot-poker, Willy lily, Snake's meat, Cows and bulls). As I discuss in the Introduction, the bizarre form of this abundant plant of woods and hedgerows (the shiny arrow-shaped leaves, often speckled with black, and the pale green sheath, sometimes streaked with purple, hooding the purple or yellow spadix, which eventually produces a spike of bright orange berries) has generated a huge number of imaginative local names over the centuries. A few (see above) still survive, and there are even some new coinings. 'Willy lily' is as splendidly ribald as anything from the first Elizabethan era; and one family's tag, 'soldier in a sentry box',[2] catches the sheathed effect of the flower structure as well as any of the traditional names. There are new interpretations of old names, too:

'I am surprised by the suggestion that Lords and Ladies is a polite Victorian convention. It seems more likely to me that it is a bit of downstairs vulgarity in the ancient tradition, i.e. "The Lord's and the Lady's".'[3]

'I understood the [common] name went back further to the days of powder and patch, when Lords and Ladies sported "beauty spots", sometimes to be seen on the leaves of the wild arum.'[4]

'My father used to pronounce "pint" to rhyme with mint, not with pint as in pint of milk. Maybe an abbreviation of pintle [slang for penis].'[5]

But, for all its bawdy associations, the plant itself is a handsome and modest one, pale and sculptural in the spring. There is a fifteenth-century carving of it in berry in Westminster Abbey, and an exceptional representation on a choir-stall in St Paul's Church at Four Elms in Kent, done by Evelyn Chambers of the Art Workers' Guild in about 1917.[6] And, perhaps with unintentional irony, wild arum is still used instead of its extravagant cousin, altar-lily (see above) at Methodist funeral services in some Cornish communities.[7]

It was also associated with St Withburga in Cambridgeshire:

'Old Fenmen in the last century ... held the traditional belief that when the nuns came over from Normandy to build a convent at Thetford in Norfolk they brought with them the wild arum or cuckoo-pint. When the monks of Ely stole the body of St Withburga from East Dereham and paused, on their way back, to rest at

Brandon, tradition has it that the nuns of Thetford came down to the riverside and covered the saint's body with the flowers. During the long journey down the Little Ouse of the barge bearing St Withburga several of the lily flowers fell into the river, where they threw out roots. Within an hour they had covered all the banks as far as Ely with a carpet of blooms, and more remarkable still, these flowers glowed radiantly at night ... The pollen of the flowers does, in fact, throw off a faint light at dusk and when the Irish labourers came in large numbers to find work on the Fens during the famines in their own country during the last century, they named the lilies Fairy Lamps. The Fen lightermen had long called them Shiners.'[8]

The baked and ground roots of lords-and-ladies were once in demand as a home-grown substitute for arrowroot (normally from the West Indian species *Maranta arundinacea*), under the name of Portland sago. But the resulting gruel tended to be bitter, and the crushed roots were more often employed as a domestic starch (especially for ruffs), though they often produced severe blistering of the launderers' hands.

Dragon arum, *Dracunculus vulgaris*, is a dramatic species from the eastern Mediterranean with a livid purple sheath, enclosing a chocolate-brown spadix, which smells of rotten meat. Edward Lear, in his Cretan journal for spring 1864, described it as 'brutal-filthy yet picturesque'.[9] Feral specimens appear occasionally in southern England from garden throw-outs or from bird-sown berries, including one in the heart of Kensington, London.[10]

A duckweed 'lawn' in a Dorset waterway.

Duckweed family *Lemnaceae*

Common duckweed, *Lemna minor*. This is the most frequent of the small water plants which can cover areas of stagnant water with mats of green. Lesser duckweed's leaves are only 2 to 4 mm in diameter, with a clove-like, three-lobed shape; packed tightly, they can make the surface of the water appear solid. In the north-west of England, a region of abundant flooded marl- and brick-pits and derelict canals, duckweed gave rise to the myth of Jenny Greenteeth, a lurking, amorphous monster that would suck naughty children into the depths if they ventured too close.

At the beginning of this century, Jenny was sometimes used as a threat to children who didn't keep their teeth clean. But most frequently she seemed to be a lurid, coded warning to children to stay away from dangerous water. The folklorist Roy Vickery found that the myth had survived in Cheshire and Lancashire at least until the 1980s. Occasionally, Jenny was described as having an actual physical form. A 68-year-old woman from Fazakerley was told as a child that Jenny inhabited two pools beside Moss Pitts Lane and 'had a pale green skin, green teeth, very long green locks of hair, long green fingers with long nails, and she was very thin with a pointed chin and very big eyes'.[1] Normally, though, she was a shapeless and invisible threat, as described to Vickery by a 34-year-old woman in 1980:

'I remember, as a very small child, being told by my mother to stay away from ponds as Ginny Greenteeth lived in them. However, I only recall Ginny living in ponds which were covered in a green weed of the type which has tiny leaves and covers the entire surface of the pond. The theory was that Ginny enticed little children into the ponds by making them look like grass and safe to walk on. As soon as the child stepped onto the green, it, of course, parted and the child fell through into Ginny's clutches and was drowned. The green weed then closed over, hiding all traces of the child ever being there. This last point was the one which really terrified me and kept me away from ponds, and indeed my own children have also been told about Ginny, although ponds aren't as numerous these days.'[2]

Least duckweed, *L. minuta*, is an even smaller-leaved species (0.5 to 2 mm across), introduced from America for use in garden ponds and aquaria. It wasn't noted in the wild until 1977, but in places it is already achieving the same lawn-like effect as its larger native cousin: 'A dense carpeting on the Kennet and Avon Canal, spreading eleven miles in three months. The Canal has been described in letters to the local papers as "the best-kept bowling green in Wiltshire". It recalls the earlier covering by *Azolla filiculoides* (water fern). In 1939, "It looked from a distance like a red asphalt road, and I heard evacuated Londoners solemnly telling newcomers that the weed was a subtle device of the Government to prevent the enemy bombers seeing the water from the air." '[3]

Rush family *Juncaceae*

Soft-rush, *Juncus effusus*. This is a thin rush, growing in tufts up to three feet high in damp woods, waterlogged ground, marshes and ditches. It has glossy green cylindrical stems and yellowish flowers, but is best known for its pith, which, with care, can be extracted in quite long strips. Well into the nineteenth century this was used in making the basic source of illumination in most country cottages – the rushlight. These ancient vegetable tapers were simply lengths of peeled rush, soaked in fat or some other inflammable substance and then burned like disembodied wicks. In his great tract on self-sufficiency, *Cottage Economy*, William Cobbett placed his entry on them between goats and mustard, and obviously regarded them as one of the staples of life: 'I was bred and brought up mostly by *Rush-light*, and I do not find that I see less clearly than other people. Candles certainly were not much used in English labourers' dwellings in the days when they had meat dinners and Sunday coats.'[1]

Fifty years previously, his near neighbour, the Revd Gilbert White of Selborne, had written the classic account of the preparation and economics of the rushlight, a tribute to the ingenuity and frugality of eighteenth-century cottagers:

'The rushes are in best condition in the height of summer; but they may be gathered, so as to serve the purpose well, quite on to autumn. It would be needless to add that the largest and longest are best. Decayed labourers, women and children, make it their business to procure and prepare them. As soon as they are cut they must be flung into water, and kept there; for otherwise they will dry and shrink, and the peel will not run. At first a person would find it no easy matter to divest a rush of its peel or rind, so as to leave one regular, even rib from top to bottom that may support the pith: but

Soft-rush. Its pith was once soaked in fat and used in household lamps.

this, like other feats, soon becomes familiar even to children; and we have seen an old woman, stone-blind, performing this business with great dispatch, and seldom failing to strip them with the nicest regularity. When these *junci* are thus prepared, they must lie out on the grass to be bleached, and take the dew for some nights, and afterwards be dried in the sun.

Some address is required in dipping these rushes in the scalding fat or grease; but this knack also is to be attained by practice. The careful wife of an industrious Hampshire labourer obtains all her fat for nothing; for she saves the scummings of her bacon-pot for this use; and, if the grease abounds with salt, she causes the salt to precipitate to the bottom, by setting the scummings in a warm oven ... A pound of common grease may be procured for four pence; and about six pounds of grease will dip a pound of rushes ...

If men that keep bees will mix a little wax with the grease, it will give it a consistency, and render it more cleanly, and make the rushes burn longer: mutton suet would have the same effect. A good rush, which measured in length two feet four inches and a half, being minuted, burnt only three minutes short of an hour: and a rush still of greater length has been known to burn one hour and a quarter. These rushes give a good clear light ...'[2]

Rushlights may not be as quaint and anachronistic as they sound. They are easy to make, as White points out, once you have the knack of peeling the rushes. They have the advantage over candles of not dripping scalding tallow, so they can be held and carried without a holder; and they burn with a clear, almost smokeless flame, which is surprisingly bright. (In the nineteenth century a single rushlight would often serve several people sitting round one table. Those doing fine work, such as lace-making, would use a globe of water as a lens, to produce a concentrated spotlight.) Their economic use of waste fat was not forgotten during the dark days of the Second World War, when they had a temporary revival in rural areas.[3]

The pith's absorbency (due to its fine cellular structure) also means that it is exceptionally lightweight, and this has been noticed – and exploited – by children:

'While walking home from school in the Lake District we frequently paused to make "cigarettes" from rushes. This called for delicate handling of the rushes, which we peeled, trying to manage as long a piece of peeled rush as possible. There was no question of trying to light the cigarette; you simply held it in your mouth as if holding a real cigarette. The texture of the plant was such that it held easily to the lips, even when you talked.'[4]

'Take a piece of rush, strip the green from it and liberate the pith with your thumbnail – producing an almost weightless white worm, up to a foot in length. Then introduce this to the updraft of an outdoor fire. The result is uncanny – it shoots heavenwards, spiralling as it goes, then gently drifts downwards only to be sucked up again by the draft.'[5]

'Make arrows out of soft-rush stems, by half-peeling a strip of the outer skin away from the pith of a section of

Great wood-rush, a typical plant of West Country woods and streamsides, decorative enough to have been taken into gardens.

rush stem, balancing the stem across the top of one's hand, then pulling sharply on the half-peeled strip – which propels the arrow at a suitable target.'[6]

'We used to peel soft-rushes to put on little leaf "plates" as "bananas", with rowan berries as "oranges".'[7]

'The girls in this area used to plait soft-rush so as to make what they called "Ladies' Handmirrors". This was done by bending the stem sharply at the middle at two points about quarter of an inch apart. The stem was then plaited by bending each side in turn sharply over the other at right angles. When this had been done along almost the whole length, the ends were pushed through the starting loop to make a small circle with a lorgnette-type handle. An adequate supply of saliva was then needed to persuade the loop to take on a saliva lens, which formed the mirror surface.'[8]

'A recent craze among local primary (and early secondary) schoolgirls in the Forest of Dean is for "friendship bracelets" usually of cotton threads, occasionally wool. My daughters have also tried soft rush, and often say "We're going for the record, plaiting three strands-long lengths with it." '[9]

Field wood-rush *Luzula campestris* (VN: Good Friday grass, Sweep's brooms), is a widespread species of all kinds of short grassland – lawns, village greens and downland. Its dark, brush-like flowers (the 'sweep's brooms') appear early in the year, nearly always in time for Easter.

'My father, who lived in his young days at Bolton by Bowland in West Yorkshire, used to tell me that the flowering of this plant in his meadows was the sign that it was time to put the wintering cattle out to start grazing.'[10]

Another conspicuous feature of the plant is the bristly hairs on its leaves: 'It has hairs on its leaves, which in hot sunshine are surprisingly active. Their motion seems to be caused by a deflection near the base of the hair of up to 45 degrees. This motion appears erratic, twisting rapidly first one way then the other. It is not the result of convection currents, but an incidental result of the irritability of the hairs in response to sunlight or heating.'[11]

Hairy wood-rush, *L. pilosa*, is a quite common species in deciduous woods and moors. The flowers are carried on the ends of loose branches at the top of the stem. **Great wood-rush**, *L. sylvatica*, is a very characteristic and ornamental plant of woods and shady streamsides on acid soils, mainly in the west and north. It forms dense tussocks of blade-like leaves up to two feet high, with loose sprays of chestnut-brown flowers.

Sedge family *Cyperaceae*

Common cottongrass, *Eriophorum angustifolium*. This is a frequent and abundant plant in wet, acid bogs throughout Britain. The seed-heads are full of silky white plumes, often dense enough to whiten whole patches of bog. Experiments have been made to see if a usable thread can be derived from the seed-plumes, but their fibres are too short. However, they have been used in Sussex for stuffing pillows,[1] and they were gathered by Scottish children for use in wound-dressings during the First World War.[2]

Common club-rush or **Bulrush**, *Schoenoplectus* (or *Scirpus*) *lacustris*. Club-rush, bulrush (as it is often called in common with *Typha latifolia* – see p. 400), or just plain 'rush' is a stout perennial found in shallow water in lakes, ponds, canals, slow rivers and dykes. It can reach up to 10 feet in height with a thickness of nearly an inch at the base.

Cottongrass – common on wet moorland.

The rounded stems are straight and jointless, which makes them ideal for plaiting and weaving into baskets, mats, chair-seats and the like, especially items that are liable to be repeatedly wettened – conditions to which rushes are, by nature, perfectly adapted. They are normally cut in June or July, whilst the ruddy-brown flower-spikes are in bloom. Later in the summer the stems tend to become too woody to work easily.

Rush also makes a good paper: 'the thinnest, finest sheets of home-made paper produced outside Japan,' believes Maureen Richardson. 'The rush is either gathered from local ponds or ditches, cadged from commercial rush-gatherers in the Fens, as discards from the bolts [bunches] they supply to rush-weavers, or obtained from chair-menders. Paradoxically, the old, worn, frayed rush seats cut away for replacement make the best paper fibre.'[3]

Club-rush is one of the species that gave the ceremony of 'rush-bearing' its title. Before the days of floor-boarding and carpets, the stone or earth floors of churches and dining-halls were strewn with a mixture of rushes, sweet-flags and scented herbs (see sweet-flag, p. 384). In churches and chapels this green carpet had to be renewed at least once a year, usually at the times of the local Wakes celebrations. Christina Hole has given the definitive description of the full ritual, which would have occurred in many parishes up to the nineteenth century:

'Every part of the parish contributed its quota of sweet-smelling rushes, sometimes carried in bundles by young women in white, but more often piled high in decorated harvest-wains, and held in place by flower-covered ropes and the high harvest-gearing. The best horses in the village were chosen to draw the carts; Morris dancers usually preceded them, and children and young people walked beside them, carrying garlands which were hung in the church after the new rushes had been laid down. Often the procession perambulated the parish in the morning, stopping outside the great houses of the district where the Morris-men danced; and then, the long round ended, the whole company came to the church, to the sound of pealing bells, and there strewed their rushes on the floor (and sometimes on the graves outside as well), and hung up their garlands in the appointed places.'[4]

Rush-bearing ceremonies still survive in a few areas, notably at Grasmere and Ambleside in the Lake District, where they take place in late July or early August. There are also signs of revival elsewhere:

'A rush-bearing procession and service is held as part of the Midsummer Rejoicing celebrated annually at Bishops Castle, but this didn't start until about 1976, and I don't think it was ever a local custom. It has however become a central part of the celebrations, with a decorated rush cart.'[5]

'The Saddleworth Rushcart Festival, in which Morris Dancers drag a cart loaded with rushes around local villages, culminates in a rush-bearing service at Saddleworth Church, Uppermill, West Yorkshire. This event takes place on the August Bank Holiday weekend and is a revival of an old tradition in the area.'[6]

'Rushes have been spread on the floor of Trinity House in Hull since time immemorial. The Norfolk Trust

Stacks of cut saw-sedge at Wicken Fen, Cambridgeshire. Sedge was once widely used as thatch in the county.

used to supply the rushes till 1969. They are now imported from Lastingham in North Yorkshire.'[7]

'Rushes are still strewn over the floor for the Liskeard [Cornwall] Mayor-making ceremony, from the room in which the councillors assemble, along the processional route to the mayor-making chamber – these days the Public Hall. The Sergeant at Arms cuts them the day before from Hendra Bridge.'[8]

Great fen-sedge or **Saw-sedge**, *Cladium mariscus*. Saw-sedge is aptly named. The tall (up to 10 feet) leaf-spears have vicious serrations at their edges and grow quite densely packed in fens and ponds. Very little in the way of either trees or mammals invades a sedge fen, and as a result they can persist for long periods, building up rich layers of peat from their old, waterlogged leafage.

Despite this, sedge has been an important economic plant in East Anglia, the only region of the country where it could be said to be common. T. A. Rowell has traced its history in the Cambridgeshire Fens, where there are at least 16 parishes which contain areas that can be identified by name as former sedge-fens. Compartments of sedge were cut on a three- or four-year rotation in summer, usually early in the morning. As with hay, dew made scything easier. R. W. Macbeth's painting of the sedge harvest at Wicken Fen (1867) shows the details of the cutting in graphic detail, including the stockings which the cutters bound round their hands and arms to protect them from the sharp teeth of the sedge.

Pendulous sedge, a native species popular in gardens.

Greater tussock-sedge. The tussocks were sometimes cut off whole and used as seats and church kneelers.

The cut sedge was chiefly in demand as a thatching material. It is more flexible and durable than reed and was used on ricks and public buildings as well as cottages – for instance to top the incomplete tower of St Mary the Great in Cambridge after the bells were rehung in 1515. It was also a popular firelighting material in Cambridge and was the only fuel purchased for the bakehouse ovens in both St John's and Corpus Christi Colleges throughout the seventeenth century. Every college had a 'sedge loft' and the servants, like the cutters, wore special gloves to protect their hands whilst handling the plant.

A subsidiary use was, ironically, in the drainage works that were to lead to the drying-out of sedge fens and its virtual extinction as a commercial harvest throughout the Fens. In some fens, drainage channels were filled with a mixture of sedge and bushes before the topsoil was replaced. It was also used for strewing in barns, on slippery bridges and, around 1900, on the gallops at Newmarket during icy weather, to improve the purchase for the horses.[9]

Greater tussock-sedge, *Carex paniculata*, is a large tussock-forming sedge of marshes, fens and wet woods throughout most of Britain. In many areas (e.g. the Norfolk Broads)[10] the tussocks were sometimes cut and trimmed to make fireside seats or kneelers for village churches. **Sand sedge**, *C. arenaria*, is quite common on sand-dunes around the whole British coast and in a few inland areas – notably the Breckland. It is one of the first colonists to arrive on open dunes, and has a striking pattern of growth – its long creeping roots spreading in straight lines just below the surface of the sand and sending up tufts of stiff leaves at regular intervals. **Pendulous sedge**, *C. pendula*, is a tall and tufted species, growing up to five feet tall, with graceful hanging flower-spikes. It is quite common by riversides and in damp woodlands (especially ancient woods on clay soils), chiefly in England and Wales. But its distribution is rather a scattered one, and specimens or colonies in odd places (churchyards, for instance) may be naturalised from gardens, where it is often grown as a decorative plant.

Grass family
Poaceae (or *Gramineae*)

Although grasses dominate the rural landscape, as crops and as the ground-cover of most open countryside, they do not in the main have sufficient individuality to have found a place in popular culture. The exception has been children's games. Here, the grass family's round-the-year accessibility and abundance have worked in its favour. And while not much discrimination is used when sucking stems for sweetness or whistling through leaf-blades, children have spotted and ingeniously exploited many of the fine differences between species.

Perennial rye-grass, *Lolium perenne*, is a tough-stemmed native of waysides, rough ground and pastures, which is also one of the most widely used species for reseeding grasslands. Its distinct, well-spaced spikelets have made it the grass of choice for a children's prediction game in the Yorkshire Dales: 'Each spikelet was pulled off in turn to fit the rhyme till it was used up. And so we chanted "This year, next year, sometime, never; silk, satin, muslin, rags; boots, shoes, slippers, clogs (this latter footwear was normal in the village); the Big House, little house, pigsty, barn", and so on.'[1]

Annual meadow-grass, *Poa annua*, **rough meadow-grass**, *P. trivialis*, and **smooth meadow-grass**, *P. pratensis*, are abundant in grassy places. Their loose heads of flowers and seeds are used to illustrate a seasonal rhyme. This version is from Somerset: 'We used meadow grass to demonstrate the four seasons, saying "Here is the tree in

Mixed meadow-grasses in the early morning, Monewden Meadow, Suffolk.

Flowering head of sweet vernal-grass.

springtime" (pushing the grass into a bunch). "Here's the tree in summer" (letting the spikes go). "Here is the tree in autumn" (brushing all the seeds off). "And here's the tree in winter ..." '[2]

A more frequent and widespread version begins with summer and is in rhyme:

Tree in summer
(Show seeds on grass stem, as it grows.)
Tree in winter
(Remove seeds with upward sweep of hand. It looks bare and twiggy like a winter tree.)
Bunch of flowers
(Show the seeds, which nestle like a bouquet in the hand.)
April showers
(Throw the seeds up into air and watch them come down on the onlooker.)[3]

Wood meadow-grass, *P. nemoralis*, is a distinctive species of old deciduous woods and hedge-banks, whose leaves stick out straight and almost horizontally from the main stem.

Cock's-foot, *Dactylis glomerata*, is an abundant perennial of rough grassland and cultivated ground throughout Britain, notable for its cluster of spikelets, which have a slight similarity to a chicken's foot (albeit a rather clubbed one), and for the sharp edge of its leaves: 'A personal name we used as children in Ayrshire was "cutting grass". If handled roughly – especially if pulled – it could give a nasty cut.'[4]

It is also one of the best grasses for sucking (though the connoisseur's favourite is **sweet vernal-grass**, *Anthoxanthum odoratum*[5]): 'Possibly the best grass for chewing. Careful traction easily breaks the stem at the node, pulling a nice clean length out of the leaf-base. Very sweet when leaves actively photosynthesising in bright weather.'[6]

Wild-oat, *Avena fatua*, is an introduced weed from continental Europe, and one of the ancestors of cultivated oats. Its rampancy in the fields (still obvious, even in these days of sprays and seed-screening) no doubt helped to underpin the euphemism 'sowing wild oats' for youthful excess and dissipation – compared, implicitly, to the wisdom of sowing 'good grain'. Wild-oat seeds are very physically active themselves and susceptible to changes of temperature or moisture, 'and if placed in a damp hand or breathed upon will instantly wriggle and move about'. This quality was once exploited by fortune-tellers and magicians. 'To cover the cheat, the magician called his magic plant the leg of an Arabian spider, or the leg of an enchanted fly, and many people were deceived by its

Wild-oats, going to seed.

use.'[7] Apparently fish can also be confused, as wild-oat seeds have been used on hooks as a fishing-lure.

Yorkshire-fog, *Holcus lanatus*, is an abundant species of rough grassland, lawns and waste ground, with distinctive grey-haired leaves and often purple-tinged plumes. The name is officially regarded as originating from the Old Norse *fogg*, meaning a 'long, lax, damp grass';[8] but it is more commonly taken as a description of the misty effect of the grass when seen at a distance, which is somewhere between the tones of Sheffield factory smoke and heather moorland. It is often used as a substitute for meadow-grasses in the seasonal tree game (see above).

Creeping soft-grass, *H. mollis*, is similar but more of a woodland grass, growing on acid soils; it is distinguished by its 'hairy knees' – little tufts of hairs on the nodes on the stems, which are otherwise smooth.

Marram, *Ammophila arenaria* (VN: Starr-grass). Anyone who has walked or picnicked on sand-dunes around the coasts of Britain will remember marram. It is a tall, tufted plant whose stiff leaves are as sharp as new cartridge paper. Marram is one of the most cleverly adapted of all seashore plants. Its root system is a convoluted mat of fibres which helps bind the sand down. Its leaves are springy enough to stand up to the most violent gales. They are covered by glossy cuticles as protection from the abrasive powers of blown sand and, in very drying winds, roll up into tubes to reduce the area of exposed leaf surface and so conserve moisture. The tufts also catch the sand, and the whole plant is a potent agent for stabilising mobile dunes, sometimes advancing as much as 30 feet a year.

Its resilience and durability made it a favourite raw material for products where indifference to weather – especially salt winds – was important: 'There was a thriving marram-weaving industry along the south-west coast of Anglesey. It used to make mats for haystacks and barn roofs and cucumber frames, and nets and cordage for fishermen; even shoes were woven out of marram. The tall grass was cut in late summer with a broad-bladed reaping hook, heaped and spread to dry in sheltered dune-slacks. Each family had their own sand-dune by tradition. Harvesting alone took a good month, and most of the village was engaged in the industry, which was of a sufficient scale in the mid-nineteenth century to supply its products throughout Wales.'[9]

Dorothy Hartley once found a marram stool, possibly a century old, underneath a layer of skulls in a derelict Irish church:

'Reasoning from the date when the church alterations were made, I reckoned that specimen of grass must have been at least 80, and more probably 120 years old. It was woven in a circular form and had been filled with peat moss, or some similar substance, and sewn up to form a hassock or stool. By its position in the depths of the ruined crypt, under stones, dust and old bones, it had probably been used as a basket to carry down rubbish when the place was cleared. It must have been damped and dried repeatedly, and then left forgotten, yet the

Marram grass on Gibraltar Point, Lincolnshire. Marram is one of the chief natural agents for stabilising sand-dunes.

grass was as elastic and firm as the grass I had seen growing on the dunes at Newborough [Anglesey], across the water. Under such conditions straw, reed, or rush must have perished, but the marram grass endured.'[10]

Dorothy Hartley believed that it was the texture of woven marram that helped give rise to the characteristic curves and scroll-work of Celtic design. Straw and reed are cylindrical and hollow, and when bent will always flatten or crack into sharp angles. But the flat blades of marram (and, to a lesser extent, the pith-filled stem of rushes) can be eased into smooth and complex curves.

Small-scale work in marram still continues here and there: 'Marram is known as starr-grass in Lancashire. It is very common on the beach, and older people refer to sandhills as "starr-hills". It is still made into serviceable brushes by us.'[11]

But its importance in stabilising dunes has made the harvesting of marram for thatch and plaiting illegal in some western coastal areas. This is perhaps an over-reaction, as marram is a true pioneer plant, needing space. Once it has completely covered a dune, it begins mysteriously to die out.

Meadow foxtail, *Alopecurus pratensis*, is common in grasslands, preferring rich, damp soils. Once the spikelets or seeds have been stripped off, this is the favourite species for giving 'Chinese haircuts'. The flowers would be stripped off the stem, leaving only their short, wiry stalks. These would then be twiddled into another child's hair (or, with older boys, hairs on the legs) – usually of the child sitting at the desk in front. A sharp pull would then remove all the hair tangled up in the stalks.[12]

Timothy, *Phleum pratense*, and **smaller cat's-tail**, *P. bertolonii*, are both common throughout Britain and notable for their long, silky heads. (Curiously, for a common native grass, *P. pratense* has acquired an American name: 'Timothy' is after Timothy Hanson, a farmer who introduced its seed to Carolina in about 1720.)[13]

'We used cat's-tails to make rabbits. The large heads would provide ears, arms and feet, with the heads wrapped round another one for the body.'[14]

Common couch, *Elytrigia repens* (VN: Squitch, Twitch, Wickens, Stroggle, Grandmother grass). This is an abundant perennial of cultivated places and rough ground, with long, complicated and obstinate roots.

'My mother, born in 1891, introduced me and my daughter (and through us her daughter) to "grandmother grass" or couch grass. One plucks off the head of the grass and sticks it in another head, still on its stem. A flip of the hand holding the stem and "Grandmother, grandmother, jump out of bed" is recited as the first head springs out of its nest.'[15]

Wall barley, one of the commonest grasses of town streets.

Couch-grass roots are an effective mild diuretic and were gathered for this purpose by the National Herb Committee during the Second World War.

Wall barley, *Hordeum murinum*, is a common annual of walls, waste and rough ground, and bare patches in dry grassland. Its bearded heads are as clinging as burdock burs and have found their way into numerous games. (Some of these are reminders that the Latin specific name, *murinum*, is derived from *mus*, a mouse, not *murus*, a wall!)

'In school we throw "flea-darts" at each other. They stick to hair and clothing. They are the flowers of wall barley and wild oats.'[16]

'We played "crawly-wallies" by picking spikes of wall

barley, holding them upside down within one's closed fist; then, by making small movements with one's fingers, the wall barley spike climbs out and escapes.'[17]

'A variation was to put a head of wall barley beneath the ribbing of the jumper of the person sitting in front of you at school and then wait for it to work its way up their back and start them scratching.'[18]

A few other species have roles as landscape plants or as the raw materials of crafts.

Arrow bamboo, *Pseudosasa japonica*, from Japan and Korea, is quite widely naturalised by river-banks and in damp plantation woodland and neglected parks, often as dense thickets. Unusually for bamboo, it flowers quite frequently. Several other bamboo species from different genera, all from south-east Asia, are also naturalised here and there, but especially in the west.

Wood millet, *Milium effusum*, is a tufted perennial growing up to four feet high in deciduous woods, chiefly in England. Its loose, delicate flower-heads floating a few feet above the ground are one of the most conspicuous features of ancient woodland vegetation once the spring flowers are over. **Giant fescue**, *Festuca gigantea*, and **hairy-brome**, *Bromopsis ramosa*, are larger, grosser and later-flowering grasses of similar woodland sites.

Crested dog's-tail, *Cynosurus cristatus*, is native and common in all grassy places, and was previously grown as a crop: 'In the nineteenth century this was used for making bonnets. The picking was done by children in Box, Wilts. The plait was reckoned to be equal in colour and durability and superior in texture to the finest Leghorn.'[19]

Quaking-grass, *Briza media* (VN: Quaker grass, Toddling grass, Totty grass, Totter grass, Tom's tottle, Dithery dock, Dothery Dicks, Wigwams, Silver spoons, Golden shekels). Locally common in old grassland, chiefly on calcareous soils, quaking-grass is very distinctive, with heads of loosely hung, heart-shaped spikelets which tremble and dance in the slightest breeze. It has, as a consequence, accumulated a host of graphic local names.

Wood melick, *Melica uniflora*, is the most beautiful of the woodland grasses, with brilliant green leaves and nodding heads of dark egg-shaped spikelets. It has a strong liking for old wood-banks. **Tufted hair-grass**, *Deschampsia cespitosa* (VN: Bull pats, Tussock grass, Hassocks, Snigglebogs, Old boys), is common in woodlands and damp rough grassland throughout Britain, forming small tussocks. **Wavy hair-grass**, *D. flexuosa*, is a fine-leaved relative with wavy stems, confined to woods and heaths on acid soils. **Canary-grass**, *Phalaris canariensis*, from north-west Africa and the Canary Islands, and **common millet**, *Panicum miliaceum*, from Asia, are both widely used in birdseed and crop up frequently as casuals in gardens, pavement cracks and rubbish-tips.

Tufted hair-grass, a common and graceful grass of woods and damp rough grassland.

Common reed, *Phragmites australis*, is a common species, often forming large beds in shallow water at the edges of lakes, canals, slow-moving rivers and brackish lagoons. It can also rapidly colonise patches of temporarily damp ground – for instance, in quarries or on new motorway verges – and persist long after they have dried out.

Large reedbeds, of the kind that are most frequent in the low-lying wetlands of East Anglia, are dramatic landscape features, especially in summer, when the tall canes are topped by a haze of purple plumes. They are also important habitats for a wide variety of birds, from small warblers to large birds of prey. But they are essentially unstable habitats and, unless they are regularly cut, will be invaded by willow scrub and eventually turn into damp woodland. Reed is still cut for thatching, chiefly in the Norfolk Broads, and there has been some resurgence in the craft with the rise in weekend-cottage conversions. But to be sufficiently long and straight, reed must be cut on a regular basis.

A promising new function for reedbeds is as water-filters. The intricate and rapidly extending root-systems of even small areas of reed seem able to extract toxic substances such as nitrates and heavy metals from water which is passed slowly through them.

Even this economic grass has a children's game attached to it: 'We made boats from the thick blades of reeds. The tip was doubled over and put through a split made in the broadest part, which formed a sail on top and a rudder beneath.' [20]

Water finger-grass, *Paspalum distichum*, is a tropical grass which has been naturalised in damp ground at Mousehole, Cornwall, since 1971 and, more mysteriously, since 1984 on the banks of the Grand Union Canal in Hackney, north London.[21] It has distinctive V-shaped flowering spikes, and I suggest 'victory grass' as a more suitable popular name.

Common reed (top) forms large beds in shallow water. In places (for example at Cley, north Norfolk, above) it is harvested for use as thatch.

Meadows

Strictly speaking, a meadow is an area of grass cut for hay. It is 'shut up' against cattle between March and June or July, mown, and then grazed until the following spring. But such places are an agricultural anachronism, and the bulk of grass is now grown in 'leys', which are regularly ploughed, sown with a single species of high-yield grass and cut for silage in the spring. Outside the network of nature reserves, true hay meadows survive in any numbers only in the Somerset Levels and the Pennine Dales. But there are anomalies which have somehow survived because they lie outside the conventional agricultural system, such as 'pub meadows':

'The Duke of York Meadow (where wild daffodils grow) [Rye Cross, Herefordshire] reminds me of another feature I have noticed in Worcestershire. There are quite a number of very nice meadows which are attached to pubs. Usually, they own just one or two fields immediately adjoining the pub and these have remained unimproved because there has been no particular need to intensify management on them. I wonder whether this could be because, in the days when beer was actually brewed within pubs and possibly distributed to one or two other local pubs, the dray horses were kept in these fields?'[1]

In common usage, 'meadow' has a much more general – and almost inverted – meaning. If there are wild flowers growing in a grassy place, then that is a meadow. It may be a clearing in a wood, a churchyard corner, even the unsprayed stubble in a cornfield. A meadow is now 'a place of the mind' as much as a precisely defined ecological system – something the Northamptonshire poet John Clare would have understood: 'It is a very old custom among villagers in summer time to stick a piece of greensward full of field flowers & place it as an ornament in their cottages which ornaments are called Midsummer Cushions.'[2]

Old hay meadow, south Somerset. Hay meadows are amongst the most rigorously managed of habitats. Yet their rich mix of flowers originated in wild, natural grasslands, such as woodland glades and river flood-plains.

Bulrush family *Typhaceae*

Bulrush, *Typha latifolia*. The name bulrush represents a rare victory for common English over botanical protocol. Up until the 1970s, the fat brown busbies that everyone knows as bulrushes (or sometimes, more graphically, as 'bull-rushes') were called 'reedmace' by botanists, who reserved the name bulrush for *Schoenoplectus lacustris* (see p. 389). The mix-up is normally traced to Sir Lawrence Alma-Tadema's picture of *Moses in the Bulrushes*, in which the infant is encased in a basket plainly hidden amongst *Typha*. (Properly it should have been papyrus, *Cyperus papyrus*.)[1] Generations of children thus saw confirmed in their Sunday School books a plant name which they probably used anyway.

Bulrushes are quite common in reedbeds and ditches and at the edges of lakes and ponds, throughout most of Britain. But surprisingly they have found very few uses here, except in the occasional flower arrangement. In Nevada (where they are called cat's-tails), the Paiute Indians based a whole economy around them, building boats with the stems and leaves and using the abundant yellow pollen as flour.[2]

Lesser bulrush, *T. angustifolia*, is a smaller, less common but similar species.

Lily family *Liliaceae*

Bog asphodel, *Narthecium ossifragum*. The yellow star-flowers of bog asphodel, ranged in short spikes on leafless stems, are often the brightest flecks of colour on peat-bogs and damp heaths. In autumn the whole plant turns tawny and can colour large patches of valley bog in places such as the New Forest. At one time it was regarded as a true, miniature asphodel, a lily of the field, and early botanical names were *Asphodelus luteus* and *A. Lancastriae*.

The modern scientific name *ossifragum* – bone-breaker – is more down-to-earth. It derives from the belief that grazing the plant made the bones of sheep brittle – though it was not bog asphodel that caused this, but the sour, calcium-poor pastures in which it occurs. It is confined to such habitats and is a species chiefly of western and northern Britain, where it was occasionally used as a substitute for saffron and as a yellow hair-dye. There are colonies on the wet heaths of west Norfolk and south-east England, especially Sussex, Surrey and Hampshire, but bog asphodel is declining throughout the lowlands as wetlands are drained.

Day-lilies, *Hemerocallis* species. These garden plants, probably from Asia originally, are naturalised in a scatter

Bulrushes, surviving the snows.

of places. **Red-hot-poker**, *Kniphofia uvaria*. Not always hardy in gardens, this spectacular species from South Africa can set seed in England and be persistent where it has become naturalised or been deliberately planted. Most of the wild colonies are in the milder south-west of Britain, but there is an incongruous slug-eaten patch established among the ancient and indisputably native limes and bluebells of Groton Wood in Suffolk.

Meadow saffron, *Colchicum autumnale* (VN: Autumn crocus, Dainty maidens). At first sight, meadow saffron looks more like a crocus than a lily. It has the same compact head of satin-sheened mauve petals, the same abrupt apearance out of the ground on top of a thin white stalk. But unlike most crocuses it comes into flower in August and September, months after its large strap-like leaves have died away. This gives it a blushing, nude appearance that has been remarked on in many old local names, from 'Naked Ladies' (West Country) and 'Strip-Jack-naked' (Norfolk),[1] to '*cul tout nu*' ('bare arse') in southern France.

It is poisonous in all its parts and has a taste for rich, damp meadowland – a combination that has always made it a problem plant for graziers. But, by way of compensation, one of its toxic constituents – colchicine – has been known since classical times as a remedy for gout, so farmers had the option of treating the crocus as a catch-crop. They would forgo spring and autumn grazing, when the leaves or flowers were out, and make do with harvesting the corms for the pharmaceutical industry. The owners of the hay meadows at Monewden (see p. 408) remember the men from the Ministry of Health coming to gather corms during the last war, when overseas supplies were cut off.

Nowadays, foreign sources and synthetic substitutes are relied on, and meadow saffron on grazing land is given short shrift, as in the Llanthony Valley and Black Mountains:

'Local farmers, if they have it growing on their land, still, with good reason, destroy it. On our neighbour's farm one of the children is sent out in the morning to have a quick look round for flowers and pick them before the cattle are sent out.'[2]

'Until seven years ago we had only two records for *Colchicum* in Powys, and I actually went into print in the local Agricultural Advisory Service bulletin, that there

Bog asphodel in the New Forest.

The autumn-flowering meadow saffron, Eades Meadow, Worcestershire. The corms are the source of the drug colchicine.

was probably not enough of it in the county to kill a sheep. I was wrong. I heard that two heifers had died of *Colchicum* poisoning at Llanmerewig, and on investigation found a field full of it. A truly marvellous sight. To cut a long story short it is now a nature reserve belonging to the Montgomeryshire Wildlife Trust.'[3]

It is now an uncommon plant, found chiefly in open woodland and a scatter of churchyards (e.g. Abbey Dore, Herefordshire), especially in its heartland on the limestones of the Welsh border country, the Cotswolds and Mendips. The largest concentration in Britain is believed to be in Wychwood Forest in Oxfordshire, where there can be up to 10,000 in bloom in early autumn.[4] *Colchicum* flowers can look bizarre in darker woodland, with their pallid stalks drawn up and twisted in the search for light and sometimes seemingly unable to bear the weight of the open bloom. They have the look of flowering toadstools or vegetable eels.

Specimens with white flowers are not uncommon and have been found recently in Great Ridge Wood and Silk Wood, Wiltshire.[5]

Snowdon lily, *Lloydia serotina*, is a delicate alpine with purple-veined white flowers and a curious disjointed distribution. It occurs widely in the European Alps, but does not crop up again until the mountains of Caernarvonshire, where a few small and fragile populations cling on to cliffs and crevices in calcareous rocks. It is named after Edward Lhwyd, the great Welsh botanist, who discovered it on Snowdon before 1696. Latter-day collectors have reduced it to just five colonies in inaccessible sites.[6]

Yellow star-of-Bethlehem, *Gagea lutea*, is a rather scarce and local plant, in damp, calcareous woods, chiefly in central England, and often not noticed even where it occurs. It has inconspicuous grass-like leaves and only rarely flowers. The blooms, when they do appear, are umbels of yellow star-flowers. A Hertfordshire Flora reports rather forlornly on a herbarium specimen: 'In 1954 a Ware Grammar School girl found it in "Broxbourne Woods" ... but could not recall the exact location.'[7]

Yellow star-of-Bethlehem, a scarce and local plant which flowers only rarely.

Wild tulip, *Tulipa sylvestris*, is a southern European species anciently introduced to Britain and naturalised in a few wooded gardens, meadows and parkland. Greatworth Manor in Oxfordshire, Ellen Willmott's old garden at Warley Place in Essex, and the site of 'Petronella's Hospital for Leprous Maidens' in Bury St Edmunds[8] are typical localities. The wild tulip produces a great many leaves but, even in cultivation, few flowers.

Varieties of garden tulips, *T. gesneriana*, occasionally persist on waste ground and waysides where they have been thrown out.

Fritillary, *Fritillaria meleagris* (VN: Snake's-head, Snake's-head lily, Crowcups, Frawcups, Leper's bells, Sulky ladies, Chequered lily, Folfalarum). The snake's-head fritillary is one of the most local of well-known British flowers, and one of the most darkly glamorous. Vita Sackville-West wrote of it in her epic pastoral poem 'The Land':

And then I came to a field where the springing grass
Was dulled by the hanging cups of fritillaries,
Sullen and foreign-looking, the snaky flower,
Scarfed in dull purple, like Egyptian girls
Camping among the furze, staining the waste
With foreign colour, sulky-dark and quaint ...[9]

But it has not been exclusively a poet's plant. Across southern and middle England, a flamboyant suite of local names – snake's-head in Oxfordshire, leper's lily in Somerset, shy widows in Warwickshire, for instance – suggested a flower that was once widespread and familiar. The anonymous namers had really *seen* those dusky, reptilian bells, with their checkered patches of mulberry and lilac seeming to overlap like scales. Even up to the 1930s the fritillary grew in its thousands in more than a hundred 10-km squares, yet always locally enough, in winter-flooded hay meadows, to be intimately known and cherished. Some individual villages in Wiltshire even had their own names for it. In Oaksey it was the Oaksey lily; in Minety, the Minety bell. It was florid, profuse, extraordinary, intensely local.

Yet the first official record of a wild fritillary in England was not made, remarkably, until 1736, and this has made it top of the list of those species whose origins have been a source of nagging curiosity amongst botanists, sometimes to the exclusion of more intriguing aspects of its social and cultural history. Is it a native? A Roman introduction? An escapee, as Geoffrey Grigson believed, from Tudor gardens?[10]

In the two and a half centuries since that first 'discovery', the fritillary has come and gone with frightening speed, its distribution savagely cut by agricultural drainage and development from 27 counties before the last war to roughly the same number of individual meadows today. But it has remained, in the best sense, a parochial plant, best glimpsed through the stories of its changing fortunes in its surviving strongholds.

It was certainly about in English gardens by the sixteenth century. Gerard calls it the 'checkered Daffodill or Ginny hen flower ... in so much that euery leafe [i.e. petal] seemeth to be the feather of a Ginnie hen' and '*Frittillaria*, of the table or boord vpon which men plaie at chesse, which square checkers the flower doth very much resemble', and he remarks that 'Of the faculties of these pleasant flowers there is nothing set downe in ancient or later writers, but are greatly esteemed for the beautifieng of our gardens, and the bosomes of the beautifull.'[11]

Philip Oswald, who has reviewed most of the writings on the fritillary in Britain,[12] thinks that the first wild record may possibly post-date Gerard's note by about only 50 years. In his manuscript for *The Natural History of Wiltshire*, John Aubrey (1626–97) wrote: 'In a ground of mine called Swices ... growes abundantly a plant called by the people hereabout crow-bells, which I never saw any where but there.' 'Crow-bell' has been a vernacular name for a number of common wild flowers, including bluebell and buttercup. But Aubrey would have known these plants, and, given the location of his estate on the upper reaches of the Avon in Kington St Michael, Wiltshire, and other vernacular names for fritillary such as 'crowcup', Oswald thinks it just possible that he was referring to fritillaries.

But the accepted first record is the one of 1736. On 11 December that year, the botanist John Blackstone wrote to a friend that it 'grows in a meadow by a wood side near Harefield [Middlesex], and has done so about forty years as a neighbouring gentleman informs me'. In the following year he published what is presumably the same record: 'In Maud-fields near Ruislip Common, observed above forty years by Mr Ashby of Brakspears.'[13] The fritillaries had vanished from Maud Fields by the end of the eighteenth century. But, in 1990, the Ruislip and District Natural History Society obtained permission from the current owner of the land to plant out 2,000 bulbs (obtained from a commercial nursery, not a wild source). So far, the survival rate has not been good, though some of the plants appear to be dividing and seeding successfully.

The best-known fritillary site in England is probably Magdalen College Meadow in Oxford, where in late April the entire north-eastern half of the meadow seems covered by a tremulous purple haze. Yet, curiously for a university city devoted to scholarship and science, this spectacular display just a few hundred yards from the oldest Botanic Garden in Britain, was not even *spotted* by botanists until 1785, when it was recorded by John Lightfoot. A century later George Claridge Druce remarked: 'It was not a little singular that the Fritillary, so conspicuous a plant of the Oxford meadows, should have so long remained unnoticed by the various botanists who had resided in or visited Oxfordshire.' Perhaps 'unnoticed by ... *botanists*' is the key. In Middlesex, it was first reported by a non-botanist, Blackstone's grandfather, Francis Ashby. By contrast, at Oxford, the Professor of Botany at the time that the Magdalen colony was belatedly 'discovered' was Humphrey Sibthorp, a man of such renowned indolence that he reputedly gave just one not very successful lecture in forty years. The men of books and laboratories were not always the most energetic and sharpest of eye in the field.

And it is just possible that the Magdalen colony was an introduction from another, more convincingly wild, Oxfordshire site. About six miles west of the city is the village of Ducklington, in the Windrush Valley. This has long had fritillaries growing in its low-lying meadowland, and in the eighteenth century the living of its church was under the patronage of Magdalen College. It has been suggested that an incumbent at Ducklington may have taken a fancy to the flowers and carried some bulbs back to his college to plant.[14] (There is a similarity between the two populations in that both have a high proportion of plants with pure white blooms.) They have certainly been in Ducklington 'beyond living memory':

'The field by the River Windrush used to flood regularly when the water table was higher. Local residents remember the fields looking purple, even black with them. The flowers used to be sent to Covent Garden, and local children used to go to Birmingham to sell fritillary

Fritillaries at first light, Magdalen College Meadow, Oxford.

posies. After the war, the water table was lowered and arable crops sown on all available fields, except for one which was purchased by Roger Peel, who lives in the Manor. It now belongs to his two sons, who own it jointly but do not live locally: it has been taken over by the local community and is legally tied up so it can't be sold except to the National Trust. Once a year on Fritillary Sunday [late April or early May] the field is open to the public.'[15]

There are many representations of fritillaries in the church, though mostly twentieth-century. They appear in a stained-glass window dating from 1934, ornament a 1973 embroidered altar frontal and pew cushions, and are carved in a semi-formal chaplet on the pulpit (where they were previously misidentified as tulips, perhaps because 'wild tulips' was once a local name for the flower). The pulpit originally came from Magdalen College. There is also a framed photograph hanging in the church of a remarkable fritillary with pale lemon flowers.

The celebration of the flower, including the holding of a 'Fritillary Sunday', when flowers can be picked (or simply admired) in return for collections for charities, has been going on in many sites for at least a century. Iffley Meadow was one of Oxfordshire's traditional picking sites, and children could reputedly sell posies of the flowers in Oxford High Street provided they were more than nine years old.[16] Iffley's fritillaries vary enormously in numbers, but probably more as a result of weather and winter flooding than picking. In 1933 there were only two blooms. In 1987 numbers dropped to 300 when bad weather had made a hay-cut impossible the previous year. Since 1983, when the Berks, Bucks and Oxon Naturalists' Trust took over the management of the meadow and began an annual and far from solemn count of blooms, the number has crept above 12,000 in both 1992 and 1994.

Iffley is one of the sites that was known to Matthew Arnold, who wrote about fritillaries in his poem 'Thyrsis':

I know what white, what purple fritillaries
The grassy harvest of the river-fields,
Above by Ensham, down by Sandford, yields;
And what sedged brooks are Thames's tributaries.

To the south-west, by some smaller tributaries of the Thames, snake's-heads survive on the Duke of Wellington's estate at Stratfield Saye. These meadows are still opened to the public (just for looking) when the fritillaries are in flower. But before the last war, the public were allowed to pick the blooms, as the poet Andrew Young has described: 'at a field-gate ... a woman sat collecting money. Paying my pence I entered the field, maroon-coloured with the drooping heads of Fritillaries. People moved slowly about, stooping to pick those flowers that looked like repentant serpents. All was so unexpected and strange that I had the feeling I was in heaven; I was even troubled to think that I was not engaged like the others. Picking flowers seemed the only occupation in heaven.'[17]

To the east of Oxford, one of the largest and most famous Fritillary Sunday sites was in the parish of Dinton (previously spelt, though probably not pronounced, Donnington) near Ford, Buckinghamshire. The first sug-

Fritillaries in a 1934 window at Ducklington church, Oxfordshire. 'Fritillary Sunday' is still kept up in the village.

gestive record that there were fritillaries in the area appears on a privately commissioned map of the area dated 1803. On it there is a large plot labelled 'Frogcup Meadow'.[18] Grigson cites the similarly pronounced 'Frockup' as a vernacular name for the plant in Buckinghamshire, suggesting that this may mean either 'frog cup', deriving from fritillary's companion creatures in the meadow, or 'frock-cup', from the shape of the flowers.[19] (Less reverent commentators have suggested 'Frock-up', from human activities in the habitat.) But Leonard Bull, who lived in Ford for many years, has no doubts about the derivation of the name, which is, he says, properly 'Frawcup':

'The name Frawcup comes from the name of the hamlet, Ford. This type of sound-change was a feature of the local dialect. Having been brought up in the area I could quote other similar examples, perhaps one which is still heard is the pronunciation of the local surname Goodearl as Goodrule.

I remember very well the origin of the mistaken derivation of Frawcup. At the end of the First World War a number of teachers came into the district who had no knowledge whatever of the local dialect and its mode of development. It was these people who mistakenly assumed the name was a slovenly corruption of "frogcup" and unfortunately their opinion became accepted as being correct. We, as children, strongly objected (for which we received corporal punishment), as we knew it really meant Fordcup.'[20]

Another local man has described the history of Fritillary Sunday in the parish:

'As a lad, in 1927, I was first taken to the place where the snake's-heads grow, and each year, with the odd exception or two, made the pilgrimage right up to 1939. Local custom made the second Sunday in May a special occasion on which a large number of people living around this area congregated at Ford to pick the snake's-heads. The snake's-head was (and is) locally called a Fraucup – Frodcup – Frawcup, and this particular Sunday was known as Fraucup Sunday.

[The meadows] were literally plastered with snake's-heads, hundreds of people picked great bunches, vying with each other as to who could pick the largest bunch. This picking did not seem to make any difference to the numbers growing each year; a lot of white forms were found, and on occasions twin-headed flowers, which were status symbols indeed.

A fence had been placed in the first hedge for people to enter, and a tin hung on it to receive donations to some charitable organisation, this tin was well filled with odd pence and ha'pence. Picnics were held among the flowers, and a great many cycles were parked by the roadside, and, as the years progressed, motor vehicles became more in evidence. Quite a few locals dug up plants, some of which or their progeny now grace some of our local gardens, orchards, lawns, etc.'[21]

A woman whose mother was brought up in Ford recalls that fritillaries also figured in local May Day celebrations: 'The children bound Fritillaries down the stem of Crown Imperial Lilies with grass and ribbon, making a pretty colourful wand. They then knocked on doors and chanted – "We've come to greet you because it's the first of May. Give us a penny then we'll run away." '[22]

Frogcup Meadow was ploughed up in the early 1950s, but a few fritillaries have clung on in a strip of damp commonland between the old meadow and the lane to Aston Mullins. I found three blooms in mid-April 1993, and met two local men who remembered the last of the Fritillary Sundays (which would have to be held in April nowadays to coincide with blooming). They confirmed the fact that several locals had taken bulbs before the meadow was destroyed. One of their neighbours' populations had increased from 25 to 250. I asked if this was in his garden, but was parried with consummate diplomacy by the reply, 'No, in a piece of ground'!

The most northern of the surviving sites for fritillaries in Britain is at Wheaton Aston in Staffordshire. They have been known here since 1787. In 1912, the local squire G. T. Hartley described them as: 'the black fields, where *Fritillaria* grow in considerable quantities ... On the first Sunday in May, an ancient wake known as Fritillary Wake, locally pronounced Falfillary, is held when people from all the villages around flock down to these moors to gather the flowers of two kinds – spotted Snakeshead and White.'[23]

Later 'Falfillary' became further corrupted to 'Folfalarum' – the '-arum' ending being quite a common suffix in vernacular English for words that are regarded as highfalutin.

Elsewhere in Staffordshire, a small population persists at Tamworth. They were rediscovered in Broad Meadow, by the River Tame, in 1958, when a local naturalist,

George Arnold, found seven flowers. In recent years, the site has been managed by the Staffordshire Trust and, 'with a combination of propagation, secrecy and security', the number had increased to over 600 in 1990.[24]

At the beginning of this century, the fritillary also occurred widely in the river valleys of Suffolk, but it gradually disappeared as meadows were ploughed or drained. One of the best, if smallest, was saved in 1938 at the very last minute: 'Part of the meadow had already been lightly ploughed, and a drainage ditch dug across the lower end ...'[25] Mickfield is now a bewitching place, a small oasis of meadow flowers in a vast arable desert. There are also fritillaries at Rookery Farm, Monewden (now known as Martins Meadows), where they grow with wild daffodils, cowslips and orchids, and at 'Fox Meadow', Framsden. Here, again, there was a tradition of a Fritillary Sunday in aid of charity. The owner, 'Queenie' Fox, told the writer C. Henry Warren in the 1960s: 'A shilling – and everybody can take home a bunch of flowers ... In aid of our local Cancer Fund ... I think people *ought* to have a chance to see the fritillaries, don't you agree?'[26]

A few years later the meadow was sprayed with a broad-leaved weedkiller. But it was summertime, the fritillaries were leafless and they survived where almost all the other meadow flowers were obliterated. In 1978, the five-acre 'Fox Meadow' was acquired by the Suffolk Trust for Nature Conservation, and since then has been opened to the public for one day a year when the fritillaries are in bloom. But as wild flower festivals go, it is a depressing one, and symbolic of the change in the fritillary's status. The surviving flowers bloom in a tangle of rank grasses and thistles in a roped-off enclosure, round which the visitors slowly parade, as if they were at a botanical zoo.

To see fritillaries in immense abundance, you must go to North Meadow, in the Thames Valley at Cricklade, Wiltshire, where in a good year there may be several million in flower. North Meadow (now a National Nature Reserve) is an ancient common, and what is known as 'Lammas Land'. Its 44 acres are shut up for hay on 13 February each year until the hay harvest (apportioned by lot)[27] some time in July. On old Lammas Day, 12 August, it becomes the common pasture of the Borough of Cricklade, and any resident of the town may put up to ten head of horses or cattle on it, or (after 12 September) 20 head of sheep. As far as is known, this system of land tenure has continued unchanged for more than 800 years, and the show at North Meadow may be the best evidence that fritillary is a native species.

But its sheer abundance here has shown how an obsession with rarity can distort our sense of value. In 1978, just after the first Open Day at Fox Meadow, a national newspaper carried a large photograph of a clump of fritillaries at the site, mentioning its recent purchase but not its location. A TV news programme picked up the story and contacted the Nature Conservancy Council (NCC) about the possibility of getting a film of this rare and apparently secret plant. The NCC suggested that it was unlikely they would be given permission to make a private site so widely and explicitly public, but that they would be welcome to come and film fritillaries at the NCC's own North Meadow. But no sooner had the number of flowers been mentioned than the reporter lost interest. How could a plant growing with millions of its own kind be described as *rare*?[28]

White-flowered fritillaries are quite frequent in some populations.

Considering the concentration of surviving fritillary meadows along river systems of middle England, most botanists now consider that the species was a native in post-glacial England. On the continent its wild populations are concentrated around the flood plain of the Rhine, which, before the opening of the North Sea channel, about 5500 BC, formed part of a single river system embracing the Thames and many rivers in the Midlands and East Anglia. Fritillaries probably then grew naturally in seasonally flooded woodland clearings throughout this region. This, alas, is now virtually an extinct habitat in Britain. The one place you may see fritillaries growing under trees, in what is probably their natural habitat, is, with some poetic justice, the damp shrubberies around Magdalen Meadow in Oxford, where the plant's nativeness was first queried.

Martagon lily, *Lilium martagon*, is a species from mainland Europe with usually pale purple, black-flecked 'Turk's-cap' flowers, naturalised in a scatter of woods, mainly in England. It may possibly be native in a few woods in the Wye Valley and Surrey. **Pyrenean lily**, *L. pyrenaicum*, is a native of the Pyrenees with greenish-yellow 'Turk's-cap' flowers, well naturalised in woods and on hedge-banks chiefly in the West Country. Two Devon villages, only five miles apart, have both claimed it as their own. In one it is known as 'the South Molton lily', and in the other as 'the Molland lily'.[29]

Lily-of-the-valley, *Convallaria majalis*, has one of the most beautiful fragrances of any of our native plants, yet demands the poorest soils to flourish. In a manner that echoes juniper (see p. 25), it occupies two distinct habitats in Britain. In the lowland south and east of England it favours ancient woods on sandy, acidic soils, such as the heathy reaches of Danbury in Essex and the greensand woods round Brickhill and Woburn in Buckinghamshire. George Claridge Druce marvelled at the 'wonderful sight' it offered there in the 1920s 'despite the raids that have been made on it in recent years'.[30] It even grew on Hampstead Heath until the middle of the nineteenth century.[31] The most celebrated (though not always the most free-flowering)[32] of these southern colonies are the 'Lily Beds' of St Leonard's Forest in Sussex, which are extensive enough to be marked on the Ordnance Survey map.

But in the west and north it is almost totally confined to limestone woods, from the Cotswolds up to the Yorkshire Dales. Lightly wooded limestone pavements such as Great Scar Close are the place to see them, with their glossy leaves clattering against the pale stone and the pure scent of their white bell-flowers drifting over the surrounding pastures. June is their peak flowering month in Britain, but in France and Germany they are one of the symbols of May Day and are known as *muguet de mai* and *Maiblume*.

All parts of the plant, including the red berries that follow the flowers, are highly poisonous, acting on the heart in a similar way to *Digitalis* (see foxglove, p. 332).

Solomon's-seal, *Polygonatum multiflorum*, is a remarkable plant to come across in a springtime copse, at any stage in its growth. Its early shoots are like narrow scrolls, which seem to expand telescopically until the full-grown stem is arching above the dog's mercury and ramsons. It can still be difficult to spot. The shelves of grey-green oval leaves are held parallel with the ground, catching the light only occasionally and obscuring the

Lily-of-the-valley at the famous 'Lily Beds' in St Leonard's Forest, Sussex.

Solomon's-seal. Its rows of udder-like flowers gave it the nickname of 'sow's tits' in Dorset.

clusters of buttery white bell-flowers. Before these are fully open they are enough like teats to have earned the local Dorset name (now obsolete) of 'sow's tits'.

It is a local plant, concentrated in old woods (especially ash-woods) on the chalk in central southern England, though there are scattered colonies in limestone woods in South Wales and the Lake District.

Angular Solomon's-seal, *P. odoratum*, is a smaller species, with fewer flowers at each leaf-junction. It is scarce, and found chiefly in limestone woods and pavements in the Pennines, and around the Severn Estuary. **Garden Solomon's-seal**, *P.* × *hybridum*, is a robust hybrid between *P. multiflorum* and *P. odoratum*, of garden origin, which is widely naturalised on hedge-banks and in woods throughout Britain.

May lily, *Maianthemum bifolium*, is a white-flowered lily with heart-shaped flowers, quite common on the continent, but a very rare plant in Britain. It once grew in Kenwood in north London, but is now known from acid woods in four locations – Durham, North Yorkshire, Swanton Novers Great Wood in Norfolk, and the same Lincolnshire lime copse where the Revd Keble Martin sketched a specimen in the 1930s.[33]

Herb-paris, *Paris quadrifolia*, is a cryptic and subtle woodland beauty. Above its four broad leaves, held flat like a Maltese cross, is a star of four very narrow yellow-green petals and four wider sepals, topped by a crown of eight golden stamens, and later a single shining black berry – the 'devil-in-a-bush' that was one of the plant's obsolete names.

The name 'paris' comes from the Latin *par*, meaning 'equal', and reflects this symmetry of the plant's parts, which also 'signed' the plant against the disorderly behaviour of witches and epileptics. But beyond its use in early herbalism, herb-paris, surprisingly, has almost no place in native folklore. This can't be altogether explained by its scarcity. Paris occurs in moist, woody places on calcareous soils through most of Britain, from Kent and Somerset up to Banffshire. In the south it seems to prefer ancient woods – in Wiltshire and Somerset on limestone, for instance, and in Essex and Suffolk on boulder clay.

But on the Cotswold plateau it is spreading into young deciduous plantations and self-sown woods. In Hampshire it flourishes in some of the damp, hollow lanes and in Derbyshire haunts streamside ash-woods. In the north-west it will tolerate more rocky terrain, and in the Yorkshire Dales it can sometimes be found marooned at the bottom of fissures in limestone pavements. Paris will spread and flower more freely when its favoured woods

Herb-paris, topped by its striking crown of eight golden stamens, is a plant of shady woods on chalk or limestone.

are thinned or coppiced. (It is the emblem of the recently formed Coppice Association.) But it is principally a plant of secluded, half-lit corners, and it is this, together with its muted coloration, that has made it such a private plant. In the dappled shade of a woodland floor in May its flat leaves often merge and overlap companion plants such as stinging nettle and dog's mercury. Searching for the clusters of even, oval leaves and for the dull glow of the golden antennae is one of the pleasures of hunting for herb-paris. Sometimes you will find plants with five leaves, and more unusually with three or six. In two woods in North Yorkshire plants with seven leaves have been found.[34] These asymmetrical varieties have always held a fascination for botanists. In the margin of his copy of Gerard's *Herball*, the seventeenth-century Welsh botanist Sir John Salusbury noted: '... Herbe Paris is found near Carewis in a place called Cadnant where a faire well springeth called St Michael's Well ... in Welsh, Fynnon Mihangell, within a boult shot of that well down the spring on that side of the water as Carewis standeth ... and by reason of the rankness of the place there I found a great store of herbe paris with five leaves apiece but the yeare 1606 I found the same with six leaves.' Five-leaved herb-paris still grows in this self-same wood at Coed Maesmyan, Flintshire, nearly four centuries on.[35]

Star-of-Bethlehem, possibly native on sandy soils in East Anglia, but more often found as an escape from gardens.

Star-of-Bethlehem, *Ornithogalum angustifolium*, is a short perennial, with a cluster of starry white flowers. It occurs scarcely in dry grassy places throughout Britain, but is possibly native only on sandy soils in East Anglia. **Drooping star-of-Bethlehem**, *O. nutans*, is a common continental species, grown in gardens and naturalised in some grassy places, chiefly in south-east England. In places it can become quite abundant, as in Bodney churchyard, west Norfolk, where it has been since at least the 1930s and is known as 'the Bodney lily'.[36]

Spiked star-of-Bethlehem or **Bath asparagus**, *O. pyrenaicum*. Bath asparagus earned its commoner name in the most straightforward of ways. Its young, unopened flower-spikes were gathered from the wild in May and sold in the markets of Bath, for cooking as asparagus. Bath is close to the centre of its distribution in Britain, which is concentrated in the counties of Avon and Wiltshire (though small colonies occur in some other localities, including Bedfordshire). In the Avon Valley, especially, and on the Bradford-on-Avon plain, it occurs in almost every hedge-bank, green lane and copse. In ancient woodland, it can be as abundant as the bluebell, and in this habitat has all the appearance of a long-established native.

Yet its somewhat odd distribution (it is happiest and most widespread in the Mediterranean region) has raised the possibility that it is an introduction. David Green has speculated that its abundance around Bath may be a legacy of the Roman occupation of this area: 'I wonder whether the bulb of *O. pyrenaicum* arrived via the earth ball of a Roman vine, or whether it was deliberately introduced for its own culinary value.'[37]

But its tall spikes of greenish-yellow flowers are a handsome and distinctive feature of the limestone countryside around Bath and look thoroughly native. It is good that commercial picking of the shoots ended some years ago – though good, too, that the tradition of eating the shoots is kept up by a few locals.

Spring squill, *Scilla verna*, is a small, native bulbous species with violet-blue flowers, often growing in great

Top: spiked star-of-Bethlehem in its heartland in Wiltshire. The young shoots were sold as 'Bath asparagus'.
Above: spring squill, abundant on many west-coast cliffs.

abundance on cliffs on the west coast, in company with thrift and sea campion. **Autumn squill**, *S. autumnalis*, is an autumn-flowering species, more or less confined to Devon and Cornwall.

Bluebell, *Hyacinthoides non-scripta* (VN: Granfer Griggles, Cra'tae – i.e. crow's toes). The sight of sheets of bluebells 'wash wet like lakes' under opening woodland leaves is one of our great wild-flower spectacles. Botanists from further east and south, used to the sweeps of colour in Alpine meadows and Mediterranean hills, still make pilgrimages to bluebell woods if they are lucky enough to be in Britain in springtime.

It is a *British* speciality, too, not just an English one. Bluebells grow in shady habitats – and in open ones in the damper west – from the cliffs of Cape Wrath in Sutherland down to Land's End. The name 'bluebell' is now almost universal, too, despite a persistent belief (held mostly by English people) that in Scotland the species is known as the wild hyacinth and that in that country 'bluebell' refers to *Campanula rotundifolia*, the summer-flowering 'harebell' of England.

But then the accepted names have always been unstable. Two centuries ago, for instance, there is not a sign of a Scots/English split. No less a champion of the Scottish language than Robert Burns uses wild hyacinth and bluebell synonymously. In 'The Song', a poem plainly set in springtime, he rejects the gaudier flowers of foreign fields:

Far dearer to me are yon humble broom bowers
Where the blue-bell and the gowan [buttercup spp.] *lurk lowly unseen:*
For there, lightly tripping amongst the wild flowers,
A-listening the linnet, aft wanders my Jean.

A few decades later, England's greatest rural poet, John Clare, was using 'blue bell' and 'harebell' for the same flower in the same poem, 'The Nightingale's Nest' (*c.* 1832).[38] Even the scientific name has veered wildly about, and in reputable British Floras has changed three times since the 1980s, from *Endymion non-scriptus* to *Scilla non-scripta* to *Hyacinthoides non-scripta* in 1991. (The traditional 'non-script' – meaning 'unlettered' – portion of the name is to distinguish the British hyacinth from the classical hyacinth, a mythical flower sprung from the blood of the dying prince Hyacinthus, on whose petals Apollo inscribed the letters AIAI – 'alas' – to express his grief.)

'Bluebell' as a name did not really come into common currency until the Romantic poets began to celebrate the flower early in the nineteenth century. And it was a later Romantic, Gerard Manley Hopkins, who caught the essence of bluebells more completely than any other writer. His words throng and ripple against each other like the flowers themselves. In his Journal for 9 May 1871, he wrote:

'In the little wood/ opposite the light/ they stood in

Suite of lilies from a late-Victorian set of hand-painted playing cards. They were used in a game resembling 'Happy Families', but with plant instead of human families.

blackish spreads or sheddings like the spots on a snake. The heads are then like thongs and solemn in grain and grape-colour. But in the clough/ through the light/ they came in falls of sky-colour washing the brows and slacks of the ground with vein-blue, thickening at the double, vertical themselves and the young grass and brake fern combed vertical, but the brake struck the upright of all this with winged transomes. It was a lovely sight. – The bluebells in your hand baffle you with their inscape, made to every sense: if you draw your fingers through them they are lodged and struggle/ with a shock of wet heads; the long stalks rub and click and flatten to a fan on one another like your fingers themselves would when you passed the palms hard across one another, making a brittle rub and jostle like the noise of a hurdle strained by leaning against; then there is the faint honey smell and in the mouth the sweet gum when you bite them ...'[39]

Two years later, Hopkins returned to the flower: 'May 11 – Bluebells in Hodder wood, all hanging their heads one way. I caught as well as I could while my companions talked the Greek rightness of their beauty, the lovely/ what people call/ "gracious" bidding one to another or all one way, the level or stage or shire of colour they make hanging in the air a foot above the grass, and a notable glare the eye may abstract and sever from the blue colour/ of light beating up from so many glassy heads, which like water is good to float their deeper instress in upon the mind ...'

The persistent allusions to water in Hopkins's descriptions are an association that is inescapable in the field. In the Chiltern beechwoods the bluebells usually open at the same time as the beech leaves, and with the filtered light dappling the trunks and the bluebells shifting in the breeze, ambling through the flowers is like walking underwater. Even the colour of the blooms has a shoal-like quality, as they bleach from sea-blue to faded denim to spindrift grey. In my own small wood, I once mistook a steep bank of end-of-season bluebells for drifting wood-smoke. Although only 15 acres in extent, it holds individual bluebell clumps which regularly flower in colours ranging from pure white (quite common), through grey, lilac and pale blue to dark cobalt. There is also a variegated form, whose flowers have the look of a white-bell, dipped

Bluebells in 'Devil's Churchyard' wood, near Checkendon, Oxfordshire.

in blue water-colour paint and then allowed to run. A few hundred yards away, in a feral colony in a friend's garden, there was in 1995 a freak bluebell with bracts almost three inches long.[40] (This is apparently a rare but regular sport. In J. M. Albright's handwritten notes in my copy of *The Flora of Oxfordshire*, he writes of May 1917: 'Found bluebell in Dean Grove, with bracts 3 inches long, tipped with blue.' Also, half a century earlier: 'May, 1857. Found bluebell at Heythrop growing through a bush with stalks over 4 feet high and 49 bells.')

Bluebells are woodland plants, but, except perhaps in East Anglia, do not need woods so much as humidity and continuity of habitat. On lowland hedge-banks and bracken-covered pastures in the uplands, drifts of bluebells may be relics of the time when woods grew on these sites. On cliffs and in ravines in the north and west, they will grow in places that have never seen a tree, let alone a wood.

'At Lowland Point on the Lizard in Cornwall, bluebells grow among rocks and boulders only feet from the surf.'[41]

Only the toxic, acid litter of conifer plantations seems able to drive them out, and even here they can slowly move back once the conifers have been removed. In the right conditions they can persist and spread seemingly indefinitely. In the woods at Ipsden in the Oxfordshire Chilterns, the bluebells still flourish, though they do not run 'Bluebell Trains' to see them as they did at the beginning of the century.[42] But the 'Bluebell Railway', which runs through five miles of wooded countryside between Horsted Keynes and Sheffield Park in East Sussex, was saved by a local preservation society after it was axed by British Rail in 1958. And, since 1972, Beaton's Wood at Arlington in East Sussex has been opened for a month of parish celebrations: 'Many local residents have a continued yearly commitment to the Bluebell Walk. The revenue from admissions ... has helped fund several major local projects, including a school swimming pool, and the construction of a new village hall, hence the bluebell motif on the wall ... [it also] pays for the upkeep of the wood, which is managed partly to enhance the aesthetic qualities of the bluebells by creating long clear vistas.'[43]

A 'Bluebell service' is held annually in Withland Wood, Leicestershire, also to raise money for charity. But recently local people report finding the surrounding wood rather bereft of flowers.[44]

The picking of bluebells can spoil woods for other visitors and is a pointless exercise, given how lifeless they look in a vase. But it does little harm to the plant itself (though one Buckinghamshire contributor was told never to pick bluebells 'with the white on' – i.e. down to the base of the stalk – as the plant would die).[45] A more serious recent development is the wholesale stripping of bluebell bulbs from woods. The cause of this, ironically, has been the fashion for 'wild' gardens, and a great increase in the demand for native bluebell bulbs. With the retail value of a single bulb currently about 10p, there is a great incentive both for the illegal 'poaching' of bulbs in public or private woods and for landowners legally to sell their bluebells to commercial diggers. (Under the Wildlife and Countryside Act of 1981, it is an offence to uproot any wild plant without the landowner's permission.) The problem is most serious in East Anglia, where bluebells are least common.[46]

But the trade is not all one way. Bluebells from local stock are being planted out in school and hospital grounds

Bluebells are popularly regarded as Britain's 'national' flower.

and in new plantations. And, just before the opening of the A41 Berkhamsted bypass in 1994, I saw a blue-stencilled board amongst the myriad of contractors' signs and directions. It read, cryptically, BLUEBELL TOPSOIL. But that was what it was, stripped off when the road was dug and stored. The following spring, spread along the new bare embankment outside a wood, it bloomed profusely.

A final point about names, to bring the story full circle. The Botanical Society of the British Isles (whose symbol is the bluebell) has organised a survey of pubs called the Bluebell or Blue Bell.[47] There is a conspicuous concentration of these in the East Midlands, but most have blue bells, not blue flowers, painted on their signs. Some have both. In Hunworth, Norfolk, 'in what appears to be a change of dubious taste', a previous 'Bluebell' has been renamed 'The Hunny Bell' and the sign changed to a barmaid clasping the flower. And in the village of Helpston, Northamptonshire, birthplace of the poet John Clare, the

Grape-hyacinth, native in grassland in the Breckland.

Spanish bluebell is a stouter, more upright species, which occurs in pink as well as white and blue. It is widely naturalised from gardens, even in ancient woods.

board outside the Blue Bell has a church bell on one side and a bluebell on the other. Clare, who enjoyed bluebells, bells and beer equally, would, I think, have relished the confusion.

Spanish bluebell, *H. hispanica*, is a much stouter plant, with less nodding and more bell-like flowers which emerge from all sides of the stem. The flowers occur in purple, pink and white as well as blue. Commonly grown in gardens and naturalised in hedge-banks, churchyards and woodland edges. But the most frequently naturalised varieties are the hybrids between this species and common bluebell, *H. non-scripta* × *H. hispanica*. These crop up wherever both parents are in moderately close proximity, even in ancient woods, and the crosses (which are fertile themselves) form a complete spectrum of colours and habits between the two parents.

Grape-hyacinth, *Muscari neglectum*, is a speciality of sandy areas in the Suffolk Breckland and chalky verges in Cambridgeshire, and not occurring wild in any other part of Britain. Its blue pearl-flowers appear in April and May on road-verges, field edges, rough grassland and the sand banks which build up around pine wind-breaks. A long-standing site is Maidscross Hill, near Lakenheath, Suffolk. Wild grape-hyacinths once grew abundantly throughout this parish, but they have now been largely obliterated by housing development and the air-base at Lakenheath. One large area of rabbit-grazed heath grassland which was covered with the flowers was levelled to make a series of concrete standings for storing bombs. When this was eventually dismantled in the 1960s, the grape-hyacinths returned 'and thousands of plants were in flower over the former area of the bomb banks up to 1976'.[48] Suffolk locals still jokingly stretch their dialect when talking of this species and refer to it as 'grey parsons'.[49] **Garden grape-hyacinth**, *M. armeniacum*, from eastern Europe, and **compact grape-hyacinth**, *M. botryoides*, from southern Europe, are both occasionally naturalised on waysides, from garden throw-outs and seed.

Ramsons or **Wild garlic**, *Allium ursinum* (VN: Stink bombs, Stinking nanny, Stinking onions, Londoner's lilies). A large stand of wild garlic in full odour is, for a couple of months a year, an impressive and unmistakable landmark. Places were named from its Old English root *hrmsa*: Ramsey Island off Pembrokeshire; Ramsbottom, Lancashire; Ramsdell, Hampshire; Ramsden and Ramsey, Essex; Ramsgreave, Lancashire; Ramsholt, Suffolk; Ramshope, Northumberland; and Ramshorn, Staffordshire.[50] In the charter for a piece of land in Berkshire granted by King Edmund to Bishop Aelfric in AD 944, one of the features used to fix the boundary in the minds of the inhabitants was a 'wild garlic wood'. A thousand years later these tangy spots, sweet-and-sour, still stick in the memory. On the Isle of Man (where there is also a 'Ramsey'): 'The motorcycle TT races are associated in some people's minds with the scent of wild garlic, as many of the places where spectators stand in mid-May are liberally covered with the plant. There are stories of fans returning from a day's racing being shunned by their friends on account of their all-pervading ambience.'[51]

'Onions were rarer than gold when we were evacuated to Ayrshire in 1940, but this was no problem, as we just went up the banks of the River Afton and picked as much wild garlic as we wanted. It seemed a kindly thought to keep posting some back to our next-door neighbour still stuck among the bombs in Liverpool. She was ever so pleased, but not so the postman. There were no polythene bags in those days, so his sack reeked permanently of the stuff till we returned three years later.'[52]

'It even made the cat smell which had been walking in it.'[53]

Perhaps it is no wonder that wild garlic's popular name is persistently mispronounced as 'ransoms'. But, for one young boy, it was not the smell, but the massed ranks of dome-shaped flowers that made him mishear their garlic tag: 'Alastair, from Bristol, aged eight at the time, confidently assured me that the white wood alliums were called Daleks. Is this how names change?'[54]

Ramsons grow in similar situations to bluebells, and often with them, but seldom actually intermixed.

Ramsons, the woodland wild garlic – unmistakable and abundant enough to figure in Old English place names.

Although they form dense and sometimes very large colonies, they rarely share space with other species – at least until their leaves begin to rot away at the end of May. In eastern England, they haunt the damper reaches of ancient woods. Further west and north, they become more tolerant, growing in hollow lanes, on stream-banks, and in scrubby scree, until in Pembrokeshire and Cornwall they can be found even on sheltered cliff-faces. In such places, and in the cracks in limestone pavements in the Pennines, the flower-stalks and leaves can grow up to two or three feet tall as they reach for the light. But nowhere can match the show they make on the banks of wooded streams in Wales (e.g. at Crafnant – 'the valley of the ramsons' [55] – in Snowdonia), with their green and white drifts echoing the dapple of ash leaves and trunks right to the water's edge.

Despite their strong smell *en masse*, ramsons are surprisingly mild to eat. The broad leaf-blades are used in salads, stews and soups, 'as the tin miners did in our area at the turn of the century when their wages were so low'.[56]

Oliver Rackham pops the leaves in peanut butter sandwiches when he is doing woodland fieldwork.[57] And, after I introduced the Italian owner of the village pub at the Lee, Buckinghamshire, to his local colonies, he added a flurry of wild garlic improvisations to his Tuscan menu: olive oil in which ramsons leaves had been steeped, for use on sun-dried tomatoes; ramsons salads; and ramsons leaves chopped and added, instead of basil, to a cold tomato sauce for pasta.

Chives, *A. schoenoprasum*, is a rare native of rocky places, usually on limestone, in a scatter of places in south-west England, South Wales, and the north of England. A long-established site is on the Whin Sill ridge near Hadrian's Wall in Northumberland. It is almost certainly native here (as it unquestionably is on wilder Whin Sill rocks a few miles to the north). But locally, as with so many unusual plants, it is regarded as a Roman relic: 'A plant famous for being left behind by the Romans is the wild chive growing by Hadrian's Wall. These plants have existed there for two thousand years, maybe nowadays the sheep nibble them down and they are less obvious to people who might have dug them up. A local man brought us, at Hexham Herbs, a small clump of chives that he had growing in his garden. His grandfather had dug up a piece of the Roman chives thirty years before (before people were generally aware of how wrong this was) and gave us a bit. It grows in our Roman garden which displays the herbs the Romans cultivated in Britain and is much smaller both in habit and flower than common chives.' [58]

A number of European onions and garlics are naturalised as throw-outs or relics of cultivation, including the **onion** itself, *A. cepa*, and varieties such as tree onion. **Garlic**, *A. sativum*, is naturalised in a saltmarsh on the River Lune near Lancaster, presumably from drifting bulbs.[59] **Rosy garlic**, *A. roseum*, an attractive pink-flowered Mediterranean species, is increasingly common on waysides and waste ground in the south-west, as is the white-flowered **Neapolitan garlic**, *A. neapolitanum*.

Three-cornered garlic, from the western Mediterranean, naturalised in hedge-banks and churchyards, and spreading.

Three-cornered garlic, *A. triquetrum* (VN: Stinking onion), is an attractive western Mediterranean species, despite its disparaging local name, with a markedly triangular stem and drooping clusters of white bell-flowers. Introduced to Britain in 1752, it began to escape into shady hedge-banks and churchyards in Cornwall in the 1860s. By the 1930s it was in Devon and seemed 'to be spreading from Cornwall'.[60] Its north-easterly march continues. I found it in a hedge-bank in the Chilterns in 1995. **Few-flowered garlic**, *A. paradoxum*, is an introduction from the Caucasus, and is spreading in waysides, rough grassland and woods throughout Britain. One contributor from Hawick calls it 'plastic grass' 'from the effect of the leaves before the flowers open along the roadsides'.[61] Of a large patch in the churchyard at Drayton Beauchamp, Buckinghamshire, in 1993, a churchwarden remarked to me, as matter-of-factly as if she were telling me the time: 'It does help to keep the Devil out of the church.' **Field garlic**, *A. oleraceum*, and **wild onion** or **crow garlic**, *A. vineale*, are wiry, medium to tall perennials, with narrow, more or less cylindrical leaves and loose umbels of greenish-white to pink flowers growing out of a compact cluster of bulbils. Wild onion is common in southern England, rarer in Wales and further north, in hedgebanks, grassland and occasionally cultivated land. Field garlic is a more local species, found chiefly in the north of England.

Summer snowflake, an exquisite flower, yet not 'discovered' in the wild until the late eighteenth century.

Round-headed leek, *A. sphaerocephalon*, is a species of predominantly Mediterranean distribution, but believed to be native at its one British site in the Avon Gorge, where it has been known on the dry limestone crags of St Vincent's Rocks since 1847. As many as 4,500 flowering heads have been counted in good years. It shares this site with three other unquestionably introduced southern European onions: rosy garlic, *A. roseum* (above), **keeled garlic**, *A. carinatum*, and the splendidly tall **honey garlic**, *Nectaroscordum siculum*, whose drooping greenish-pink flowers smell not of garlic but of honey. All these three species are now believed to have been brought to the Avon Gorge by a Bristol schoolmaster, G. H. Wollaston, who had collected their bulbs and seeds whilst on a visit to Sicily and spread them around the clifftop of St Vincent's Rocks in about 1897.[62]

Summer snowflake, *Leucojum aestivum* (VN: Loddon lily, Snowflakes). As with the fritillary (see p. 403), there has been a lingering suspicion as to why this exquisite and conspicuous lily should have remained apparently 'undiscovered' until the late eighteenth century. It is two or three feet tall, with gracious sprays of iris-like leaves and clusters of white bell-flowers as big as acorn-cups. They hang at the end of long stalks, each petal daubed with an emerald beauty-spot near the tip. When the flowers nod towards you in the breeze, you catch a flash of gold from the stamens. In its favoured habitats it grows in great, sweeping beds. And Gerard, writing at the end of the sixteenth century, remarked: 'These plants do grow wilde in Italy and the places adiacent, notwithstanding our London gardens haue taken possession of them all, many yeeres past.'[63] Not a plant, in short, that was likely to be overlooked.

Yet it was not until the 1780s that William Curtis found the first wild colony, 'betwixt *Greenwich* and *Woolwich* ... close by the Thames side, just above high water mark, growing ... where no garden, in all probability, could ever have existed'. He went on to ask how 'so ornamental a plant, growing in so public a place, could have escaped the prying eyes of the many Botanists who have resided in London for such a length of time'.[64] A fair question, to which Geoffrey Grigson (though not Curtis himself) responded with a resounding 'Impossible', suggesting instead that the seeds had floated down-river from gardens – presumably in the snowflake's current heartland beside the tributaries of the Thames in Oxfordshire, Berkshire and Wiltshire.

Yet in these places it grows in identical situations to those of indisputably wild populations on the continent, in winter-flooded, wooded swamps. To see the plant *en masse* in damp willow carr by the River Loddon in Berkshire, or the Wiltshire Avon, north of Salisbury, is to be both strongly persuaded of its native status and to understand why early botanists might have missed it in the wild. By the Loddon south of Twyford, for instance, the isolated clumps at the corners of riverside gardens have all the look of escapes, quite possibly the Mediterranean variety (ssp. *pulchellum*) favoured by gardeners. Then, south of Sandford Mill, the river-edge grows wilder. There are bigger, taller stands of snowflake growing amongst marsh marigolds under the trees. Follow their trail through the dark alders and nettles, and you enter a quite different habitat, a shifting, humid swamp, caked with a flood-wrack of willow branches and leaf litter. And amongst this debris are sheaves upon sheaves of snowflake, in patches sometimes hundreds of yards square. It is an astonishing sight, but an inhospitable place and not one likely to tempt an early botanist in a wet April.

At the close of the nineteenth century George Claridge Druce noted that summer snowflakes were used for decorating the altar in Wargrave Church, Berkshire, and that 'large quantities from Shillingford have been sold in the Oxford streets recently'.[65] In Long Wittenham, Oxfordshire, they were anciently included in May Garlands (perhaps circumstantial evidence of their being indigenous in the Upper Thames region): 'The ceremony of crowning the May Queen, which includes a parade through the village in period costume and maypole dancing, was revived in 1968. The posies carried are mainly of wild flowers, and by custom they included Loddon lilies. They are known locally as snowflakes.'[66]

Summer snowflakes – Loddon lilies – in Berkshire.

Spring snowflake, *L. vernum*, is possibly native by shady streams and in damp scrub in a couple of localities in Somerset and Dorset. It flowers in February and March.

Snowdrop, *Galanthus nivalis* (VN: Candlemas bells, Mary's taper, Snow piercer, February fairmaids, Dingle-dangle). The snowdrop has always seemed an ambiguous, paradoxical flower. It is a species of winter as much as spring, and the eighteenth-century poet Thomas Tickell called it 'vegetable snow'. We look on it as a wild flower, yet most of its colonies probably began as garden escapes. It may not even be a British native, despite its seemingly ancient pedigree.

The whole of the snowdrop's history is fascinatingly contrary, and it is quite possibly both native and naturalised in Britain. It grows in wild habitats on the continent, in damp woods and meadows up to 1,600 metres. (Though, despite its alpine aura and leaf-tips specially hardened for breaking through frozen ground, it does not occur at all in Scandinavia or the colder northern reaches of Europe.) It is regarded as a native in northern Brittany, not that far from the most persuasively wild colonies in south-west England. And where it turns the ground white, as in 'Snowdrop Valley' near Timberscombe in Somerset or along the shelving banks of the Coundmoor Brook in Shropshire, it is difficult to believe that the plant is not anciently native here too.[67]

Yet snowdrops were not recorded as growing wild in Britain until the 1770s, when they were found in Gloucestershire and Worcestershire. Even as garden plants they seemed slow to attract attention. The first mentions of the name which *The Oxford English Dictionary* can trace are in 1664, when it is listed in John Evelyn's *Kalendar of Horticulture*, and Robert Boyle referred to 'Those purely White Flowers that appear about the end of winter, and are commonly call'd Snow drops'. But thirty years previously, in his revised edition of Gerard's *Herball*, Thomas Johnson adds a footnote to the entry on the 'Timely flouring bulbous Violet': 'Some call them also Snowdrops.' There is no doubt that this is the snowdrop, and, as in Gerard's original edition of 1597, there is an unmistakable drawing and description.[68]

Yet 'Timely flouring bulbous Violet' is too mannered a name for a plant which was already loose and familiar in the countryside, and Gerard and Johnson make it plain that they knew it only from London gardens. The manner in which snowdrops spread also suggests that they may have been originally introduced from an area with warmer winters. Most colonies in Britain, whether in gardens or the wild, reproduce almost exclusively by division of the bulbs rather than by seed. This is partly because cultivated populations (and those naturalised from them) usually come from the same genetic stock, and are sterile; and partly because there is rarely enough insect activity in February to cross-pollinate them. But where different species or varieties grow close together, and there is mild weather at flowering time, seedlings appear.

Despite the lateness of both garden and wild records, it is likely that snowdrops were growing in this country much earlier than the eighteenth century, albeit rather locally. Their pure white blooms have long been accepted by the Catholic Church as a symbol of Candlemas (2 February), the Feast of the Purification of the Virgin Mary, and the link with monastic sites is striking right across Britain:

'I spent six months at the Benedictine Priory of our Lady at Burford, and remember that on the Feast of Candlemas great bunches were picked to decorate the chapel for the day.'[69]

'When I retired to the Dorset–Wilts border I was charmed by the sheets of snowdrops round the village of Donhead. At first I thought of them as garden escapes, but came to realise that the reverse might be the case, and that those in the gardens had come from the hedgerows and wooded areas from a considerable distance around. There are some areas in this county where the snowdrops do appear to be native, but a remark by a villager prompted me to think of Shaftesbury Abbey, founded by Alfred the Great for the Benedictine Nuns with his daughter the Lady Aethelgiva as abbess. This was on a scale of some importance and its wealth increased with subsequent monarchs, with the abbey gaining vast areas of land as far as Tisbury [three miles north of Donhead] ... This seems to support the theory that snowdrops, symbols of purity, were originally planted by the nuns.'[70]

'There is a wonderful display near the remains of Ankerwyke Priory, Middlesex. This is a vast area of Crown Lands, and surrounds the priory where Henry VIII courted Anne Boleyn under the yew tree [this tree, which still survives, is one of the oldest in Britain].'[71]

'In Newbury, there is a church called St Mary's of Speen. It is in a beautiful setting, and very old. The churchyard, and a lot of the surrounding wood, is full of

snowdrops; they are so thick they look like a bluebell wood. They lie in woods that belong to a Dominican monastery right by the churchyard.'[72]

Snowdrops thrive amongst the ruins of St Rachunds Abbey Farm in Kent. The present owner thinned the bulbs and sold them for the Methodist Church restoration fund: 'I probably sold over £150 worth in two years. We were pleased that from the ruins of the old abbey money was made and used to enlarge a twentieth-century church. I think the old monks would have been pleased.'[73]

There are also large surviving colonies at the site of the twelfth-century Cistercian Roche Abbey, near Maltby in Yorkshire; within the skirting walls that are all that remain of the Grey Friars priory in Dunwich, Suffolk; in the ruins of Walsingham Priory, Norfolk; in a cemetery in Abbotskerwell in Devon, which was created by the Augustines in the nineteenth century; at the twelfth-century priory at Brinkburn near Rothbury, Northumberland; and around a farm at Copdock, Suffolk, built 400 years ago on the site of a monastery.[74]

The likelihood that many colonies of feral snowdrops originated with ecclesiastical plantings is supported by records from more workaday church grounds:

'The whole village [Reighton, North Yorkshire] is teeming with snowdrops. The churchyard is scattered with them. There are several small wooded areas in which they are massed, and a large tumulus and a very long drive leading to Reighton Hall are white at this time of year [February]. They are scattered along all the old hedgerows and tumbling out of most gardens.'[75]

In Hampshire, there is a celebrated colony in St Wilfrid's Church at Warnford, which seems to have been quite indiscriminate in its spread: 'The churchyard is full of them, they grow alongside the park boundary railings ... There is also a small clump growing by the side of a footpath leading to an area called Betty Munday's Bottom. Apparently Betty Munday was a lady of the night many years ago.'[76]

There is a Snowdrop Open Day at Castle Hedingham, Essex, and in Kirk Bramwith, South Yorkshire, an annual 'Snowdrop festival' with a brass band concert in the churchyard.[77] Colonies are also cherished by local inhabitants at churches, for instance, at Lamberhurst in Kent, Lower Bourne in Surrey, Burton in South Wirral, Falkenham near Ipswich, Wherwell in Hampshire, Newchurch on the Isle of Wight, Damerham in Hampshire, at the tenth-century Church of Merthyr Issui, Gwent, and at an

Above: snowdrops – 'February fairmaids' – are used to celebrate the Feast of Candlemas (2 February) and are often found associated with churches and monastic foundations. Left: Damerham churchyard, Hampshire.

old people's home in Bocking, Essex, which is on the site of a monastic foundation dating back to AD 995. One of the most northerly sites is in an old graveyard outside Cromarty on the Black Isle, Easter Ross.[78]

In Monkton in Wiltshire the churchyard colony is a living memorial to the Revd J. Brinsdon, an eighteenth-century incumbent: 'Among other things he tried his best to teach the children to read and write. To make them more eager to learn the alphabet, he planted snowdrops in the churchyard in the shape of letters. After his death the snowdrops spread all over the churchyard.'[79]

The churchyard of St Mary the Virgin at Drayton Beauchamp in Buckinghamshire has more earthy associations. Between 1860 and 1883, the incumbent was Henry Harpur Crewe, a distinguished botanist and plantsman, with a penchant for snowdrops, after whom a green cultivar was named earlier this century. There is no trace of this in the churchyard (or of an improbable pink variety some villagers claim to have seen), but it is full of more common varieties of single and double snowdrops, set amongst aconites and ground-ivy. They cluster around the grave of Harpur Crewe himself, which is marked by a simple Victorian cross wreathed with carved ivy leaves.

Snowdrops will linger wherever there have been gardens and cottages. Isolated roadside colonies often mark the sites of demolished cottages:

'One of my favourite walks at Wern-y-wil was along a cart track between a field hedge and an area of scrubland. It was only when I went one February and found flourishing patches of snowdrops, that I thought to scrabble amongst the winter-thin brambles. I found the remains of two cottages – stone-built but less durable than the snowdrops in their gardens.'[80]

'My grandsons have a den which occupies a table-sized clearing in a vicious thicket of blackthorn scrub. In February they bring me a token snowdrop that I cannot reach.'[81]

'Snowdrops occur on Salisbury Plain Army Ranges, on sites of old or destroyed cottages, e.g. Framstead on the Larkhill Artillery Ranges, and in what were the gardens of the now vanished village of Imber (taken over by the Army in the Second World War).'[82]

'Another memory from childhood is that of making regular winter trips to a burnt-down cottage at Shadwell, near Thetford, Norfolk, to pick snowdrops. This place seemed mysterious to me as a child and I always wondered if the occupants had died in the fire. The derelict garden was white with snowdrops growing from the surrounding undergrowth into the charred remains of the cottage's timbers and foundations.'[83]

At Bourne End, Hertfordshire, I have found them lining the leats of what was once a mill (also burnt down) where they were first recorded in 1849.[84] The Oxfordshire writer Mollie Harris has noted that they were sometimes planted to mark the line to outside privies, for a few weeks in winter anyway.[85]

Yet the bulbs clearly travel further and faster than can be explained by vegetative spread alone, and being ferried by water is one plausible explanation: 'A large colony extends for about half a mile through the entire length of a wood facing onto the River Mole, Surrey, and reaches up to 100 yards back from the river. Interestingly most of the plants are submerged when the river floods, and if a flood coincides with flowering, it rather spoils the plants. However I think that on balance the flooding is beneficial. There are several large houses in the area with grounds

Probable native snowdrops in woods at Timberscombe, Somerset.

bordering the river, and there are clumps of snowdrops in the grounds, some along the banks. I am inclined to think that the river may spread the plants from one area to another.'[86]

In Wiltshire, which could well claim to be the top snowdrop county in England, the pattern of snowdrop spread is clearly visible. In the valleys of rivers such as the Nadder and Ebble, which thread their way between the chalk downs, snowdrops grow profusely on streamsides, village greens and road-verges. They seep perceptibly into lower or damper patches and runnels, so that even the most subtle contours of the low ground seem to be marked out with white tracery.

Despite its sanctification for Candlemas, the snowdrop is one of the white blossoms that are still regarded as being unlucky if brought into a house. In parts of Northumberland, Westmorland and Hampshire, for instance, single flowers especially are still viewed as 'death-tokens'. (One Victorian explanation for this was that the flower 'looked for all the world like a corpse in its shroud'[87] – though the belief may have anti-Catholic roots, as with the similar fear of bringing may blossom indoors: cf. hawthorn, p. 211.) Yet the diary of a young girl from Harpenden, Hertfordshire, for February 1901 notes: 'Our February Fair-Maids were up very early this year to grace the mourning for our Queen, and we have had them on our table every day since the first Sunday after she died.'[88]

A more secular Victorian role (though echoing the ancient associations with religious purity) was as an emblem for virginity in the 'language of flowers'. A few blooms in an envelope were often used to warn off over-ardent wooers. In Yorkshire 'there was an old custom, celebrated on February 2nd, for village maidens to gather bunches of snowdrops and wear them as symbols of purity ... In some parts here they are known as "snow-piercers", like the French "perce-neige".'[89]

Yet this history of deliberate introduction and cultivation does not mean that snowdrops are not authentic wild natives in some parts of the west and south. And even where their origins are doubtful, they always have a wild cast about them. Perhaps more than any other garden bulb they are planted haphazardly, in untended corners of lawns and shrubberies, dared almost to break bounds – and loved most of all when they do, edging in scattered clumps away into the shade, as one walker saw them in Durham:

'No white sheets billowing in the breeze here, as often seen in well-manicured churchyards, or secluded corners of equally secluded country houses; these were in *their* element, sharing their habitat with fallen branches, curled-up brown leaves, rotting vegetation and the sounds from the beck below. Some were even growing in the water's edge and on miniature islands, heads nodding the way towards the Greta half a mile away. They accompanied me for about 300 yards, by which time the beck had reached open fields, and there were no more. Looking at the map I now notice that there is the remains of an ancient church only 100 yards from where the beck enters the river; also a site of the medieval village of Brignorth close by.'[90]

Daffodil, *Narcissus pseudonarcissus* (VN: Daffys, Daffy-downdilly, Lent lily, Lenten lily, Easter lily, Glens). Whatever happened to the wild daffodil? It is now a rare plant across great stretches of England and Wales, a flower that people make pilgrimages to see in a 'host'. Yet in the late sixteenth century John Gerard regarded it as growing 'almost euery where through England' and 'so well knowne to all, that it needeth no description'.[91] When the Belgian botanist Charles de l'Écluse visited England in 1581 he found that 'It grows in such profusion in the meadows close to London that in that crowded quarter commonly called Cheapside in March the country women offer the blossoms in great abundance for sale, and all the taverns may be seen decked out with this flower.'[92]

To judge from contemporary Floras, it continued to be one of the most widespread, common (and commonly picked) spring flowers until the middle of the nineteenth century. Then, across much of central and eastern England especially, it seemed to slip into a rapid and mysterious decline. It was not picking that was responsible (wild daffodils are no more harmed by this than cultivated varieties) even though this was deplored by, among others, the Revd Keble Martin in Devon in 1939: 'In spite of the ruthless tearing up of the plants in many districts, it still flourishes and is generally distributed.'[93]

But Devon was one of the few areas where the wild daffodil did continue to flourish. By the 1930s it had already acquired what is now one of the most curiously disjointed distributions of any once-widespread species. It occurs in widely separated zones, for instance in south Devon, pockets of the Black Mountains in Wales, stretches of the Gloucestershire–Herefordshire border country, the Sussex Weald, Farndale in Yorkshire and the Lake District. Elsewhere, even in seemingly suitable places (e.g. the lanes of south-east Cornwall),[94] it occurs in small patches many miles apart. There is a slight westerly tendency in the pattern, and a shift in climate may be involved in its decline. But for once, the loss of old woods and meadowland can't be seriously implicated, since in favoured regions daffodils can be abundant and quite indiscriminate in their choice of habitat. In the Blackdown Hills in Somerset, for instance, I have seen them growing around the foot of road signs, at the edges of pig-sties and on newly canalised river-banks.

There is probably no single or easy explanation for the scattering of the colonies. But the history of the trade and celebrations that have surrounded the wild daffodil in recent years does give some clues as to how its status as a 'popular' flower may have influenced its fortunes.

Nowhere has it been more popular than in the area around Newent, Dymock and Ledbury on the Gloucestershire–Herefordshire border. This was already famous as a focus for 'Lent lilies' before the First World War. The poet Lascelles Abercrombie came to live at Ryton, close by, in 1910, and celebrated the local landscape in his best-known poem, 'Ryton Firs':

> *From Marcle way,*
> *From Dymock, Kempley, Newent, Bromberrow,*
> *Redmarley, all the meadowland daffodils seem*
> *Running in golden tides to Ryton Firs ...*[95]

The region acquired the nickname of 'The Golden Triangle', and in the 1930s, the Great Western Railway began running 'Daffodil Specials' from London for the sake of weekend tourists who came to walk amongst the 'golden tides' and to buy bunches at farm gates. And also for the casual workers who helped local people harvest them for the city markets. John Masefield, who was born in Ledbury, wrote of these piece-workers: 'And there the pickers come, picking for the town ... Hard-faced women, weather-beaten brown.'

This was the first place I ever saw wild daffodils myself, in the late 1970s, and I can understand very well how Londoners would travel 100 miles to see the

The colony of wild daffodils that Wordsworth commemorated in his poem 'I wandered lonely as a cloud ...' still survives on the banks of Ullswater in the Lake District.

Wild daffodils: 'taut, pert, two yellows'.

Dymock hosts. My companion on that trip, Francesca Greenoak, had not seen them before either and her field-notes of that first impression caught their lightness and vivacity perfectly: 'The flowers themselves are lovely – taut, pert, two yellows; the yellows of the trumpet not turning green but arching right back to the papery bud-cover; the surrounding petals appear to be forward-pointing in some (perhaps the early ones?) and stretched out in a pale star in others. In fields it looks as if they are growing in bunches.'[96]

But it was not just the dazzle and daintiness of the flowers themselves; it was their extraordinary, almost impertinent profusion. They ramped in ditches, across orchards, around the edges of arable fields, up the banks of underground reservoirs, along the Ross Spur of the M50, in amongst the bracken in young conifer plantations, and in dazzling clumps (neatly mown around) in churchyards. At the old church in Kempley, the vicar told us a little of the important role the daffodils had played in the local economy. Many farmers and orchard-owners used to regard them as a catch-crop, harvested on a 'pick-and-pay' basis. At the end of April the meadows went back to their normal business of growing hay, or fruit. But the extra income must have helped preserve many from premature 'improvement' and, our informant suspected, led to daffodil bulbs being transplanted onto holdings where they were not so common.

A 1930s guide to the village of Preston (two miles north of Dymock) fills in more detail about the spring harvest:

'They are picked by the local women and children and sold to an agent, who sends them to South Wales, where they find a ready sale for Mothering Sunday and also for Palm Sunday, for in many places the old [Welsh] custom of "flowering the graves" on that day still persists. The wild daffodils are also sent to northern industrial towns, where they find a ready sale too as they can be sold so much more cheaply than the cultivated daffodil. Some of the local boys also augment their supplies of pocket-money considerably by selling huge bunches of these flowers to passers-by at Preston Cross, where there is a continual stream of traffic to Birmingham and the Midlands and to South Wales.'

The 'Daffodil Line' to Dymock was closed in 1959 and the local daffodils faded into obscurity for a while. But in the 1980s, a new kind of daffodil consciousness began to emerge in the district, based not on picking, but on the distinctive flavour the flowers give to the local landscape: 'Many people, both local and from further afield, come to see the daffodils, and "Daffodil Teas" are held in various parish halls, e.g. Oxenhall and Kempley, to coincide with this. The flowering time is notoriously difficult to judge! ... The wild daffodil is a significant local emblem and appears in all sorts of places. It used to be on the Newent CND group banner. One of the bakeries in Newent is called The Daffodil. There is a wrought-iron gate in Newent with a row of daffodils forming part of the design.'[97]

In 1988 a 10-mile-long 'Daffodil Way' was opened between the villages of Dymock, Kempley and Four Oaks. It was created on the initiative of the Windcross Public Paths Project, with support from national and local authorities, and the co-operation of local people and farmers. It follows a roughly circular route along existing rights of way, and passes through orchards, woods, meadows and alongside lanes and streams, and is rarely without both close-up and long-distance vistas of hosts of Lent lilies.

An 1890s greeting card. The child's posy includes true wild daffodils – perhaps then still regarded as common wild flowers.

It is clear from the survival of daffodils around Dymock and elsewhere that even commercial picking had little effect on their populations (cf. primroses, p. 165). The harvest was a consequence of local abundance, not an erosion of it, and throughout Britain, the areas in which there was extensive picking in recent memory still have the greatest concentrations of flowers. In Cornwall, Devon, Gloucestershire, Herefordshire, Hampshire and North Yorkshire they were also one of the seasonal crops of gipsies, up until the mid-1980s:

'From the woods at Tehidy, gipsies used to gather daffs to sell in Camborne, outside Woolworths on a Saturday morning.' (Cornwall)[98]

'In the 1970s the gipsies used to turn up with large baskets to pick them and take them to towns to sell.' (Hampshire)[99]

'Up to 1985 gipsies still sold the wild daffodils from door to door. I had one call at my door in Minchinhampton, Glos, a long way from Newent.'[100]

'When I was a child living at Redcar, I remember women coming round the houses selling bunches of wild daffodils from Farndale, at one old penny a bunch.' (Yorkshire)[101]

Some of these well-known local picking grounds are named after the plant: 'Wild daffodils are in and around Plymouth. Across the Tamar to the west of the city is situated the Cornish village of Cawsand. A large population of daffs has attracted local people for many years. It is called "Goldie Bank".'[102]

The prefix 'Gold' also occurs, perhaps coincidentally, in the names of a number of daffodil sites in Hertfordshire (Golden Parsonage, Goldingtons, etc), and in the same county there is also a well-endowed 'Dilly Wood' at Sacombe.[103] Other woods are known locally as 'Daffy Copses' – e.g. at Washford, near Somerset[104] – but this nickname is almost always informal, even personal:

'During the war a small airfield was located here [Blandford St Mary, Dorset], known as the Tarrant Rushton Airfield. Nissen huts were put up in the Daffy Coppice and the servicemen lived in them for several years ...

About 10 years ago the land was returned to the owners. We were alarmed when we saw that the hazel wood was being cleared out and the ground ploughed up to plant some kind of fir trees. Each spring we would visit the Daffy Coppice, and for several years the flowers were very scant. They have increased in the last few years, and I am thrilled to tell you that this year [1992] it is a host of golden daffodils, or as I've always known them, Lent Lilies.'[105]

'There is a small, open wood close to where I live, about three miles north of Chelmsford in Essex, with wild daffodils. It looks as if it was once a coppiced wood but now has only standards. Individual wild daffodil plants are widely separate, but occur throughout. The name "Daffy Wood" attracted my attention and I must admit that at first I doubted any connection with wild daffodils until I went there and saw them for myself ... I am studying the history of this village [Broomfield, Essex] and recently came across a reference to the wood in a deed dated 1658. Then, it was called "Daffadille Groves", so it appears to be a genuine wild daffodil wood.'[106]

After Dymock and Devon, the Weald – especially in Sussex – probably provides the most extensive wild daffodil territory, though here the flowers are more restricted to ancient, undisturbed habitats and every individual patch seems to be known. Well-loved colonies can be found, for instance, in Kent at Lesnes Abbey Wood (one of the nearest colonies of authentically wild flowers to London) and Elchin Wood near Elmsted; in East Sussex around Ashdown Forest and at Heathfield, Uckfield, Mayfield and Staplecross; in West Sussex, in pockets in St Leonard's Forest and on lane-banks between Horsham and Dial Post, at Fittleworth and Plaistow. Further south at Thakeham it grows in the tall, double hedges known locally as 'shaws' and in woods near Frenchlands, Steyning, where (shades of Dymock) 'my mother used to dig up clumps of the bulbs and plant them in other surrounding woods, where they have now spread'.[107]

There are local celebrations and walks in other areas. Butley Woods in Suffolk are opened to the public at daffodil time[108] (and contain a small colony of a fully double sport of *N. pseudonarcissus*, quite different from garden doubles).[109] At Dunsford, in Devon, there is a famous two-mile walk along the River Teign, where daffodils grow in profusion on the bankside and in the woods and meadows. And there are still a fair number in Gowbarrow Park, on the shores of Ullswater ('Beside the lake, beneath the trees'), whose ancestors inspired what is probably the best-known line in English poetry, Wordsworth's 'I wandered lonely as a cloud ...' But for the real flavour of these doughty spring messengers in the wet and wind-swept English Lakes, read William's sister Dorothy's diary record of 15 April 1802, the 'threatening, misty morning' when they first glimpsed the host:

'When we were in the wood beyond Gowbarrow Park we saw a few daffodils close to the water-side. We fancied that the lake had floated the seeds ashore, and that the little colony had so sprung up. But as we went along there were more and yet more; and at last, under the boughs of the trees, we saw that there was a long belt of them along the shore, about the breadth of a country turnpike road. I never saw daffodils so beautiful. They grew among the mossy stones about and about them; some rested their heads upon these stones as on a pillow for weariness; and the rest tossed and reeled and danced, and seemed as if they verily laughed with the wind, that blew upon them over the lake; they looked so gay, ever glancing, ever changing. This wind blew directly over the lake to them. There was here and there a little knot, and a few stragglers higher up; but they were so few as not to disturb the simplicity, unity and life of that one busy highway.'[110]

Tenby daffodil, *N. pseudonarcissus* ssp. *obvallaris*. Some kind of daffodil is, of course, the national flower of Wales and St David. But its precise identity is uncertain. It may be the wild Lent lily. It may simply be a notional daffodil (cf. the Scotch thistle – see p. 455). It may even be a leek by another name, which would neatly resolve the apparent anomaly of a country having two floral symbols: 'The distinction between the daffodil and the leek is much smaller in Welsh than in English, and may be the source of the confusion as to which is the national plant: leek = Cennin; daffodil = Cennin aur (golden leek). [A botanically satisfying explanation, too, since both are lilies.]'[111]

More recently, attempts have been made to pin the honour on the Tenby daffodil. This is certainly a distinctive variety, with short, stiff stems and small, beautifully proportioned, uniformly yellow flowers in which the petals are held almost at right angles to the trumpet. It also appears to be unique to South Wales. But it was not discovered until the end of the eighteenth century, and its status as a species is still uncertain. Nevertheless it has had a fascinating history in the two centuries since, which has been reviewed by David Jones for the Tenby Museum.[112]

It was first reported by the Welsh botanist R. A. Salisbury in 1796, when it was apparently abundant in fields and pastures between Tenby and the Preseli Hills in Pembrokeshire. As it became better known, it became highly fashionable amongst horticulturalists, much to the delight of some Tenby farmers. In 1893, the curator of the Tenby Museum reported: 'From enquiries, I gather that up to 1885, a steady trade had been done by people here in the bulbs, men being sent into the country districts by one man here systematically to hunt for them. On some fields belonging to Holloway Farm, which you no doubt remember is just outside Tenby on the Marsh Road, the daffodil grew very abundantly; the owner, a man named Rees, learning of the value of the flowers at Covent Garden, sold the bulbs, the entire crop on fields, for £80.'[113]

There was also a steady trade in colourful and improbable historical myths about the flower's origins, which doubtless helped to raise its curiosity value – and price – among dealers. The bulbs had been traded by Phoenician sailors for a cargo of anthracite. They had been brought over by Flemish settlers in the early twelfth century, or to the physic gardens of French or Italian monks, perhaps in the monastery on Caldy Island, just off Tenby. Unfortunately for all these theories of exotic introduction, nothing resembling the Tenby daffodil grows wild anywhere else in Europe.

The old double daffodil variety van Sion growing on a grave in Monewden churchyard, Suffolk *(see p. 433)**.*

But they helped raise the profile of the plant, and the profits of its most ruthless plunderer, Mr Shaw. According to a paper read to the Cardiff Naturalists' Society by Charles Tanfield Vachell in 1894, this ambitious nurseryman was responsible both for popularising the name 'Tenby Daffodil' and for driving the plant virtually into extinction in the wild:

'So delighted were the wholesale bulb dealers with the new flower that orders for the bulbs arrived in rapid succession. Mr Shaw was enabled to engage a staff of collectors who scoured the greater part of South Pembrokeshire for several seasons in a vigorous attempt to meet the phenomenal demand.

Considerable quantities were found by the south side of the Haven, even as far as Castlemartin, but by far the largest quantities were obtained around Narberth, Clynderwen, Llanycefn and Maenclochog.

As a rule the farmers on whose land they grew regarded them as little better than weeds and readily parted with them for a trifle and sometimes for nothing, though Mr Shaw's men, as a result of three days' excursion, often brought him a heaped cart load, which he sold for £160 or more. So well did he keep his secret that he had a complete monopoly of the trade until the supply was practically exhausted.'[114]

The ploughing of grassland during the two wars, and the intensive agriculture since, destroyed most of the colonies of putatively wild Tenby daffodils that the nineteenth-century exploitation failed to uproot. But the flower has been extensively grown in gardens around Tenby, in neighbouring Carmarthenshire and in the Aeron Valley, Cardiganshire, and is now widely naturalised in hedge-banks, churchyards and closes near to farms and cottages.[115] And in the early 1970s, a bizarre sequence of events meant that it began to spread even wider. A young boy from Essex, on holiday in the town, walked into the local tourist office and asked where he could get some Tenby daffodil bulbs to take home to his aunt. Neither Tenby's director of tourism, J. E. Evans, nor any of his staff, had ever heard of the flower and thought it was either a mistake or a practical joke. But a plumber who was working in the office heard the discussion and brought in *The Reader's Digest Book of British Flowers*, which showed and described the flower quite clearly. Mr Evans, realising the daffodil's potential prestige value, persuaded a local nurseryman to search out cultivated stocks from specialist suppliers, and featured it in the forthcoming 'Tenby in Bloom' celebrations.

Interest in the flower built up, and large quantities were planted out in the town itself. A few years later, the Tenby daffodil was given the Royal imprimatur when the Prince of Wales wore one in his buttonhole on a visit to the Principality. And in 1992, 10,000 bulbs were planted out as one of the key features of the first National Garden Festival, held at the site of a former steelworks in Ebbw Vale.

The area around Tenby is now awash with Tenby daffodils, on the verges of approach roads, on roundabouts and increasingly in gardens. Perhaps it has simply gone back to its roots, as its most likely origin centuries ago was as a spontaneous and hardy hybrid between the Lent lily and an unknown cultivar. But in remote corners of the Preseli Hills there are still a few defiantly wild clumps, floral guerrillas whose identity cannot be so tidily explained away.

But increasing numbers of daffodil cultivars are becoming naturalised as a result of widespread municipal plantings like those around Tenby, especially 'Primrose-peerless', 'Nonesuch' and varieties of Pheasant's-eye, *N. poeticus*. Ironically it was in Wales that worries about saturating country towns with cultivated daffodils were first expressed at an official level. In 1992, the Welsh Office told Abergavenny Council that it was becoming over-narcissistic and that 'there were too many daffodils in Wales'.

One much more local naturalisation has an intriguing story behind it. Amongst the fritillaries and green-winged orchids of the ancient hay meadow at Rookery Farm, Monewden in Suffolk there is a colony of the van Sion daffodil, a double-flowered, single-coloured variety first bred by a Flemish gardener in 1620. It is exceptional for

Butcher's-broom has a liking for growing round the base of trees. The clumps may be older than the trees themselves.

garden plants to penetrate the tight sward of wild species that develops in old, undisturbed grassland and just how the van Sion effected its entry had been something of a mystery. But in 1973, John Trist tracked the plant back, along the paths and hedge-banks where it also grows, to Monewden churchyard, one mile away. There he found the daffodil growing on three graves (dated between 1830 and 1857) all belonging to members of the Garnham family, who owned Rookery Farm for generations prior to 1899. So the van Sion's source is plain enough. And perhaps it found its own way from churchyard to meadow along the footpaths and field edges. It was clearly a favourite of the Garnhams, a kind of family emblem, and they may have fancied seeing it not just as a grave ornament but in the setting of one of their working meadows.[116]

Wild asparagus, *Asparagus officinalis* ssp. *prostratus*, is a very scarce and local native subspecies, which grows more or less prostrately on grassy sea-cliffs in Cornwall and Pembrokeshire, though it has been locally common enough to have one small Cornish island named after it. *A. officinalis* ssp. *officinalis* (VN: Sparrow-grass), the familiar garden asparagus, is much more widespread, though it is a southern European plant, brought to Britain some time in the sixteenth century as a vegetable and quickly naturalising. It is frequent on the sandy soils of the Breckland, where there is one of the largest acreages of asparagus under cultivation, especially in disturbed soil at the edges of conifer plantations. Pigeons are very fond of the red berries and no doubt help in its dispersal. It also occurs on rubbish-tips, in churchyards and by railways. I have found clumps of garden asparagus more than three feet across, surviving as relics of wartime railside allotments in west London and still bearing good spears.

Butcher's-broom, *Ruscus aculeatus*. It is hard to credit that butcher's-broom is a member of the lily family. It is like a dwarf holly, a spiny woodland shrub growing knee-high, with spiny leaves (though in fact these are not leaves but flattened stems) that carry scarlet berries in their centres. One of its obsolete vernacular names is 'knee-holly'. It is confined as a native plant to old woods and hedgerows in the south of England and in Wales and East Anglia.

The name butcher's-broom originated because bundles of the spiny stalks were once used to scour butchers' blocks. In the mid-seventeenth century, a few writers mention another butchers' gadget, a miniature indoor hedge made of its branches and placed around meat to keep mice at bay. By the nineteenth century both uses had fallen out of fashion. But old practices have a habit of surviving as symbolic customs and there may have been an echo of that original, utilitarian broom in a custom amongst Victorian butchers on the Isle of Wight of decking 'their mighty Christmas sirloins with the berry-bearing twigs'.[117]

Iris family *Iridaceae*

Yellow iris, *Iris pseudacorus* (VN: Yellow flag, Segg, Jacob's sword). Common by streams, rivers and ponds, and in fens, ditches and damp meadows throughout Britain, this is a robust plant, and the yellow flowers, out from mid-May, are one of the great spring ornaments of wetlands. It is sometimes suggested as the origin of the 'fleur-de-lis' of heraldry. 'Segg' (a variant of 'sedge') is from the Anglo-Saxon for a short-sword, a reference to the blade-like character of the leaves.

The yellow flag, the commonest native iris.

Stinking iris, *I. foetidissima* (VN: Roast-beef plant, Gladdon, Bloody bones). 'Stinking' is an unjust epithet for our other native iris. The smell, such as it is, comes from the leaves, but only if rubbed; and it is far from stinking, resembling slightly stale, raw (but not roast) beef. But scent associations are notoriously personal. One mother's vegetarian child told her it smelt of nut roast.[1] A thoroughly carnivorous Devon family call it 'steak and kidney pudding'.[2]

Gladdon, or gladwyn (another Old English word for a sword), grows in a wide variety of habitats. It is quite common in open wood and on banks and cliffs in southern England (especially in Dorset and Devon) and parts of Wales, but also grows sparingly, often in company with herb-paris and twayblade, in shady corners of ancient woods in East Anglia.

It is also widely naturalised in churchyards. It will not have been introduced for its modest flowers, which are a pallid purple or blue (or very occasionally yellow) and faintly cobwebbed with dark lines. Gladdon is taken into graveyards for the same reason it is grown in gardens, for its brilliant orange seeds, which lie in rows, like peas in a pod, on the open segments of the seed capsules all winter, gleaming against the evergreen leaf spears. They are frequent ingredients in winter gravetop bouquets and church flower arrangements. Bird-sown seeds from these sprays are the usual source of the naturalised clumps in churchyard hedges, as, for example, at Tilbury-juxta-Clare in Essex, and Drayton Beauchamp, Buckinghamshire.

Many species and varieties of European and American irises have escaped from gardens. **Snake's-head** or **widow iris,** *Hermodactylus tuberosus* (from the Mediterranean), with its curious and sombre green flowers topped with blackish-purple epaulets, has been naturalised on sandhills and wayside banks in the West Country since the middle of the nineteenth century. **Bearded irises,** the *Iris germanica* group, are the commonest garden varieties. They are frequently well established on railway banks and waste ground, often forming quite large clumps. In the wild they usually revert to the basic form of violet-blue flowers, with darker 'falls'. **Siberian iris,** *I. sibirica*, from Asia, **purple iris,** *I. versicolor*, from North America,

Stinking iris is commoner in Wales and south-west England. The crushed leaves smell slightly of stale beef.

blue iris, *I. spuria*, from Europe, all have a few scattered but established wild colonies in reed-swamps and other damp places. The white-flowered **Turkish iris**, *I. orientalis*, is naturalised on calcareous banks in Somerset, Kent and north Hertfordshire; and the so-called **English iris**, *I. latifolia* (from the Pyrenees, in fact), is naturalised in grassy places in Kent and Shetland.

Spring crocus, *Crocus vernus* (VN: Nottingham crocus), and **autumn crocus**, *C. nudiflorus*. Towards the middle of the nineteenth century the flood meadows of the River Trent to the south of Nottingham were covered in March with the soft lilac spears of spring crocuses. It was a spectacular display that supported something close to a local festival:

'This harbinger of the vernal season is at the present moment enlivening the Nottingham meadows with thousands of its purple blossoms. Hundreds of "young men and maidens, old men and children" may be seen from the Midland Railway picking the flowers for the ornamentation of their homes. Any stranger fond of flowers who visited Nottingham now for the first time would feel surprised to see large handfuls of Crocuses in the windows of the poorer and middle-class inhabitants. Crocuses in mugs, in jugs, in saucers, in broken teapots, plates, dishes, cups – in short, in almost every domestic utensil capable of holding a little fresh water; and very beautiful they look, even amidst these incongruities, still they look far fresher when seen nestling among the fresh green herbage of these oft-inundated meadows.' (1872)[3]

But by the 1970s, its colonies were greatly reduced: 'In 1974, I went to visit a friend who lived at Nuthall, about five miles from Stapleford. He showed me a pot of spring crocus which had come from an old lady's garden. This lady had originally dug up the plants from the meadow areas of Nottingham. In the 1970s the crocus was found growing in the General Cemetery, Beeston Field Golf Course, The Old Bramcote Church grounds, and Moorgate Congregational Church grounds.'[4]

In October the meadows also carried clumps of the

Spring crocus in Nottingham, where it has been naturalised since the Middle Ages.

purple-flowered autumn crocus, *C. nudiflorus*. Neither of these species is native to Britain. Both come from hills of southern and central Europe, and their successful naturalisation in the improbable settings of a Midlands city has always been a puzzle. But a plausible explanation has been pieced together by a local botanist, Steve Alton.[5] He saw that the crocus populations seem to have radiated out from a central point – the Priory of Lenton, which was founded in the early twelfth century by monks of the Cluniac Order. Cluny is in Burgundy, where the crocus grows wild, and the Clunian monks were Benedictines, celebrated as gardeners and herbalists. Steve Alton believes that the crocuses may have been brought to England as additions to monks' herb gardens, most probably to provide a cheap substitute for saffron (the stigmas of the closely related *C. sativus*). There is circumstantial evidence – for the autumn crocus at least – in the pattern of its distribution elsewhere in the Midlands. No less than nine of its anciently naturalised sites occur along the well-used road between the Lenton priory and one of its outlying hermitages, across the Pennines at Kersall in Lancashire.

Whatever their origins, the Nottingham crocuses were able to flourish in the Trent meadows because of the way they were managed. They were 'Lammas land' – hay meadows 'shut up' to allow grass to grow between Candlemas (2 February) and midsummer; then again between Old Lammas Day (13 August) and 3 October: 'It so happened that while the fields were closed to allow the hay to grow ... the spring crocus flowered, produced its leaves and set seed. In addition, the autumn crocus produced leaves and perhaps small amounts of seed. While the fields were closed in August and September, to allow the grass to grow for winter grazing, the autumn crocus flowered. It is this fortunate compatibility between the traditional management of British meadows and the life-cycle of Central European crocuses that allowed the two species to flourish and spread for hundreds of years.'[6]

This century, they have fared less well. Many of the sites have been developed or flooded. The autumn crocus has a site in a more remote Trent valley meadow, but seems to have left the city itself. But *C. vernus* has clung on, and keeps a tenuous link with its past. There is a colony on the Nottingham University campus, close to the former site of the Priory Grange of Farnstead (part of the priory demesne, which may have had its own herb garden). Steve Alton believes the small colony in the General Cemetery (see above) may be a remnant of a much larger population which grew all over 'the Forest' – a large area of open space which hosts the annual Goose Fair. This area was formerly known as the Sand Field, and was one of the chief areas of grazing land around the city. When the Burgesses of Nottingham removed their livestock from the Lammas meadows down by the Trent, they moved them to the Sand Field. The crocus corms may well have been transported there stuck to the feet of cattle. The Sand Field was divided up in 1865 to give various areas of public open space (including a cemetery).

In Lancashire, Allan Marshall has done an independent study of the sites of the autumn crocus, and reached a similar conclusion – though in this case the agents of distribution seem to have been the Knights of the Order of St John of Jerusalem, returning from the Crusades. All the persuasively 'wild' sites of autumn crocus in the county – at Middleton, Chadderton, Crompton, Oldham,

Wild gladiolus, found only in the New Forest, and almost invariably under bracken.

Milnrow, Healey, Rochdale Milhouse and Hollingworth – are on land owned by the Knights of St John Hospitallers in the thirteenth century.[7]

Elsewhere in Britain both spring and autumn crocus are quite widely naturalised in churchyards, parks and roadsides, the latter being the most thoroughly established (though fine displays of *C. vernus* can be seen at Inkpen in Berkshire and at Ellen Willmott's old garden, now a nature reserve, in Essex). Other species that occasionally escape and naturalise include the lavender-flowered **early crocus**, *C. tommasinianus*, **Kotschy's crocus**, *C. kotschyanus*, and **Bieberstein's crocus**, *C. speciosus* (both pale mauve with dark outside veining), which occur in churchyards and along waysides in Surrey and Suffolk.[8]

And in a wide grass verge alongside Church Lane, Tottenham, Brian Wurzell has found a remarkable colony of crocuses – anciently planted, of course, but now thoroughly established – which includes not only most of the species above but also a 'hybrid swarm' made up of crosses between the pale lilac *C. biflorus* and the yellow *C. chrysanthus*. He has distinguished at least ten varieties – including pure white flowers; a form with white segments above and dark purple 'birds' wings' pencilled below; and a dusky hybrid with mixed yellow and lilac flowers.[9]

Wild gladiolus, *Gladiolus illyricus*. The ancestors of our blowsy cultivated gladioli were first introduced to Britain from southern Africa in the mid-eighteenth century. They could hardly be more different from the wild European species, the 'corn flags' of seventeenth-century gardens, slender, well-proportioned plants, only a foot or so tall with magenta, lily-like flowers. **Eastern gladiolus**, *G. communis* ssp. *byzantinus*, is one of the vivid floral memories many people bring back from Mediterranean holidays. Occasionally, naturalised specimens of this species turn up in southern Britain, usually as relics of cultivation on old bulb fields. (In the Isles of Scilly, where feral gladioli are common, they are known as Slippery or Whistling Jacks.)

These are plainly all escapes. But in 1856, a clergyman discovered a closely related gladiolus species, *G. illyricus*, growing wild in the New Forest.[10] It is still there, in more than 50 sites, and has the look of a genuinely wild species. On the European mainland it grows in scrub and light woodland, and it is a woodland edge species in the Forest

Montbretia naturalised amongst Michaelmas-daisies on South Stack, Anglesey.

too – though always under a canopy of bracken fronds. It seems that only under such unpalatable cover can the gladioli escape grazing by the local ponies and cattle.[11]

Montbretia, *Crocosmia × crocosmiiflora*, is a horticultural cross between two southern African species. It did not even exist, let alone grow wild, until the 1870s. In its 'home' country, it is now a commonly cultivated plant, but unknown outside gardens. Yet in Britain, 6,000 miles to the north, it naturalises so readily and adaptably that you can see its spikes of orange flowers all through the western reaches of Britain, from Cornish cliffs to stream-banks in Argyllshire. It deserves a more welcoming and expressive vernacular name than the gardeners' Latin tag it is stuck with at present.

Black bryony family *Dioscoreaceae*

Black bryony, *Tamus communis*, is the only member of the yam family to grow in the British Isles. It is a common species in England and Wales, scrambling over hedges and wood margins, with heart-shaped, strikingly glossy leaves and bright shiny-red berries, which often remain on the plant into midwinter. It is no relative of its namesake white bryony (see p. 131), though both derive their names from the same classical root (the Greek *bruein* means 'to be full to bursting') and both are poisonous, acting in almost identically irritant ways.

Black bryony, a common wayside scrambler.

Orchid family *Orchidaceae*

The orchids make up the most glamorous and mysterious of our wild plant families. They have extraordinary life-cycles, sometimes blooming only once in a decade. A little imagination can make their flowers resemble insects, reptiles, ball-gowned ladies, even monkeys, and a little more can bridge the gap between modest English woodlands and distant rain-forests. Yet there is surprisingly little mythology or cultural association attached to them, beyond the rather negative (and not particularly accurate) belief that they are all rare, endangered and highly sensitive. Although this is true of some species, others are proving themselves highly adaptable and capable of moving into the most improbable habitats. This has a lot to do with their being a youthful family in evolutionary terms,

Common spotted-orchids, commonest of the tribe in Britain.

still throwing up new forms and hybrids (see Young's helleborine, below); and also with the fact that many orchids produce enormous quantities of exceptionally lightweight seed, which can be blown long distances. These days this often fetches up on artificially open habitats, low in nutrients and free of competition (quarries, for example), which replicate orchid-rich natural habitats such as sand-dunes and cliff-tops. It is this paradoxical, opportunistic quality of many orchids – the exquisite bloom transforming the spoil tip – that has become the basis for the true modern myth of the family, a botanical version of Beauty and the Beast. Increasingly, orchids are found on the lime-rich waste-tips of old chemical factories, on abandoned colliery land, reclaimed airbase runways and bunkers, urban recreation grounds, motorway verges, roundabouts, and power-station fly-ash tips.[1] Orchids invading garden lawns are a source of particular satisfaction:

'We had 198 blooms of spotted orchids on our lawn this year [1993] and we are very proud of them. We originally had two plants – one came in on its own and the other we rescued from someone who had dug one up and put it in a pot, and after talking to us felt thoroughly ashamed. They have increased to this number over about 15 years. We mow them in the early autumn, and in the spring, we go down on our hands and knees and mark the orchids as they come up, even the tiny seedlings ... We also rescued some lady's tresses from a friend who had a ruthless gardener and feared for their safety. We swapped a deep cut turf. We now have 10 plants.' (Hampshire)[2]

'I have observed an abundance of pyramidal orchids growing quite freely in my front garden. Unfortunately, this is not the result of any strongly held conservationist views, but is a direct consequence of a somewhat precarious financial state. Early in 1992, I was made redundant. This was followed by a whole series of home appliance failures, i.e. the washing machine, the cooker, the fridge and yes, the lawn mower. Unable to purchase a replacement mower I decided to neglect the garden and let nature take its course. In the June of this year [1993] I observed a number of the purple conical-shaped flowers appearing on the now overgrown lawn.' (Clwyd)[3]

Lady's-slipper, *Cypripedium calceolus*, is one orchid that is genuinely rare, and now reduced to a single, heavily guarded site in Yorkshire. Collectors have played a role in its decline, though it was probably never a common plant in Britain. It formerly grew wild in open woods on the Pennine limestones, from Derbyshire[4] to Cumberland and Durham. Its best-known stations were in the valleys of the upper Wharfe in the West Riding, especially between Litton and Grassington; on the southern slopes of the Cleveland Hills in the North Riding; and in the craggy ravine known as Castle Eden Dene, near Hartlepool in Durham.[5]

The remarkable flowers – claret-coloured petals crowning a large, bright-yellow pouch, rather like a garden calceolaria – make the lady's slipper the only native orchid that, even to an amateur, plainly belongs to the same group as the tropical species sold by florists. So it is perhaps worth meditating on the fact that the flowers most frequently mistaken for this rare prodigy by non-botanists are the gaudy but not dissimilarly shaped blooms of that much maligned and impetuous immigrant, the Indian balsam (see p. 274)!

Lady's-slipper orchid, the rarest species, known now at just a single site in Britain.

White helleborine, *Cephalanthera damasonium*, is a shade-loving species with white, egg-shaped flowers that rarely open fully. It chiefly haunts beechwoods on chalky soil in the south of England; but: 'Large white helleborine has grown in a patch of ground next to my bungalow since 1985. I leave the parent plants to die off so that the seeds can drop, and this year [1993] three came up in the cracks in a concrete path.' (Wiltshire)[6]

Broad-leaved helleborine, *Epipactis helleborine*, is an uncommon but widespread orchid found chiefly in old woods across most of Britain – except in Glasgow, where it prefers suburban and wasteland habitats. Seventy-five per cent of all the city's colonies are in parks, cemeteries, golf-courses, gardens, railway embankments and roadsides, spoil heaps and quarries. Jim Dickson, senior lecturer in botany at Glasgow University, has done a tally of the populations close to the heart of the city:

'John Lyth found two plants growing under a tree in Carlton Place, less than one kilometre from George Square. It grows among shady gravestones at the Necropolis, in the campus of Glasgow University, in the Botanic Gardens, at the fences of the playing fields along Anniesland Road, in Jordanhill College and many other places. Broad-leaved helleborine does not just have a predilection for small private gardens in Pollokshields. It grows in gardens in Bearsden, Milngavie, Kelvindale, Newlands, Netherless, Williamwood and Low Blantyre. It is open to being called a snob because all these gardens are owner-occupiers. There is not a single record from the gardens of the large council house developments, in particular those to the north-east of the city.'[7]

There is no obvious explanation for this bizarre distribution, which is reproduced in no other British city. But it does help to explain the presence on two Glasgow 'bings' (coalmine spoil-tips) of **Young's helleborine**, *E. youngiana*, of which broad-leaved helleborine may be one parent. Young's helleborine was first discovered in northern England in 1982 and is believed to be a new, 'stabilised' hybrid, that is a cross which has become a self-perpetuating species. It was found in Glasgow three years later. Its other putative parent is **dune helleborine**, *E. leptochila* var. *dunensis*, which in Glasgow also occurs on wooded bings. It is endemic to northern Britain and has appeared nowhere else in Europe.

Violet helleborine, *E. purpurata*, is similar to *E. helleborine*, but later-flowering and with a distinct purple sheen to stem and leaves. It is chiefly a plant of ancient woods in southern and central England, but is often found on road-verges close to woods. In Bentworth, Hampshire, it grows under a beech tree in the churchyard.[8] One population in Hockeridge Wood, Berkhamsted, has a few specimens with white, green and violet variegated leaves, like a hosta; and others have pushed their way through a concreted drive to a house next to the wood.

Ghost orchid, *Epipogium aphyllum*, is the most mysterious and secretive of all our orchids, prone to sudden manifestations and disappearances, and spending most of its life underground. The ghost orchid is entirely lacking in chlorophyll, and the flowers and stem (it has no leaves) look as if they are made of yellow wax, dashed with violet. It relies on nutrients from rotting vegetation, and is con-

Violet helleborine does not flower till high summer and tolerates heavy shade.

fined to deep, moist leaf litter in the most shaded parts of old beechwoods. Once every ten years or so, the underground rhizomes may form a flower-bud, which, depending on the weather, may or may not put up a flower-spike the following summer. Only a score or so have been seen in Britain since records began, and the Chiltern beechwoods are now its most reliable stronghold. But it has also been found on the Herefordshire–Shropshire border, where, true to tradition, it was discovered by a woman.[9]

Twayblade, *Listera ovata*, is widespread and quite common in woods, scrub and grassy places, chiefly on calcareous soils. One colony in Lincolnshire grows by the Kinema-in-the-Wood at Woodhall Spa (surely one of the most romantic botanical addresses in Britain).[10] Twayblade is named from the two large oval leaves that sit at the base of the flower stem – the 'twa blades' in Scots. ('Twayblade' itself was seemingly coined by Henry Lyte in the late sixteenth century.)[11] The small, yellowish-green flowers are vaguely humanoid in shape, though John Gerard, somewhat fancifully, thought they resembled gnats or newly hatched goslings.[12]

Autumn lady's-tresses, *Spiranthes spiralis*, is a remarkable and delicate species, whose flowers grow in a near-perfect spiral. When the individual white blooms are just opening, packed tightly against one another round the short stem, the spike has the look of a sea-creature, or an ivory ornament, turned on a lathe.

Autumn lady's-tresses crops up locally in short grassland as far north as Yorkshire, and in places – for instance, in the damp slacks at Dawlish Warren, south Devon – it can grow as thickly as grass. Its preference for very short turf makes it one of the orchid species which is often found on garden lawns. In parts of the chalk country of Hampshire, Sussex and Kent it has almost become a suburban plant: 'Since 1983 Lady's tresses orchids started to appear on lawns on two estates in Seaford Fairways and Golden Key Estates [East Sussex]. They were very limited at that time, but in 1992 in July, August and September, the lawns were completely covered with them, and also some of the flower borders. On one lawn alone my mate and myself counted over three thousand orchids.'[13]

1992 was one of the species' boom years, and that same summer a front lawn in Barnham, West Sussex, sported 672 spikes in an area approximately 230 yards square.[14]

Creeping lady's-tresses, *Goodyera repens*, is a native of pine and birch woods in northern England and Scotland. An odd outlying population occurs in the pine plantations of East Anglia, where it was reputedly introduced in the roots of saplings brought from Scotland.

Autumn lady's-tresses, often found in lawns on the southern chalk.

Greater butterfly-orchid, *Platanthera chloranthera*, and **lesser butterfly-orchid**, *P. bifolia*, are white-flowered species whose blooms are strongly fragrant at night. They are sometimes hard to tell apart in the field (and occasionally grow together); but the greater is more a plant of woods and calcareous soils and is much the commoner in the south, while the lesser tends to favour open, heathy and acidic sites.

Pyramidal orchid, *Anacamptis pyramidalis* (often inexplicably pronounced py*ram*idal orchid by many botanists), is quite widespread on chalk and limestone areas throughout Britain. It has a preference for old, semi-natural grassland, including churchyards, but is also beginning to appear in much more artificial habitats. Colonies have been noted, for instance, on the verges of

Pyramidal orchids on chalk downland in Dorset.

both the A66 and the M66 in Lancashire; on the Oxford ring-road and the banks of the Marina at Maryport, Cumbria; and 'at the famous Mucky Mountains, St Helens, alongside the Sankey canal ... this is waste from the Alkali industry forming grasslands resembling limestone grassland'.[15]

'A large population at Stansted airport, growing on rubble left after the Second World War, was successfully "translocated" in 1986 into a new specially designated "wildlife area" close to terminal access roads. In 1986, when the grassland was moved, it had a population of 60 pyramidals. Now there are more than 660, with a scatter of bees and common spotted.'[16]

'Several years ago a road improvement scheme was carried out on the A420 between Giddeahall and Ford, Wilts. Because of the unstable nature of the limestone the cutting was extended well beyond the road surface, leaving a flat area which remained sterile for many years. However, soil has gradually formed, and a few years ago a small group of pyramidal orchids appeared. This year [1993] they have spread along the entire length of the cutting, with many hundreds of flowers present. The colour ranged from deepest magenta to two specimens which were a lovely pure albino.'[17]

The most famous roadside sites, which have both become traffic-slowing landmarks in June, are the Claydon and Coddenham roundabouts on the A14 in Suffolk: 'The roundabouts were constructed in the late 70s. By the early 1980s a few pyramidal orchids were growing on both roundabouts. By the summer of 1992 the numbers had increased markedly to between 5,000 and 6,000. At this stage both sites were designated as roadside nature reserves under the scheme operated by the Suffolk Wildlife Trust and the County Surveyor's Department. Management entails a late summer cut with the cuttings being raked and removed. Some thinning of the trees has also occurred to allow the orchids to further colonise each site. Possibly because of this the total number of flowers increased to 11,000 in 1993.'[18]

Pyramidal orchids can occur in many shades of pink, and occasionally white. Sometimes the individual flowers in the spike occur upside down.

Fragrant orchid, *Gymnadenia conopsea*, is a quite frequent species of grassland sites in England and Wales, generally preferring dry sites on chalk or limestone, though forms do occur on heaths and even in fens (ssp. *densiflora*). The flowers are pink to purple, occasionally white, and are carried in the form of a cylindrical spike. Their fragrance is similar to mock-orange (though, like *Philadelphus* scents, not to everyone's taste) and is strongest in the evening.

Frog orchid, *Coeloglossum viride*, is more like a diminutive twayblade than a frog; or perhaps, with its flattened head and long legs, a tadpole. It grows locally in short grassland mainly on the southern chalk and northern limestones.

The marsh- and spotted-orchids (*Dactylorhiza* spp.) are generally the most widespread and most catholic in their choice of habitats. They are also extremely variable, and there are many subspecies and hybrids as well as simple colour variants. **Common spotted-orchid**, *Dactylorhiza fuchsii*, is the most frequent British orchid and can be found in open woods, scrub, fens and grassland

Marsh-orchids growing at Saltfleetby, Lincolnshire.

(usually on chalky soil) as well as spoil-tips, railway embankments and old quarries. It can vary in colour from almost pure white to deep pink, and in height from a few inches to two feet. Large populations, including hybrids with **southern marsh-orchid**, *D. praetermissa*, are frequent on damp ground at old industrial sites, including the fly-ash tip at Aberthaw Power Station near Barry in Glamorgan, and on the spoil around British Steel's deep ore mines at Scunthorpe.[19]

Heath spotted-orchid, *D. maculata*, is similar, but prefers damp places in bogs, marshes and acid grassland. It is the commoner of the two in northern and western Britain, and is plentiful enough on peaty roadsides in parts of Scotland to be picked for vases.

Early-purple orchid, *Orchis mascula* (VN: Adder's meat, Blue butchers, Bloody butchers, Red butchers; Goosey ganders, Kecklegs, Kettle cases, Kite's legs). The variety of local names (Grigson lists more than 90)[20] suggests that the early-purple orchid was once both abundant and well known. As late as 1950, Jocelyn Brooke described it as 'one of the few orchids that can fairly be called common in this country'.[21] It could hardly be called that now, although it is still widespread throughout Britain, occurring on most non-acidic soils, and in a great variety of habitats: ancient woodland (especially coppice), hay meadows, chalk downland, old banks, cliff-top grassland, limestone pavements. Although it still occurs in some numbers in such places, they are precisely the kind of habitats that have suffered most from development and modern farming over the past 50 years, and where they have gone the orchid has usually vanished with them.

The early-purple orchid is a handsome plant, with spikes of pink to purple (or very occasionally white) flowers, and bold, blade-shaped leaves. The flowers usually appear at the same time as the bluebells (and often in their company), and have a distinctive, fugitive scent:

'Early-purple orchids smell wonderful when first opened, like lily of the valley, but this is soon tinged with blackcurrant. When they go over they reek of tom cat.'[22]

'This was one non-poisonous plant we never picked, and this was because locally it was called Adders' Meat. We were told by grown-ups that adders lived nearby and fed on them. As they were not plentiful perhaps this was a cunning way of preserving them, but I think the real reason was far less sophisticated, and was in fact a practical warning because the colouring of the splotched leaves bears some resemblance to a coiled adder.'[23]

Lizard orchid, painted in Bath in 1784. The illustration shows the testicle-like root tubers which gave plants of this family the Latin name Orchis.

Orchis means testicle, and beneath an early-purple orchid is a pair of root-tubers – one new and expanding, filling up for next year's growth, and one withering, as it supplies the plant this season. (John Ruskin was shocked when he learned the derivation of the word, and suggested that orchids be renamed 'wreatheworts'.)[24] The symbolism of the tubers' form could hardly be ignored, and some old vernacular names, such as 'dog-stones', make explicit reference to it. Concoctions of orchid root were given as aphrodisiacs back into classical times. Robert Turner, in *Botanologia* (1664), wrote that enough early-purple orchids grew in Cobham Park, Kent, to pleasure all the seamen's wives in Rochester. And they are almost certainly the 'long purples' of Shakespeare's ambiguous description of Ophelia's garland in *Hamlet* (see purple-loosestrife, p. 234).

An unusual use for the tubers was in a drink called Saloop, or Salep. This was popular amongst manual workers in the nineteenth century, and was made by grinding the dried tubers into a flour, and mixing with hot milk or water, honey and spices. It probably originated (along, perhaps, with bulk supplies of the tubers) in the Middle East, where a similar drink is called *sahleb*.

Green-winged orchid, *O. morio*, is somewhat like a small early-purple, but the 'hood' of the flower is shot through with delicate green veining. It is a declining species of undisturbed short grassland, especially damp meadows. It made an unexpected (and unprecedented) appearance 'at short mid-wicket' on the cricket ground at Stansted Park in West Sussex in May 1992.[25]

Burnt orchid, *O. ustulata*, is now a nationally scarce species, which is found especially on chalk downland in Wiltshire. The unopened buds at the top of the flower-spike are a dark purplish-brown, giving it a scorched look – hence the name.

Lady orchid, *O. purpurea*, is a tall and elegant species, sometimes reaching two and a half feet in height, with a large, broadly cylindrical flower-spike. The individual flowers have something of the look of women in crinoline ball-gowns. They are generally pale pink or rose, with a 'head' of deeper purple. It is a plant of scrub, woodland edges and rides, and confined these days almost exclusively to sites on the chalk in Kent. Previously it was more widespread across south-east England, and in the late 1940s Jocelyn Brooke saw its flowers being sold on Kent street corners for sixpence a bunch. He regarded it as perhaps the most beautiful of all our native orchids: 'Coming upon it suddenly, on the fringe of some Kentish woodland, its showy spikes standing out so vividly against the dingy thickets of dog's-mercury, one is struck, above all, by its "exotic" quality; there is something curiously alien about these tall pagodas of purple and white blossom.'[26]

Military orchid, *O. militaris*. The story of the military orchid's decline, fall and subsequent resurrection in England could be a parable for the fortunes of all our wild flowers. Up to the nineteenth century, to judge from the records, it had been comparatively widespread in the

Green-winged orchid, a declining species of damp, short, undisturbed grassland. Bratoft Meadows, Lincolnshire.

chalk country of southern England. There had been a colony at the bottom of the slope beneath the Duke of Bridgewater's monument at Ashridge, Hertfordshire.[27] In Essex, in 1738, it was found in 'Belchamp Walter Parish, on a little hillock in the corner of a ploughed field adjoining the way leading from Goldington Hall by the lime kiln towards Gestingthorpe'.[28]

But 'Souldiers Cullions' as Gerard called it (literally, 'soldiers' testicles'),[29] is a plant of the warm south, on the edge of its range here, and it has always been temperamental. Some years it would flower, other years not even show itself above ground, and during the cool summers of the second half of the nineteenth century it began to disappear from one district after another. By about 1914 it seems to have been extinct.[30] For the next 30 years the orchid was not seen, or at least not reported, by a single soul. But the possibility that an isolated specimen of this

Burnt orchid, named from its dark, 'scorched' tip. A nationally scarce species of chalk downland.

Lady orchid, so called from the flowers' 'crinoline gown' appearance. Now virtually confined to Kent.

handsome flower – it is a little like a compact lady orchid – might be blooming in some remote corner of the Chiltern Hills turned searching for lost 'soldiers' into an insatiable quest for some orchid lovers (and probably some collectors, too).

In the end, the military orchid was found again almost by chance. The botanist J. E. Lousley had gone to the southern Chilterns in May 1947 ostensibly for a picnic. But, as he put it, 'I selected our stopping places on the chalk with some care, and naturally wandered off to see what I could find. To my delight I stumbled on the orchid just coming into flower.'[31] But, aware of the threat collectors might pose to the plant, he never made the location of his find public. In 1948 the writer Jocelyn Brooke (see above) made the fruitless search for military orchids the theme of the first volume of his trilogy of autobiographical novels.[32] It was not until the 1960s that the colony (or possibly another in the same region) was refound in Buckinghamshire. The site was Homefield Wood near Marlow, and the occasion was announced by a now legendary coded telegram from its discoverers: 'The soldiers are safe in their home field.'

In 1975, the naturalists' trust that managed the site decided to go public, and announced the mysterious orchid's return amidst high security and not a little melodrama. I wrote about it at the time:

'This time local naturalists took steps to ensure that it would not vanish again because of any human agency by setting an electric fence around it. There were rumours of round-the-clock watches and of a warden who carried a shotgun with his sandwiches. Souldiers Cullion was about to be restored to the public, but on rather different terms from those it had enjoyed in Gerard's day. When the press were finally told about the return of this prodigal to an idyllic woodland glade only 50 miles from London, they knew they had a story. For the *Daily Mirror*'s photographer it meant "a pledge of secrecy, a rendezvous in a car park off a lonely country road ... A long walk, an electrified fence, a last few careful steps." And there, "The Beauty that Must Blossom in Secret", the headline in the shadows. More people must have seen the rather smudgy black-and-white picture of *Orchis militaris* the following morning than had seen the plant in its whole history in this country. Yet it seemed a far cry from "the little hillock in the corner of a ploughed field" and the chance of finding it for yourself on a spring picnic.'[33]

Happily, this over-protective stance was abandoned at the end of the 1980s. A summer warden was installed at Homefield Wood and the general public welcomed. Far from leading to their immediate demise at the hands of rapacious pickers, the scheme has been a huge success. The population has grown to over 50 plants (with more than 20 flowering in 1993), and thousands of people have

More anthropomorphic orchids: military orchid (left), known from just two sites, in Suffolk and the Chilterns. Right: man orchid, rather more widespread.

seen them and the 11 other species of orchid that grow close by. The only fencing seen these days is to protect the plants from trampling and rabbit-grazing.

But there is a certain amount of low-key 'gardening' at the site. Trees have been thinned and patches of chalk downland grazed by sheep, in order to create the dappled mosaic of scrub and grass that the military orchid prefers. And, since no pollinating insect has yet been firmly identified, visitors in May stand a chance of witnessing a touching act of symbiosis between orchid-lovers and needy plants: the voluntary wardens, on their knees, delicately extracting pollen from the flowers with proboscis-like stalks of grass, and ferrying it to other clumps.[34]

Military orchids also grow in some numbers in a chalk-pit near Mildenhall in Suffolk, where they were first found in 1955. They are larger and more vigorous than the Chiltern flowers and seem to have more in common with continental plants.[35] As there is no history of the orchid in Suffolk, this does suggest that they may be introductions or windblown colonists.

Monkey orchid, *O. simia*. The individual flowers of this very rare species resemble monkeys rather more than those of the military orchid resemble soldiers. The 'lip' is thick, the 'arms and legs' are curved and long, though more prehensile-looking than the short 'tail'. The monkey orchid, once comparatively common on open chalk hills on both sides of the Thames, is now reduced to two sites in Oxfordshire and two in east Kent.

Man orchid, *Aceras anthropophorum*, is a species of grassland and scrub on chalk and limestone soils, found in scattered localities in the south and east of England. The stem, which normally grows to about a foot in height, carries quite closely packed yellowish-green flowers of a vaguely humanoid shape, and sometimes rust-tinged at the edges.

On the continent, where both man and monkey orchids are much commoner, they occasionally hybridise, producing what has been nicknamed the 'missing-link orchid'. One specimen of this was found at an east Kent monkey orchid site in 1985.

Lizard orchid, *Himantoglossum hircinum*. The lizard orchid has the most bizarre appearance of any of the 'mimic' species. It can grow up to three feet tall, and the spike carries sometimes as many as 80 closely packed flower-heads. The flowers begin to open from late June, and passable imitations of small lizards begin to uncoil, with the petals and sepals forming the head, and the lip dividing into two legs and an impressively long (up to two inches) and twisted tail. They are greenish purple in colour, spotted with darker purple on the tail, and smell distinctively of goat.

The lizard orchid is a sun-loving, continental species, and though it may take many years to grow from seed to flower, it is abundant in road-verges, field edges, vineyards and rough grassland throughout southern Europe. It has, till recently, been a rare plant in Britain, occurring sporadically on chalk and limestone grasslands in the south and east of England. Its populations have fluctuated notoriously. It was thought to be extinct in 1900, then expanded rapidly in the warm spell between 1920 and

Bee orchid, often appearing in large numbers on disturbed roadsides, quarries and mine spoil-heaps.

1940, then went into another decline.[31] But in the six years between 1988 and 1994 – responding perhaps to global warming – the number of flowering plants tripled from 962 to more than 3,000.

Bee orchid, *Ophrys apifera*. The first sight of a bee orchid is an experience few flower-lovers ever forget. There is nothing quite like the sculptured *oddity* of the blooms, perched like sun-bathing, pink-winged bumble-bees on the stalk. They are beautiful, bizarre (the brown 'body' is even furry to the touch) and exotic. Yet bee orchids sometimes behave like rampant weeds. They appear in huge numbers on disturbed chalk soils, linger for a few years and then vanish. Large populations have exploded like this on, for example, a new roundabout near Hitchin, Hertfordshire; shingle banks bordering the Telephone Exchange car park in Milton Keynes, Buckinghamshire; and at the old limestone quarries near Worksop, Nottinghamshire.[36] They can also appear on lawns, sometimes miles from the nearest wild colony.[37]

Fly orchids haunt scrub and dappled woodland edges on chalk and limestone soils in England.

Bee orchids can be frequent on short, calcareous grassland, sand-dunes, spoil-heaps, railway embankments and woodland rides. They also haunt damp places, such as shallow fens (where the spikes can often grow immensely tall, bearing as many as ten separate flowers) and damp, clayey meadows: 'Three years ago orchid leaves were noticed on the running track and hockey pitch of our playing fields, which are on heavy clay soil and permanently waterlogged in winter. These plants were moved in large turves to our meadow area, where they should escape mowing and trampling. I expected some sort of marsh orchid, but to our surprise they turned out to be bee orchids.' (Oxfordshire)[38]

The bee orchid flower is believed to have evolved originally as a decoy, to trick real bees into attempts to mate with successive flowers and so help with pollen transfer. But this has never been reliably observed, and in Britain the flowers are self-pollinating. Some human enthusiasts, however, remain optimistic about the flower's capacity as a lure: 'One of my very favourite flowers from an early age has always been the bee orchid. We used to visit an uncle and aunt in Dorking and often walked on Box Hill, so I got to know the orchids well as a child. When I was at university I went to a fancy-dress party *as* a bee orchid, with the outer perianth represented by a head-dress which I fashioned out of wire, coloured tissue paper and odd bits of cloth, the lip represented by a sort of bib made in the same way. I had a green pullover and borrowed some green trousers.'[39]

Variants of the bee orchid crop up from time to time, for instance the **wasp orchid** (var. *trollii*), in which the tip of the lip is continued downwards, into a sharp sting-like point. In var. *chlorantha*, the whole flower is coloured in various shades of yellowish-green.

Fly orchid, *O. insectifera*, is one of the most exciting and testing orchids to hunt down. The inflorescences – velvet-bodied, with shiny metallic waists – resemble wingless bluebottles impaled on the stalk, and hide in the shade of dogwood and dog's mercury at the edge of beechwoods and chalk scrub, chiefly on chalk and limestone areas of England.

Scottish vernacular plant names

Peter Marren

The following list is derived from numerous sources: correspondence, current Scottish Floras, library sources (notably the card index of the School of Scottish Studies, Edinburgh), conversations, and my memory of spoken names in north-east Scotland. These are the names which I believe are still in use by some people, somewhere. Those I have often heard in use I denote with the standard 'seen it' code of the Floras – an exclamation mark. There are almost certainly many other names in use locally. The *Scottish National Dictionary*, published in instalments from *c.* 1938 to 1976, contains several hundred. However, many of them seem to be extracted from ballads and other ancient matter, and I have omitted these unless there is some evidence that the names are still current in everyday speech. There has never been a scholarly study of country plant names in Scotland. It would be a difficult undertaking now, the researcher needing to spend much time living in remote areas talking to elderly country folk and taping conversations. Since local names are part of an oral tradition, they may have no fixed spelling. Scholars wrote them down phonetically, as they were pronounced.

The most up-to-date source of Gaelic names is *Flora of the Outer Hebrides* (1991) by Pankhurst and Mullin. Most plants in this Flora have been given a Gaelic name by Dr Margaret Bennett: i.e. there are many hundreds. Many are poetic. Others seem to be translations from the English, and it is open to question how many are in genuine use. Margaret Bennett explained to me that Hebrideans were and are good field botanists, but it is hard to believe that they could name most wild plants in their native tongue. I have played safe and included Gaelic names only when they can be verified from recent books about dyeing, folk medicine and the like. Orkney and Shetland have their own distinctive folk flora, which includes many Norse names of probable ancient origin. Fortunately, two scholars collected Orkney and Shetlandic flower names in the 1940s, and the latter have been recorded in the recent Flora of those islands.[1]

In general, Scottish names are less fanciful than the English. They are characteristically 'pawkie' and down-to-earth, and may convey a sense of a plant's use (blackin' girse) or its appearance (gowk's thimmles; locken gowan), or even the nature of its growth (quickens). Many or most derive from the old rural Scotland of Burns and Hogg, the world of the kailyard and the kirk, of ploughman and shepherd, of intermittent famine and half-concealed superstition. That world has vanished, except along the fringes and in parts of the Isles, and it is there, and almost only there, that flowers are still widely known by their pre-scientific names.

A century ago a large proportion of the flora was useful, as herbal medicines, natural dyes, fuel, kitchen plants and children's playthings. At that time these plants were everyday objects and each community had its own names for them. Today the names are falling from the language, helped by the quest for standardisation which seems to be a goal of the scientific botanist. There are two modern Floras (Shetland and the Outer Hebrides) that have restored folk names to their rightful place in the botanical community, an example I hope future Scottish Floras will follow.

One problem for the Flora compiler is that farmers and crofters were not scientists. Folk names often group together plants of similar appearance or habit and occasionally even of similar colour, taste or smell. Here are some that wildly breach the botanical rules:

Blawort, blavers or **blue bonnets**. Blue flowers: cornflower was the favourite (though obviously not now); but also harebell, sheep's-bit, scabious, speedwells, etc.

Bogin. A general name for a marsh plant.

Curl doddie. The name used for devil's-bit and other scabiouses, ragwort and (especially in Shetland) orchids – derives from a fancied resemblence to the curly head of a child. The stems can also be twisted so that the flower-head will spin like a little top.[2] Not to be confused with *carl* doddies, the ribwort plantain.

Deid man's bellows. Bugle, foxglove, red-rattle and *Boleti* (though oddly enough not puffballs, which are deid man's e'en). Presumably related to the puff-mechanism of the seed-capsule or pores.

Duck's meat. Chickweed, duckweed – both are tiny and succulent.

Ekrabung. A Shetland word meaning 'field grass', that is, any grass that invades arable land, e.g. false oat-grass, Yorkshire-fog or couch. Elsewhere invasive grasses are often 'knot-grass', which seems to suit couch the best. Any grass in seed is 'dog's corn'.

Gowan! The *Scots Word Dictionary* says a daisy, but it seems also to be a general name for yellow or white field flowers: buttercups, marsh-marigold, celandine, dandelion, etc. There are also special gowans: witches' gowan and milk gowan (dandelion), dog's gowan (feverfew), horse gowan (ragwort), white gowan (oxeye daisy), etc.

Gollan. Probably the same as gowan, but *SND* applies it particularly to daisy-like plants. There are also yellow gollans: corn marigold, ragwort, etc. Possibly cognate with guld (yellow). The name sounds Irish (cf. the whitefish: 'powan' in Scotland, 'pollan' in Ulster) and, if so, applies to west not east Scotland.

Gowk's ... A gowk is literally a cuckoo, more generally a fool, as in gowk's clover (wood-sorrel), gowk's shilling (yellow-rattle), gowk's thimmles (harebell), gowk's gillyflower (bitter-vetch). Only a fool would mistake these for the real thing.
Hemlock or **kex**. Umbellifers generally, but especially hogweed. Seldom the botanical hemlock, which is not a common plant in Scotland. 'Skeeters' is another general name, referring to the hollow stems.
Mappie's lugs. 'Monkey-faced' flowers: *Mimulus, Antirrhinum*, etc. These are also called frog's mouth.
Murrik. Any edible bulb, root or tuber.
Peasair nan Luch. Gaelic – literally 'mice peas' and probably applying to any vetch bearing small pea pods.
Souraks or **soouricks**. Sour, sharp-tasting leaves, most often of sorrel but also of wood-sorrel.
Spink. Bright and pretty flowers, most often lady's-smock (cuckooflower in Scotland), but also primrose, cowslip, etc. A 'spinkie den' is a flowery dell (a nice Scots/English contrast).
Windlestrae. Grasses with long thin stalks that sway in the wind, e.g. rye-grass, crested dog's-tail and Yorkshire fog.

Plant names in botanical order

Horsetails are *earball* in Gaelic. *Equisetum fluviatile* has been called paddock's pipes in the north-east and faerie's spindles in Shetland.
Bracken is breckan, or sometimes simply 'fern' or 'great fern', even in the Gaelic *raineach mhòr* ('great fern'). *Bun rainnich* are bracken roots. 'Great brake' is archaic. Few other ferns attract much attention, but I would be inclined to credit 'sword-fern' for hard fern (*an raineach chruaidh*). Fraser Darling called it 'spleenwort'.[3]
White water-lily is cambie-leaf. This is the common water-lily of the north and west and there are several Gaelic names: *lili bhan* is merely 'white lily'; more poetic is *duilleagbhaite bhan*, which translates either as 'the drowned white leaf' or 'the white leaf of drowning'. Margaret Bennett suggests the latter contains a note of warning – its beauty the lure to a watery grave. The roots are *ruamalach*, a source of black dye. *SND* has 'lily-can' for yellow water-lily.
Marsh-marigold is a famous flower with many names: gowan, gollan or yellow gollan (general); blugga (Shetland); Gaelic: *lus buidhe Bealltainn* ('the yellow plant of Beltane', for its appearance on 1 May – much later than in England). Also *a carrach-shod* ('the clumsy one of the marsh'). But its mystique is declining: 'nearby is a field covered every spring in a carpet of golden marsh marigolds. This field is now being drained for grazing and I mentioned to some very intelligent children from the north that it would not be there for their children to see. They pass it twice a day on the school bus, and not one of them had even noticed it!'[4]
Globeflower is also a gowan, but usually prefixed as 'locken gowan', 'lockety gowan', 'lapper gowan' or variations, all referring to the 'locked up' petals. In the past it was also 'bull gowan' or 'witches' gowan'. In the Borders it is 'stocks', from the resemblance to a closed cabbage stock.
Creeping buttercup is distinguished from other kinds as sit sikker! (i.e. it sits sure – is hard to get rid of); In Shetlands it is kraatae (crow's toes), from the shape of the leaves; Gaelic: *buidheag*. Other buttercups are harmless and so are just 'gowans'.
Lesser spearwort is wil fire (i.e. wild fire), doubtfully current; Gaelic: *glaisleum*.
Lesser celandine is solandene or pilewort. A rarely-used Deeside name is herb of St Ternan (Ternan being a native of Banchory).
Water-crowfoot. The old Gaelic names *fleann uissage* (water-follower) or *lion na h'aibhue* (river flax) may no longer be in use. Andrew Spink comments that 'along with most aquatics it is most frequently just named "weed" or, if given a name, water buttercup is much more frequent than water-crowfoot. It is also called *Ranunculus* by people who would not dream of calling meadow buttercups by that name.'[5]
Meadow-rue has an unusually succinct Gaelic name: *ru*. This rue is *Thalictrum minus*. But the rarer *T. alpinum* is common enough in Shetland to be known as redshanks.
Barberry is barruba (*SND*).
Poppy is blavers (a name normally used for blue flowers). Gaelic: *lus a' chadail* ('the sleep plant'). *Papaver dubium* is the Scottish poppy, but Gaels might have had *P. somniferum* in mind.
Fumitory has the Gaelic name *lus deathach-thalmhainn* ('earth-smoke plant'), which is also what fumitory means – a mist of leaflets drifting from the soil.
Stinging nettle is just plain 'nettle'. Gaelic: *feanntag*.
Bog-myrtle is usually just myrtle; also gall, gale or Scotch gale. Gaelic: *roid*!, an ancient name still current; also *miortal*.
Orache is iron-root.
Chickweed is chickenweed or occasionally 'the hen's inheritance'. It is also a duck's meat. It is still called arvi or ervi in Shetland.
Greater stitchwort is 'snap stalks' and 'break bones' according to McCallum Webster.[6]
Ragged-Robin is rag-a-tag (*SND*); Gaelic: *caorag leana* ('marsh spark').
Red campion is sweet William or sweet Willie in Shetland, where the flowers are larger than usual and often a deep magenta. A Gaelic name is *cirean coilich* ('cockscomb').
Sea campion has a Norse name in Shetland: *sholgirse*, from

hjol (a wheel), shared with yarrow and sneezewort, which refers to the circular flowers. Also buggie-flooer (also referring to wheels). In Moray, where it grows on steep, treacherous cliffs, it has been called dead man's bells or devil's hatties.
Redshank has leaves spotted the colour of dried blood. In the Roman Catholic west, this is *am boinne-fola* ('blood spot') or *lus chrann ceusaidh* ('herb of the Cross-tree'). It is also known as flower-lady's thumb – it bears the Virgin Mary's thumb-print – and as 'useless', from what she had to say about it.
Amphibious bistort. is yallowin' girse in the northern isles, where it is remembered as the source of a dye.
Water-pepper is spotted knotweed; Gaelic: *lus an fhogair* ('weed that banishes') – it is rumoured to drive away pain, flies, etc.
Knotgrass is recorded as rabbit sugar in McCallum Webster's Flora,[7] but, since she uses old names, this may not be in current usage. The folk 'knotgrass' is just as likely to be couch.
Sorrel is sooraks! and variations; sour dock(en); soordock; soorik (Shetland). All are still widely used. The same names are sometimes applied to other sharp-tasting plants, e.g. wood-sorrel, redshank. Gaelic: *sealbhag* (from *searbh*, 'sour').
Dock leaves are still dockans! everywhere; Gaelic: *copag*. *Rumex pseudoalpinus*, monk's-rhubarb, is also known as smeird docken and butter blades.
Thrift is gressie-spink or sea-daisy.
St John's-worts are specially honoured in Gaelic-speaking areas as *caod achlasen Chaluim Chille* ('the flower carried in the arms of St Columba'). Also *lus na Maighdinn Muire* (milking plant) and Virgin Mary's herb. The former is applied by some to *Hypericum pulchrum* alone.
Mallow seeds are a wayside snack and have been called 'biscuities'. As everywhere else, the botanical common mallow is the country 'marsh mallow'.
Sundews Gaelic: *lus na feàrnaich* ('plant of *earnach*', a cattle disease reputed to be caused by eating sundews; another interpretation is 'plant with the shields': the leaves of *Drosera anglica* are the shape of the shields on the Bayeux Tapestry).
Common rock-rose is solflower, apparently no longer in use (*SND*).
Violets and pansies. *Viola tricolor* is often heartsease as well as wild pansy.
Garlic mustard is sticky Willie.
Water-cress is well-girse or wild skirret.
Lady's-smock is usually cuckooflower (-flooer) in Scotland. Also Mayflower or spink.
Scurvygrass is screebie-grass, screevie-girse or variations. Gaelic: *maraiche* (a mariner) or *carran* ('the thing for scurvy'). Its antiscorbutic properties are still well-known in the north-west.
Shepherd's-purse. Gaelic: *an sporan* (i.e. a sporran, from the little heart-shaped seed-pods); also *lus na fala*.
Charlock is skelloch! Also skellies, runch (shared with other yellow cresses) or wild mustard. (In Orkney, a gritty substance made from charlock seeds was reuthie bread.)
Wild radish shares runch (meaning crunchy) with charlock and wild mustard and other cressy plants. In Shetland it is runchie. It is also ryfarts (for obvious reasons).
Heather is often ling; Gaelic: *fraoch*, still used as the battlecry of the MacDonalds.
Bilberry is always blaeberry; Gaelic: *lus nan dearc* or *lus nan braoileag*, a dye plant as well as a feast.
Bearberry. I have only known it called bearberry. Other names, possibly no longer in use, are dogberry (Grigson)[8], gnashacks (McCallum Webster)[9] and rapperdandie (*SND*). Gaelic: *grainnseag*. The names all imply uselessness.
Primrose is May flooer or buckie-faulie.
Stonecrops, as a group, are suckles (from their teat-like leaves?) or dog's kail (*SND*). *Sedum anglicum* is supposed to have been a Highland delicacy, *biadh an t' sionnaidh* ('lord's food').
Roseroot. An outing of the Banff Field Club in 1918 referred to it by the apt and delightful name of sea-dollies (it still grows on sea-cliffs in Buchan).
House-leek is foose, guardian of homes.
Meadowsweet is queen of the meadow or blackin' girse (from its former use as a dye); Gaelic: *lus chneas Chuchulainn* (Cu Chulainn's plant). A good Norse name used in Orkney and Shetland is *yölgirse* or *yülegirse*, not from Christmas but from an Old Norse word for angelica.
Cloudberry is aivrons, averin and variations, noop (Borders) or knoutberry. Gaelic: *lus nan eighreag*.
Stone bramble is roebuck-berry. It is sufficiently familiar to be a clan badge.
Raspberry is rasp, wood-rasp or siven, but usually raspberry.
Blackberries are always brambles in Scotland (e.g. bramble bush), sometimes pronounced brammles. Gaelic: *an druise bennaichte* ('blessed bramble' – as Christ chastised the moneylenders with it).
Silverweed is or was mascorns in the north-east; Gaelic: *brisgean*. Once important as a famine-plant, and one of the plants known as 'bread and cheese', it is now ignored.
Tormentil is ewe-daisy in the Borders; Gaelic: *leannartach* or *braonan nan con*.
Dog-roses. Gaelic: *ros nan con* is just a translation. Hips are puckies or dog-hippens. The sharp bits are breers or buckies. I have been shown a 'genuine' Jacobite white rose near Stirling; there is also an Ayrshire rose, a form of field-rose.
Kidney vetch has been called cat's claws, presumably from the leaves rather than the flower-heads.

Bird's-foot-trefoil is craw's taes or knifes an' forks, from the spiky seed-pods, or cockies and hennies, from the vari-coloured flowers. A Shetland name is horse yakkels ('horse teeth'). Shamrock is used around Glasgow, possibly fitting any trefoil plant.
Common vetch is fitchacks! as well as mice peas.
Bitter-vetch is *cairmeal* (literally 'dig and enjoy'), corrupted to carmile. Also gnapperts and gowk's gillyflower (referring to its attractive multi-coloured flowers).
Clover is souk or soukies! – the flowers can be sucked like sweets. White clover alone has been called milkies in Moray.
Broom is the brume of the ballads; Gaelic: *bhealaidh*.
Dyer's greenweed is usually weld in Scotland, an ancient dye.
Gorse is usually whin(s), though sometimes furze or gorse; Gaelic: *coras*; *rusg conuisg*.
Purple-loosestrife. Gaelic: *lus na sith chainnt* ('the peace-making plant').
Dwarf cornel is glutton-berry (sharp-tasting, it made an aperitif before a Highland banquet), still occasionally in use as a name in the north-east.
Sun spurge is or was little guid, deil's apple trees (excellent!) or deil's kimstaff (*SND*).
Wood-sorrel is a wayside titbit called gowk's meat, gowk's clover or sookie-sooriks. It is also widely called shamrock.
Marsh pennywort is sheep-rot or shilling grass.
Cow parsley. In the 1970s, Alison Rutherford knew it as 'dog's flourish' – it flourished on road-verges around Helensburgh 'where the dogs have been'.[10]
Sweet cicely is sometimes myrrh; Gaelic: *cos uisge*.
Pignut is arnuts! (usually lousy or lucy arnuts), referring to the ground-nut. Other names are knotty meal, cronies and knot-girse. It was the tasty nut which distinguished this otherwise 'ordinary' plant.
Ground-elder is usually bishopweed, but also goutweed.
Hemlock water-dropwort has the Gaelic name *fealladh bog* ('gentle deceit').
Spignel is badminnie, from the circular seeds; older books have 'Highland Mickim'.
Lovage is usually just that, not 'Scottish lovage'. It is also occasionally called sea parsley.
Wild angelica is ait-skeiters; Gaelic: *lus-nam-buadha*. 'Skeiters' or 'skite' is used for any umbellifer with a hollow stalk suitable for a pea-shooter.
Hogweed is kex, kecksie (Shetland) or bear-skeiters.
Autumn gentian is dead man's mittens in Shetland, from the half-open buds 'like livid finger nails protuding through the green'.[11] (Contrariwise, centaury has been called 'gentian' in Scotland.)
Sea bindweed. Gaelic: *flur a' Phrionnsa*, 'the Prince's flower'!, from the well-known legend that Bonnie Prince Charlie sowed it on landing in Eriskay in 1745. It still grows there and in a few other places in the Hebrides.
Bogbean is water triffle. Its Shetland name gulsa-girse recalls its former use as a cure for jaundice. Gaelic: *ponair chapull* (horse's or mare's bean) or *an tri bhileach* ('three-leaved plant').
Oysterplant is sea gromwell in McCallum Webster's Flora[12] (though this has a bookish ring).
Forget-me-nots are mammy-flooers in Shetland (which suits them, somehow).
Greater plantain is warba; wayburn-leaf (probably from waybread); wavverin leaf (Shetland); warba blades (NE); wabret (Borders); bird's meat. Gaelic: *cuach Phadraig* ('[St] Patrick's bowl').
Ribwort plantain is sodgers!; carl doddies!; rib-grass; and Johnsmas-flooer (Shetland). All are still in use. Carl doddies is the name of a game that has transferred itself to the plant.
Foxglove has numerous names. The flowers are dead man's bells or witches' (or dead woman's) thimbles. The tubers are dead man's thoombs. The whole plant is gensie pushon (Orkney) or in Gaelic *lus nam ban-sith* ('the fairy woman's plant').
Germander speedwell is cat's e'en! or Jenny's blue e'en.
Brooklime is bekkabung in Shetland, which is similar to Linnaeus's Swedish-Latin name beccabunga. They both mean 'brook plant'.
Yellow-rattle is gowk's shillins, from the seeds, or doggins (*SND*).
Lousewort is still hinney-flooer or bee-sookies in Shetland, where it is common.
Harebell is of course Scottish bluebell, but not as often as is claimed. It is known as aul' man's bells in Buchan;[13] and elsewhere occasionally as gowk's thummles (thimbles). Confusingly, the real bellflower, *Campanula latifolia*, has been called harebell in Scotland. Gaelic: *currac* (or *brog*) *na cubhaig* ('cuckoo's hood').
Lady's bedstraw. Gaelic: *lus an leasaich* ('the rennet plant' – it helped to curdle milk for cheese). It was once a greatly valued source of red dye and called *rugh*.
Cleavers is sticky Willie! Other names are stickers!; sticky-grass; catchweed; loosy-tramps; Lizzie-in-the-hedge; bleedy tongues.
Moschatel is wee toon clock.
Devil's-bit scabious is a curl doddie; Gaelic: *ura bhallach*.
Burdock is burrs; burdog in Shetland; Gaelic is better: *suirichean suirich* ('the foolish wooer').
Thistles. The Gaels distinguished between *cluran leàna* (marsh thistle), *cluran deilgneach* (spear thistle) and *fothannan achaidh* (creeping thistle). In Old Scots it is always thrissle.

Common knapweed. Gaelic: *cnapan dubh* ('the black knob').
Cornflower is now so rare that country names must be historic, but once it was the archetypal blue flower – blaver, blawort, blue bonnets, etc.
Dandelion is pee-the-bed; dentylion; clocks; horse gowan. In Moray it has been 'burn pipes' (you rub the leaves on the affected part), also 'denties', from the dainty meal it made for pet rabbits.[14] Gaelic: *bearnan Bride* ('St Bride's plant' – it came into flower on her day, 1 February).
Mountain everlasting is Scotch edelweiss; Gaelic: *spòg cait* ('cat's paw').
Daisy is a gowan or wallie; in Shetland it is also a kokkeloori; muckle (or big) kokkeluris are oxeye daisies and mayweeds.
Tansy is stinking Tam (suggested by the English name?); and occasionally 'yellow buttons'. Gaelic: *lus na Fraing* ('French weed') or *lus an righ* ('king's weed').
Mugwort is muggart or muggins.
Yarrow is pronounced yarra. It is also supposedly doggie's brose (*SND*); Gaelic: *lus chasgadh na fala* ('the plant that staunches bleeding' – as yarrow was a well-known styptic).
Sneezewort is adder's tongue; Gaelic: *cruaidh lus* ('hard weed') or *meacan ragaim* ('stiff plant').
Scentless mayweed. Gaelic: *buidheag an arbhair* ('corn daisy').
Pineappleweed has been called 'apple ringin' in the Borders. This name properly belongs to the cultivated and very unalike southernwood *Artemisia abrotanum*, a case of kinship by taste.
Oxeye daisy is muckle gowan, horse gowan or just plain 'gowan'. It is also wild marguerites in the Glasgow area, where a genteel older name is patio marguerites.[15]
Corn marigold is the once notorious gule (variations: gool, guld), meaning gold. 'The Gule, the Gordon and the Hoodie Craw are the three warst things that Moray ever saw.' Gaelic: *bile-bhuidhe* ('yellow blight').
Marsh ragwort is the common ragwort of the Outer Hebrides and northern isles. On Orkney it has two names: tissics in the east but tirsoos in the west. On Shetland it is gowan or horse gowan (cf. 'as yellow as da gowan').
Ragwort was commonly called ragweed and also tansy. As the foulest of weeds, it was and is stinking Willie (the Duke of Cumberland, of course) but also stinking Davie (who was he?). Other names are curly doddies, weebie, benweed. Gaelic: *buaghallan*. Historically it was sometimes guld or gowan, though the former is normally the corn marigold.
Groundsel is swally! and grunny-swally.
Colt's-foot is dishilago or tushylucky gowan, both recalling its former use for chest complaints, commemorated in its Latin name, *Tussilago* (*tussis* is a cough). Also son before father, meaning a back-to-front plant, whose flowers appear before the leaves. Gaelic: *gormag-liath*.
Butterbur, another back-to-front plant, is usually wild rhubarb!; also Paddy's rhubarb in the Borders; as well as flapper-bags and sheep-root.
Rushes. A clump of rushes is a sprot and is tolerable; a field-full is called rashes and means trouble. Crofters distinguish between soft-rush (*Juncus effusus* or *J. conglomeratus*) and the coarse, hard heath rush (*J. squarrosus*). The former is floss. Heath rush is stuil-bent (after the basal leaves) or, in Shetland, burra. Burra-stikkels are pot brushes or besoms made from rushes – the next best thing on treeless islands.
Cottongrass is draw-moss in the north-east and Lukki's oo in Shetland. The Gaelic name *cannach* was still in use in Glasgow a century ago.
Grasses. Only a few grasses are distinguished by particular names: couch is still quickens, or kwigga in Shetland – *kvika* means quick, vital. A rammock is a generic term for a creeping root, characteristic of the dreaded quickens. Wrack and whigga are alternative old names. Nell Hardie used to gather a grass called 'jingling silver' in the south-west, possibly *Deschampsia cespitosa*.
Reed (*Phragmites australis*) is star-reed! (with emphasis on *reed*, hence sometimes 'streed'); stower (Orkney); the Gaelic *cuilc*! (pronounced culker) probably applied to any reed-like grass, e.g. *Phalaris arundinacea*. On the Ythan (NE) star-reed is also the name of *Schoenoplectus lacustris*.
Bog asphodel. Gaelic: *bliochan*; still sometimes called limrek or limmerik (limb + break), especially in Shetland.
Scottish asphodel (*Tofieldia pusilla*) has been called lamb-lily.
Herb-paris is glamourie-berrie or deil in a bush (*SND*), doubtfully extant.
Bluebell is in Scotland the wild hyacinth, or so we are for ever being told. But as a boy in Glasgow Jim Dickson learned to call it 'bluebell' and it grew in what were known as 'bluebell woods'.[16] Alison Rutherford believes bluebell was the name of the urban working class and wild hyacinth the name of the rural middle class. In the Borders it is still called crawtraes from the shape of the petals. As a wild plant, bluebell is rare or absent in the north of Scotland, where the paler, more hyacinth-like Spanish bluebell (or the hybrid) is more likely to be found. Could this be the real 'wild hyacinth' of Scotland?
Wild garlic (ramsons) is usually plain 'garlic'; 'ramsons' or 'ramps' are not widely used. Gaelic: *gairgean* ('irritable person').
Orchids. In Shetland, orchids are still curly doddies, possibly from the twisted 'stems' (really ovaries) of the flowers. Elsewhere they are, or were, beldairy or puldary (buldeeri in the northern isles). *SND* has gentle Janet, hen's kames, deid man's thoombs and dog's dubbles, the last three of course referring to the tubers. I have never heard any of them in use.

The Scottish thistle

The thistle emblem of Scotland originated in the late Middle Ages as a personal badge of the Stuart Kings. It is the subject of several recent articles, most notably one by Jim Dickson and Agnes Walker entitled 'What was the Scottish Thistle?'.[17] The topic was also worked up into an illustrated booklet, *The Thistle of Scotland*, published by Glasgow Museums and Art Galleries in 1963. This is a summary, with one or two ideas of my own.

Fifteenth-century kings had personal emblems and badges which were worn on chains and symbolised in heraldry and the decorative arts – the white stag of Lancaster, the white rose and sun-burst of York and the white boar of Richard III. The portcullis was part of the iconography of the House of Tudor and survives on modern coinage. The thistle seems to have been borrowed by the Royal Stuarts from their French allies: an Order of the Thistle was founded in France in 1370 and is depicted on contemporary tapestries. The thistle (then pronounced thrissle) first appears in Scotland a century later (1470) on a silver groat of James III. That king left 'a covering (coverlet) browdin with thrissillis' to his heir, and accounts of James V mention 'thissillis of gold to put upoun the Kingis bonat'. The portrait of the latter king shows him wearing a collar adorned with thistle heads, and the marriage of his father James IV to Margaret Tudor of England was celebrated in a contemporary carol, 'The Thrissle and the Rose'.

Why a thistle? No doubt the plant embodied qualities that the Stuarts (and by extension the Scots) saw as their own – tough and durable, proud and fearless, defiant against aggressors. Possibly the thistle also emblemised the Scots' chosen arm of defence – a prickly fence of massed spears, the *schiltrom*. Equally important, the thistle makes an unusually decorative, versatile and satisfactory symbol with its bold outline and what Geoffrey Grigson called its 'gallant colour'.[18] It is surely no coincidence that the Scottish monarchy adopted the thistle at about the same time that the English kings were turning to roses. They symbolised a growing sense of nationhood in both countries at a time when symbols and allegories were potent.

By the sixteenth century, the thistle had become a national emblem. The coinage often depicted a thistle, and Mary Queen of Scots went so far as to place it on the Great Seal of Scotland. By the turn of the century the thistle was being widely carved in wood and moulded in masonry. But it was not until 1687, long after the Union of the Crowns, that King James VII (James II of England) founded the Order of the Thistle, with its chapel in the Abbey Church of Holyroodhouse. Its famous motto *Nemo me impune lacessit* is a pun, and evokes the prickly aggressiveness of the thistle.

Which thistle was it? Grigson dismisses the question as academic: 'The kings of Scotland chose a thistle, not a particular thistle ... They were not botanists.'[19] No doubt they weren't, but the question is of practical interest. The thistles planted each year in the gardens of Holyroodhouse before the Queen's visit are the cotton thistle, *Onopordum acanthium*. This for many is *the* Scottish thistle, despite the fact that it hardly grows wild in Scotland. (The 'official' BSBI name is now cotton thistle, but many older Floras call it Scottish or Scotch thistle.) The attribution dates back to the visit of George IV to Edinburgh in 1822, when the plant was chosen by Sir Walter Scott to be carried in procession. Cotton thistle is in many cases the best match for the thistle of heraldry, decoration and some coins, with a tightly splayed head and a neat, round, prickly capitulum. However that is not because artists copied cotton thistles from life, but because this species just happens to match the stylised thistles of heraldry more closely than other thistles. Heraldic art sacrificed scientific accuracy for effectiveness of design. If the Stuarts had in mind any particular thistle it would have been the one they saw every day, all over Scotland, the spear thistle, *Cirsium vulgare* (known by Burns as the 'burr thistle'). They would probably not have known the cotton thistle. Spear thistle is also the best match for the early coins. The only other contender would be melancholy thistle, but, although some coins depict similarly splendid 'cossack' heads, the melancholy thistle has few prickles and you can sit on it with impunity. The well-known legend of the Viking who trod on a thistle and whose bellow of pain alerted the Scots implies dwarf thistle. But that is found no further north than Yorkshire, and the tale seems to date from no earlier than 1829. It may be significant that *Cirsium vulgare* is known as the Scottish Thistle just across the water in Northern Ireland.

Among the display of thistlry one now sees everywhere in Scotland, from the breakfast table to the Lyon College of Arms, are the symbols of the National Trust for Scotland and Scottish Sports Council, the dustwrapper of *Chambers Scots Dictionary*, the peculiar logo of Scottish Natural Heritage, and Britoil's Thistle Field. Far more moving than any of these are the cotton thistles on the castle mound of Fotheringhay, Northamptonshire, on the site of Queen Mary's execution. They are said to have been planted there by Queen Mary's attendants and are still called Queen Mary's thistles locally. If so, there is something rather miraculous about this small corner of Scotland in a 'foreign' field where the thistles still bloom defiantly above the slow drift of the Nene.

Source notes

SELECT BIBLIOGRAPHY

Allen, David Elliston, *The Victorian Fern Craze*, 1969

Aubrey, John, *Memoires of Naturall Remarques in the County of Wilts*, 1685 (edited by John Britton and first published in 1847 as *The Natural History of Wiltshire*)

Bates, H. E., *Through the Woods*, 1936

Beales, Peter, *Roses*, 1992

Betjeman, John, *Collected Poems*, 3rd edn, 1970

Brewis, Anne *et al.*, *The Flora of Hampshire*, 1995

Bromfield, W. Arnold, *Flora Vectensis*, 1856

Brooke, Jocelyn, *The Wild Orchids of Britain*, 1950

Bulleyn, Willyam, *Bulleins Bulwarke of Defence Againste All Sicknes, Sornes, and Woundes ...*, 1562

Bunce, R. G. H. and Jeffers, J. N. R., *Native Pinewoods of Scotland*, 1977

Burton, Rodney, M., *Flora of the London Area*, 1983

Cave, C. J. P., *Roof Bosses in Mediaeval Churches*, 1948

Chambers, E. K. and Sidgwick, F., *Early English Lyrics*, 1907

Clapham, A. R., Tutin, T. G., and Warburg, E. F., *Flora of the British Isles*, 1952; 3rd edn, 1987

Clare, John, *The Midsummer Cushion*, Ann Tibbles (ed.), 1978

———, *The Shepherd's Calendar*, Geoffrey Summerfield and Eric Robinson (eds), 1964

Cloves, Jeff, *The Official Conker Book*, 1993

Cobbett, William, *The Woodlands*, 1825

Coles, William, *Adam in Eden*, 1657

Cooke, M. C., *A Fern Book for Everybody*, 1867

Cornish, Vaughan, *Historic Thorn Trees in the British Isles*, 1941

———, *The Churchyard Yew and Immortality*, 1946

Culpeper, Nicholas, *The English Physician*, 1699 edn

Curtis, William, *Flora Londinensis*, 1777-99

Dakers, Caroline, *The Countryside at War*, 1987

Dallimore, William, *Holly, Yew and Box*, 1908

Davies, Jennifer, *The Wartime Kitchen Garden*, 1993

Deering, C., *Catalogus Stirpium*, 1738

Dickson, Camilla and James, 'The diet of the Roman army in deforested central Scotland', in *Plants Today*, July-August, 1988

Dickson, J. H., *Wild Plants of Glasgow*, 1991

Dony, John, *Flora of Hertfordshire*, 1967

Druce, George Claridge, *The Flora of Oxfordshire*, 1886; 2nd edn, 1927

———, *The Flora of Berkshire*, 1897

———, *The Flora of Buckinghamshire*, 1926

Ekwall, Eilert, *The Concise Oxford Dictionary of English Place-names*, 4th edn, 1960

Ellis, E. A., *The Broads*, 1965

Evelyn, John, *Sylva, or a Discourse of Forest-trees*, 1664; 2nd edn, 1670; 3rd edn, 1679; 4th edn, 1706

———, *Acetaria: A Discourse of Sallets*, 1699

Ewen, A. H. and Prime, C. T. (trans. and ed.), *Ray's Flora of Cambridgeshire*, 1975

Fisher, John, *The Origins of Garden Plants*, 1982

Fitter, R. S. R., *London's Natural History*, 1945

Forsyth, A. A., *British Poisonous Plants*, 1968

Fowles, John, *The Tree*, 1979

Fowles, John and the Kenneth Allsop Trust, *Steepholm: A Case of History in the Study of Evolution*, 1978

Friend, Revd Hilderic, *Flowers and Flower-lore*, 1883

Gelling, Margaret, *Place Names in the Landscape*, 1984

Genders, Roy, *The Scented Wild Flowers of Britain*, 1971

Gerard, John, *The Herball*, 1597

———, *The Herball*, 2nd edn enlarged and amended by Thomas Johnson, 1633

Gibbons, Euell, *Stalking the Wild Asparagus*, 1962

Gilbert, O. L., 'The ecology of an urban river', in *British Wildlife*, 3(3), 1992

———, *The Flowering of Cities*, 1993

Gillam, Beatrice (ed.), *The Wiltshire Flora*, 1993

Gilpin, William, *Remarks of Forest Scenery*, 1791

———, *Observations on the Western Parts of England*, 1808

Gimingham, C. H., *Ecology of Heathlands*, 1972

Goody, Jack, *The Culture of Flowers*, 1993

Grainger, Margaret, ed., *The Natural History Prose Writings of John Clare*, 1983

Greenoak, Francesca, *God's Acre*, 1985

Grigson, Geoffrey, *Gardenage*, 1952

———, *The Englishman's Flora*, 1955 and 1987

———, *A Herbal of All Sorts*, 1959

——— (ed.), *Dictionary of English Plant Names*, 1974

Grindon, L. H., *The Trees of Old England*, 1868

———, *The Shakspere Flora*, 1883

Hadfield, Miles, *British Trees*, 1957

Hall, P. C. (ed.), *Sussex Plant Atlas*, 1980

Harington, Sir John (trans.), *The Englishman's Doctor, or the School of Salerne*, 1607

Hartley, Dorothy, *Made in England*, 1939

———, *Food in England*, 1954

Hepburn, Ian, *Flowers of the Coast*, 1952

Hickin, Norman E., *The Natural History of an English Forest*, 1971

Hole, Christina, *British Folk Customs*, 1976

Hoskins, W. G., *The Making of the English Landscape*, 1955

———, *The Common Lands of England and Wales*, 1963

Jackson, Kenneth (ed.), *A Celtic Miscellany*, 1971

Jefferies, Richard, *Wild Life in a Southern County*, 1879

———, *Nature near London*, 1883

———, 'Wild Flowers', in *The Open Air*, 1885

Jermyn, Stanley T., *Flora of Essex*, 1974

Johns, Revd C. A., *Flowers of the Field*, 1853

Keble Martin, Revd W., *Sketches for the Flora*, 1972

——— and Fraser, Gordon T. (eds), *Flora of Devon*, 1939

Kent, Douglas H., *The Historical Flora of Middlesex*, 1975

Keyte, Hugh and Parrott, Andrew (eds), *The New Oxford Book of Carols*, 1992

Krussmann, Gerd, *Roses*, 1982

Lavin, J. C. and Wilmore, G. T. D. (eds), *The West Yorkshire Plant Atlas*, 1994

Leather, E. M., *The Folk-lore of Herefordshire*, 1912

Lees, Edwin, *The Botany of Worcestershire*, 1867

Lightfoot, John, *Flora Scotica*, 1777

l'Obel, Matthias de, *Stirpium Observationes*, 1576

Loudon, J. C., *Encyclopaedia of Plants*, 1836

———, *An Encylopaedia of Trees and Shrubs*, 1842

Lousley, J. E. *Wild Flowers of Chalk & Limestone*, 1950

Lyte, Henry, *A Nievve Herball*, 1578

Mabey, Richard, *Food for Free*, 1972

———, *Plants with a Purpose*, 1983

——— (ed.), *The Frampton Flora*, 1985

——— (ed.), *The Flowers of May*, 1990

——— and Evans, Tony, *The Flowering of Britain*, 1980

——— and Greenoak, Francesca, *Back to the Roots*, 1983

Marren, Peter and Rich, Tim, 'Back from the brink. Conserving our rarest flowering plants', in *British Wildlife*, 4(5), 1993

McClintock, David, *Companion to Flowers*, 1966

Miles, Roger, *The Trees and Woods of Exmoor*, 1972

Miller, Philip, *The Gardeners Dictionary*, 3rd edn, 1737; 6th edn, 1752

Milner, J. Edward, *The Tree Book*, 1992

Ministry of Food, *Hedgerow Harvest*, 1943

Mitchell, Alan, *A Field Guide to the Trees of Britain and Northern Europe*, 1974

Morris, M. G. and Perring, F. H. (eds), *The British Oak*, 1974

Morton, John, *The Natural History of Northamptonshire*, 1712

North, Pamela and the Pharmaceutical Society of Great Britain, *Poisonous Plants*, 1967

Opie, Iona and Peter, *The Lore and Language of Schoolchildren*, 1959

———, *Children's Games in Street and Playground*, 1969

Parkinson, John, *Theatrum Botanicum*, 1640

Payne, R. M., 'The flora of walls in south-eastern Essex', in *Watsonia*, 12(1), 1978

Pechey, John, *The Compleat Herbal of Physical Plants*, 1694

Perring, F. H. and Walters, S. M. (eds), *Atlas of the British Flora*, 1962

Perring, F. H., Sell, P. D., Walters, S. M., and Whitehouse, H. L. K., *A Flora of Cambridgeshire*, 1964

Perring, F. H. and Farrell, L. (eds), British Red Data Books: 1. *Vascular Plants*, 2nd edn, 1983

Petch, C. P. and Swann, E. L., *Flora of Norfolk*, 1968

Pevsner, Nikolaus, *The Leaves of Southwell*, 1945

Pollard, E., Hooper, M. D., and Moore, N. W., *Hedges*, 1974

Porta, John Baptista, *Natural Magick*, 1558

Pratt, Anne, *The Poisonous, Noxious and Suspected Plants of Our Fields and Woods*, 1857

Preston, T. A., *The Flowering Plants of Wilts*, 1888

Rackham, Oliver, *Hayley Wood: Its History and Ecology*, 1975

———, *Ancient Woodland*, 1980

———, *The History of the Countryside*, 1986

Ratcliffe, D. A. (ed.), *A Nature Conservation Review*, 1977

Raven, John and Walters, Max, *Mountain Flowers*, 1956

Ray, John, *Catalogus Plantarum circa Cantabrigiam nascentium*, 1660

Rendall, Vernon, *Wild Flowers in Literature*, 1934

Richens, R. H., *Elm*, 1983

———, 'Studies on *Ulmus*'. The first paper, 'The range of variation of East Anglian elms', is in *Watsonia*, 3(3), 1955; the remaining six are in *Forestry*. A full list is given with 'Essex elms' in *Forestry*, 40, 1967

Riddelsdell, H. J. *et al.*, *Flora of Gloucestershire*, 1948
Robinson, William, *The Wild Garden*, 1870
Roden, D., 'Woodland and its management in the medieval Chilterns', in *Forestry*, 41, 1968
Rodwell, J. S. (ed.), *British Plant Communities*, 1991-5
Roper, Patrick, 'The British Service Trees', in *British Wildlife*, 6(1), 1994
Ruskin, John, *Proserpina*, 1874-86
Salisbury, Sir Edward, *Weeds and Aliens*, 1961
———, *Weeds and Aliens*, 2nd edn, 1964
Salmon, Charles Edgar, *Flora of Surrey*, 1931
Simpson, Francis, *Simpson's Flora of Suffolk*, 1982
Sinker, C. A. *et al.*, *Ecological Flora of the Shropshire Region*, 1985 and 1991
Smith, J. E. and Sowerby, J., *English Botany*, 1790-1814; 3rd edn, Boswell Syme (ed.), 1849-54
Soyer, Alexis, *The Culinary Campaign*, 1857
Spencer, Jonathan, 'Indications of antiquity. Some observations on the nature of plants associated with ancient woodland', in *British Wildlife*, 2(2), 1990
Stace, Clive, *New Flora of the British Isles*, 1991
Stewart, A., Pearman, D. A., and Preston, C. D. (eds), *Scarce Plants in Britain*, 1984
Stevens, H. M. and Carlisle, A., *The Native Pinewoods of Scotland*, 1959 and 1996
Summerhayes, V. S., *Wild Orchids of Britain*, 1951
Swan, George A., *Flora of Northumberland*, 1993
Threlkeld, Caleb, *Synopsis Stirpium Hibernicarum*, 1727
Tibble, J. W. and Anne (eds), *The Poems of John Clare*, 1935
Tubbs, Colin R., *The New Forest*, 1986
Turner, William, *The Names of Herbes*, 1548 (facsimile reprint 1965)
———, *The Herball*, 2nd edn, 1568
Vickery, Roy (ed.), *Plant-lore Studies*, 1984
———, *A Dictionary of Plant Lore*, 1995
Wallis, J., *The Natural History and Antiquities of Northumberland*, 1769
Walters, Max, *Wild and Garden Plants*, 1993
Webb, R. H. and Coleman, W. H., *Flora Hertfordiensis*, 1849
White, Florence (ed.), *Good Things in England*, 1932
White, Gilbert, *The Natural History of Selborne*, 1789
———, *A Naturalist's Calendar ... ; extracted from the papers of the late Rev. Gilbert White M.A.*, J. Aiken (ed.), 1795
———, *The Antiquities of Selborne*, Sidney Scott (ed.), 1950
———, *Journals*, Francesca Greenoak (ed.), 1986-89
White, John, *The Way to the True Church*, 1608
Wilkinson, Gerald, *Trees in the Wild*, 1973
———, *Epitaph for the Elm*, 1978
Wolley-Dod, A. H., *Flora of Sussex*, 1937
Wynne, Goronwy, *Flora of Flintshire*, 1993

NOTES

Introduction

1 Barbara Tocher, Exhibition Florist, Tamworth, Staffs.
2 Grigson, 1955
3 David Underdown, *Revel, Riot and Rebellion: Popular Politics and Culture in England, 1603–1660*, 1985
4 See for example Mary Roberts, *Flowers of the Matin and Even Song*, 1845; James Neil, *Rays from the Realms of Nature, or Parables of Plant Life*, 1879
5 Goody, 1993
6 John Vidal, *Guardian*, 10 December 1993
7 Caroline Smedley, Newton Regis, Staffs.
8 Alison Rutherford, Helensburgh, Strath.
9 Ronald Blythe, 'An inherited perspective', in *From the Headlands*, 1982
10 M. K. Farmer, Petersfield, Hants.

Horsetails, *Equisetaceae*

1 Elizabeth and Rachel Stevens, Fleetwood, Lancs.
2 Tony Baker, Winster, Derby.
3 Mabel Turner, Writtle, Essex
4 Sue Goss, Bledlow Ridge, Bucks.
5 Grigson, 1959
6 Aubrey, 1685
7 Peter Casselden, Chesham, Bucks.
8 Burton, 1983

Ferns, *Pteropsida*

1 Allen, 1969
2 Culpeper, 1699
3 Len and Pat Livermore, Lancaster, in *BSBI News*, 61, 1992
4 Caroline Male, Halesowen, W. Mids
5 M. H. Beard, Little Wilbraham, Cambs.
6 Ida Turley, Ty Gwyn, Clwyd
7 Bill Chope, Baden Powell Scouts Association, King's Heath, Birmingham
8 Chris Walker, Condover, Shrops.
9 Lightfoot, 1777
10 'Stack silage made from bracken', in *Transactions of the Highland Society of Scotland*, 20, 1988
11 Whybrow, *c.* 1920
12 C. Walker, Condover, Shrops.
13 Dorothy Mountney, Harleston, Norf.
14 Martin Spray, Ruardean, Glos.
15 A. J. Cherrill and A. M. Lane, 'Bracken ... infestation of rough grazing land in the catchment of the River Tyne, northern England', in *Watsonia*, 20(2), 1994
16 S. A. Rippin, Fforest Coalpit, Gwent
17 P. M. and Madeline Reader, Horney Common, E. Susx
18 Dr Jack Oliver, Lockeridge, Wilts.
19 S. M. Walters, 'Cambridgeshire ferns – ecclesiastic and ferroviatic', in *Nature in Cambridgeshire*, 12, 1969
20 C. C. Babington, *Flora of Cambridgeshire*, 1860; Perring, Sell, Walters and Whitehouse, 1964
21 Charlotte Chanter, *Ferny Combes. A ramble after ferns in the glens and valleys of Devonshire*, 1856
22 Cooke, 1867
23 Rickard, Martin, 'Ferns – a case history', in *The Common Ground of Wild and Cultivated Plants*, A. Roy Perry and R. Gwynn Ellis (eds), 1994
24 *ibid.*
25 Cooke, 1867

Pines, *Pinaceae*

1 Hole, 1976
2 Ayers Cleaners, per Common Ground, London
3 Peter Webb, Suffolk County Council
4 Stevens and Carlisle, 1959; Bunce and Jeffers, 1977
5 Max Sinclair, Avoncroft Museum of Building, Bromsgrove, W. Mids
6 E. Morris, Minsterley, Shrops.
7 S. Healey, Hoarstone Farm, Wribbenhall, Here.
8 Barbara Barling, Yatton, Here.
9 Andy Patmore, Forest Ranger, Salcey Forest, Northants.; P. C. McE., Berkhamsted, Herts.
10 G., Stonegrave, N. Yorks.
11 Dorothy Halliday, Wantage, Oxon.
12 Kenneth Watts, 'Scots Pine and droveways', in *Wiltshire Folk Life*, 19, 1989
13 T. J. Flemons, Luston, Here.
14 D. I. H. Johnstone, *In Search of Scotch Ale*, 1984
15 Elizabeth Bartlett, St Endellion, Corn.

Junipers, *Cupressaceae*

1 J. B. Foster, Arnside, Lancs.; also Richard Simon, Cullen, Banff.
2 O. L. Gilbert, 'Juniper in Upper Teesdale', in *Journal of Ecology*, 68, 1980
3 Ian Findlay, National Nature Reserve Warden, Upper Teesdale, Durham
4 Jennifer Raven, Dorking, Surrey
5 John R. Akeroyd, Dereham, Norf., in *BSBI News*, 59, 1991
6 Ratcliffe, 1977
7 Dr Brian Moffat, Fala Village, Lothian
8 Vickery, 1995
9 Grigson, 1955
10 Wendy Morgans, Kington, Here.
11 Clare Mahaddie, Milton Keynes, Bucks.

Monkey-puzzles, *Araucariaceae*

1 Wilfrid Blunt, *In for a Penny: A Prospect of Kew Gardens*, 1978

Yew, *Taxaceae*

1 Wilkinson, 1973
2 Cornish, 1946
3 Robert Turner, *Botanologia: The Brittish Physician*, 1664
4 Aubrey, 1685
5 Sir Thomas Browne, *Hydriotaphia*, 1658
6 White, 1789; also White, 1950
7 Robert Hardy, *Longbow*, 1976
8 Friend, 1883
9 Leather, 1912
10 Cornish, *op. cit.*
11 Milner, 1992
12 White, 1950
13 Allen Meredith, Bushey, Herts.; also Allen Meredith, *Touchwood* (in preparation); Milner, 1992
14 Wordsworth, 'Yew-Trees', composed 1803
15 L. Harris, Low Lorton, Cumbr.; Scott Henderson, Keswick, Cumbr.
16 Jill Burton, *The Element Iw in Cheshire Place-Names*, 1987
17 Ruth Ward, Culham, Oxon.
18 Madeline Reader, Horney Common, E. Susx; also C. M. Maudslay, Duddleswell, E. Susx
19 Susan Cowdy, The Lee, Bucks.
20 Julia Upton, Derbyshire County Council
21 Professor G. I. Ingram, Hastingleigh, Kent
22 Ida Turley, Ty Gwyn, Clwyd
23 James Anderson, *The Selborne Yew*, 1993
24 David Allen and Sue Anderson, *The Great Yew Excavation: February 1990*, Hampshire County Museums Service Interim Report, 1990
25 White, 1789

Churches and churchyards

1 Pevsner, 1945
2 Greenoak, 1985
3 John Spencer, Churchwarden, Studley, Warw.

Birthworts, *Aristolochiaceae*
1 Primrose Warburg, Oxford, Oxon.
2 Kevin and Susie White, Hexham, Northum.
3 Ruth Ward, Culham, Oxon.; Julie Meech, Lower Broadheath, Worcs.
4 Preston, 1888

Water-lilies, *Nymphaeaceae*
1 Cave, 1948, quoted by Grigson, 1955
2 William Cowper, 'The Dog and the Water-Lily: No Fable'

Buttercups, *Ranunculaceae*
1 Dr Larch Garrad, Manx Museum, Douglas, I. of M.
2 Mr Wilkinson, Clitheroe, Lancs.
3 Simpson, 1982
4 White, 1789
5 Martin Cragg-Barber, Chippenham, Wilts.
6 Mrs Cawthorne, Stanton, Suff.
7 P. Hill, Clifton, S. Yorks.; D. W. Hingley, Rotheram, S. Yorks.
8 Pratt, 1857
9 *Independent*, 29 September 1993
10 Grigson, 1955
11 Spencer, 1990
12 K. Martin, Stanley Common, Derby.
13 Smith and Sowerby, 1849–54
14 Robinson, 1870
15 Grainger, 1983
16 Lavin and Wilmore, 1994
17 Gerard, 1597
18 White, 1986–89
19 Mrs Gibson-Poole, High Salvington, W. Susx
20 Jill Lucas, Fixby, W. Yorks.; also Pevsner, 1945
21 Mary and Eric Humphries, Woodford, Northants.
22 *Transactions of the Suffolk Naturalists' Society*, 30, 1994
23 Grigson, 1955
24 Sonia Holland (ed.), *Badgeworth Nature Reserve Handbook*, 1977; Edgar Milne-Redhead, Great Horkesley, Essex
25 Perring and Farrell, 1983
26 White, 1795
27 Lyte, 1578
28 Stephen Gill (ed.), *William Wordsworth*, 1984. 'To the Small Celandine' and 'To the Same Flower' were composed in the spring of 1802 and first published in *Poems in Two Volumes*, 1807
29 Frances Macdonald, Stratford upon Avon, Warw.
30 Andrew Spink, Dept of Botany, University of Glasgow
31 Curtis, 1777–99
32 Gillam, 1993
33 Jill Lucas, Fixby, W. Yorks.

Barberries, *Berberidaceae*
1 Peter Webb, Suffolk County Council; Gillam, 1993
2 Hartley, 1954
3 Caroline Giddens, Exmoor Natural History Society, Minehead, Somer.
4 Jane Arnold, Bishopstone, Wilts.

Poppies, *Papaveraceae*
1 *The Times*, 28 March 1992
2 H.W., London W1
3 Salisbury, 1961
4 Friend, 1883
5 Grigson, 1955
6 Clement Scott, *Poppy-land Papers*, facsimile edn, Christine Stockwell, 1993
7 Dakers, 1987, quoting Ivor Gurney
8 Edmund Blunden, *Undertones of War*, 1928
9 William Orpen, *An Onlooker in France 1917-19*, 1924
10 Salisbury, 1964
11 Pamela Francis, Letterston, Dyfed
12 Ruskin, *Proserpina*, 1874–86
13 Hazel Sumner, St Weonards, Here.; and many other contributors, including S. A. Rippin, Fforest Coalpit, Gwent, and Margaret Pilkington, Lindfield, W. Susx
14 John Presland, Winsley, Wilts., per Martin Cragg-Barber, Hullavington, Wilts.
15 *Gardeners' Chronicle*, March 1889
16 P. Tickner, Shirley, Croydon, Surrey
17 John R. Palmer, South Darenth, Kent, in *BSBI News*, 59, 1991
18 Grigson, 1955
19 Gerard, 1597
20 Dr Sue Thompson, Hawstead, Suff.; Susan Cowdy, The Lee, Bucks.; Marion Dadds, The Lee, Bucks.; Helen Weidell, Newbury, Berks.; T. J. Flemons, Luston, Here.; Jacqueline Seaborn, Evesham, Worc.

Fumitories, *Fumariaceae*
1 Coles, 1657
2 Payne, 1978

Planes, *Platanaceae*
1 Mitchell, 1974; also Hadfield, 1957
2 Peter Webb, Suffolk County Council
3 Robin Hamilton, Hitcham, Suff.

Elms, *Ulmaceae*
1 Betjeman, 1970
2 Wilkinson, 1978
3 Sheila Beosham, Greenstead Green, Essex; Roy Fussell, Chirton, Wilts.; David M. Norfitt, Coventry, W. Mids; Barbara Penman, Hever, Kent; Mark Powell, Riseley, Beds.
4 Rackham, 1986
5 Richens, 1967
6 Rackham, 1986
7 Janey Rimington, Brighton, E. Susx
8 Humberside County Council, Kingston upon Hull, Humbs.; Faith Moulin, Yatton, Avon
9 Richens, 1983
10 The Revd Edward Houston, Paulesbury, Northants.
11 Canon T. Barnard, Lichfield Cathedral, Staffs.
12 David Wall, The Cathedral School, Lichfield, Staffs.
13 J. N. Rounce, Great Walsingham, Norf.
14 Rackham, 1986

Hops, *Cannabaceae*
1 Brian Moffat, Soutra, Lothian
2 *Guardian*, 13 September 1993; Stuart Carpenter, Bishops Stortford, Herts.; Pooran Desai, Bioregional Development Group, Carshalton, Surrey
3 Madeline Reader, Horney Common, E. Susx; Sinker *et al.*, 1985 and 1991
4 Kevin and Susie White, Hexham, Northum.
5 Roy Fussell, Chirton, Wilts.
6 Frank Penfold, Arundel, W. Susx, in *Sussex Trust Newsletter*
7 David Wall, Lichfield, Staffs.
8 Pevsner, 1945
9 Gerard, 1597
10 Philip Henry Stanhope, 5th Earl Stanhope, *Life of the Rt Hon. William Pitt*, 1861–62
11 Ursula Bowlby, Ullinish, Isle of Skye

Mulberries, *Moraceae*
1 Grown by and presented to me by that great geographer David Lowenthal
2 Dony, 1967; also S. A. Rippin, Fforest Coalpit, Gwent; Kathleen Coleman, Castleford, W. Yorks.
3 Vanessa Baker, Colliers Wood, Gtr London
4 Audrey Allan, Netherthong, W. Yorks.
5 Jill Lucas, Fixby, W. Yorks.; also John Ackroyd, Dewsbury, W. Yorks.
6 Trevor Moxom, Huddersfield, W. Yorks.
7 H. G. B. Coast, Chatham, Kent; Dickson, 1991
8 Gilbert, 1992

Nettles, *Urticaceae*
1 Ekwall, 1960
2 K. A. H. Cassels, Wimbish Green, Essex
3 Rackham, 1980
4 Jack Oliver, Lockeridge, Wilts., in *BSBI News*, 63, 1993
5 Peter Marren; Mary Beith, Melness, Sutherland
6 Betty Don, Frampton Cotterell, Avon
7 Mark Powell, Riseley, Beds.
8 Antony Galton, Exeter, Devon
9 B. Johnson, Northwich, Ches.
10 D. C. Fargher, Port Erin, I. of M.
11 Christine Ashworth, Rochdale, Lancs.; also Philip Hodges, Ewloe Green, Clwyd
12 Sheila Llewellyn, Burcot, Oxon.
13 B. W. and M. Wilson, Ulverston, Cumb.
14 Margaret Bown, Lindfield, W. Susx
15 Grigson, 1955
16 Davies, 1993
17 Barbara Last, Berwick St James, Wilts., in *BSBI News*, 68, 1995; Alec Bull, East Tuddenham, Norf., in *BSBI News*, 69, 1995; Alan Showler, High Wycombe, Bucks., in *BSBI News*, 69, 1995
18 D. P. Stephens, Quethiock, Corn.

Bog-myrtles, *Myricaceae*
1 M. Haines, Linwood, Hants.
2 Gelling, 1984
3 George Y., Stonegrave, N. Yorks.
4 M. J. Yates, Whitby Naturalists' Club, Saltburn, Cleve.
5 Penny Bennett, Littleborough, Lancs.
6 Mary Beith, Melness, Sutherland; also Bridie Pursey, Elphin, Sutherland
7 Kathleen MacLeod Rodger, Elphin, Sutherland

Beeches, *Fagaceae*
1 D. Flitney, Barnet, Herts.
2 Evelyn, 1664
3 Nick Delaney, Honorary Warden, Sladden Wood, Dover District Council, Kent
4 Mrs Jackson, Caton, Lancs.
5 Jean Sharples, Eardisley, Here.
6 Susan Telfer, East Cowes, I. of W.
7 Peter Marren
8 Bob Mills, ATS Tree Surgeons, per *West Sussex County Times*
9 Michael Sumpster, Farley Hill, Berks.
10 Andy Patmore, Ranger, Salcey Forest, Northants.
11 David Morfitt, Coventry, W. Mids
12 Sue Paice, Great Barrington, Northants.
13 Zoë Upchurch, Hartley Wintney, Hants.
14 Geoff Marsh, Lytchett Matravers, Dorset
15 Evelyn Smith, Norfolk Society, Norwich
16 Brewis, 1995
17 Rackham, 1980

18 Fowles, 1979
19 Cave, 1948
20 George Hayward, Oak Apple Club, Great Wishford, Wilts.; also C. C. G. Ross, *The Story of Oak Apple Day in Wishford Magna*, 1987; Hole, 1976
21 Friend, 1883
22 Pamela Michael, Lostwithiel, Corn.
23 Glenys Lund, Moreton, Lancs.
24 Robin Ravilious, Chulmleigh, Devon
25 Morris and Perring, 1974
26 Rackham, 1980
27 Miles Hadfield, 'The oak and its legends', in Morris and Perring, 1974
28 The Revd R. M. Robertson Stone, *A Short History of the Bale Oak*, 1993
29 Roden, 1968
30 White, 1789
31 Gilpin, 1791
32 Peter Webb, Suffolk County Council
33 Anon., Ingleby Greenhow, Cleve.
34 Glenys Lund, Moreton, Lancs.
35 R. H. Mills, Regional Director, National Trust, High Wycombe, Bucks.
36 Joan Poulson, per Ruth Ward, Culham, Oxon.
37 Anthony Bayfield, Eastbourne, E. Susx
38 Phil Gates, Crook, Durham
39 John Chinery, Hayes, Middx; A. McRae, Holmbrook, Cumbr.; Roger Deakin, Mellis, Suff.; G. D. Bridges, Wiveliscombe, Somer.; Kevin Pyne, Burley, W. Yorks.; B. Stewart, Hastings, E. Susx
40 Sheila Evans, Llanfwrog, Clwyd
41 Brian Cave, Longhope, Glos.
42 John A. Dolwin, Crowborough, E. Susx, in *Quarterly Journal of Forestry*, January 1994
43 F. L. Forbes, Watford, Herts.; A. V. B. Flecchia, Croydon, Surrey
44 Dorothy Mountney, Harleston, Norf.
45 Hazel Sumner, St Weonards, Here.
46 Vera Gleed, Wooton Bassett, Wilts.

Birches, *Betulaceae*
1 Madeline Reader, Horney Common, E. Susx
2 Peter Marren
3 Loudon, 1842
4 Peter Marren
5 Barbara Mellish and Helen Beet, Handcross WI, W. Susx
6 Rackham, 1980
7 Jackson, 1971
8 Una Cosgrove, Balmaclellan, Kirkud.
9 Rackham, 1980
10 Sheila Davies, Cookham Dean, Berks.
11 Anthony Bayfield, Eastbourne, E. Susx
12 Peter Casselden, Chesham, Bucks.
13 P. J. Corben, Frome, Somer.
14 per Judy Powell, Education Officer, Suffolk Wildlife Trust
15 Dr Larch Garrad, Manx Museum, Douglas, I. of M.
16 Ian and Victoria Thomson, Bentworth, Hants.; Roy Fussell, Chirton, Wilts.
17 Tony Hare, London
18 Mark Powell, Riseley, Beds.
19 Adrian Harris, Dragon Project, London
20 Meg Game, 'Cobnuts and conservation', in *British Wildlife*, 6(6), 1995
21 Meg Game, Kent
22 Aubrey, 1685
23 quoted in Rackham, 1980
24 Alma Pyke, Great Houghton, Northants.
25 M. J. Yates, Saltburn, Cleve.
26 The Revd Peter Gilks, Upper Clatford, Hants.
27 per C. I. P. Denyer, Chief Clerk to the Queen's Remembrancer, Royal Courts of Justice, London
28 Nigel Ashby, Greatford, Lincs.
29 Maggie Colwell, Box, Glos.

Hedges
1 Rackham, 1986
2 David Morfitt, Coventry, W. Mids.
3 Pollard, Hooper and Moore, 1974
4 Ann Tate, 'Squatters' hedges break the rules', in *Countryman*, 1993
5 Vikki Forbes, Bushey, Herts.
6 Mark Purdey, Lydeard St Lawrence, Somer.

Goosefoots, *Chenopodiaceae*
1 Ekwall, 1960; also Gelling, 1984
2 Milton Luby, The Hemp Patch, Shawbury, Shrops.
3 Grigson, 1955
4 Hepburn, 1952
5 P. J. Spicer, Chichester, W. Susx
6 Roberta Blattner, Lewes, E. Susx
7 the late Jack Bishop, Blakeney, Norf.
8 'Plants, People, Places' project, Liverpool Museum
9 Judith Swarbrick, County Library, Preston, Lancs.
10 Mary Coote, Heacham, Norf.
11 Stace, 1991

Pinks, *Caryophyllaceae*
1 Barbara Penman, Hever, Kent
2 Grigson, 1955
3 Jonathan and Wendy Cox, Kingston St Mary, Somer.
4 Carol Bennett, Sprowston, Norf.
5 *Oxford English Dictionary*, 1933
6 Gerard, 1597
7 Mike Coyle, Stoke, Devon
8 H. G. B. Coast, Chatham, Kent
9 Jo Darrah, Victoria & Albert Museum, per Maureen Patterson, Enfield, Middx; Margaret Pilkington, Lindfield, W. Susx
10 Pamela Michael, Lerryn, Lostwithiel, Corn.
11 Perring and Farrell, 1983
12 Gerard, 1597

Knotweeds, *Polygonaceae*
1 Trevor Smith, Mytholmroyd, per Mrs Ellison, Mytholmroyd, W. Yorks.
2 A. Lee, Seascale, Cumb.
3 Thelma Farrer, Carlisle, Cumb.
4 Joan Nichols, Beetham, Cumb.
5 W. E. Foster, Epping, Essex
6 Robinson, 1870
7 R.W.R., Sway, Hants.
8 'Village green', in *Devon Community Council Newsletter*, December 1992
9 Robin Ravilious, Chulmleigh, Devon
10 Martin Spray, Ruardean, Glos.
11 Judith Cheney, University Botanic Garden, Cambridge
12 Gibbons, 1962
13 Maura Hazelden, Crymych, Dyfed; Sheila Evans, Ruthin, Clwyd
14 P. Lock, Kingsdown, Bristol
15 Burton, 1983
16 *ibid.*
17 David Bevan, Warden, Railway Fields Nature Reserve, Harringay, Essex
18 Hans Helbaek, 'Early crop plants in southern England', in *Proceedings of the Prehistoric Society*, 1952
19 Clare, 1964
20 L. Clay, Paignton, Devon
21 White, 1932
22 Nigel Mussett, Head of Biology, Giggleswick School, Settle, N. Yorks.
23 Kevin and Susie White, Hexham, Northum.
24 Jane Arnold, Bishopstone, Wilts.
25 B. Phillips, Romsey, Hants.
26 Ian and Victoria Thomson, Bentworth, Hants.
27 Margaret Bown, Lindfield, W. Susx

Thrifts, *Plumbaginaceae*
1 Roy Maycock, Milton Keynes, Bucks.
2 Stace, 1991
3 Grigson, 1955
4 Gerard, 1597
5 Jean Kington, Leyburn, N. Yorks.

Peonies, *Paeoniaceae*
1 Gerard, 1597
2 Gerard, 1633
3 W. L. Bowles, *Banwell Hill*, 1829, quoted by Grigson, 1955
4 Fowles, 1978
5 Gerard, 1597

St John's-worts, *Clusiaceae*
1 Hole, 1976
2 Friend, 1883
3 A. P. Mead, Kingston St Mary, Somer.
4 Burton, 1983; Gillam, 1993

Limes, *Tiliaceae*
1 Rackham, 1980
2 Edgar Milne-Redhead, Great Horkesley, Essex
3 A. J. L. Fraser, Conservation Manager, Worcestershire Nature Conservation Trust, Worcs.
4 Rackham, 1980
5 C. D. Pigott, 'Factors controlling the distribution of *Tilia cordata* at the northern limits of its geographical range, IV', in *New Phytologist*, 1989
6 David Esterley, 'Out of the ashes', *Independent Magazine*, 4 July 1992
7 Bob and Margaret Marsland, Hallwood Green, Glos.
8 Dr Donald Pigott, University Botanic Garden, Cambridge
9 Lloyd James, Stratton St Margaret, Wilts.
10 N. Owens, Corton Denham, Somerset
11 Peter Lewis, Ewell, Surrey
12 Anne McKean, Forty Hill, Middx
13 Rackham, 1976
14 Alistair Scott, Forestry Commission, Edinburgh
15 Pam Gorman, Dartington, Devon

Plants, places and names
1 Gelling, 1984; Ekwall, 1960
2 Ruth Ward, Culham, Oxon.
3 Caroline Giddens, Minehead, Somer.

Mallows, *Malvaceae*
1 Dickson and Dickson, 1988
2 J. N. Rounce, Great Walsingham, Norf.
3 Marren and Rich, 1993
4 Jean Humphrey-Gaskin, Sudbury Hill, Middx
5 Alexander Dean, Melton Constable, Norf.

Sundews, *Droseraceae*
1 Gerard, 1597
2 Dr Larch Garrad, Manx Museum, Douglas, I. of M.

Rock-roses, *Cistaceae*
1 Gillam, 1993
2 R. Wilson, in Stewart, Pearman and Preston (eds), 1994

Violets, *Violaceae*
1 Keble Martin and Fraser, 1939
2 Riddlesdell *et al.*, 1948
3 Tibble and Tibble, 1935
4 Rackham, 1975
5 Mark Powell, Riseley, Beds.
6 Gerard, 1597
7 Dorothy Mountney, Harleston, Norf.
8 Sinker *et al.*, 1985 and 1991
9 Raven and Walters, 1956
10 Mrs Seeley, Catton, Northum.
11 Mabey and Greenoak, 1983

Tamarisks, *Tamaricaceae*
1 Hepburn, 1952
2 R. M. Wickenden, Staplecross, E. Susx

White bryony, *Cucurbitaceae*
1 Forsyth, 1968
2 Pratt, 1857
3 Porta, 1558
4 Caroline Male, Halesowen, W. Mids

Willows, *Salicaceae*
1 Peter Levi, Frampton on Severn, Glos.
2 Edgar Milne-Redhead, Great Horkesley, Essex
3 *ibid.*
4 Charles Watkins, University of Nottingham, Notts.
5 Desmond Hobson, Wantage, Oxon.
6 Edgar Milne-Redhead, Great Horkesley, Essex
7 David Bleasdale, Trawden, Lancs.
8 E. Milne-Redhead, 'The B.S.B.I. Black Poplar survey, 1973–88', in *Watsonia*, 18(1), 1990
9 John Kirkpatrick, Aston on Clun, Shrops.
10 Dr Daffydd Huws, Caerffili, Clwyd; Hazel Harrison, Llanbedr, Gwynedd; Judith Fearnall, Hope, Clwyd
11 Riddelsdell *et al.*, 1948
12 Sonia C. Holland, *The Black Poplar in Gloucestershire*, 1992
13 Graham King, Norfolk County Council, Norwich
14 Peter Webb, Suffolk County Council, Ipswich
15 Barbara Wilson, Marlesford, Suff.
16 Mary Taylor, Bardwell, Suff.
17 B. Kelsey, Worstead, Norf.
18 Margaret Wingrove, Aylesbury, Bucks.
19 Susan Cowdy, The Lee, Bucks.
20 G. F. Peterken and F. M. R. Hughes, *Restoration of Floodplain Forests*, 1994 (from ICF Discussion Meeting on Forests and Water); Jonathan Spencer, *The Native Black Poplar in Britain: An Action Plan for its Conservation*, 1994
21 Mark Powell, Riseley, Beds.
22 Janet Brunswick, Weston Turville, Bucks.
23 Gerard Manley Hopkins, *Selected Poems and Prose*, ed. W. H. Gardner, 1953
24 Rackham, 1976
25 Alan Bennett, *Writing Home*, 1994
26 O. L. Gilbert, 'Regenerating Balsam Poplar ... × Black Poplar ... at a site in Leeds', in *Watsonia*, 19(3), 1993
27 Anita Jo Dunn, Charlbury, Oxon., in *BSBI News*, 66, 1994
28 Daniel Keech, Common Ground, London
29 Ida Turley, Ty Gwyn, Clwyd
30 Geoff Locke, Rockhampton, Glos.
31 William Berry, Brook Farm, Chediston, Suff.
32 Jill Goodwin, Ashmans Farm, Kelveden, Essex
33 Andrew Brockbank, Irby, Wirral, Ches.
34 Frances MacDonald, Stratford upon Avon, Warw.

Cabbages, *Brassicaceae* (or *Cruciferae*)
1 Fitter, 1945
2 Peggy Bridges, Bath, Avon
3 Julie Meech, Lower Broadheath, Worcs.
4 Ekwall, 1960; Gelling, 1984
5 Grigson, 1952
6 W. E. Foster, Epping, Essex
7 J. B. Hurry, *The Woad Plant and its Dye*, 1930
8 Edwin Lees, *Pictures of Nature*, 1856
9 Dorothy Hilton, Tewkesbury, Glos.
10 Jill Goodwin, Kelvedon, Essex; Jane Wise, Stogumber, Somer.; R. M. Wickenden, Staplecross, E. Susx
11 The Revd R. Addington, Charsfield, Suff., per Suffolk Naturalists' Society
12 Colin Jerry, Peel, I. of M.
13 Genders, 1971
14 Susan Tyler Hitchcock, *Gather ye Wild Things*, 1980
15 Ekwall, 1960
16 Evelyn, 1699
17 Mayhew, *London Labour and the London Poor*, 1851
18 Elizabeth Roy, Greatford, Lincs.
19 N. Morris, Andover, Hants.
20 Turner, 1548; Gerard, 1597
21 C. Pickett, Norfolk Society, Terrington St Clement, Norf.
22 Katherine Luto, London
23 Soyer, 1857
24 A. J. Showler and T. C. G. Rich, '*Cardamine bulbifera* ... in the British Isles', in *Watsonia*, 19(4), 1993
25 Rosemary Reynolds, St Albans, Herts.
26 Mike Coyle, Stoke, Devon; Robin Ravilious, Chulmleigh, Devon
27 Grigson, 1955
28 Eva M. Cole, Mayfield, E. Susx
29 Bessie Hancock, Farnham, Surrey; M. Marsland, Dymock, Glos.; V. P. Helme, Lugwardine, Here.; E. A. Shuck, Cutnall Green, Worcs.; M. J., Upton-by-Chester, Ches.; Judith A. Webb, Kidlington, Oxon.; Alma Pyke, Great Houghton, Northants.; H. M. Edbrooke, Shrewsbury, Shrops.; Stan Farmer, Dumf.; A. MacLean, Gourock, Renfrewshire
30 Tony Bayfield, Eastbourne, E. Susx
31 Anne Proctor, Edlesborough, Beds.
32 Account based on contributions to *BSBI News*, 56, 64 and 65, 1990, 1993 and 1994, by Nick Scott, Amble, Northum.; Trevor G. Evans, Chepstow, Gwent; Felicity Woodhead, Bournemouth, Dorset; and Simon J. Leach, Taunton, Somer.; with additional localities from Philip Oswald, Cambridge.
33 Dr Alan Baker, Sheffield University, and Professor Steve McGrath, the Institute of Arable Crop Research, Rothamstead.
34 Dony, 1967
35 Nick Sturt, West Wittering, W. Susx, in *BSBI News*, 58, 1991
36 John R. Palmer, South Darenth, Kent, in *BSBI News*, 66, 1994
37 Doris Thompson, Ramsgate, Kent
38 N. D. Mitchell, 'The status of *Brassica oleracea* ... (Wild Cabbage) in the British Isles', in *Watsonia*, 11(2), 1976
39 Roy Fussell, Chirton, Wilts.
40 Keble Martin and Fraser, 1939
41 Colin Jerry, Peel, I. of M.
42 Petch and Swann, 1968
43 Mrs Lloyd-Williams, per R. Lewis, Ty'n-y-groes, Gwyn.
44 H. G. B. Coast, Chatham, Kent
45 White, 1986–89
46 Dr Larch Garrad, Manx Museum, Douglas, I. of M.

Mignonettes, *Resedaceae*
1 Lesley Davey, Sible Hedingham, Essex
2 Ruth Ward, Culham, Oxon.
3 Jill Goodwin, Kelvedon, Essex

Heathers, *Ericaceae*
1 James Robertson
2 Gimingham, 1972
3 Hoskins, 1963
4 N. W. Norman Moore, 'The heaths of Dorset and their conservation', in *Journal of Ecology*, 50, 1963
5 Phyllis Somes, Fawley, Hants.
6 James Robertson
7 Colin Jerry, Peel, I. of M.
8 Madeleine Reader, Horney Common, E. Susx
9 M. R. Newman, St Leonards on Sea, E. Susx
10 Mary Beith, Melness, Sutherland
11 Joyce Dunn, Bridge of Allan, Stirling
12 Bruce Williams, Broomhill, Glasgow
13 *ibid.*
14 Ratcliffe, 1977
15 Liza Goddard, Farnham, Surrey; Margaret Evershed, Ewhurst, Surrey; A. P. Mead, Kingston St Mary, Somer.
16 Robin Ravilious, Chulmleigh, Devon
17 T. T. Freeston, Wellington, Somer.
18 Colin Jerry, Peel, I. of M.
19 C. Walker, Condover, Shrops.
20 Janet Preshous, Lydham, Shrops.
21 S. A. Rippin, Fforest Coalpit, Gwent
22 E. Woolrich, Trentham, Staffs.

Primroses, *Primulaceae*
1 Dorothy Gibson, Tunstall, Cranforth, Lancs.; also A. Dawson, Lochgilphead, Argyll.
2 Linda Ridgley, Warwickshire Rural Community Council, Warw.
3 Maureen Bayliss, Ollerton, Ches.
4 Christine Butcher, Holt, Wilts.
5 Lily Kelly, Handcross WI, W. Susx
6 Maggie Colwell, Box, Glos.
7 Margaret Trevillion, Germoe Churchtown, Corn.
8 Susan Cowdy, The Lee, Bucks.
9 J. Borough, Orleton, Shrops.
10 Rackham, 1980
11 P. Nash, Hele, Devon
12 T. Hull *et al.*, 'Primrose Picking in South Devon: The Social, Environmental and Biological Background', in *Nature in Devon*, 3, 1982
13 R. D'O. Good, 'On the distribution of the primrose in a southern county', in *Naturalist*, 809, 1944
14 Rackham, 1980
15 John Richards, University of Newcastle upon Tyne, in *BSBI News*, 60, 1992
16 H. M. Porteous, Leafields Farm, Shut Green, Staffs; also Jo Pasco, Tarewaste, Corn.

17 Gerard, 1597
18 Mrs Benyon, The Lee, Bucks.
19 Ray, 1660
20 Sowerby, 1790–1814
21 Henry Doubleday, in *The Phytologist*, 1, 1842; see also Jermyn, 1974
22 Mabey and Evans, 1980
23 Ian Hickling, Compton, Berks.
24 Susan Telfer, East Cowes, I. of W.
25 Anne Peyton, Lambley, Notts.
26 Genders, 1971
27 Ian Hickling, Compton, Berks.
28 Grigson, 1959
29 Hugh McAllister, Ness Gardens, Wirral, Ches.
30 E. Chaplin, Sunbury-on-Thames, Middx; P. Carter, Shaftesbury, Dorset
31 Dorothy Mountney, Harleston, Norf.

Spring festivals
1 Hole, 1976
2 E. H. W. Crusha, Charlton-on-Otmoor, Oxon.
3 Ruth Wheeler, 'The wild flower garlands', in *The Bampton Beam*, April 1993; also Ruth Ward, Culham, Oxon.
4 George Herbert, *A Priest to the Temple*, 1652

Gooseberries, *Grossulariaceae*
1 Richard Simon, Cullen, Banff.
2 Mabey and Greenoak, 1983
3 John Dossett-Davies, Witney, Oxon.
4 Ellis, 1965
5 Penny Hands, Weasenham St Peter, Norf.; also John R. Turner, Ely, Cambs.

Stonecrops, *Crassulaceae*
1 Robin Ravilious, Chulmleigh, Devon
2 Gerard, 1597
3 Gerard, 1633
4 Margaret Evershed, Ewhurst, Surrey
5 Mike Pratt, Cleveland Community Forest, Cleve.

Saxifrages, *Saxifragaceae*
1 Parkinson, 1640
2 *ibid.*
3 A. Wainwright, *Walks in Limestone Country*, 1970
4 Curtis, 1777–99
5 Swan, 1993
6 l'Obel, 1576; Lyte, 1578

Roses, *Rosaceae*
1 James Robertson
2 Marie Mitchell, Bolton, Lancs.
3 N. Owens, Corton Denham, Somerset
4 Raven and Walters, 1956
5 Jean Kington, Leyburn, N. Yorks.
6 Grigson, 1955
7 Richard Simon, Cullen, Banff.
8 Derek McLean, Stenton, E. Lothian
9 Mabey, 1972
10 Caroline and Peter Male, Halesowen, W. Mids
11 Dave Earl, Southport, 'Plants, People, Places' project, Liverpool Museum
12 Mr and Mrs Heard, Othery, Somer.
13 Ruth Ward, Culham, Oxon.
14 Kath Edwards, Bowdon, Ches.; Mrs N. Beresford, Bathley, Notts.
15 Hartley, 1954
16 *Evening Standard*, 8 October 1957
17 Jill Hill, Lower Peover, Ches.
18 Mike Palmer, 'Plants, People, Places', Liverpool Museum
19 Carol Bennett, Sprowston, Norf.
20 Simon Leatherdale, Forest Enterprise, Woodbridge, Suff.
21 Deering, 1738
22 Pevsner, 1945
23 Hilary Forster, Sedbury, Gwent
24 Hilda Evans, New Tredegar, Gwent
25 Dorothy Gibson, Tunstall, Lancs.
26 White, 1986–89
27 Jill Lucas, Fixby, W. Yorks.
28 Ruth Ward, Culham, Oxon.
29 Roy Fussell, Chirton, Wilts.
30 Mrs Gibson-Poole, High Salvington, W. Susx
31 White, 1986–89
32 T. J. Flemons, Luston, Here.
33 Susie White, Hexham, Northum.
34 Gerd Krüssmann, *Roses*, 1982
35 Pam Sinfield, Milton Keynes, Bucks.
36 per Ronald Blythe, Wormingford, Essex; also William T. Stearn, 'The Five Brethren of the Rose: An old botanical riddle', in *Huntia*, 2, 1965
37 Krüssmann, *op. cit.*; Beales, 1992
38 M. J. Yates, Whitby Naturalists' Club, Saltburn, Cleve.
39 Gerard, 1597
40 Dorothy Hinchclife, Murton, Cumbria, quoted in Davies, 1993
41 Dr Diane Bannister, Senior Information Officer, Boots Contract Manufacturing, Nottingham
42 Ministry of Food, 1943
43 Elizabeth Mellor, Haverhill, Suff.; S. Robson, Ripley, Surrey
44 J. N. Rounce, Great Walsingham, Norf.
45 Colin McLeod, Dundee
46 Evangeline Dickson, Westerfield, Suff.
47 Margaret Crichton, Helensburgh, Dunbarton.; G. Tinkler, Flimby, Cumb.; Michael Bradford, Penwithick, Corn.
48 A. E. Burrows, Dringhouses, York
49 Ray Veerman, Horticultural Adviser to the Crime Prevention Office, Essex Police
50 John Prince, Milton Keynes, Bucks.
51 Goody, 1993
52 Allan Harris, Colinsburgh, Fife
53 Catherine Bennett, Llys-y-frân, Dyfed
54 Mabey and Evans, 1980
55 Jean Kington, Leyburn, N. Yorks.
56 Cobbett, 1825
57 Pamela Michael, Lostwithiel, Corn.
58 Jack Oliver, Lockeridge, Wilts.
59 Mark Powell, Riseley, Beds.
60 Hazel Brecknell, Quarndon, Derby.
61 Edith Boxall, Handcross WI, per Janet Masters, Nymans, W. Susx
62 Jonathan and Wendy Cox, Kingston St Mary, Somer.
63 H. G. B. Coast, Chatham, Kent
64 Ray Tabor, Hundon, Suff.
65 L. J. Day, Harlow, Essex; also Cherry Chapman, Great Yeldham, Essex
66 Anne Proctor, Edlesborough, Beds.
67 Hamish Eaton, Weston Turville, Bucks.
68 Sue Benwell, Weston Turville, Bucks.
69 E. Woolrich, Trentham, Staffs.; also G. Greaves, Eccleshall, Staffs.
70 Peter Marren
71 Nigel Slater, 'On the damson trail', in *Gardens Illustrated*, August/September 1995
72 Barbara Fell, Quainton, Bucks.
73 Rackham, 1980
74 Tom Smith, Whatcote, Warw.
75 Gordon Maclean, Frilford, Oxon.
76 Andy Jackson, 'The Plymouth Pear', in *British Wildlife*, 6(5), 1995
77 Rackham, 1986
78 Grindon, 1883
79 Dorothy Hinchcliffe, Murton, Cumb.
80 A. Garfitt, Wells, Somer.
81 Grigson, 1955
82 Peter Marren
83 Gavin Maxwell, *Raven Seek Thy Brother*, 1968
84 Dr Larch Garrad, Manx Museum, Douglas, I. of M.
85 Colin Jerry, Peel, I. of M.
86 Jackson, 1971
87 Clare, 1978
88 Patrick Roper, 'The distribution of the wild service tree ... in the British Isles', in *Watsonia*, 19(4), 1993
89 Geoff Locke, Rockhampton, Glos.
90 D. Blissett, Solihuill, W. Mids
91 Grindon, 1868
92 Richard Jackson, Brooks Green, W. Susx
93 Bromfield, 1856
94 E. M. Crampton, Tenterden, Kent
95 Howarth Greenoak and Alice Kilpatrick, Wigginton, Herts.
96 H. G. B. Coast, Chatham, Kent
97 Barbara Penman, Hever, Kent
98 Patrick Roper, Sedlescombe, W. Susx; see also Roper, 1994
99 Simon Leatherdale, Halstead, Essex
100 Nigel Ashby, Greatford, Lincs.
101 Daniel Keech, Common Ground, London
102 Judith Marshall, Gedney Dyke, Lincs.
103 B. Heath-Brown, Welwyn Garden City, Herts.
104 Roper, 1994
105 Hickin, 1971
106 Roper, 1994
107 Quentin Kay, Llanmadoc, W. Glam.
108 Evelyn, 1679
109 Stace, 1991
110 David Cann, Copplestone, Devon
111 Roper, 1994
112 Stace, 1991
113 Gerard, 1597
114 Burton, 1983; Salmon, 1931
115 Wolley-Dod, 1937
116 Bates, 1936
117 Hoskins, 1955
118 Caroline Giddens, Minehead, Somer.
119 *ibid.*; also M. Pearl Todd, Burgess Hill, W. Susx; Beryl Murfin, Rayleigh, Essex; Robert Hockley, Much Marcle, Here.; Dorothy Dixon, Cresswell, Staffs.; Margaret Fox, Tadcaster, N. Yorks.
120 Judith Allinson, Settle, N. Yorks.
121 Pauline Harris, Hagley, W. Mids
122 Anne Morris, Clwyd; also B. Bristow, Southport, Lancs.
123 Suzanne Royd-Taylor, Folkestone, Kent
124 Gillian Craig, Cambridge; Janie Clifford, Frampton on Severn, Glos.
125 Mother Mary Garson, Grace and Compassion Convent, Turners Hill, W. Susx
126 Veronica Holliss, Lumley, Hants.
127 Therise Christie, Folkestone, Kent
128 Sheila Dennison, Nympsfield, Glos.; also Dorothy Bell, Petts Wood, Kent; Peter Thornton, Eastwood, Notts.; C. M. Newman, Salisbury, Wilts.
129 Marina Warner, *Alone of All Her Sex*, 1978
130 Jennifer Westwood, Norton Subcourse, Norf.; also, Runcorn Hill Visitors' Centre, Ches.
131 Goody, 1993
132 Peter Rollason, Hitchin, Herts.

133 Rackham, 1986
134 Rackham, 1980
135 Cornish, 1941
136 Mavis Abley, Salcombe Regis, Devon
137 Cornish, *op. cit.*
138 Vickery, 1995
139 B. M. Fell, Quainton, Bucks.; Clare Mahaddie, Milton Keynes, Bucks.; H. Wilson, Houghton-le-Spring, Tyne and Wear
140 Vickery, 1995
141 *ibid.*
142 Cheshire Landscape Trust, Chester
143 Jim Beynon, Saltwells Local Nature Reserve, Dudley, W. Mids
144 Genevieve Leaper, Stonehaven, Kincard.
145 Elizabeth Telper, Selkirk, Borders
146 Ruth Watson, Haverfordwest, Dyfed
147 Dr Larch Garrad, Manx Museum, Douglas, I. of M.
148 J. A. Moulton, Ravenshead, Notts.
149 Dr John J. Evans, Secretary, West Bromwich Albion FC, W. Mids
150 Gwen Everitt, Torpoint, Corn.; also Mrs Gibson-Poole, High Salvington, W. Susx
151 Hartley, 1954
152 D. E. Allen, 'A possible scent difference between *Crataegus* species', in *Watsonia*, 13(2), 1980
153 A. D. Bradshaw, 'The significance of hawthorns', in *Hedges and Local History*, 1971
154 E. Woolrich, Trentham, Staffs.

Wild foods
1 Gailann Keville-Evans, Shirley, Hants.

Peas, *Fabaceae* (or *Leguminosae*)
1 Cobbett, 1825
2 Hadfield, 1957
3 Gerard, 1597
4 Martin Wainwright, *Guardian*, 16 May 1992; Sharron Cocker (ed.), *Talking Spanish*, 1992
5 Donald Rose, *The Flora of Wiltshire*, 1957
6 Grigson, 1955
7 Jefferies, 1885
8 Alison Rutherford, Helensburgh, Dunbarton.
9 Mick Jones, Dancers End Reserve Manager, BBONT, Oxford
10 Peter Marren
11 Loudon, 1836
12 Petch and Swann, 1968
13 Grigson, 1955
14 Salisbury, 1964
15 Joan Lancaster, Bletchley, Bucks.
16 Edward and Helene Wenis, Leonia, New Jersey, USA, in *BSBI News*, 56, 1990
17 Harry Richardson, Shoreham-by-Sea, W. Susx
18 Petch and Swann, 1968
19 North, 1967
20 E. Morris, Minsterley, Shrops.; also Janet Preshous, Lydham, Shrops.
21 Anne Sandford, Stoke St Milborough, Shrops.; Jonathan and Wendy Cox, Kingston St Mary, Somer.
22 Geoff Brown, Cockermouth, Cumb.
23 Rose Macdonald, Llangain, Dyfed; Maura Hazelden, Crymych, Dyfed
24 James Robertson
25 Diana Stannus, South Zeal, Devon
26 Mrs J. B. Foster, Arnside, Lancs.
27 Dorothy Mountney, Harleston, Norf.
28 Tony Baker, Winster, Derby.
29 Gill Perkins, Halesworth, Suff.
30 Mrs Jean Williamson, Whitby, Cleve.
31 Mike Coyle, Stoke, Devon; A. Hosier, Northchurch, Herts.
32 John Fishenden, Bolton-le-Sands, Lancs.
33 James Robertson
34 Ruth Ward, Culham, Oxon.
35 Mary A. Coburn, Harpenden, Herts.
36 George H. Whybrow, *The History of Berkamsted Common*, *c.* 1920
37 Geoff Brown, Cockermouth, Cumb.
38 Ursula Bowlby, Ullinish, Isle of Skye
39 Colin Jerry, Peel, I. of M.; also D. C. Fargher, Port Erin, I. of M.
40 per Danny Hughes, Beaford Arts Centre, Devon

Sea-buckthorns, *Eleagnaceae*
1 Petch and Swann, 1968

Purple-loosestrifes, *Lythraceae*
1 Grigson, 1955
2 Grigson, 1974

Mezereons, *Thymelaceae*
1 Miller, 1752
2 White, 1789
3 Sue Goss, Bledlow Ridge, Bucks.
4 Forsyth, 1968
5 Peter Casselden, Chesham, Bucks.

Willowherbs, *Onagraceae*
1 Wallis, 1769
2 Webb and Coleman, 1849
3 Mabey, 1985
4 Johns, 1853
5 Preston, 1888
6 Lees, 1867
7 Berkhamsted Citizens Association, Herts.
8 Riddelsdell *et al.*, 1948
9 T. G. Tutin, 'Natural factors contributing to a change in our flora', in *The Changing Flora of Britain*, J. E. Lousley (ed.), 1953
10 Salisbury, 1961
11 Susan Cowdy, The Lee, Bucks.
12 Colin Twist, A. Culverhouse, Chris Poulson, Sam Hallett, Peter Gateley, 'Plants, People, Places' project, Liverpool Museum
13 Stace, 1991
14 Portia Fincham, Todmorden, W. Yorks.
15 Helen Mason, Toddington, Beds.
16 Colin Jerry, Peel, I. of M.
17 M. Trayner, Knaresborough, N. Yorks.
18 Philip and Janet Oswald, Cambridge
19 Gerard, 1597
20 l'Obel, 1576

Dogwoods, *Cornaceae*
1 Ray Tabor, Hundon, Suff.
2 Maurice Young, Marlow, Bucks.

Mistletoe, *Viscaceae*
1 *Transactions of the Woolhope Naturalists' Field Club*, 1864
2 Hole, 1976
3 Leather, 1912
4 Grigson, 1952
5 Miller, 1737
6 Alfred Lord Tennyson, 'Morte D'Arthur', 1842
7 Blyton, 1944
8 *Transactions of the Woolhope Naturalists' Field Club*, 1870
9 Bryan Hay, Chesterfield, Derby.
10 Jill Lucas, Fixby, W. Yorks.
11 G. Wildon, Malvern, Worcs.
12 D. McKirgan, Weston-super-Mare, Somer.; Maggie Colwell, Box, Glos.
13 Graham Talbot, Maidenhead, Berks.
14 Mary Gay, Ledbury, Here.; Joan Dickinson, Twyning, Glos.; Peter Hill, Ross-on-Wye, Here.; T. J. Flemons, Luston, Here.; P. Powell, Hereford; J. Madeley, Monkhopton, Shrops.; A. P. Radford, West Bagborough, Somer.
15 Morton, 1712
16 Helen Mayo, Chepstow, Gwent; Jean Churchill, Blandford, Dorset; Jo Dunn, Charlbury, Oxon., in *BSBI News*, 70, 1995; Joy Barnes, Taunton, Somer.; R. Booty, Galleywood, Essex
17 Sinker *et al.*, 1985 and 1991
18 S. A. Rippin, Fforest Coalpit, Gwent
19 Mrs Holbrow, Yeovil, Somer., in *Plant-lore Notes and News*, April 1992
20 Simon Leatherdale, Halstead, Essex; M. Tucker, Warborough, Oxon.; Mrs Copper, Cowes, I. of W.; Constance Swain, Lostwithiel, Corn.
21 Jermyn, 1974
22 Angus Idle, Wycombe Urban Wildlife Group, High Wycombe, Bucks.
23 Pam Tickner, Shirley, Surrey
24 C. M. Langstaff, Home Farm, Winkburn, Notts.

Spindles, *Celastraceae*
1 Turner, 1568
2 Simon Leatherdale, Halstead, Essex

Hollies, *Aquifoliaceae*
1 Daniel Keech, Common Ground, London
2 Stone House Recall Group, Bishops Castle, Shrops.
3 Hazel Brecknell, Quarndon, Derby.
4 Grace Wheeldon, Bakewell, Derby., in *BSBI News*, 63, 1993
5 Simon Leatherdale, Forest Enterprise, Woodbridge, Suff.
6 Ekwall, 1960
7 Mary Jannetta, Cullen, Banff.
8 Dianne Harvey, Haddiscoe, Norf.
9 Dorothy Beck, Lydd, Kent; also G. F. Peterken and J. C. E. Hubbard, 'The shingle vegetation of southern England: the holly wood on Holmstone Beach, Dungeness', in *Journal of Ecology* 60, 1972
10 Ruth Ward, Culham, Oxon.
11 Nature Conservancy Council, 'The Food and Feeding Behaviour of Cattle and Ponies in the New Forest', 1983
12 Martin Spray, 'Holly as a fodder in England', in *Agricultural History Review*, 1981; see also J. Radley, 'Holly as a winter feed', in *Agricultural History Review*, 1961
13 Lorna Gartside, Saddleworth, Oldham, Lancs.
14 Mary Aitkenhead, Greenfield, Oldham, Lancs.
15 Geoff Boswell, Oldham, Lancs.
16 Bob and Margaret Marsland, Hallwood Green, Glos.
17 P. H. Nicholls, 'On the evolution of a forest landscape', in *Transactions of the Institute of British Geography*, 56, 1972
18 B. R. Oakley, High Yewdale Farm, Windermere, Cumb.
19 Tubbs, 1986
20 E. V. Reece, Eardiston, Worcs.
21 A. Rich, Whitnell Farm, Binegar, Somer.
22 Mrs Mary Taylor, Botton Head, Taham Fells, Lancs.

23 Paul Ottard, Walford College of Agriculture, Shrops.
24 Norman Chalk, North Walsham, Norf.
25 Stan Tanner, Dumf. and Galwy
26 P. J. Rollason, Hitchin, Herts.
27 'The Ratcatcher', Basingstoke, Hants.
28 Susan Gogarty, Nettlecombe, I. of W.
29 Norma D'Lemos, Truro, Corn.; also Jill Hill, Lower Peover, Ches.
30 G. MacKay Smith, Rothesay, Isle of Bute
31 T. T. Freston, Wellington, Somer.
32 Ken Procter, Wilmslow, Ches.
33 Bill Shephard, Newport, I. of W.
34 Justin Brooke, Marazion, Corn.
35 Tess Baker, Legant, Corn.; Ronald Blythe, Wormingford, Essex; Geoff Marsh, Lytchett Matravers, Dorset
36 Dallimore, 1908
37 Margaret Stump, Berkeley, Glos.
38 Simon Leatherdale, Forest Enterprise, Woodbridge, Suff.; C. Walker, Condover, Shrops.
39 Susan Cowdy, The Lee, Bucks.
40 Madeline Reader, Horney Common, E. Susx
41 The Revd F. Vere Hodge, Glastonbury, Somer.
42 Rosemary Teverson, Corn.
43 P. Mead, Kingston St Mary, Somer.
44 B. P. Major, Huntington, N. Yorks.
45 Patty Jackson, Mold, Clwyd
46 Elspeth Wrigley, Holden Clough, Lancs.
47 D. G. Grant, Lee on Solent, Hants.
48 Alwen Byer, Shrewsbury, Shrops.
49 Beth Howell, Walkden, Lancs.
50 Rosy Berry, Marlborough, Wilts.
51 Sue Jenkins, Leeds, W. Yorks.

Midwinter greenery

1 Paul Jackson, Aberystwyth, Dyfed
2 Colin Jerry, Peel, I. of M.

Box, *Buxaceae*

1 Ekwall, 1960; Gelling, 1984
2 John Bullman, Parish Council, Boxford, Suff.; J. F. Wilkinson, Box, Glos.
3 Rackham, 1986
4 Tony Harwood, Boxley Parish Councillor, Maidstone, Kent
5 Evelyn, 1706
6 Gilpin, 1808
7 C. J. L. Farmer, Petersfield, Hants.; the Vicar of Boxgrove, W. Susx
8 Internal documents, Nature Conservancy, Monks Wood, 1972
9 Aubrey, 1685
10 Bernard and Carla Phillips, Wells, Norf.
11 Shirley Warnes, Lillington, Warw.
12 Friend, 1883
13 'Plants, death and mourning', in *Daily Telegraph*, 1 December 1868, quoted in Vickery, 1984

Spurges, *Euphorbiaceae*

1 F. Rugman *et al*., '*Mercurialis perennis* (dog's mercury) poisoning: a case of mistaken identity', in *British Medical Journal*, 287, 24–31 December, 1983
2 'Part of a Letter from Mr. T. M. in Salop, to Mr. William Baxter, concerning the strange effects from the eating Dog-mercury ...', in *Philosophical Transactions of the Royal Society*, 203, VIII, September 1693
3 Rackham, 1980
4 Trevor James, Hertfordshire Environmental Records Centre, Hitchin
5 Ben Norman, Watchet, Somer.
6 Dr Larch Garrad, Manx Museum, Douglas, I. of M.
7 Grigson, 1955
8 Elizabeth Ringe, Graffham, W. Susx; Mary Hignett, Oswestry, Shrops.
9 Sinker *et al*., 1985 and 1991
10 Grigson, 1955; Fisher, 1982

Buckthorns, *Rhamnaceae*

1 Gerard, 1597; Lyte, 1578

Flaxes, *Linaceae*

1 Gerard, 1633

Milkworts, *Polygalaceae*

1 Grigson, 1955

Horse-chestnuts, *Hippocastanaceae*

1 Gareth Pearce, Woodlavington Primary School, Somer.
2 Cloves, 1993
3 Lyte, 1578
4 Gerard, 1597
5 Gerard, 1633
6 Mrs Healey, Wribbenhall, Here.
7 Ernest Law, *The Chestnut Avenue*, 1919
8 Helen Steinlechner, Hampton, Middx
9 Opie and Opie, 1969
10 Cloves, 1993
11 Opie and Opie, 1969
12 Nigel Mussett, Head of Biology, Giggleswick School, N. Yorks.
13 Opie and Opie, 1969
14 *ibid.*
15 Ernest Worthington, Calverton, Notts.; A. Baxter, Ilkley, W. Yorks.
16 Eleanor Nesbitt, Coventry, W. Mids

Maples, *Aceraceae*

1 Roger Lines, High Ham, Langport, Somer.
2 Mrs Linford, Watford, Herts.
3 Rackham, 1980
4 Pevsner, 1945; T. C. Burnard, Corsham, Wilts.
5 Evelyn, 1706
6 Peter Casselden, Chesham, Bucks.
7 Gerard, 1597
8 Thomas Sharp, *Oxford Replanned*, 1948
9 George Tinkler, Carlisle, Cumb.
10 D. Parsons, Countryside Officer, City of Bradford Metropolitan Council
11 Paul Lipscombe, Beaminster, Dorset
12 V. Smethurst, Penistone, S. Yorks.
13 Peter Branney, Cumbria Broadleaves, Bowness, Cumb.
14 Julia Upton, Youlgreave, Derby.
15 Miles, 1972
16 Edna Mallett, Wigginton, N. Yorks.
17 J. Morton Boyd, 'Sycamore and conservation', in *Tree News*, Summer 1993; also R. C. Steele, 'Sycamore in Britain', in *Quarterly Journal of Forestry*, April 1992
18 John Taylor, Causewayhead, per Joyce Dunn, Bridge of Allan, Stirling.
19 James Robertson; Paul Jackson, Aberystwyth, Dyfed
20 White, 1932
21 E. Crouch, Liskeard, Corn.
22 J. N. Rounce, Great Walsingham, Norf.; also J. Lawmon, Cyncoed, Cardiff; Ida Turley, Ty Gwyn, Clwyd
23 Peter Marren
24 Robin Ravilious, Chulmleigh, Devon
25 Heather Paul, Royal Botanic Gardens, Edinburgh
26 Valerie Hetherington, North Berwick

Trees-of-heaven, *Simaroubaceae*

1 Hadfield, 1957
2 Burton, 1983

Wood-sorrels, *Oxalidaceae*

1 P. Mead, Kingston St Mary, Somer.
2 Mabey, 1985
3 Evelyn, 1699
4 Dorothy Gibson, Carnforth, Lancs.
5 Michael Braithwaite, Hawick, Roxburghshire; Ida Turley, Ty Gwyn, Clwyd

Crane's-bills, *Geraniaceae*

1 Michael Braithwaite, Hawick, Roxburghshire; Alison Rutherford, Helensburgh, Dunbarton.
2 Vera Gordon, Sefton, Lancs.
3 Gerard, 1597
4 William Hudson, *Flora Anglica*, 1762
5 Ian Brown, Blythe Bridge, Staffs.; Paul Jackson, Aberystwyth, Dyfed; Ida Turley, Ty Gwyn, Clwyd
6 Graham Rice, Oundle, Northants., in *BSBI News*, 60, 1992
7 Greenoak, 1985
8 Patrick Roper, Sedlescombe, E. Susx

Balsams, *Balsaminaceae*

1 Sinker *et al*., 1985 and 1991
2 Salmon, 1931
3 Salisbury, 1964
4 A. O. Hume, in *Journal of Botany*, 1901
5 Diana Harding, Dulverton, Somer.
6 J.M., Sheffield
7 R. Butler, Maidenhead, Berks.
8 Brett Westwood, Cookley, Worcs.
9 Jo Pasco, Tarewaste, Corn.
10 Anne Emmett, Warrington, Ches.
11 Sue Goss, Bucks.; Barbara Penman, Hever, Kent.
12 Godfrey Nall, Shirley, W. Mids
13 Deirdre Barrett, Wetheringsett, Suff.

Ivies, *Araliaceae*

1 Duncan Ross, Newton Poppleford, Devon
2 Michael Morgan, Canvey Island, Essex
3 Jennifer Sandy, Gunnislake, Corn.
4 N. J. Burrell, Wareham, Dorset
5 Simon Russell, Sevenoaks, Kent
6 R. Chilver, Shaftesbury, Dorset
7 The Revd E. Pearson, Penrhos, Gwent
8 Matthew Edmonds, Midland Tree Surgeons, Sudbury, Derby.
9 Major D. A. Pudsey, The Hampshire Honey Farm, Sutton Scotney, Hants.; also M. L. Somers, Surrey Beekeepers' Association, Churt, Surrey
10 Thomas Whately, *Observations on Modern Gardening*, 1770
11 Jane Burse, Bentham, Lancs.
12 S. Barlow, Watersheddings, Lancs.
13 E. J. Saynor, King's Lynn, Norf.
14 Jacqueline Brook, Selly Park, Birmingham
15 Irene Tait, Caversham, Berks.
16 Walters, 1993
17 Glynys Morgan, Ammanford, Dyfed
18 K. W. Gilbert, Otterbourne, Hants.
19 Jill Betteridge, Cottenham, Cambs.
20 Ernie Marshall, per Michel Hughes, Chudleigh, Devon
21 Beth Veale, Herne Bay, Kent
22 Mary Sutch, Bardwell, Suff.
23 Wyn Lawrence, Farnborough, Hants.
24 Jennie Lancaster, St Helens, Lancs.
25 C. Walker, Condover, Shrops.

26 Peter Marren
27 E. Morris, Minsterley, Shrops.
28 John Pike, Copplestone, Devon
29 J. Ellis, Ventnor, I. of W.
30 C. H. Williams, Shrewsbury, Shrops.
31 P. Mann, Bristol
32 Eluned Davies, Tremeirchion, Clwyd
33 Gwen Redshaw, Rhewl, Clwyd
34 Alison Rutherford, Helensburgh, Dunbarton.
35 Opie, 1959
36 Keyte and Parrott (eds), 1992
37 Amanda, Hessle, Humbs.
38 Jill Lucas, Fixby, W. Yorks.
39 Chambers and Sidgwick, 1907
40 H. A. McAllister and A. Rutherford, '*Hedera helix* ... and *H. hibernica* ... in the British Isles', in *Watsonia*, 18(1), 1990
41 Peter Marren; also Bruce Philp, West of Scotland College of Agriculture
42 E. O. Holdsworth, Blackburn, Lancs.
43 M.L.D., Overstrand, Norf.
44 D.H., Scorton, N. Yorks.
45 Mandy Barwell-Parker and Simon Rogers, Coventry, W. Mids
46 Denis Malsher, Hanworth, Middx
47 Y. Howe, Stockbridge, Hants.
48 M. H. O. Hoddinott, Chester
49 M. Murray, Wolverhampton, W. Mids
50 Richard Stewart, Ipswich, Suff.
51 J. B. Foster, Arnside, Lancs.
52 Stephen Alton and Sarah Ausberger, Letchworth, Herts.; Jane Coles, Berrynarbor, Devon; K. Davis, Saltash, Corn.; Audrey Harrison, Launceston, Corn.; B. J. White, New Ollerton, Notts.; Penny Anderson, Chinley, Lancs.
53 Dickson, 1991
54 Peter Marren
55 Roy Fussell, Chirton, Wilts.
56 Rachel Hamilton, Hitcham, Suff.

Carrots, *Apiaceae* (or *Umbelliferae*)

1 Carol Bennett, Sprowston, Norf.
2 Mrs Gibson-Poole, High Salvington, W. Susx
3 From Witham, Essex, May 1983, per Roy Vickery, in *BSBI News*, 69, 1995
4 Frances MacDonald, Stratford upon Avon, Warw.
5 Barbara Penman, Hever, Kent
6 N.M., Datchworth, Herts.
7 M. H. Beard, Little Wilbraham, Cambs.
8 Gerard, 1597
9 The Revd Richard Addington, Charsfield, Suff.
10 Nigel Mussett, Head of Biology, Giggleswick School, N. Yorks.
11 Mike Pratt, Cleveland Community Forest, Cleve.
12 Mary Hignett, Oswestry, Shrops.
13 Colin Jerry, Peel, I. of M.; Dr Larch Garrad, Manx Museum, I. of M.
14 Peter C. Horn, Kempstone, Beds.
15 R. A. Roberts, Ironbridge, Shrops.
16 Rackham, 1980
17 per Evelyn Simpson, Weavering, Kent
18 Gerard, 1597
19 Lyte, 1578; S. A. Rippon, Fforest Coalpit, Gwent
20 'Part of a Letter from Mr. Ray, F.R.S. to Dr. Sloane, giving an Account of the Poysonous Qualities of Hemlock-Water-Drop-Wort', in *Philosophical Transactions of the Royal Society*, 238, V, March 1698
21 Lightfoot, 1777
22 Harington, 1607
23 Hilary Forster, Sedbury, Gwent
24 John Cameron, *Gaelic Names of Plants*, 1883
25 Pratt, 1857
26 Katrina Porteous, Beadnell, Northum.
27 Turner, 1548
28 Brian Wurzell, Tottenham, London, in *BSBI News*, 58, 1991
29 Clapham, Tutin and Warburg, 1952; confirmed by Anthony Galton, Exeter, Devon
30 Ruth Ward, Culham, Oxon.
31 Burton, 1983
32 S. G. Terry, Yelverton, Devon
33 Tony Hare, London
34 Kathleen Frith, FRCOG, Romford, Essex
35 The *Gardener's Magazine*, 12, 487, 1836
36 James Fenton, Elgin, Moray, in *BSBI News*, 58, 1991
37 per E. Charles Nelson, National Botanic Gardens, Dublin, in *BSBI News*, 57, 1991
38 Robinson, 1870
39 Peter Morris, Hampton-in-Arden, W. Mids
40 Allan Harris, Colinsburgh, Fife

Gentians, *Gentianaceae*

1 Druce, 1926
2 Vera Harding, Lyndhurst, Hants.

Nightshades, *Solanaceae*

1 McClintock, 1966
2 Pratt, 1857
3 Roy Maycock, Milton Keynes, Bucks.
4 Burton, 1983
5 Doris Page, Barrow-in-Furness, now of Victoria, British Columbia
6 In a note to the poem 'To the Lady Fleming', composed 1823
7 Simpson, 1982
8 Gerard, 1597
9 Barabara Last, Berwick St James, Wilts.
10 Alison Rutherford, Helensburgh, Dunbarton.
11 Peter Casselden, Chesham, Bucks.
12 S. H. R. Jackson, Broughton, Humbs.
13 *Rail News*, October 1970
14 Pechey, 1694
15 Burton, 1983
16 Julian M. H. Shaw and Belma Konuklugil, Dept of Pharmaceutical Sciences, Nottingham University, in *BSBI News*, 65, 1994

Bindweeds, *Convolvulaceae*

1 A. S. Monckton, in *Quarterly Journal of Forestry*, January 1993
2 Nancy Small, Lingwood, Norf.
3 H. Carter, Keeper of Natural History, Reading Museum, Berks.; also T. Freeston, Wellington, Somer.

Jacob's-ladders, *Polemoniaceae*

1 C. E. Raven, *John Ray, Naturalist*, 1950 and 1987

Borages, *Boraginaceae*

1 Carl Linnaeus, *Skånska Resa*, 1751, quoted in Wilfred Blunt, *The Compleat Naturalist*, 1971 and 1984
2 Mike Pratt, Cleveland Community Forest, Cleve.
3 Barbara Last, Berwick St James, Wilts.
4 Mike Coyle, Stoke, Plymouth, Devon
5 Mrs Williamson, [n.a.]
6 Mark Powell, Riseley, Beds.
7 Dr Gavin Ewan, Aylesbury, Bucks.
8 R. Lewis, Ty'n-y-groes, Gwyn.
9 Peter Crossland, Upper Denby, W. Yorks.
10 Alice Payton, via her daughter Helen Turner, Goring-on-Thames, Berks.
11 K. Copestake, The Grange Rest Home, Staffs.; Kathleen Simpson, Wilmslow, Ches.
12 C. Coiffait, Welton, Humbs.
13 Norma de Smet, Isfield, E. Susx
14 Mark Powell, Riseley, Beds.
15 Evelyn, 1699
16 Jefferies, 1883
17 Richard Simon, Cullen, Banff.
18 collected in *Sibylline Leaves*, 1817
19 quoted in Rendall, 1934
20 Marjorie Wilson, Wymington, Northants.
21 Pechey, 1694

Vervains, *Verbenaceae*

1 Richard Folkard, *Plant Lore*, 1892
2 John White, 1608
3 Colin Jerry, Peel, I. of M.
4 Francesca Greenoak, Wigginton, Herts.

Dead-nettles, *Lamiaceae*

1 Gerard, 1633
2 White, 1986–89
3 Druce, 1886
4 P. R. Marren, 'The past and present distribution of *Stachys germanica* L. in Britain', in *Watsonia*, 17(1), 1988
5 Ray Tabor, Hundon, Suff.
6 Peter Marren
7 Nicolette Goddard, Rushden, Northants.
8 Hall, 1980
9 Gerard, 1597
10 T. Law, London
11 Threlkeld, 1727
12 Mary Beith, Melness, Sutherland
13 William Sole, *Menthae Britannicae*, 1798
14 Margaret Evershed, Ewhurst, Surrey

Plantains, *Plantaginaceae*

1 John Josselyn, *New-Englands Rarities Discovered*, 1672
2 J. H. G. Grattan and C. Singer, *Anglo-Saxon Magic and Medicine*, 1952. Version in Grigson, 1955
3 T. T. Freeston, Wellington, Somer.
4 Elizabeth Telper, Selkirk, Borders
5 The Church of Christ's Women's Fellowship, Selston, Derby.
6 Pam Gorman, Dartington, Devon; University of Sussex Natural History evening class, per Margaret Pilkington, Lindfield, W. Susx
7 H. G. B. Coast, Chatham, Kent
8 Peter Marren
9 Elizabeth Telper, Selkirk, Borders
10 Margaret Bennett, School of Scottish Studies, Edinburgh
11 C. Walker, Condover, Shrops.

Butterfly-bush, *Buddlejaceae*

1 Walters, 1993
2 M. H. O. Hoddinott, Chester
3 Gilbert, 1993

Urban commons

1 Richard Mabey, *The Unofficial Countryside*, 1973
2 Gilbert, 1993

Ashes, *Oleaceae*

1 White, 1789
2 Rackham, 1980
3 Katrina Porteous, Beadnell, Northum.
4 Alistair Scott, Forestry Commission, Edinburgh

5 Franklyn Perring, Oundle, Northants.; Roger Deakin, Mellis, Suff.
6 Eleanour Brown, Coulter, Lanark.
7 J. Roberts, Rhiw, Gwyn.
8 Roy Fussell, Chirton, Wilts.
9 Simon Leatherdale, Forest Enterprise, Woodbridge, Suff.
10 Jane Allen, Hawick, Roxburghshire
11 Nancy Roberts, Weybridge, Surrey
12 Terry West, Cranbrook, Kent
13 Rackham, 1980
14 Lynn Fomison, Ropley, Hants.
15 Desiree Merican, Shoreham-by-Sea, W. Susx
16 Mabey, 1985
17 Ellen Slade, Eastbridge, Kent

Figworts, *Scrophulariaceae*
1 Catherine Bennett, Llys-y-frân, Dyfed
2 Parkinson, 1640
3 Friend, 1883
4 William Baxter, *British Phaenogamous Botany*, 1833–43
5 Anon., Runcorn Visitors' Centre, Ches.
6 Hazel Sumner, St Weonards, Here.
7 Gwen Redshaw, Rhewl, Clwyd; also Diana Harding, Dulverton, Somer.
8 Davies, 1993
9 T. J. Flemon, Luston, Here.
10 per Northamptonshire Wildlife Trust, Oundle Branch
11 M. E. Braithwaite, Hawick, Roxburghshire
12 Anne McKean, Forty Hill, Enfield, Middx
13 quoted in Rendall, 1934
14 Mabey (ed.), 1990
15 Grigson, 1955
16 Perring and Farrell, 1983
17 John Vaughan, *Flowers of the Field*, 1906

Broomrapes, *Orobanchaceae*
1 Sheila Beosham, Greenstead Green, Essex
2 White, 1986–89

Bellflowers, *Campanulaceae*
1 Friend, 1883
2 Tony Bayfield, Eastbourne, E. Susx
3 Grigson, 1955
4 Gerard, 1633; W. Legge, Donhead St Andrew, Dorset
5 Hall, 1980
6 Richard Fitter, Chinnor, Oxon., in *BSBI News*, 66, 1994
7 Roy Maycock, Milton Keynes, Bucks.

Bedstraws, *Rubiaceae*
1 Genders, 1971
2 Rendall, 1934
3 James Burnett, Reading, Berks.
4 Betty Dow, Frampton Coterell, Avon
5 Peter Marren
6 Vera Burden, Great Missenden, Bucks.
7 Anne Berens, Shipton-under-Wychwood, Oxon.
8 M. Burden, Winshill, Staffs.

Honeysuckles, *Caprifoliaceae*
1 Thomas Hale, *A Compleat Body of Husbandery*, 1756
2 Evelyn, 1706
3 Robert Chambers, *Popular Rhymes of Scotland*, 1847
4 Christine Ashworth, Rochdale, Lancs.
5 Barbara Penman, Hever, Kent
6 *Folklore*, 104, 1993
7 A. P. Mead, Kingston St Mary, Somer.
8 D. C. Fargher, Port Erin, I. of M.
9 Maura Hazelsdean, Crymych, Dyfed
10 T. J. Flemons, Luston, Here.
11 Joyce Dunkley, Market Harborough, Leic.
12 Duff Hart-Davis, *Independent*, 26 June 1993
13 S. A. Rippin, Fforest Coalpit, Gwent
14 Lady Statham, Reigate, Surrey
15 Jane Belsey, Rock, Worcs.
16 Katrina Porteous, Beadnell, Northum.
17 F. H. Barratt, Cottingham, Humbs.
18 Colin Jerry, Peel, I. of M.
19 S. G. Terry, Yelverton, Devon
20 Alastair Scott, Forestry Commission, Edinburgh
21 Martin Spray, Ruardean, Glos.
22 G. H. Knight, 'Tree with a future', in *Countryman Wildlife Book*, 1969
23 Parkinson, 1640
24 K. G. Messenger, Uppingham, Leic.
25 *Sussex Life*, January 1985
26 Daphne Cooper, Shrewsbury, Shrops.
27 Tony Bayfield, Eastbourne, E. Susx
28 Bulleyn, 1562
29 Grindon, 1883

Valerians, *Valerianaceae*
1 George Griffin, per F. J. Cockersole, Cockermouth, Cumb.
2 Hudson, 1762

Teasels, *Dipsacaceae*
1 Kevin and Susie White, Hexham, Northum.
2 Mrs and Mrs Heard, Othery, Somer.; Betty Don, Frampton Cotterell, Avon; Rollo and Janie Clifford, Frampton on Severn, Glos.; Anne Jeffery, Stroud, Glos.; Ruth Ward, Culham, Oxon.
3 The Revd Lynne Mayers, Anfield, Liverpool
4 Hilary Forster, Sedbury, Gwent
5 Jefferies, 1879
6 Deering, 1738
7 Hazel Wilson, Abergele, Clwyd
8 *BSBI News*, 63, 1993
9 Brian Wurzell, Tottenham, London, in *BSBI News*, 62, 1992

Daisies, *Asteraceae* (or *Compositae*)
1 Tony and Faith Moulin, Yatton, Bristol
2 Peter Gateley, Maghull, Liverpool, per 'Plants, People, Places' project, Liverpool Museum
3 The Curator, City of Edinburgh Museum
4 Hole, 1976
5 Mark Powell, Riseley, Beds.
6 A. G. Barr (Soft Drinks) plc
7 Hilda Evans, New Tredegar, Gwent
8 School of Scottish Studies, Edinburgh
9 Elaine R. Bullard, Kirkwall, Orkney, in *BSBI News*, 60, 1992
10 J. Mowbray, Canonbie, Dumf.
11 Caroline Smedley, Newton Regis, Staffs.
12 Mike Coyle, Stoke, Devon
13 Clare, 1964
14 Peggie Pittkin, Pershore, Worcs., in *BSBI News*, 58, 1991
15 Salisbury, 1961
16 Dorothy Knowles, Liverpool
17 Gillam, 1993
18 Godfrey Nall, Shirley, W. Mids
19 John Keats, 'I Stood Tip-toe Upon a Little Hill', 1817
20 Children of Needham Market School, Suff.
21 Julia Upton, Youlgreave, Derby.
22 Margaret Evershed, Ewhurst, Surrey
23 C. J. Peat, Carlton, Warw.
24 Jack Boyce, Soham, Cambs.
25 Julia Sterling, Chelmondiston, Suff.
26 Mike Coyle, Stoke, Devon
27 Roy Maycock, Milton Keynes, Bucks.
28 Jane Arnold, Bishopstone, Wilts.
29 Pechey, 1694
30 Mary Hignett, Oswestry, Shrops.
31 Tony Hare, 'Lesser Fleabane: a plant of seasonal hollows', in *British Wildlife*, 2(2), 1990
32 Gerard, 1597
33 Robinson, 1870
34 Peter J. Cook, Withernsea, Humbs., *BSBI News*, 66, 1994
35 B. Wurzell, '*Conyza sumatrensis* naturalized in southern England', in *Watsonia*, 15(4), 1985; Mick Crawley, Imperial College, Ascot, Berks., in *BSBI News*, 68, 1995
36 Grainger, 1983
37 Seren Hathaway, Gwernaffield, Clwyd; also Hilary Foster, Sedbury, Gwent
38 S. A. Rippin, Fforest Coalpit, Gwent
39 John, Herne Bay, Kent
40 Opie and Opie, 1959
41 Seren Hathway, Gwernaffield, Clwyd
42 E. Reeves, Sidcup, Kent
43 William Hazlitt, *Lectures on the English Poets*, 1819
44 Beryl Haynes, Handcross, W. Susx
45 Geoffrey Chaucer, Prologue to *The Legend of Good Women*, Text F, lines 40–8 and 179–87
46 E. S. Johnson *et al.*, 'Efficacy of feverfew as prophylactic treatment of migraine', in *British Medical Journal*, 31 August 1985
47 J. Farrow, Martham, Norf.; Anne Fowler, Llanbedr, Gwyn.
48 Dr J. D. S. Birks, Malvern, Worcs.
49 Marie Mitchell, Bolton, Lancs.
50 Thomas Tusser, *Fiue Hundreth Points of Good Husbandry*, 1573
51 Pattie Barron, Bath
52 Val Gateley, 'Plants, People, Places' project, Liverpool Museum
53 Margaret Trevillion, Germoe, Corn.
54 Nancy Girdler, Hurst, Berks.; also Christine Butcher, Holt, Wilts.
55 Dr Larch Garrad, Manx Museum, Douglas, I. of M.
56 Colin Jerry, Peel, I. of M.
57 Friend, 1883
58 Sue Thompson, Hawstead, Suff.
59 Thomas Hill, *The Gardeners Labyrinth*, 1577 (edited with an introduction by Richard Mabey, 1987 and 1988)
60 Heather R. Winship, 'Chamomile – The herb of humility in demise', in *British Wildlife*, 5(3), 1994
61 Ekwall, 1960
62 Q. Kay and G. Hutchinson, in *BSBI News*, 63, 1993; Salisbury, 1961
63 Salisbury, 1964
64 Kathleen Simpson, Wilmslow, Ches.
65 Ekwall, 1960
66 quoted in Rendall, 1934
67 Meg Stevens, Llanfrynach, Powys
68 W. Teasdale, Corby Hill, Cumb.
69 Janet White, Great Totham, Essex
70 Ray, 1660
71 P. D. Sell, 'The rediscovery of the Fen Ragwort in Cambridgeshire', in *Nature in Cambridgeshire*, 16, 1973; S. M. Walters, 'The re-discovery of *Senecio paludosus* L. in Britain', in *Watsonia*, 10(1), 1974
72 Forsyth, 1968
73 Bryn Celan, ADAS, Newtown, Powys
74 C. Walker, Condover, Shrops.; Ursula Bowlby, Ullinish, Isle of Skye

75 Robin Page, *Daily Telegraph*, 10 September 1993
76 'The Ragwort', in Clare, 1978
77 Colin Jerry, Peel, I. of M.; D. C. Fargher, Port Erin, I. of M.
78 Druce, 1886
79 Druce, 1927
80 Peter Casselden, Chesham, Bucks.
81 S. A. Rippin, Fforest Coalpit, Gwent
82 J. E. Lousley, 'A new hybrid *Senecio* from the London area', in *Report of the Botanical Society and Exchange Club of the British Isles*, 12, 1946
83 Ray, 1660
84 Bulleyn, 1562
85 A. P. Mead, Kingston St Mary, Somer.
86 Margaret Lovel Graham, Headley, Hants.
87 R. M. Wickenden, Staplecross, E. Susx
88 Gerard, 1597
89 Jane Allan, Hawick, Roxburghshire; Chris Alsop, Buckley, Clwyd
90 Pamela Michael, Lostwithiel, Corn.
91 Edward Deville, Hertford, Herts.
92 E. Woolrich, Trentham, Staffs.
93 Norma D'Lemos, Truro, Corn.
94 Brian Morris, 'The pragmatics of folk classification', in Vickery, 1984

Flowering-rushes, *Butomaceae*
1 Grigson, 1955

Water-plantains, *Alismataceae*
1 See discussion in Peter Fuller, *Theoria*, 1988
2 Salmon, 1931
3 Marren and Rich, 1993

Frogbits, *Hydrocharitaceae*
1 Druce, 1886

Lords-and-ladies, *Araceae*
1 Letter to Christopher Merret, 1668, in *Letters of Sir Thomas Browne*, G. Keynes (ed.), 1931
2 D. Moore, Solihull, W. Mids
3 Paul Jackson, Aberystwyth, Dyfed
4 Carol Bennett, Sprowston, Norf.
5 Sheila Evans, Llanfwrog, Clwyd
6 Barbara Penman, Hever, Kent
7 Rosemary Teverson, Cornwall Trust for Nature Conservation
8 E. M. Porter, *Cambridgeshire Customs and Folklore*, 1969
9 Edward Lear, *The Cretan Journal*, Rowena Fowler (ed.), 1984
10 Mary Miller, Kensington, London

Duckweeds, *Lemnaceae*
1 Mrs Ethel Kerry, Irby, Wirral, Lancs., in Roy Vickery, in *Folklore*, 94(2), 1983
2 Mrs D. Wakeham, Bebington, Lancs., in Roy Vickery, *op. cit.*
3 Dr Jack Oliver, Loveridge, Wilts.

Rushes, *Juncaceae*
1 William Cobbett, *Cottage Economy*, 1823
2 White, 1789
3 Mabey, 1983
4 Ida Turley, Ty Gwyn, Clwyd
5 Simon Leatherdale, Halstead, Essex
6 Jonathan and Wendy Cox, Kingston St Mary, Somer.
7 Philip Oswald, Cambridge
8 I. O. Jones, Ashley Heath, Shrop.
9 Martin Spray, Ruardean, Glos.
10 Edgar Milne-Redhead, Great Horkesley, Essex
11 T. E. C. Bird, Alderley Edge, Ches.

Sedges, *Cyperaceae*
1 Margaret Pilkington, Lindfield, W. Susx
2 Bridie Pursey, Elphin, Sutherland
3 Maureen Richardson, Whitney on Wye, Here.
4 Hole, 1976
5 Janet Preshous, Lydham, Shrop.
6 Nick Barnes, Tideswell, Derby.
7 Mrs Russell, Trinity House, Hull, Humbs.
8 Sally Hall, Herodsfoot, Corn.
9 T. A. Rowell, 'Sedge in Cambridgeshire: its use, production and value', in *Agricultural History Review*, 34, 1986
10 Richard Hobbs, Norfolk Naturalists' Trust; also Ellis, 1965

Grasses, *Poaceae* (or *Gramineae*)
1 Dorothy Mountney, Harleston, Norf.
2 Sandra Woodman, Clutton. Avon
3 Jane Hall, Romsey, Hants.; also Caroline Smedley, Newton Regis, Warw.; Harry Triggs, Gwern-y-Brenin, Shrop.; Anne Newcombe, Churchstoke, Powys; Jonathan Curry, Crosspool, Sheffield
4 Hugh McAllister, Deputy Director, Ness Gardens, Ches.
5 Francesca Greenoak, Wigginton, Herts.
6 Bill Chope, Baden Powell Scouts Association, King's Heath, Birmingham
7 Friend, 1883
8 Grigson, 1974; C. E. Hubbard, *Grasses*, 3rd edn, revised by J. C. E. Hubbard, 1984
9 James Robertson
10 Hartley, 1939
11 Elizabeth and Rachel Stevens, Fleetwood, Lancs.
12 Cameron Crook, Preston, Lancs.; Aaron Woods, Milton Keynes, Bucks.; also Ray Woods, Llandindrod Wells, Powys, per Mary Briggs, in *BSBI News*, 60, 1992
13 Grigson, 1974
14 Pauline Conder, West Cross, Swansea
15 Joan Johnson, Guildford, Surrey
16 'John', Herne Bay, Kent
17 Jonathan and Wendy Cox, Kingston St Mary, Somer.
18 Caroline Giddens, Exmoor Natural History Society, Minehead, Somer.
19 A. P. Mead, Kingston St Mary, Somer.
20 Caroline and Peter Male, Halesowen, W. Mids.
21 Tony Hare, London

Meadows
1 Andrew Fraser, Conservation and Education Manager, Worcester Nature Conservation Trust, Hindlip, Here.
2 Preface to Clare, 1978

Bulrushes, *Typhaceae*
1 McClintock, 1966
2 Margaret Wheat, *Survival Arts of the Primitive Paiutes*, 1967

Lilies, *Liliaceae*
1 Grigson, 1955
2 S. A. Rippin, Fforest Coalpit, Gwent; also T. J. Flemons, Luston, Here.
3 Bryn Celan, ADAS, Newtown, Powys
4 Ruth Ward, Culham, Oxon.
5 Gillam, 1993
6 Perring and Farrell, 1983
7 Dony, 1967
8 Simpson, 1982
9 Vita Sackville-West, *The Land*, 1926
10 Grigson, 1955
11 Gerard, 1597
12 Philip Oswald, 'The Fritillary in Britain. A historical perspective', in *British Wildlife*, 3(4), 1992
13 Kent, 1975; J. Blackstone, *Fasciculus Plantarum circa Harefield sponte nascentium*, 1737
14 Derek Wells, Hilton, Cambs.
15 Ruth Ward, Culham, Oxon.
16 *ibid.*
17 Andrew Young, *A Prospect of Flowers*, 1945
18 E. Duffey *et al.*, *Grassland Ecology and Wildlife Management*, 1974
19 Grigson, 1955
20 Leonard Bull, Bainbridge, N. Yorks.
21 A. T. Woodward, Stone, Bucks., per Peter Knipe, Dorchester, Oxon.
22 Elizabeth Evans, Kinoulton, Notts.
23 G. T. Hartley, *Some Notes on the Parish of Lapley-cum-Wheaton Aston*, per Ken Harris, Wheaton Aston, Staffs.
24 Caroline Smedley, Newton Regis, Staffs.
25 F. W. Simpson, 'Mickfield Fritillary Meadow, East Suffolk', in *SPNR Handbook*, 1938
26 C. Henry Warren, *Content with What I Have*, 1967
27 Mabey and Evans, 1980
28 Philip Oswald, *op. cit.*
29 Robin Ravilious, Chulmleigh, Devon; D. J. and K. O'Connor, Exeter, Devon
30 Druce, 1926
31 Kent, 1975; Gerard, 1597
32 James Bolam and Susan Jameson, Wisborough Green, W. Susx
33 Keble Martin, 1972; Perring and Farrell, 1983
34 Jean Kington, Leyburn, N. Yorks.
35 Wynne, 1993
36 Greenoak, 1985
37 Gillam, 1993
38 Clare, 1978
39 *The Journals and Papers of Gerard Manley Hopkins*, ed. H. House, completed by G. Storey, 1959
40 Francesca Greenoak, Wigginton, Herts.
41 Caroline and Peter Male, Halesowen, W. Mids
42 Ruth Ward, Culham, Oxon.
43 Helen Proctor, Upper Dicker, E. Susx
44 Jane Thompson, Markfield, Leics.
45 Joan Lancaster, Bletchley, Bucks.
46 Martin Jenkins and Sara Oldfield, *Wild Plants in Trade*, 1992
47 F. H. Perring, in *BSBI News*, 69, 1995; *Bluebell Signs on British Inns*, BSBI report [1994]
48 P. J. O. Trist (ed.), *An Ecological Flora of Breckland*, 1979
49 Ronald Blythe, Wormingford, Essex
50 Gelling, 1984
51 Colin Jerry, Peel, I. of M.
52 N. McArdle, Winterley, Ches.
53 John and Anna Taylor, Kirmond le Mire, Lincs.
54 Jane Halle, Romsey, Hants.
55 William Condry, *The Natural History of Wales*, 1981
56 Barbara Grover, Penzance, Corn.; also Pamela Michael, Lostwithiel, Corn.; Philip Hodges, Ewloe Green, Clwyd
57 Rackham, 1980
58 Susie White, Hexham, Northum.
59 Len and Pat Livermore, Lancaster, in *BSBI News*, 60, 1992
60 Keble Martin and Fraser, 1939
61 Michael Braithwaite, Hawick, Roxburghshire

62 L. C. Frost *et al.*, '*Allium sphaerocephalon* L. and introduced *A. carinatum* L., *A. roseum* L. and *Nectaroscordum siculum* (Ucria) Lindley on St Vincent's Rocks, Avon Gorge, Bristol', in *Watsonia*, 18(4), 1991
63 Gerard, 1597
64 Curtis, 1777–99
65 Druce, 1897
66 Ruth Ward, Culham, Oxon.
67 Denny Ingram, Huntspill, Somer.; P. M. Sharp, Upton, Somer.; Hilary Stephenson, Bicton Heath, Shrops.
68 William Withering, *A Botanical Arrangement of British Plants*, 1776; *Oxford English Dictionary*, 1933; Gerard, 1633 and 1597
69 Eileen Plume, Leominster, Here.
70 Mrs Legge, Donhead St Andrew, Dorset
71 Eddie Smith, Wraysbury, Middlesex
72 Irene Payne, Donnington, Berks.
73 Grace Moynan, Dover, Kent
74 P. Hill, Clifton, N. Yorks.; Patrick King, Walsingham, Norf.; G. E. Watts, Abbotskerswell, Devon; Eva James, Carmlington, Northum.; G. B. and A. M. Maddison, Hainault, Essex
75 A. Knapton, Reighton, N. Yorks.
76 V. M. Cook, Park Gate, Southampton, Hants.; also Joan D. Owen, Fareham, Hants.
77 Valerie Burgess, Sible Hedingham, Essex; Sallie Duckitt, Kirk Bramwith Parochial Church Council, Moss, S. Yorks.
78 M. W. B. Hunter, Lower Bourne, Surrey; Jacquie Moon, Puddingstone, Ches.; Mandy Archer, Kirton, Suff.; L. Mould, Southampton, Hants.; Margaret Hale, Sandown, I. of W.; Patricia D. Newnes, Chepstow, Gwent; E.D., London; Margaret Long, Wadsley, S. Yorks.
79 Ingrid Foster, Monkton, Wilts.
80 Mary Hignett, Oswestry, Shrops.
81 Pamela Michael, Lostwithiel, Corn.
82 Roy Fussell, Chirton, Wilts.
83 Clare Jones, Norwich, Norf.
84 Webb and Coleman, 1849
85 per Ruth Ward, Culham, Oxon.
86 H. Whiting, Redhill, Surrey
87 Friend, 1883
88 Mary A. Coburn, Harpenden, Herts.
89 Viv Street, Hemsworth, W. Yorks.
90 M. O'Sullivan, Middleton St George, Durham
91 Gerard, 1597
92 Carolus Clusius, *Rariorum Plantarum Historia*, 1601
93 Keble Martin and Fraser, 1939
94 Chris Scott, Saltash, Corn.
95 Sean Street, *The Dymock Poets*, 1994
96 Mabey and Evans, 1980
97 Margaret and Bob Marsland, Hallwood Green, Dymock, Glos.
98 Dr Susan Warr, Mutley, Devon
99 Mr and Mrs T. M. Cave, Wiveliscombe, Somer.
100 Maggie Colwell, Box, Glos.
101 Ida Mullins, York; also Ursula Bowlby, New Forest, Hants.; Susan Cowdy, The Lee, Bucks.
102 Mike Coyle, Stoke, Devon
103 Trevor James, Hertfordshire Environmental Records Centre, Hitchin
104 E. Small, Washford, Somer.
105 G. Hurst, Bladford St Mary, Dorset
106 Doug Shipman, Broomfield, Essex
107 Michael Adey, Welling, Kent; A. Russell, Clevedon, Avon; Heather Hastings, Nutley, E. Susx; Doreen Farrant, Waldron, Heathfield, E. Susx; Rosemary Gilbert, Uckfield, E. Susx; E. Cole, Mayfield, E. Susx; Joan Dunlop, Staplecross, E. Susx; P. C. Mosby, Plummers Plain, Horsham, W. Susx; Deirdre Howe, Dial Post, W. Susx; Mary Middleton, Plaistow, W. Susx; Liza Goddard, Boundstone, Surrey; J. B. Haslam, Danhill Farm, Thakeham, W. Susx; I. MacDonald, Wiston, W. Susx
108 Alan Paine, Trimley St Mary, Suff.
109 Simpson, 1982
110 Dorothy Wordsworth, *Journals*, in *Home at Grasmere*, Colette Clark (ed.), 1960
111 Paul Jackson, Aberystwyth, Dyfed
112 David Jones, *The Tenby Daffodil*, 1992
113 *ibid.*
114 Charles Tanfield Vachell, *A Contribution Towards an Account of the Narcissi of South Wales*, 1894
115 T. A. Warren Davis, *Plants of Pembrokeshire*, 1970
116 Mabey and Evans, 1980
117 Bromfield, 1856

Irises, *Iridaceae*

1 Faith Moulin, Yaton, Bristol
2 Robin Ravilious, Chulmleigh, Devon
3 *Gardeners' Chronicle*, March 1872
4 L. Sharp, Stapleford, Notts.
5 Steve Alton, Nottinghamshire Wildlife Trust
6 *ibid.*
7 Allan Marshall, Newhey, Lancs., in *BSBI News*, 58, 1991
8 Simpson, 1982
9 Brian Wurzell, Tottenham, London, in *BSBI News*, 60, 1992
10 William A. Clarke, *First Records of British Flowering Plants*, 2nd edn, 1900
11 Tubbs, 1986

Orchids, *Orchidaceae*

1 Barry Hunnett, Scunthorpe, Humbs.; Carol Woodward, Hucknall, Notts.; Ian Hartland, Worksop, Notts.; F. A. Niker, Frogpoool, Truro, Corn.; J. Groves, Chepstow, Gwent; B. Huggins, Tunley, Bath; Dr Peter Shaw, Dept of Environmental Sciences, Roehampton Institute, London
2 Madge Goodall, Otterbourne, Hants.; also Lilian Endsell, Horley, Surrey
3 David Monks, Llanddulas, Clwyd
4 W. R. Linton, *Flora of Derbyshire*, 1903
5 Summerhayes, 1951
6 V. M. Catton, Salisbury, Wilts.
7 Dickson, 1991
8 Vicky and Ian Thomson, Bentworth, Hants.
9 T. J. Flemons, Luston, Here.
10 E. Barker, Solihull, W. Mids
11 Lyte, 1578
12 Gerard, 1597
13 D. P. Godwin, Seaford, E. Susx
14 Brian Hopkins, Barnham, W. Susx; also Liz Bell, Tonbridge, Kent; R. Goodchild, Fareham, Hants.
15 Mary Bradbury, Cockermouth, Cumb.; Derek Sodo, Burnley, Lancs.; Nigel Northcott, Sandford-on-Thames, Oxon.; Anne Lee, Seascale, Cumb.; Sally Edmondson, Lancashire Wildlife Trust
16 Penny Anderson, Chinley, Ches.
17 John Tucker, Chippenham, Wilts.
18 Ian Dunnett, Gipping Valley Countryside Project, Suffolk County Council; also Mrs Doney, Needham Market, Suff.; C. M. Twyman, Claydon, Suff.
19 David Binstead, Sully, S. Glam.; Barry Hannett, Scunthorpe, Humbs.
20 Grigson, 1955
21 Brooke, 1950
22 Robin Ravilious, Chulmleigh, Devon
23 Margaret Trevillion, Germoe Churchtown, Corn.
24 Ruskin, 1874–86
25 *The Times*, 13 May 1992
26 Brooke, 1950
27 Webb and Coleman, 1849
28 quoted in Jermyn, 1974
29 Gerard, 1597
30 Ronald Good, 'On the distribution of the Lizard Orchid (*Himantoglossum hircinum* Koch)', in *New Phytologist*, 35, 1936
31 Lousley, 1950
32 Jocelyn Brooke, *The Military Orchid*, 1948
33 Mabey, 1980
34 Belinda Wheeler, Homefield Wood Warden, BBONT, Oxford; see also the annual reports on the site published by the Chiltern Military Orchid Group
35 Perring and Farrell, 1983; also Summerhayes, 1951
36 S. A. White, Leighton Buzzard, Beds.
37 A. Chowns, Berkhamsted, Herts.; Rhoda Downes, Poole, Dorset
38 J. A. Webb, Kidlington, Oxon.
39 Antony Galton, Exeter, Devon

Scottish vernacular plant names

1 W. Scott and R. Palmer, *The Flowering Plants and Ferns of the Shetland Islands*, 1987
2 Grace Corne, quoted in *Plant-lore Notes and Views*, 38, 1995
3 F. Fraser Darling, *Natural History in the Highlands and Islands*, 1947
4 Janet Lawson, Elphin, Sutherland
5 Andrew Spink, Dept of Botany, University of Glasgow
6 Mary McCallum Webster, *Flora of Moray, Nairn and East Inverness*, 1978
7 *ibid.*
8 Grigson, 1955
9 McCallum Webster, *op. cit.*
10 Vickery, 1995
11 Scott and Palmer, *op. cit.*
12 McCallum Webster, *op. cit.*
13 David Welch, *Flora of North Aberdeenshire*, 1993
14 Vickery, 1995
15 Alison Rutherford, Helensburgh, Dunbarton.
16 Jim Dickson, Dept of Botany, University of Glasgow
17 J. H. Dickson and A. Agnes Walker, 'What was the Scottish Thistle?', in *Glasgow Naturalist*, 20(2), 1981
18 Grigson, 1955
19 *ibid.*

Acknowledgements

To the Countryside Commission, English Nature, the Ernest Cook Trust and Reed Books for so generously supporting the research stage of the project.

To Common Ground – and Sue Clifford and Angela King especially – who acted as charitable host to the project and who were unfailing in their support and encouragement whenever my own enthusiasm showed signs of wilting. And to Daniel Keech and John Newton, who worked as full-time information and research officers, principally from Common Ground's office, but who also did invaluable and original fieldwork beyond the call of duty.

To Peter Marren, James Robertson and Ruth Ward, who co-ordinated research for us in Scotland, Wales and Oxfordshire respectively.

To Roz Kidman Cox, editor of *BBC Wildlife* Magazine, and Patrick Flavelle, producer of BBC TV's *CountryFile*, who gave us space and time (and encouragement) to recruit contributors.

To the many societies and associations, national, regional and local, whose members and staff were amongst the principal contributors:

Botanical Society of the British Isles, British Bryological Association, British Association for Nature Conservation, British Naturalists' Association, British Trust for Conservation Volunteers, Butterfly Conservation Society, Churchyard Conservation Project, John Clare Society, Council for Protection of Rural England, Countryside Council for Wales, Folklore Society, Forest Enterprise, Help the Aged, Herb Society, Learning through Landscapes, Local Agenda 21, National Association for Environmental Education, National Association of Field Studies Officers, National Farmers' Union Scotland, National Federation of Women's Institutes, National Trust, Open Spaces Society, Plantlife, Poetry Society, Ramblers' Association, the Royal Botanic Gardens at Kew and Edinburgh, Royal Forestry Society, Royal Society for Nature Conservation, Scottish Environmental Education Council, Scottish Natural Heritage, Tree Council and the Parish Tree Wardens network, Watch, Youth Hostels Association.

Arthur Rank Centre, Association of Leicestershire Botanical Artists, Bioregional Development Group, Bolton Museum, Cheshire Landscape Trust, Cleveland Community Forest, Cobtree Museum, the County Wildlife Trusts of: Berkshire, Buckinghamshire and Oxfordshire; Cambridge and Bedfordshire; Cleveland; Cornwall; Derbyshire; Gloucestershire; Hampshire and Isle of Wight; Hertfordshire and Middlesex; Kent; Lancashire; London; Norfolk; Northamptonshire; North Wales; Nottinghamshire; Scotland; Suffolk; Surrey; Wiltshire. Cymdeithas Edward Llywd, Derbyshire Ranger Service, Deeside Leisure Centre, Flora of Dunbartonshire Project, Groundwork Trusts of Merthyr and Cynon, Kent Thames-side and Amber Valley, Humberside County Council Planning Department, Liverpool Museum, Manchester Herbarium, Mid-Yorkshire Chamber of Commerce, Montague Gallery, Norfolk Rural Life Museum, Norfolk Society, North York Moors National Park, Social and Education Training Norfolk, Ted Ellis Nature Reserve, South-east Arts, University of Sussex Centre for Continuing Education, Warwickshire Rural Community Council, Wildplant Design.

Boxley Parish Council, Bradford City Council Countryside Service, Buchan Countryside Group, Thomas Coram School, Cumbria Broadleaves, Dragon Environment Group, Embsay with Eastby Nature Reserve, Giggleswick School, Great Torrington Library, Greenfield Valley Heritage Centre, Mike Handyside Wildflowers, Harehough Craigs Action Group, Hedingham Heritage Group, Hexham Nursery, Horsham Natural History Society, Ingleby Greenhow Primary School, Graham Moore Landscape Works, Lee Parish Society, Little Wittenham Nature Reserve, Oakfield Methodist Church, Paulersbury Horticultural Society, Pytchley Parish Sunday School Group, St Mary's Church Kirk Bramwith, Spiral Arts, Wealden Team Conservation Volunteers, West Bromwich Albion Football Club, White Cliffs Countryside Management Project, Whitegate and Marton Parish Council.

My personal thanks for their support to Charles Clark, Gren Lucas, Sir Ralph Verney and the late Sir William Wilkinson, and to the Leverhulme Trust for their generous award of a Research Fellowship to help fund my own researches.

To the friends and colleagues who gave me hospitality, company and much stimulating information during my own research trips: Elizabeth Roy and Nigel Ashby, Ronald Blythe, Hilary Catchpole and the pupils of Thomas Coram School, Berkhamsted, Rollo and Janie Clifford, David Cobham, Mike and Pooh Curtis, Roger Deakin, the late Edgar Milne-Redhead, Robin and Rachel Hamilton, Anne Mallinson, Richard Simon, Jane Smart, Jonathan Spencer, Ian and Vicky Thomson. And to Liza Goddard, who acted as guinea-pig contributor whilst I was still refining the idea of the project and who accompanied me on some of the early field-trips, and to Pattie Barron, for her patience.

To Penelope Hoare, my publisher at Sinclair-Stevenson, and the directors of Reed Books, who gave unflagging support to the project over what, in the modern book world, was a very long gestation period.

To Robin MacIntosh, my personal assistant, who helped in so many ways, especially with the awesome task of putting the contributions into some order. To Vivien Green, my agent, especially for bolstering me up during the low periods of the writing. To Douglas Matthews, for the speed and accuracy with which he produced the lengthy indexes. To Philip Oswald, whose vast background knowledge and meticulous attention to detail not only gave the text botanical respectability but added a wealth of anecdotes, historical notes and stylistic improvements. And to Roger Cazalet, my editor, for his diligence, dedication, patience and care.

And finally, the warm thanks of all of us go to the many thousands of people who contributed to the book and without whom it would not exist. Those of you whose stories and notes have found their way into the text are individually acknowledged in the Source notes. But every single contribution helped form the entries and the overall flavour of the book. Please keep contributing, as we hope that this first edition of *Flora Britannica* will not mark the end of the project so much as the beginning of a new phase.

Picture credits

Bob Gibbons: frontispiece, 8, 9, 11, 15, 18 (both), 19, 20r, 22, 24, 25r, 26, 27, 28, 33, 36, 38, 40, 41 (both), 42, 43, 44, 47, 48, 49, 50 (both), 55 (both), 62, 63r, 64l, 66, 67, 69, 70, 76t, 82, 84, 87, 88, 89, 93, 94, 95, 96l, 97, 98, 101r, 102, 103, 104, 105, 106 (both), 108, 111, 112, 114 (both), 115, 116 (both), 117, 121, 122, 123 (both), 125, 126, 127, 129t, 130 (both), 136, 139, 140, 142t, 143, 144, 145, 147l, 148, 149 (both), 151, 152 (both), 153, 154, 155, 156 (both), 158, 159, 160, 161, 162, 163, 164, 165, 169, 172 (all), 173b, 175, 177, 178b, 179, 180, 181, 182 (both), 183, 185 (both), 187t, 188 (both), 189, 190 (both), 191, 192, 193, 194l, 195t, 198, 199, 201 (both), 203, 204, 205, 207, 208, 209, 213, 216r, 217 (both), 218 (both), 219, 220, 221 (both), 222, 223, 224, 225, 226t, 228, 229r, 232, 233, 235, 236, 238, 239, 240, 241, 243, 244, 245, 246, 248, 252-3, 255, 258 (both), 259, 260, 261b, 263, 264, 266, 267l, 268, 269, 270, 271 (all), 272, 273, 274l, 275, 281 (both), 283 (both), 284, 285 (both), 287, 288, 290, 291, 293, 294, 297, 298, 299tl and bl, 300, 301 (both), 302 (both), 303, 304t, 305 (both), 307, 308, 309t, 310l, 311, 312, 313 (both), 316 (both), 317 (both), 320, 322, 325, 327, 328l, 330, 331 (both), 333, 335, 336, 338, 339, 340, 341 (both), 342, 345r, 346, 347, 349 (both), 351, 352, 353, 356 (both), 357, 358l, 359 (both), 360, 362 (both), 363, 364, 365, 366, 368, 372, 373, 374, 375, 378, 379b, 381, 382, 383, 386, 387, 388, 389, 390, 391 (both), 392, 393 (both), 394, 396, 397 (both), 398-9, 400, 401, 402 (both), 406, 408, 409, 410 (both), 411, 412, 416 (both), 417, 418, 419, 420, 422, 423, 424, 426-7, 428, 431, 432, 434, 436, 437, 438 (both), 439, 440, 441, 442, 443, 446 (both), 447 (both), 448, 449.

Gareth Lovett Jones: 13, 21, 29, 30-1, 37, 45, 51, 56r, 59, 61, 64r, 73, 76b, 78, 80l, 83, 85, 86, 96r, 107, 109, 110, 120, 124, 131, 132, 133, 134, 141, 142b, 146, 147r, 157, 178t, 186, 195b, 197, 202 (both), 215, 216l, 226b, 227, 229l, 231, 234, 247, 251, 254, 257, 261t, 265, 276, 277, 278, 280, 289, 299r, 306, 309b, 315, 323, 344, 345l, 350, 354, 385, 395, 414, 433.

Heather Angel: 72; Ashmolean Museum: 267r; Bodleian Library: 25l, 63l, 184; Bridgeman Art Library: 90, 176, 310r, 321; British Museum: 279l; Carlisle City Museum and Art Gallery: 79; Ian Collins: 7; Common Ground: 10, 174, 413, 435; Derby Museum and Art Gallery: 358r; Mary Evans Picture Library: 429; Tony Evans ©: 39, 46, 53, 58, 168, 171, 210, 282, 318, 367, 376, 415; Sonia Halliday and Laura Lushington: 304b; Hulton Getty Picture Collection: 20l, 57; Hunterian Art Gallery, University of Glasgow: 113; A. F. Kersting: 65; Andrew Lawson: 404-5; London Transport Museum: 262; Richard Mabey: 16, 80r, 81, 324; Marlborough Photographic Library: 187r; Margaret Marsland: 119; Mercury Gallery: 54; National Gallery, London: 379t; National Trust Photographic Library: 14t (John Bethell), 348 (Andreas von Einsiedel); Natural Image/Peter Wilson: 74, 100, 128, 173t, 196, 292, 337, 369, 445; Nature Photographers/E. A. Janes: 371; Norfolk Museum Service (Norwich Castle Museum): 211; Katrina Porteous: 328l; Royal Horticultural Society: 444; Eddie Ryle-Hodges: 14b, 34, 56l, 101l, 129b, 132t, 150, 237, 274r, 279r, 370; Edwin Smith: 118, 194r, 214; Homer Sykes: 355; by courtesy of the Board and Trustees of the Victoria & Albert Museum (Bridgeman Art Library): 60, 135, 329.

Index of places

General index